Neglected Diseases and Drug Discovery

RSC Drug Discovery Series

Titles in the Series:

1: Metabolism, Pharmacokinetics and Toxicity of Functional Groups: Impact of Chemical Building Blocks on ADMET
2: Emerging Drugs and Targets for Alzheimer's Disease; Volume 1: Beta-Amyloid, Tau Protein and Glucose Metabolism
3: Emerging Drugs and Targets for Alzheimer's Disease; Volume 2: Neuronal Plasticity, Neuronal Protection and Other Miscellaneous Strategies
4: Accounts in Drug Discovery: Case Studies in Medicinal Chemistry
5: New Frontiers in Chemical Biology: Enabling Drug Discovery
6: Animal Models for Neurodegenerative Disease
7: Neurodegeneration: Metallostasis and Proteostasis
8: G Protein-Coupled Receptors: From Structure to Function
9: Pharmaceutical Process Development: Current Chemical and Engineering Challenges
10: Extracellular and Intracellular Signaling
11: New Synthetic Technologies in Medicinal Chemistry
12: New Horizons in Predictive Toxicology: Current Status and Application
13: Drug Design Strategies: Quantitative Approaches
14: Neglected Diseases and Drug Discovery

How to obtain future titles on publication:
A standing order plan is available for this series. A standing order will bring delivery of each new volume immediately on publication.

For further information please contact:
Book Sales Department, Royal Society of Chemistry, Thomas Graham House, Science Park, Milton Road, Cambridge, CB4 0WF, UK
Telephone: +44 (0)1223 420066, Fax: +44 (0)1223 420247, Email: books@rsc.org
Visit our website at http://www.rsc.org/Shop/Books/

Neglected Diseases and Drug Discovery

Edited by

Michael J. Palmer
Pfizer, Sandwich, Kent, UK

Timothy N. C. Wells
Medicines for Malaria Venture, Geneva, Switzerland

RSC Publishing

RSC Drug Discovery Series No. 14

ISBN: 978-1-84973-192-8
ISSN: 2041-3203

A catalogue record for this book is available from the British Library

Published by The Royal Society of Chemistry,
Thomas Graham House, Science Park, Milton Road,
Cambridge CB4 0WF, UK

Registered Charity Number 207890

For further information see our web site at www.rsc.org

Foreword: Introduction to Neglected Diseases

Improvements in social conditions and healthcare have transformed life in the developed world over the last century where, for example, life expectancy has increased dramatically. Indeed, some babies born today can expect to live into the next century! These advances are undoubtedly due to a combination of improved hygiene, social conditions and lifestyle, but some 40% is estimated to result from new medicines. A clear example is heart disease: hypertension is now well controlled and cardiovascular risk reduced using safe and effective medicines. Similar breakthroughs have been seen in other areas: for instance, deaths due to HIV/AIDS have been brought under control in most developed countries. These advances have involved a significant investment, not only by the pharmaceutical industry (US biopharmaceutical companies spent over US$65bn on R&D in 2009) but also by academic groups in science and clinical research. Even with the extremely high prices for some medicines in the West, such healthcare expenditure is not sustainable and is under increasing pressure, so how can we address serious diseases in less advanced countries where safe food and water supplies may be more pressing priorities?

Nowhere is the divide more apparent than for HIV/AIDS, where 70% of the 33 million sufferers are located in sub-Saharan Africa, with limited access to effective drugs that are priced at US$10,000/year in the US. Even generic alternatives that may be more than 10-fold cheaper are beyond many patients' reach and in some countries, life expectancy has dropped to below 40 years. However, all is not lost and "where there is a will, there is way" as seen by previously successful worldwide collaborations to eradicate smallpox and polio. There is a now a similar sense of social responsibility in some developed nations that high mortality and unnecessary suffering, particularly amongst

RSC Drug Discovery Series No. 14
Neglected Diseases and Drug Discovery
Edited by Michael J. Palmer and Timothy N. C. Wells

Published by the Royal Society of Chemistry, www.rsc.org

children, cannot continue, and that we should mobilise international collaborations to control neglected diseases, reduce mortality and improve quality of life.

The George Institute for International Health estimates that total R&D spending on neglected diseases was US$3.1 billion in 2008, although AIDS, malaria and TB consumed almost 75%. Funding was strongly led by the US NIH ($1.1 billion) and the generosity of the Bill and Melinda Gates Foundation ($0.64 billion), followed perhaps surprisingly by the biopharmaceutical sector ($0.39 billion). Pharmaceutical companies have long recognised their responsibility to shareholders rather than non-profitable markets, but management has become increasingly receptive to persuasive lobbying from research scientists that neglected diseases must be addressed. Consequently, the tide has turned over the past decade where, for example, GSK and Novartis have established dedicated research centres to tackle diseases of the developing world while various bio- and pharmaceutical-companies have also invested directly, or in kind. Innovative Public Private Partnerships (PPPs) have been established, including MMV, TB Alliance, OneWorldHealth and DNDi, and are starting to bear fruit.

I first became involved with neglected diseases in 1999 by answering an advertisement in *Nature* and offering my services to the nascent WHO/TDR malaria initiative which I recall had only 2 or 3 staff members. However, we created an Expert Scientific Advisory Committee, where I served as the inaugural chair, and our first Call for Proposals overwhelmed us with more than 100 responses. Ten were selected for funding and we were on our way! A couple of years later, the Medicines for Malaria Venture was formally established as a Swiss charity, which today has over 50 employees with an annual budget of $55 m. More importantly, this investment is now making a significant impact with one new medicine launched, 2 more in registration and an R&D portfolio of over 20 innovative projects. When these achievements are scaled across the overall research activities against neglected diseases, we can be proud and confident that we are taking important strides against some of today's scourges that cause so much suffering, deteriorating social conditions and economic disaster. It has been estimated that in 2008, Africa lost over $12 billion GDP to malaria, and yet control of the disease was possible in 1914 in Panama, and was a key factor in completing canal construction. Surely, such game-changing scenarios must instil an even greater sense of urgency into our own research and access efforts.

Encouragingly, participation by academia, PPPs, pharmaceutical and biotech companies, together with other research organisations in the fight against neglected diseases has led to an enhanced sense of collaboration and camaraderie across the sector. Probably for the first time, pharmaceutical companies have provided their compound files for third-party screening, and in some cases hit structures have been placed in the public domain to encourage community exploitation. Projects with similar targets have joined forces to avoid duplication, prioritise objectives and identify development candidates as rapidly and efficiently as possible. Provision of resource in kind from large and small organisations has been outstanding.

Of course, there are still major economic, scientific and political challenges that must be faced and overcome. Despite continued generosity from many donors, world economies face increasing pressures, which we must appreciate, but which should not deter continued and focussed fundraising. We have an unusual responsibility for the transparent use of funds and clearly communicating objectives, progress and issues to our stakeholders. It is essential we help funders appreciate the importance of maintaining a robust discovery pipeline even as successful projects move into the more expensive development phase. Scientifically, we are dealing with dangerous organisms where attack, evasion, and resistance are the norm, such that a continuous pipeline of novel agents suitable for combination therapies will be required for most of the diseases we are addressing. Some pose additional challenges, such as drug penetration and 6-month compliance for TB, while a \$1, 3-day treatment for malaria has significant cost of goods issues. Target validation is a continuing problem where genomic approaches have not yet blossomed, but whole cell screening is enjoying considerable success by providing attractive lead series for innovative medicinal chemistry follow on. Political will and stability will obviously have a major effect on drug distribution and treatment campaigns, where a key issue is how patients who subsist on a \$1/day could ever access and afford modern, effective medicines.

I am delighted to offer an introduction to this excellent volume in the RSC series, which I am sure will be well received by researchers and lay folk alike. I am particularly impressed by the breadth and depth of topics covered by such acknowledged scientific experts. Surely, this volume is a fitting testament to the dedication and commitment of the editors, authors, colleagues, funders and everyone else involved in the collective fight against neglected diseases. We have started to make meaningful progress, but the best is yet to come!

Simon F. Campbell, Kingsdown

Preface

Arguably, neglected disease drug discovery has a number of hurdles that differentiate the arena relative to conventional programmes. Notwithstanding the very challenging targets that kill and debilitate large numbers of people across the world, resistance to treatment is often a recurring problem, and the funds available for research, development and to pay for treatment are limited. Such thinking is the basis of this volume.

This book seeks to aid both researchers and lay folk, by summarising the key learning to date and also by giving a clear overview of the challenges that remain. In *Neglected Diseases and Drug Discovery*, most of the key diseases that fall into this class are detailed. The book sets out to make a critical appraisal of ongoing research with a focus on the key science that has led to breakthroughs, especially from a medicinal chemistry perspective. Our intention is that the following chapters can serve as a useful guidebook to ongoing and new drug discovery efforts in this field.

The book seeks to cover in some depth current efforts in the malaria, trypanosomatid, flavivirus and tuberculosis fields, wherein an upsurge in research efforts has been evident in recent times. Additionally, there is some focus on the 'neglected' neglected diseases, notably diarrhoea, helminths, HIV (in worldwide terms) and lower respiratory tract infections. In these latter areas we seek to highlight the pressing need for better and more widely available treatments.

Our utmost appreciation goes to the authors who have given their precious time in order to share their experiences and make this account a reality.

Mike Palmer
Timothy Wells

RSC Drug Discovery Series No. 14
Neglected Diseases and Drug Discovery
Edited by Michael J. Palmer and Timothy N. C. Wells

Published by the Royal Society of Chemistry, www.rsc.org

Contents

RSC Drug Discovery Series No. 14
Neglected Diseases and Drug Discovery
Edited by Michael J. Palmer and Timothy N. C. Wells

Published by the Royal Society of Chemistry, www.rsc.org

CHAPTER 1

Malaria: New Medicines for its Control and Eradication

TIMOTHY N. C. WELLS AND
WINSTON E. GUTTERIDGE

Medicines for Malaria Venture, 20 Rte de Pré-Bois, 1215 Geneva, Switzerland

1.1 Introduction

Malaria is caused by protozoan parasites of the genus *Plasmodium* that infect and destroy red blood cells, leading to fever, severe anaemia and, if untreated, cerebral malaria and death. *Plasmodium falciparum* is the dominant species in sub-Saharan Africa, and is responsible for almost one million deaths each year.[1] The disease burden is heaviest in sub-Saharan African children under 3 years old (who have frequent attacks and little immunological protection), and also in expectant mothers.[1] Malaria is both a cause and a consequence of poverty: in countries with intense malaria transmission, the economic impact of the disease results in a slowing of economic growth of 1.3% per year[1], translating to a reduction of the Gross Domestic Product in sub-Saharan Africa estimated to be US$12 billion per year.[2]

The global fight to control malaria requires a multifaceted approach. At present, we have a wide range of effective tools. Medicines can be used to prevent as well as to cure, especially in vulnerable populations such as infants[3] or pregnant women.[4] Insecticides and larvicide spraying and the use of insecticide-impregnated bed nets to protect against infection by mosquitoes have dramatically increased in recent years. This success brings with it the need for

RSC Drug Discovery Series No. 14
Neglected Diseases and Drug Discovery
Edited by Michael J. Palmer and Timothy N. C. Wells

Published by the Royal Society of Chemistry, www.rsc.org

the development of the next generation of insecticides, since resistance to the current gold standard, the pyrethroids, is already an issue.[5] Developing a vaccine is proving especially challenging, as the parasite has sophisticated mechanisms for avoiding the host immune system. The best candidate currently is GSK 257049, known as RTS,S/AS202, where phase III trials are expected to finish in July 2013.[6] Phase II studies suggest that it will reduce risk of clinical malaria, and decrease mortality in severe malaria by 50%.[7]

In November 2007, the Bill and Melinda Gates Foundation set an agenda with the final goal of completely eradicating malaria[8], an objective supported by both the World Health Organization (WHO) and its Roll Back Malaria (RBM) partnership.[9] Commentators have described this objective as 'worthy, challenging, and just possible', but one which must be pursued with balance, humility, and rigorous analysis.[10]

The addition of this new goal has implications for the global malaria R&D agenda, which have been discussed in a variety of different working groups.[11] Future antimalarial medicines must not only be able to treat the asexual blood stages of *P. falciparum*, but also to block the transmission of the parasite to other persons via the mosquito vector and, in the case of *P. vivax* infection, to target the dormant liver-stage of the parasite. In this chapter, we discuss the current pipeline of antimalarial medicines, and the target product profiles for the generation of such products.

1.2 The Challenges of the Different *Plasmodium* Species

Four main species of the malaria parasite infect humans. *P. falciparum* is responsible for the vast majority of the malaria-linked deaths in sub-Saharan Africa and is therefore the most important target. *P. vivax* constitutes as much as 25–40% of the global malaria burden,[12] particularly in South and Southeast Asia, and Central and South America. It does not normally progress to cerebral malaria, and has been traditionally labelled benign. However, *P. vivax* causes a greater host inflammatory response than *P. falciparum* at equivalent parasitaemia. Mortality from *P. vivax* is most likely underreported, as recent analyses in Papua (Indonesia), have shown similar mortality figures in children to those found with *P. falciparum*.[13]

Medicines that are active on the asexual erythrocyte stages of *P. falciparum*, such as the artemisinin-based combination therapies (ACTs), are assumed to be fully active against the other species. The formal clinical database supporting this assumption is relatively thin but is well supported by empirical observation. Historically, mixed infections of *P. falciparum* and *P. vivax* are rarely reported, possibly because *P. falciparum* suppresses the development of *P. vivax,* but PCR detection methods have shown that these can be as high as 30%.[14] The other two species are *P. malariae* and *P. ovale*. Currently, these are diagnosed by microscopy, and represent a small percentage of infections. Diagnosis based on Polymerase Chain Reaction (PCR) will undoubtedly lead to a re-evaluation of the presence of mixed infections, since it is able to quantify low parasite numbers, and is often more definitive as a diagnostic.

From a treatment and eradication perspective, there are 3 key differences between the species. The first difference occurs in the liver. Following infection of the patient, parasites rapidly progress to infect hepatocytes, undergo asexual schizogony, and release large numbers of merozoites into the host bloodstream. In *P. vivax* and *P. ovale*, some of the liver parasites become dormant (a form known as the hypnozoite).[15,16] These forms can be reactivated after periods that vary from 3 weeks to several years, dependent on the strain of parasite and the status of the host. Unless the hypnozoites are eliminated, malaria will continue to relapse periodically. Since *P. vivax* transmission is rarely intense, activation of hypnozoites is thought to be a major contributor to disease frequency.

The second difference between the species is in the time taken for the parasite to replicate in the host. The time between febrile paroxysms varies from around 48 hours for *P. falciparum* and *P. vivax* to 72 hours for the more benign *P. malariae*. There has been a recent interest in a fifth species, *P. knowlesi*, a parasite of Old World monkeys, now known to infect humans.[17] PCR methods show that it is often misdiagnosed as *P. malariae* infection, which is usually uncomplicated and has low parasitaemia. However, *P. knowlesi* replicates every 24 h, and so is potentially life-threatening if not treated expeditiously.[18] Therapeutically, the challenge in this case is to have a therapy with a rapid onset of action.

Third is the timing of the appearance of gametocytes in the blood stream.[19] In *P. falciparum,* the gametocytes do not appear until several days after the initial parasitaemia and fever, whereas in *P. vivax* they appear concurrently or even before asexual parasites. An ideal treatment for blood stages of *P. vivax* must be able to kill existing gametocytes, rather than simply preventing them from differentiating (see Figure 1.1).

1.3 Currently Available Antimalarials

The roots of most antimalarial treatments are based on three natural products: quinine, lapinone and artemisinin.[20] In each case the natural product was known to have some activity from traditional medicine, and was isolated, shown to have some activity and then this activity was improved by classical medicinal chemistry.

The first widely used antimalarial drug was quinine, a natural product extracted from the bark of the tree *Cinchona calisaya*. It causes parasite death by blocking the polymerisation of the toxic by-product of haemoglobin degradation, haem, into insoluble and non-toxic pigment granules, resulting in cell lysis and parasite cell autodigestion.[21] This means that the parasite is not able to generate resistance at the target site: the molecular target is a non-mutatable chemical reaction. Quinine itself is active when given 3 times a day for 7 days. Initial attempts to synthesise quinine led to the synthesis of dye

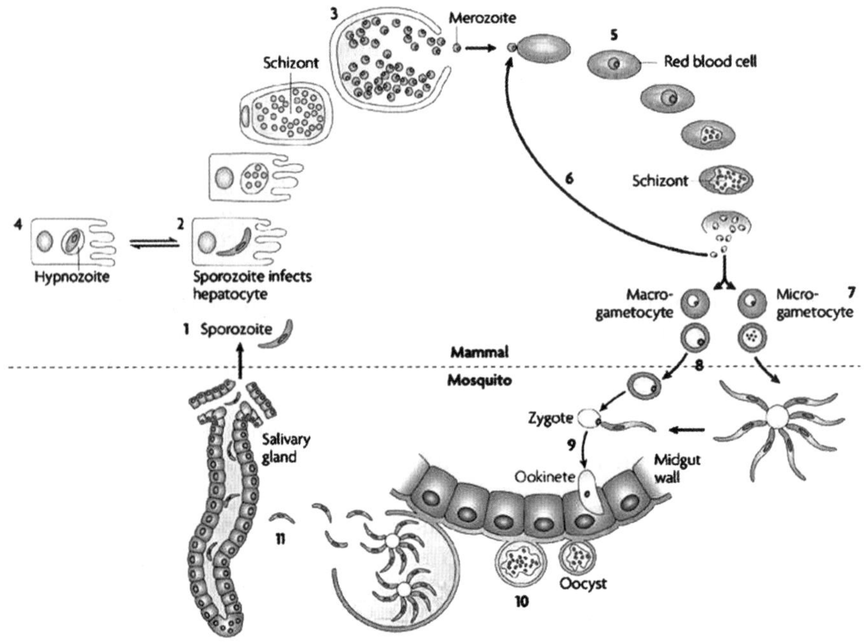

Figure 1.1 Life cycle of the malaria parasite.
1 Sporozoites are injected into humans with the saliva of a female *Anopheles* mosquito. **2** They are rapidly taken up into the liver, passing through Kupffer cells to hepatocytes. **3** Here the parasites develop to form several thousand merozoites. **4** In *P. vivax* and *P. ovale* only, some liver-stage parasites remain as a dormant form, or hypnozoite,[11] characterised histopathologically as a small uninucleate parasite, which remains dormant for a few weeks, or up to several years.[1] These species can therefore start a new cycle of asexual infection even without a mosquito bite. **5** The liver cells rupture and the merozoites are released into the blood, rapidly invading erythrocytes. **6** The intra-erythrocytic parasites replicate synchronously, leading to the classical cycle of fever observed clinically. **7** Some merozoite-infected red cells develop into male and female gametocytes. In *P. falciparum,* these are formed in the later stages of infection, whereas, in *P. vivax*, they are formed at the same time as the asexual stages. **8** Gametocytes are taken up into the female mosquito gut during the blood meal. **9** The male gametocytes are activated (exflagellation) and fuse with the female gametocytes to form diploid ookinetes. These ookinetes migrate to the mid-gut of the insect, pass through its walls and form the oocysts. **10** Meiotic division occurs and sporozoites are formed. **11** These then migrate to the salivary glands of the mosquito. Taken from Wells *et al.*[91]

substances, some of which are actually antimalarials in their own right, such as methylene blue. Later work produced chloroquine, a 4-aminoquinoline, which was the mainstay of malaria prophylaxis and treatment for the second half of the 20th century. This also has the advantage that its electronics allow it to selectively concentrate into the food vacuole. Further synthetic work has yielded many more aminoquinolines and related amino-alcohols such as amodiaquine, mefloquine, halofantrine, lumefantrine, piperaquine and pyronaridine. These medicines are characterised by a large volume of distribution and a long half life (terminal half lives of over 5 weeks are reported for choloroquine). They also require reasonably high doses (total dose of between 1250 and 2500 mg for adults, normally split into three daily doses. They have been linked to cardiac safety issues: at high doses prolongation of the QTc interval has been seen with some medicines, and indeed this led to the withdrawal of halofantrine by SmithKlineBeecham. Two more recent medicines in this class are Ferroquine (SSR97193, by sanofi-aventis, in phase II) and naphthoquine (launched as ARCO, by Kunming Pharmaceutical Coroporation). Both have long half lives, and the clinical question is whether their therapeutic window is large enough to support a single dose as a cure. Naphthoquine is administered as a 400 mg single dose with artemisinin and is reported to be safe[22] and effective in small scale clinical trials,[23] although the key data on QTc prolongation are currently not available. Ferroquine is currently in phase II trials with a lowest adult dose of 100 mg. An alternative approach has been to produce chloroquines linked to molecules known to reverse the CQ resistance transporter.[24] Although this is synthetically interesting, the challenge is still to show superiority in cardiovascular safety, which is far from trivial. In addition, molecules such as azithromycin have been shown to reverse chloroquine resistance clinically,[25] and so the challenge for a new molecule will always be to demonstrate additional benefit over known medicines.

Lapachol is a hydroxynaphthoquinone used to treat malaria and fevers in South America.[26] It was initially reported in the 19th century. As part of the American war effort it was tested in *P. lophurae* infected ducks in 1943, and showed weak activity. A close synthetic derivative, lapinone, was also active. This was subsequently confirmed in patients with *P. vivax*[27] by intravenous administration for 4 days. Solving the bioavailability issues led to the development of the orally bioavailable, metabolically stable molecule atovoquone,[28] one of the active ingredients for Malarone, the current mainstay of antimalarial prophylaxis for travellers. Further work focussed on the 4-pyridones and led to the development of GSK932121.[29] Work on this molecule was stopped in phase I after safety concerns with a pro-drug formulation. Hydroxynaphthoquinone and 4-pyridines target the cytochrome bc1 complex, and so in addition to the solubility/bioavailability and drug metabolism issues, there is clearly with such molecules a need to show selectivity against inhibition of the host electron transport. Recently, new inhibitors with selective bc1 inhibition have been reported based on an acridinone template WR 249685.[30]

The third approach is built on the discovery of the sesquiterpene lactone artemisinin (known as *Qing hao su*) in 1972 by Chinese scientists.[31] It is an endoperoxide-containing natural product isolated from the leaves of the sweet wormwood, *Artemisia annua*. Derivatives of artemisinin were subsequently shown to be more potent than the parent molecule, including dihydroartemisinin (DHA, believed to be the main active metabolite of all the derivatives), artemether, artemotil and artesunate (see Figure 1.2). Artemisinin itself is highly insoluble: chemical modification to artesunate increases oral bioavailability, and also makes it suitable for intravenous administration in severe malaria. The artemisinin derivatives are fully active against all existing drug-resistant strains of *P. falciparum*. Unlike all other antimalarial drugs, they act on all stages of the parasite intraerythrocytic life cycle and therefore rapidly kill all the blood stages of the parasite, resulting in the shortest fever and parasite clearance times of all such medicines.[32] Furthermore, the artemisinins also kill gametocyte stages – thereby reducing transmission from humans to mosquitoes.

(i) The 4-aminoquinolines and amino-alcohols used in the treatment of uncomplicated malaria.

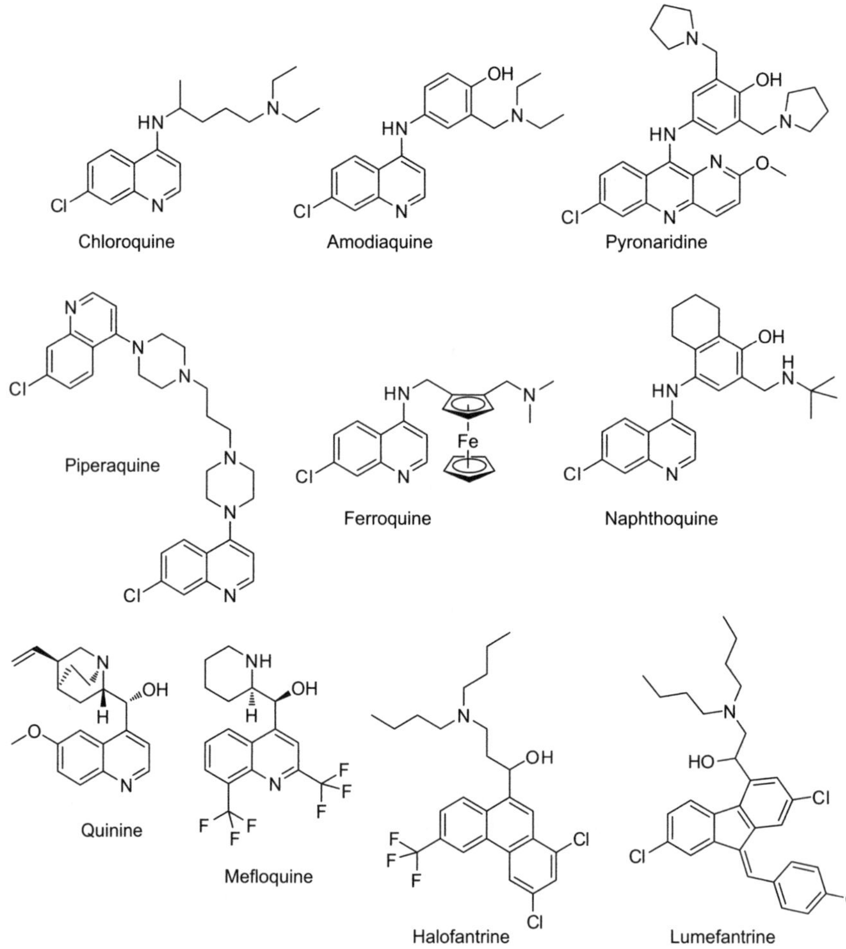

Figure 1.2 Structures of key antimalarial compounds.

(ii) The development of electron transport inhibitors. The natural product Lapinone was used as the design for Lapichol, which has limited oral bioavailability. Further work on this has yielded Atovaquone, which is part of the successful prophylactic, Malarone. Work to further improve the series has led to compounds such as WR 249685 from the Walter Reed Institute of Army Research, but so far no compounds have entered clinical development.

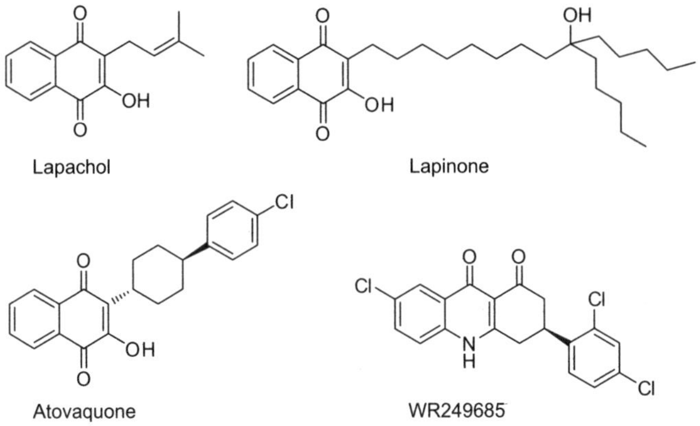

(iii) The artemisinins: artemisinin, dihydroartemisinin (DHA) the principle metabolite, the methyl ether, artemether, and the ethyl ether, artemotil; and artesunate.

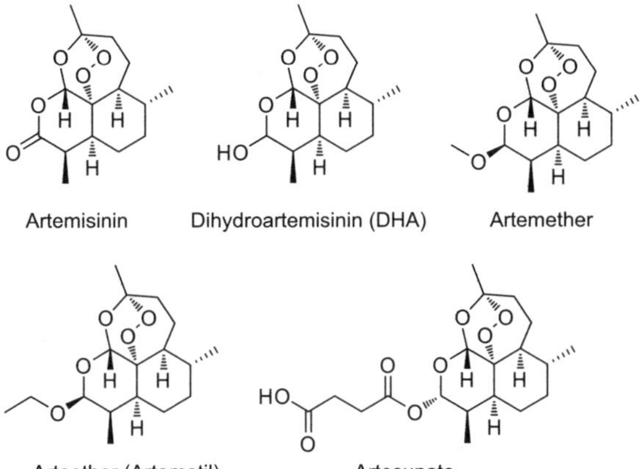

Figure 1.2 Continued.

Not all antimalarial drugs can be traced back to natural products. Some were rationally designed following an antimetabolite approach. For example, malaria parasites are unable to salvage folate, but need this cofactor to synthesise tetrahydrofolate for methylation reactions. Inhibitors of dihydropteroate synthase (sulphonamides such as sulphadoxine) and dihydrofolate reductase (2,4-diaminopyrimidines such as pyrimethamine) are potent antimalarial drugs, especially when administered in combination.

For most of the second half of the 20th century, control of acute uncomplicated malaria caused by all four species of *Plasmodium* relied on chloroquine for first-line treatment and a combination of sulphadoxine and pyrimethamine (SP) as second-line treatment.

(iv) The pathway beyond artemisinin. The semi-synthetic derivative artemisone; the endoperoxide/4-aminoquinoline fusion compound Trioxaquine (SAR116242, also known as PA1103); the first generation endoperoxide OZ277 (now under development as Rbx11160); the next generation endoperoxide OZ439, an endoperoxide CDRI 97/98 (under development by IPCA); and a synthetic endoperoxide RKA182, with a distinctive tetraoxane.

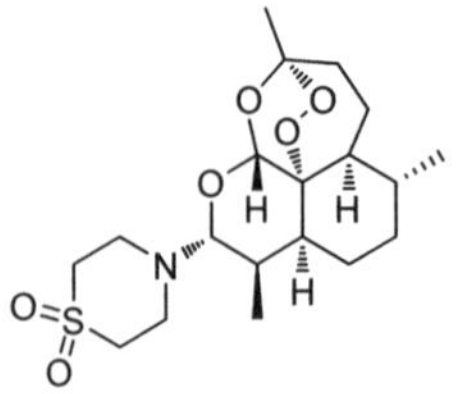

Artemisone

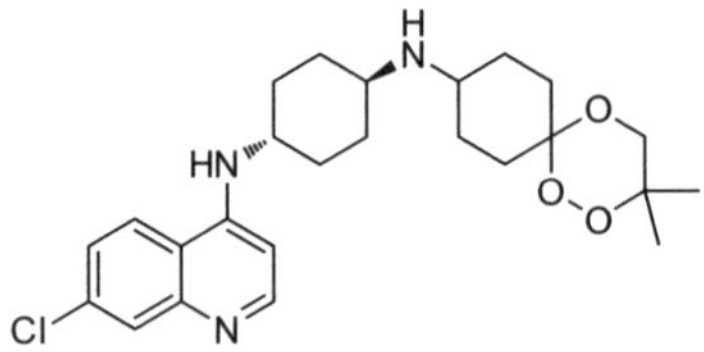

SAR116242 (Trioxaquine)

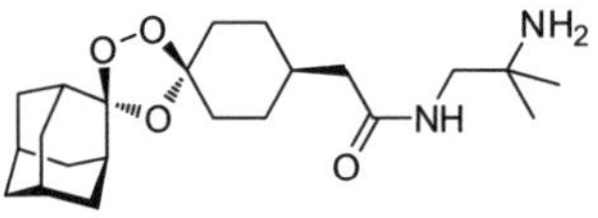

OZ277/RBx11160 (Arterolane)

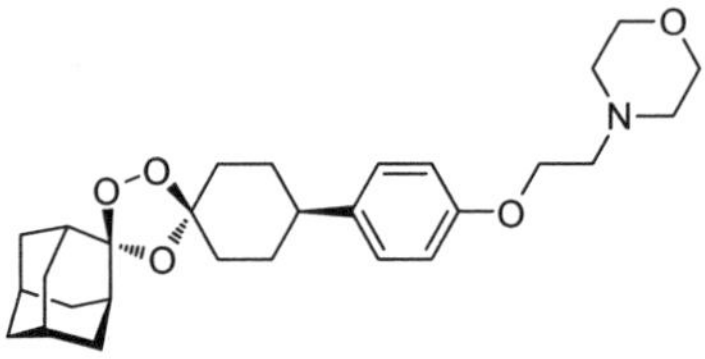

OZ439

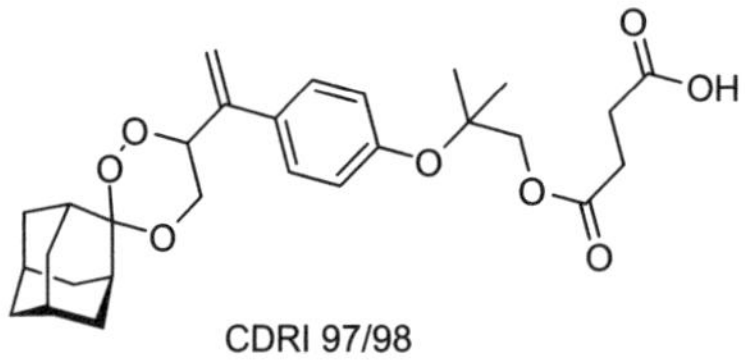

CDRI 97/98

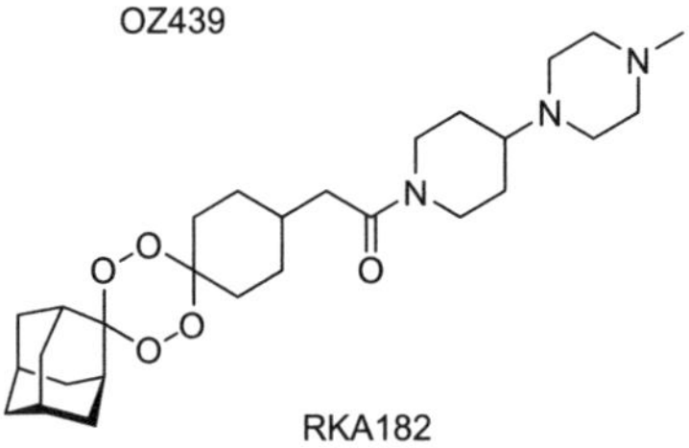

RKA182

(v) The current non-artemisinin containing combinations (NACTs): atovaquone-proguanil (Malarone), the proguanil is metabolised into cycloguanil, which acts synergistically and sulphadoxine-pyrimethamine (SP, Fansidar).

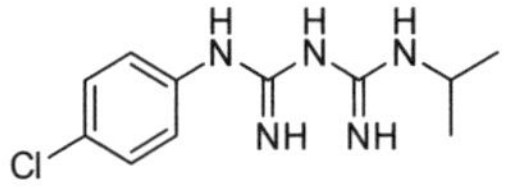

Proguanil

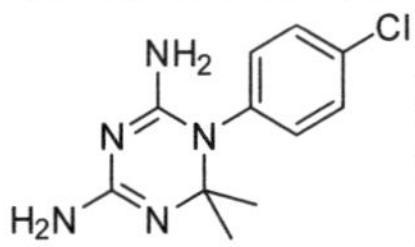

Cycloguanil (active metabolite)

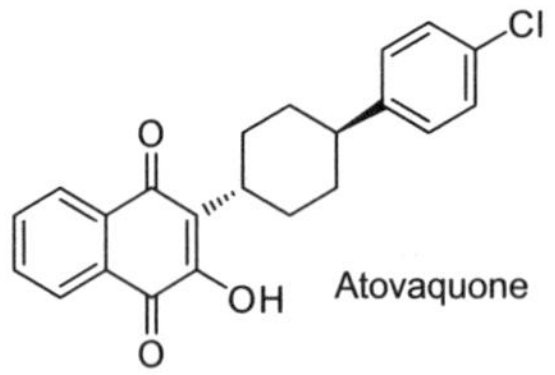

Atovaquone

Malarone

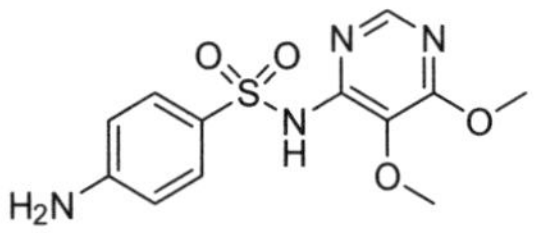

Sulphadoxine

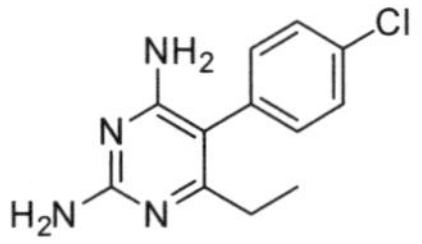

Pyrimethamine

Fansidar (SP)

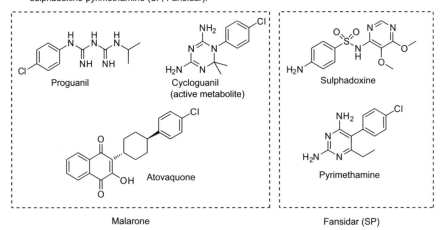

Figure 1.2 Continued.

(vi) The 8-aminoquinolines targeting *P. vivax* hypnozoites: pamaquine, primaquine and tafenoquine.

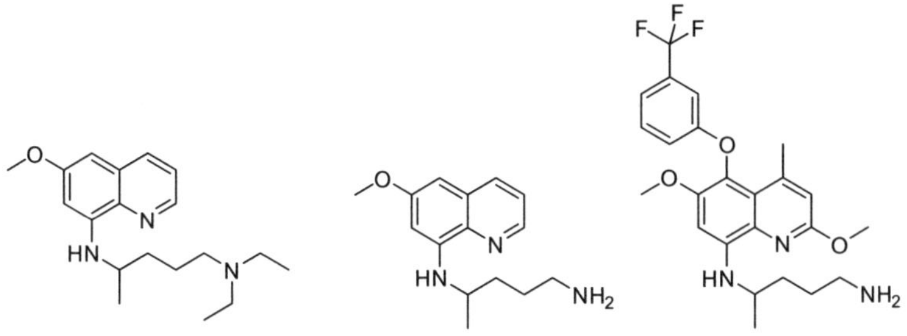

Pamaquine Primaquine Tafenoquine

(vii) Spiroindolone NITD 609, the first clinical fruits of the whole cell screening strategy.[66]

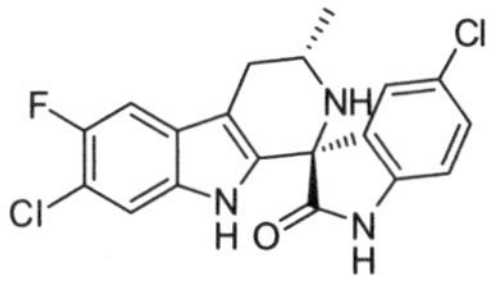

NITD609

(viii) Natural products: identified from extracts which have been demonstrated to have activity in patients with clinically defined malaria (according to WHO guidelines).

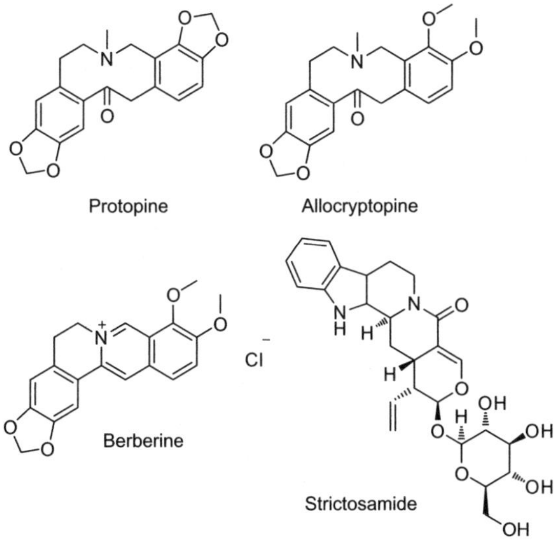

Figure 1.2 Continued.

1.4 Resistance

Resistance is a fact of life with antimalarial drugs, but the danger can be reduced by combination therapy. The frequency of mutations that might lead to resistance to drugs in *P. falciparum* is estimated at 1 in 10^{10} parasites compared with a parasite burden of 10^{12}–10^{13} in humans with severe malaria. Combining two medicines with different mechanisms of action lowers the probability that a resistant parasite will emerge and become established.[33] Against this background, it is not surprising that in all cases, *Plasmodium* strains resistant to current antimalarial drugs emerged. Resistance against drugs such as sulphadoxine-pyramethamine, where there are clear biological targets, arose more rapidly than with drugs such as chloroquine where the target is an immutaable chemical reaction and therefore other mechanisms, such as blockade of drug uptake, must be selected. Also, for reasons that are still not clear, resistance arose more rapidly in *P. falciparum* than in *P. vivax*. This might simply be an expression of less drug pressure, but this is by no means certain.

With the demise of chloroquine and SP because of drug resistance, in 2006 the WHO produced new treatment guidelines for uncomplicated *P. falciparum* malaria: they recommended that the treatment of choice should be a combination of two or more antimalarials with different mechanisms of action.[34] More than this, they suggested that artemisinin monotherapy should be withdrawn, to protect the class against the emergence of resistance – the first time such a suggestion had been made. The standard treatment rapidly became artemisinin-based combination therapies (ACTs), moving towards fixed dose (with both drugs in the same tablet, to prevent the artesunate being used as monotherapy). Artemether-lumefantrine, the first such fixed-dose artemisinin combination therapy developed to international standards of good practice, was launched by Novartis in 2001. Amodiaquine-artesunate was developed as a fixed-dose combination by the Drugs for Neglected Diseases Initiative (DNDi), and launched in 2008. The same year, a paediatric-friendly version of artemether-lumefantrine was launched as the result of a collaboration between Novartis and Medicines for Malaria Venture. In 2010, 82 million people were treated with Coartem or Coartem-D artemether-lumefantrine, and an additional 21 million treatments of generic artemether-lumefantrine were sold, mainly subsidised by the Affordable Medicines for Malaria Facility AMFm. The second most important combination by volume was amodiaquine-artesunate produced mainly by sanofi-aventis, with around 45 million courses of treatment supplied in 2010. So far, three generic producers of artemether-lumefantrine have been set up in Africa, and are able to supply drugs at the same price as Novartis ($0.30 for the smallest children, through to $1.20 for adults). Two other fixed-dose combinations are due to be launched over the next year: DHA-piperaquine (a collaboration between Sigma-Tau and Medicines for Malaria Venture) and pyronaridine-artesunate (a collaboration between Shin Poong and Medicines for Malaria Venture). Another version of DHA-piperaquine is available from Holley-Cotec. A fixed-dose combination of mefloquine-artesunate is available in Brazil from Drugs for Neglected Diseases initiative/Farmanguinhos/Fiocruz and in Europe from Mepha. A final combination of artemsinin (not the soluble artesunate) with naphthoquine has been marketed by Kunming Pharmaceutical Corporation: although this has the advantage on paper of being a treatment given in a single dose, or two doses in the same day, there are few data around the long -term safety, and no clinical studies have been carried out to determine good clinical practice. In summary, in 2010, over 150 million treatments for malaria were produced, enough for 60% of the cases of malaria identified globally. This is a tremendous step forwards compared with 5 years ago, when less than 10% of malaria patients were getting the best-quality treatments. (See Table 1.1).

One question we should ask is whether there are too many fixed-dose artemisinin combination medicines available or in late clinical development.[35] The clear answer is 'no': it is already apparent that there are clear differences between these medicines: safety profiles are not the same, costs of goods sold can vary considerably, shelf lives are different, not all have paediatric versions and some have longer half lives, and so would give an advantage in terms of post-treatment prophylaxis (how long after you are treated before you fall ill again). This is important in some parts of Africa where children may have 10 episodes of malaria a year; it becomes less important in parts of Asia where the transmission rate is much lower. Most importantly, cross-resistance patterns to the non-artemisinin partner drug are not the same. If artemisinin resistance becomes an issue, ACTs will still be clinically effective, but the resistance pressure on the partner drugs will also increase. A wider range of partner drugs is useful; but for now any new partners must bring significant advantages in terms of cost, dose or safety, properties that are difficult to predict ahead of pivotal studies.

1.5 Drugs for *Plasmodium vivax*

P. vivax malaria is generally still treatable with chloroquine, and so remains WHO's recommendation for first-line use against this species of *Plasmodium*. However, some areas in South and Southeast Asia now harbour chloroquine-resistant parasites. In these areas, WHO recommends the use of an ACT. Pyronaridine-artesunate is the only fixed-dose ACT to date where *P vivax* erythrocytic stages have been included in the label request, and pyronaridine appears to be the most potent anti-vivax agent *in vitro*.[36]

For *P. vivax* malaria, the particular challenge remains prevention of relapse. Patients who have had all blood stages of the parasite killed can still be re-infected by activation of the hypnozoite.[37] Amongst the first generation of synthetic drugs for malaria was the 8–aminoquinoline, pamaquine (Figure 1.2), introduced in 1926.[38] It has anti-relapse activity but causes haemolysis in patients who have a deficiency in the enzyme glucose 6-phosphate dehydrogenase (G6PD), meaning that ideally it should not be prescribed without patient screening. This is a big problem in disease-endemic areas where up to 20% of the patients might have this deficiency.[39] A second-generation molecule, primaquine, was used initially on returnees from the Korean war, where long-term relapses were extremely common. It is now the anti-relapse drug of choice for *P. vivax* malaria. However, primaquine also causes

Table 1.1 Fixed-dose artemisinin combination therapies for treatment of malaria already available or in late-stage development.

	Artemether Lumefantrine	*Artesunate Amodiaquine*	*Dihydroartemisinin Piperaquine*	*Artesunate Pyronaridine*	*Artesunate Mefloquine*	*Artemisinin Naphthoquine*
Partnership	Novartis, MMV	Sanofi-aventis, DNDi,[b] MMV[c]	Sigma Tau, MMV, Pfizer[a]	Shin Poong, MMV	Farmanguinos, DNDi, Mepha, Cipla,[d]	Kunming Pharmaceutical Corporation KPC
Trade Name	Coartem®/Coartem-D®	Coarsucam®	Eurartesim®[f]	Pyramax®	—	ARCO
Launch Date	1Q'01/1Q'09[e]	4Q'08	3Q'11	1Q'12	2Q'08	?
Key Strengths	Market leader with over 350 million treatments given. Excellent safety data. Paediatric formulation. WHO Prequalified	Once a day dosing. First-line therapy in francophone Africa. WHO prequalified	Once per day therapy. Long terminal half life of piperaquine. 80% patients are protected against reinfection at 42 days	Once per day therapy. Clinical data & registration also for *P. vivax* malaria. Potential to combine with primaquine for radical cure. Paediatric formulation available	Once a day therapy. Satisfactory safety record in Thailand of non-fixed combination. Successful treatment of *P. vivax* malaria in chloroquine-resistant areas	Single dose of 1 g artemisinin, and 400 mg Naphthoquine; maybe split over one day
Key Weaknesses	Twice per day treatment. Low bioavailability, potential dependence on fatty foods	Resistance (to amodiaquine) can compromise efficacy. Reputation for significant nausea. No approval by stringent Regulatory Authority	Stability: DHA least stable of the artemisinins; Paediatric formulation still in development	In registration.	Mixture of two diastereoisomers. Psychiatric and GI adverse events. No approval by stringent Regulatory Authority. Difficult to use in countries where Mefloquine is used as prophylaxis, currently expensive ($2.50).	No GCP clinical studies, little long-term safety data. No approval by Stringent regulatory authority

Market size (Number of treatments sold in 2010)	82 million total, (37 million CoartemD). Estimated 21 million additional generic Artemether Lumefantrine	45 million (fixed dose) combination	2 million	None (launch date in 2011)	400'000	Estimated to be 1 million
Stability	24 months	36 months	24? months	24 + months	No data	No data
Formulation	Tablet (adult) Dispersible (child)	Dispersible tablet for all ages	Tablet (adult) Crushed tablet (child) Dispersible (child, 2012)	Tablet (adult) Sachet (child)	Crushed tablet for all ages	Tablets

The table includes recently launched products or products which have completed phase III clinical studies and are expected to be launched over the next 2 years. Currently, the market share for ACTs is approximately 160 million treatments, of which over 100 million are artemether–lumefantrine (Novartis plus generic versions), and most of the rest are are a fixed-dose combination of artesunate-amodiaquine.

[a]Product developed by Sigma Tau, in collaboration with MMV.

[b]Drugs for Neglected Disease *initiative*.

[c]This medicine was developed by sanofi-aventis and DNDi, and MMV is partnering with Sanofi aventis and DNDi for Phase IV studies.

[d]Product developed at Farmanguinos, but to be manufactured by Cipla for some markets outside Brazil; a fixed-dose combination has also been produced by Mepha, but only at premium prices.

[e]Coartem®D is child-friendly dispersible form of artemether–lumefantrine launched this year; the original tablet form, Coartem®, was launched in 2001.

[f]Dihydroartemisinin-piperaquine is sold by Chongqing Holley as Artekin®: this has not been prequalified by WHO, or approved by a Stringent Regulatory Authority.

haemolysis in G6PD-deficient patients and must be given daily for 14 days to be effective; such compliance is unachievable in the field, and some countries even have 5–7 days in their treatment guidelines (even though clinical evidence says this does not work). A third-generation 8-aminoquinoline, tafenoquine, is in clinical development, and could be given as a single dose.[40] A phase II study to determine the safe dose of tafenoquine in G6PD patients is ongoing, sponsored by GSK. The parallel efficacy study is planned to start in 2011. The challenge in this field is the identification of new chemical entities which can kill the hypnozoite without the liabilities of the 8-aminoquinolines to G6PD patients. The most significant barrier to success here is undoubtldely that the currently available animal models cannot accurately predict the safe and the effective exposure needed in man. Mouse models of G6PD dependent haemolysis are being validated, but will still carry the considerable risk that clearance of deformed erythrocytes may be different between humans and mice. There are two additional barriers: first, it is currently impossible to maintain stable long-term cultures of *P. vivax* – so there are no standard 'relapsing' clones to study. This is important since relapse rates can be highly variable. Second, there are no cell-based models of hypnozoite formation in human cell lines. Model systems are available where the related *P cynomolgi* infects primary rhesus cells, which suggests that a human cell line with the right cell surface antigens or the right architecture will need to be developed (Clemens Kocken, *personal communication*).

Drugs targeting severe malaria: This is a challenging area for drug development. Intravenous artesunate is replacing quinine as the reference medicine, based on superior efficacy and tolerability.[41,42] Until recently, the major challenge was to ensure a supply of drug manufactured to international standards of Good Manufacturing Practice (GMP), so that it can be prequalified by the WHO, allowing it to be purchased by donors. In December 2010, WHO prequalified Guilin Pharmaceuticals, supported with regulatory advice from Medicines for Malaria Venture, who can provide artesunate for injection at a price that is affordable in disease endemic countries (around $1 per 60 mg vial). Any new medicine for this indication would have to show superiority over artesunate, (or quinine if artesunate fails) which would require large clinical studies, of more than 5000 patients. In any case, the development path for a new severe malaria treatment would need to initially demonstrate rapid activity in uncomplicated malaria in man, since the animal models of severe malaria are not predictive.[43] One interesting proposal has been for the use of a rectal artesunate formulation for the emergency treatment of severe disease. In a sub-group analysis, patients more than 6 hours from hospital showed a significant benefit to being given such a suppository.[44] This clearly illustrates the dichotomy of severe malaria: the patients who are most at risk, are by definition the ones furthest from hospital, and so access to medicines becomes the key question. This treatment is under development by the WHO TDR, and is under regulatory discussion with the UK authorities.

1.6 Prophylaxis

Prophylaxis for pregnant women and children has been promoted in many sub-Saharan African countries. In children under five, it increases haemoglobin levels, reduces the frequency of clinical episodes and reduces overall mortality. The effects in pregnant women, particularly among primigravids (first-time pregnancy), decrease the number of premature births, with a subsequent increase in the life expectancy of the newborn. A regimen of 2 or 3 doses of sulphadoxine-pyramethamine, chosen because of its long half-life, delivered during pregnancy (IPTp) have been shown to provide significant benefit both to the mother and the foetus, and is currently implemented routinely as part of comprehensive malaria control strategies throughout most endemic countries. A similar approach has been followed with young infants (IPTi).[45] Unfortunately, sulphadoxine-pyramethamine is being increasingly compromised by drug resistance, even for such indications. The drug development challenge for new products here, however, is enormous. Before medicines can be used for IPT, then they need to be demonstrated to be active in curing malaria. However, most countries are unhappy about the idea of having the same medicine used for prophylaxis and for treatment. Furthermore, the benefit/risk ratio in IPT will fall as the incidence of malaria falls, and so it is most unlikely that new medicines will be developed for this indication.

Antimalarial medicines are used not only for treatment but also for prophylaxis of high-risk groups. Non-immune travellers to malaria endemic areas, mostly tourists, constitute an important group. Of all the drugs developed, only pyrimethamine, proguanil and atovaquone are known to have causal prophylactic activity, and stop the infection reaching the blood stage. All other chemoprophylactics work solely as blood schizonticides. The drug of choice is GSK's Malarone, a fixed combination of atovaquone-proguanil, which has causal prophylactic, blood schizonticidal and transmission blocking activities.[46] Mefloquine is less expensive and has a long half-life, which provides good prophylaxis even when administered weekly, but it is associated with gastrointestinal and psychiatric side effects. Both molecules are expensive, partly because of synthetic issues, but also because, since they are used in commercial markets, there is less of a price pressure on the active pharmaceutical ingredient.

1.7 Development Challenges

The development of a medicine with two active ingredients brings special challenges. First, there is the need to select two active ingredients with different mechanisms of action, which preferably are therapeutically synergistic, without adversely affecting the uptake, distribution or safety of each other. The optimum dose for each ingredient needs to be confirmed in patients. Second, co-formulation is often a challenge, since the drugs must not interact with each other chemically. The final product has to be stable, preferably for 2–3 years with less than 5% degradation, under conditions of high temperature

(30 ± 2 °C) and relative humidity of between (65 ± 5 and 75 ± 5%).[47] With a cost limitation of $1 per adult treatment, formulation cannot add significantly to the expense. Third, medicines may have to be dosed and formulated differently for small children and pregnant women.

1.8 The Next Generation of Antimalarials: Developing a Target Product Profile

Successful drug discovery means starting out with a clear idea of what the final product should look like, the Target Product Profile (TPP). Since the time taken to develop and launch a medicine is over a decade, this Target Product Profile must be able to address issues in the future, and show how a new medicine would be superior over the treatment that is current, and the expected treatment paradigm in 10 to 15 years time.

The push for new medicines to drive the eradication of malaria has led to the development of the concept of SERCaP: a medicine that, with a single expoure, can produce radical cure and prophylaxis against all plasmodium species.[48] Obviously, this is an aspirational goal, and it should always be remembered that the medicine will be a combination. A single molecule need not tick all boxes, but the more needed, the higher the cost of goods and, for example, if some only contribute to transmission-blocking, there will be ethical issues surrounding their deployment. It could be that this is a 'goal too far', similar to desiring a new type of transport which can travel on land, in air and under water, but with the appeal of a sports car. Alternatively, it could be seen as aspirational 30 years ago as hoping for a device that sends telephone calls, shoots and replays video clips, and acts as an airline boarding pass. These examples underline that prior to the design of a perfect medicine, we must first understand the properties of the individual components.

Four sets of properties are required in the next generation of antimalarials. Putting them together in appropriate permutations would give the ideal medicine, but they are worth persuing initially in their own right. Details of the target product profiles have been published by Medicines for Malaria Venture.[49]

(a) Non-artemisinin-based combination therapy (NACT) for treatment of acute uncomplicated malaria in children and adults. This must have activity against the blood parasites of the 4 key species that cause malaria in humans, including organisms resistant to existing therapies. The big debate here, of course, is whether synthetic endoperoxides will be active in areas of artemisinin resistant malaria.

(b) A single agent for the radical cure of malaria caused by *Plasmodium vivax* and *P. ovale* and thus prevent relapse. Ideally, this must have a sufficiently large safety/efficacy window to be used in patients who are G6PD deficient.

Table 1.2 High level Target Product Profiles for antimalarial drugs.

Product For	*Uncomplicated Malaria*	*Severe Malaria*	*Radical Cure of Relapsing Malaria*	*Intermittent Preventive Treatment (IPT)*
Technical name	Oral blood schizonticide	Parenteral, rapidly acting blood schizonticide	Anti-hypnozoite	Causal prophylactic – prevents the initial infection of liver and stops replication in liver and in erythrocytes
Gold standard	Artemether–lumefantrine (chloroquine for *P. vivax*)	*i.v.* artesunate replacing i.v. quinine as a result of the recent clinical studies	Primaquine (tafenoquine)	Sulphadoxine-Pyramethamine
Key weakness to be overcome	Twice per day treatment; low bioavailability, with second dose after 6 h. Potential dependence on fatty foods, because of low bioavailability. Mismatch between pharmacokinetics, exposing products to resistance	Availability of GMP material for *i.v.* or *i.m.* (artesunate). Guilin now prequalified by WHO; other suppliers submitting in 2011	Compliance: current drug requires 14 days of dosing. Dose-limiting haemolysis in G6PD-deficient patients.	Propensity for development of resistance – a failing treatment in most countries
Added value for new treatments	New mechanism of action (no cross resistance with artemisinins or 4-aminoquinoline/amino-alcohols)	None – until artemisinin resistance becomes common.	Maintain effects against hypnozoites and gametocytes without toxicity and with 3 days of dosing.	Lower propensity for development of resistance.
Cost	Less than US$1	Less than US$5 per patient	Less than US$10	Less than US$1

(c) A medicine that can block the transmission of *Plasmodium* infection, preventing re-infection of mosquitoes and thus transmission to other people. There has been much debate about how, and even if, such a medicine could be developed. The best conceptual starting point here is primaquine, which has anti-gametocyte activity. It is not clear whether a 'pure' anti-gametocyte compound exists. Proof of concept studies (phase IIa) would probably involve volunteer infections, and clinical confirmatory trials (phase III) would use a combination with a schizonticide.

(d) Causal prophylactic activity. Clearly, for travellers and other migrant populations such as the military, this has always been an important area, and is one of the few places in malaria where there is a significant commercial market. However, this also includes new agents for intermittent preventative treatment of malaria in pregnant women (IPTp), children (IPTc) and infants (IPTi). Compounds would have to be shown to cure uncomplicated malaria before clinical studies on prophylaxis could be undertaken. It would also be important to have a long half-life (several weeks), which could be effected by slow release formulation provided the compound was potent (less than 10 mg/day).

Treatment of severe malaria is a difficult issue. A combination product is not required, but a compound with pure antimalarial activity would need to be shown to be active clinically in uncomplicated malaria. A compound with adjunctive activity would need to be shown to be active in a different disease, given the extreme vulnerability of these patients, and the fact they are generally small children.

As well as defining the clinical parameters, such as efficacy and safety as part of the target product profile, several other factors come into play which can be used to aid the early selection of potential candidates. The first is cost. Current antimalarials are available in disease endemic countries for prices of around \$1.20 for an adult treatment, and as low as \$0.30 for an infant dose. This is a tremendous achievement, but is still too expensive for many families, and will always require subsidy. Further reductions in cost for a combination product could come from selecting more potent compounds in man (many of the 4-aminoquinolines are used at doses of 10 mg/kg/day). Alternatively, lowering the cost of the raw materials is a possibility: recent work by Medicines for Malaria Venture and our partners has shown it is possible to significantly lower the bulk cost of some active ingredients. The second area is half-life. In the malaria field we have become used to medicines with a half-life of over 7 days, but this is far from the norm in pharmaceuticals. If a new generation of medicines with long half-lives is to emerge from research, this will be either as a result of focusing research on these classes of molecules, or working on slow-release formulations. Here, potency is an issue. Once per month, formulations have been made for other medicines, but are limited at a dose of 0.2 mg/kg/day. Third, one key factor to look at is the Parasite Reduction Rate in man. Artemisinins have a tremendous benefit for the patient in clearing parasites in as short a period as 24 hours, since they are active against all stages of the parasite lifecycle. New medicines with this potential are clearly at a premium. Finally, safety is a critical issue. Primaquine is not sufficiently safe to be given to the general population because of concerns about haemolysis in G6PD subjects, which make up 10–20% of the population in disease endemic countries.[39] Similarly, dapsone has been withdrawn from therapy because of inadequate safety in these populations.[50] Rarer adverse events may not arise until later in clinical development, and are usually hard to predict from preclinical data.

1.9 Finding New Molecules: Genes and Screens

Although artemisinin-based therapies have changed the lives of hundreds of millions of patients, the risk of resistance emerging should always be at the back of our minds. Early warnings have come from the Thai–Cambodian border.[51] Before resistance can be declared clinically, 3 criteria have to be met. First, a change in the clinical outcome – this is clearly seen. Second, there needs to be confirmation that these patients had a reasonable plasma concentration of the drug, which is the case. Third, that there is a significant change in the IC_{50} of the parasite when studied *ex vivo*. This has been so far hard to show conclusively, with very small IC_{50} shifts. There are suggestions that the parasite merely stays dormant for 48 hours, avoiding the drug pressure. Whatever the mechanism, the WHO is working hard to contain the outbreak in Southeast Asia, and the artemisinin combination therapies are still able to produce the required adequate clinical and parasitological response. However, given that it can take 15 years to bring a drug from discovery through to the market place, and that the success rates are not high, then it is prudent to invest heavily in finding new molecules and new scaffolds. Two approaches are being followed.

Fast followers: the first approach is 'fast followers'. Of the 4 clinically active families (quinine, mitochondrial inhibitors, artemisinin derivatives and anti-folates), Not sure that quinine is covered, might need a review type reference here all are being worked on, and are covered in later chapters. The anti-folate question – a pyrimethamine-type agent, which looks better than pyrimethamine *and* works against all the resistant strains – is a challenge. Such molecules are being identified,[52] but they face a high hurdle in development. Synthetic endoperoxides have been developed and are discussed in a later chapter, but will need to demonstrate activity against artemisinin-resistant parasites. Currently, the only generally accepted model would be changes in parasite clearance time in phase II clinical studies, so again this is a large hurdle to overcome.

New molecular targets: in malaria drug discovery, the sequencing of the whole genomes of *P. falciparum*[53] and *P. vivax*[54] has been important in identifying the full range of potential targets against which a drug can be expected to interact. Moreover, it has provided the basis for comparisons between the gene expression patterns at different stages of the life cycle and between different parasite species. This data set allows searches for new target classes – never before pursued in drug discovery.[55] It is still possible to prioritise novel targets based on the likelihood of finding a small molecule inhibitor[56] and based on their similarity to targets for which a drug has already been found. This opens up the possibility of 'orthologue' searching. Here, compounds in a company's database are selected from medicinal chemistry programmes that were directed against a human target that is a close orthologue of the parasitic target. A small collection of compounds, including structurally related, active and inactive compounds against the human target can then be tested against the parasite enzyme. This has been successful with other parasites, with hit rates as high as 10–25% and can lead to compounds that are selective against the parasite target.[57] However, it must be stressed that these targets are not 'validated' in the strictest sense of the word – that inhibitors working against the target have not been shown to work in man. This is a big risk: across all therapeutic areas, two-thirds of medicines fail the first time they are tested in man. Although there are many reasons to be more optimistic about the track record for anti-infectives, we still need to proceed cautiously. There is a concern that resistance generation for protein-based targets has historically been relatively rapid.[58]

Cell-based screening: overall, in anti-infectives, target-based screening has not fulfilled its promise. Analysing the results of screening against validated molecular targets for antimicrobials, Payne *et al.*[59] concluded that the molecular-based approach was not efficient at all. Target validation was an imperfect science, and going from enzyme hits to killing the microbe, was not simple. The membranes surrounding most microbes and the cell walls external to them prevent access in all too many cases. The recommendation was two-fold, either work on clinically validated targets or screen against the whole microorganism (Figure 1.3).

Advances in image processing and automation means that this is now possible. Over the last few years, several million compounds have been screened against the parasite in the intra-erythrocytic stages, resulting in over 20 000 compounds with activity below the micromolar level – a hit rate of 0.5%.[60–62]

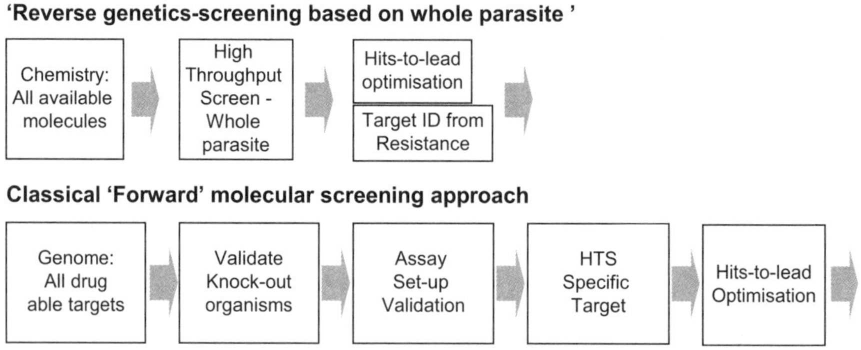

Figure 1.3 The advantages of performing the screening of compound libraries against whole parasites and identifying targets in parallel, once a chemical series has been elaborated. This process led to the extremely rapid identification and development of the spiroindolone NITD-609. From Ref 63.

Not only was this hit rate much higher than most of us predicted, but it is also much higher than that seen when screening molecular targets. This has also led to a sea change in the way drugs can be discovered. GSK, Novartis, and the academic consortium led by St Judes Hospital agreed to deposit their screening hits in a public domain archive.[64] This opens up two possibilities: first, that chemists with an interest in a particular structural class or type can see what other information is available in terms of substructure analysis. Second, it helps to focus ongoing screening on truly novel scaffold types: there is little point in rescreening compound libraries with similar content. Early prioritisation of scaffolds can be based on half-life, speed of killing, dose in man and potential low cost of goods. How feasible is it to develop the compound in the absence of a target? It can be argued that the real target of many drugs is unknown and that for registration, the authorities are more interested in knowing that a molecule is effective and safe. Rottman *et al.*[65,66] show that this can be done rapidly; their hit was optimised into a preclinical candidate in 3 years, this is turbocharged even by non-neglected disease standards. Target identification can come later, and in parallel with development. In the Rottman *et al.* work, the first hints of the target came from resistance studies – in this case, the identification of PfATP4, and the impact on protein synthesis.

Natural products: The Rottman paper has a final twist: although the original hit was identified in a natural product screening set, it is not a natural product, but a fully synthetic compound. Molecular hits from natural product collections have not been high, but where they occur, for example, quinine and artemisinin, they have been highly productive. Three hundred new anti-malarial compounds have been isolated from plants used in traditional medicine over the period 2005 to 2008.[67] However, this promising number uses a cut-off of 11 micromolar in cellular assays. If the same cut-off used in pharmaceutical screening is applied, only 20 new structures pass – one thousandth the yield obtained from screening random pharmaceutical diversity. The barrier for what is defined as interesting suddenly goes up. This raises a fundamental question about natural products work in malaria. Either compounds should be exquisitely potent (single digit nanomolar hits have been seen in oncology, for example) or they should have some other major advantage. In malaria, many natural products have actually been tested out on patients, and therefore, if it could be verified in an observational study that these patients actually had malaria (rather than a non-specific fever) – then we might be on the road to another artemisinin.

Of all the natural products literature so far, only 2 clinical studies stand out. First, a traditional treatment from the Democratic Republic of Congo, PR 259 CT1, containing strictosamide, which has under gone a phase I safety study[68] and a phase IIb controlled study in 65 patients treated for 7 days, and the patients had significant parasitaemia and fever (according to WHO research criteria). The proportion of patients cured at day 14 was 90.3% compared with amodiaquine artesunate at 96.9%.[69] This is marginal; WHO guidelines are to look 28 days after treatment and change first-line treatment once cure rates are <90%, but it is unlikely that either a *cinchona* or *artemesia* decoction would do better. Interestingly, the extract is active in murine models,[70] but not *in vitro* suggesting that deglycosylation may be required for activity. The second example is a decoction of *Argemone mexicana* where 199 patients were treated with the extract containing 3 active ingredients, compared with 102 in the control arm with amodiaquine artesunate.[71] Here, the median age was only 5 years (showing that any effect seen does not necessarily require the patient to have some immune protection. Again, the effect seen was borderline: 89% success at day 28, compared with a 95% success rate for amodiaquine-artesunate, but extremely exciting for an unpurified fraction. There may be other such studies in the literature, but these two serve to illustrate (a) that good clinical observational studies can be done, (b) natural products which are active in man can be identified and (c) that it is unlikely that these scaffolds would have been found by traditional screening routes. The important thing is now to use these molecules as the basis of medicinal chemistry programmes to make new medicines that better fit the target product profiles.

Drug repositioning: screening of almost 2000 of the known approved drugs and molecules which were in phase II, yielded some interesting results.[72] Astemizole was identified as a positive, which makes sense based on the similarity of its structure to the chloroquine series. Several other molecules from oncology were shown to be active. This raises the question that all of these molecules would have been interesting if they were already known to be inactive in man for their primary indication. Any molecule that has already been into man but failed for reasons of efficacy, and yet reached a reasonable plasma concentration would be a prime candidate for screening. There are several hundred of these compounds scattered across the pharmaceutical industry, and these will be a major focus for us over the next years.

1.10 Eradication: Moving Beyond the Erythrocytic Stages

The success of high-content screening in finding new chemotypes with activity against the erythrocyte stages suggests that this could be applied to the other stages of the parasite lifecycle. These assays exist for gametocytes for *P. falciparum*; the challenge is to automate them and make them suitable for 384 well assays. For the hypnozoite stages of *P. vivax*, the problem is more complex.[45] Infective sporozoites are needed, which means having access to mosquito breeding facilities: new technologies for cryopreservation of sporozoites may facilitate this process.[73] The cell culture challenges are more complex, as *P. vivax* requires reticulocytes for culture. Hepatocytes are available but they have very low infection rates.[74] Most work is still focused on primary human hepatocytes or infecting primate hepatocytes with the related species, *P. cynomolgi*. The recent spotlight on eradication has served to increase the amount of funding for this area, and there are some initial results suggesting a cell-based assay for hypnozoites might be possible.

Having consistent tests for all of the stages of the parasite life cycle will be a major step forward in the agenda to find new medicines. A short-term objective would be to test all the molecules that are currently in the malaria drug development pipeline in the available cell assays for all the significant stages of the parasite lifecycle discussed above, and generate malaria 'life cycle fingerprints' for each one. This will enable side-by-side comparison of molecules at a relatively early stage, and could be key for positioning the right medicines into the right early clinical assays.

1.11 The Malaria Research Pipeline

Over 90% of the malaria drug discovery projects underway target the asexual blood stages of *P. falciparum*. The first group are those which attempt to improve on drugs that have validated activity in human disease (traditionally known as 'fast followers'). Strategies include improving inhibitors of dihydrofolate reductase to develop drugs that are active against resistant strains,[75] and new inhibitors of mitochondrial electron transport to overcome atovaquone resistance.[76] Here, the TPP is relatively simple and the key advantage of a new-generation inhibitor will be action against resistant strains of the parasite. Improvements in the potency of 4-aminoquinolines or amino alcohols, is another potential approach.[77] In this case, there is the opportunity to make a radical change in the way the drug is presented (for example, combining a 4-aminoquinoline with an endoperoxide or combining an antimalarial with a resistance blocker).[78] However, with the prospect of 5 fixed-dose artemisinin combination therapies being available to patients over the next 2 years, any new 4-aminoquinoline-containing compound must demonstrate a significant benefit in terms of cost and safety (Figure 1.4).

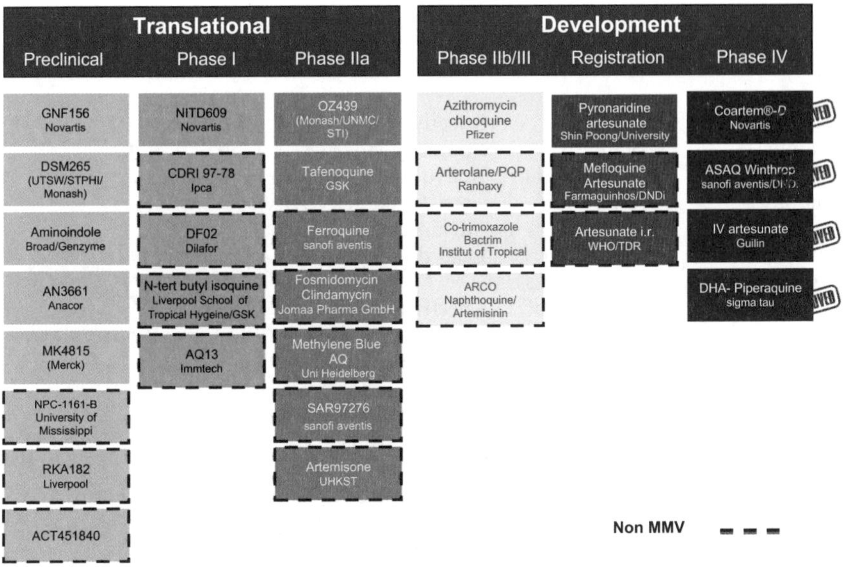

Figure 1.4 Global antimalarial drug portfolio July 2011: projects which are in collaboration with MMV are in open boxes, projects with no active collaboration with MMV are shown with a dashed border.

Another strategy is to develop drugs based on new molecular-based targets, for which there are supporting data that an inhibitor will have an effect on the parasite, but as yet there is no clinical validation (Table 1.3). Here, the project in question is usually supported by strong *in vivo* data, such as the failure to be able to rescue a viable gene knock-out in *P. berghei* or *P. falciparum*. Such validation is always useful but is subject to the concern that confirmation is based on a negative result and that there are significant differences between *P. berghei* and *P. falciparum*. For many of these targets, medicinal chemistry is aided by high-resolution structural information. Pathways such as nucleoside biosynthesis have been highlighted by genome sequence analysis,[41,46] leading to the identification of three potential drug targets. Dihydroorotate dehydrogenase (DHODH) has been targeted, as the parasite and mammalian forms differ significantly. Screening against the parasite DHODH has yielded potent and selective compounds that can now be optimised with the availability of high-resolution structural data of the enzyme–ligand complex.[79,80] The power of the rational design approach has also been underlined by the design of inhibitors of adenosine deaminase.[81] Also, transition state analogues of purine nucleoside phosphorylase have been shown to be active against

Table 1.3 Drug discovery projects in malaria: molecular targets.

Molecular Target	*Mechanism*	*Key Objective*	*Development Stage*	*Reference*
Dihydrofolate reductase	Folate Biosynthesis	Overcome existing resistance	Preclinical	49
DHODH	Pyrimidine synthesis	Selectivity and potency *in vivo*	Preclinical	53,54
Purine nucleoside phosphorylase	Nucleoside synthesis	Clinical proof of concept	Phase II for autoimmune disease	57
Adenosine deaminase	Nucleoside synthesis	Selectivity and potency *in vivo*	Preclinical	55
Cytochrome bc1 complex	Mitochondrial respiration	Active against existing resistance, selectivity against host target	Preclinical	50
Subtilisin-like protease	Egress from erythrocytes	Lead discovery	Discovery	63
Falcipains	Proteolysis of haemoglobin	Selectivity against host proteases, pharmacophores which do not cross react with host thiols	Lead optimisation	62
Fab I	Apicoplast lipid synthesis	Lead discovery	Discovery	59
Histone deacetylase	DNA replication	Active against existing resistance, selectivity against host target	Discovery	61
Kinases	Signal transduction	Active against existing resistance, selectivity against host target	Discovery	60

P. falciparum.[82,83] Some of these compounds are ready for testing in humans and have been shown to be safe, so it should be possible to rapidly evaluate the target in human malaria.

Another set of key pathways to be utilised as drug targets occur in the apicoplast in *Plasmodium*. This organelle is not found in the human host, and contains metabolic pathways and enzymes which are similar to those seen in plant choloroplasts.[84] These pathways should offer targets that have a high degree of selectivity. The best studied is the 1-deoxy-d-xylulose 5-phosphate (DOXP) pathway for isoprenoid biosynthesis, which is the target of fosmidomycin, now in Phase II clinical trials in combination with other antimalarials. As might be expected, its safety profile is excellent, it clears parasites as rapidly as chloroquine and it shows good activity against existing drug-resistant strains but an adequate clinical and parasitological response has yet to be demonstrated in all appropriate patient populations. In addition, the apicoplast has unique fatty acid biosynthesis pathways that can be targeted,[85] suggesting that other parasite targets that are inhibited by herbicides should be considered. Unfortunately, these pathways can in some cases lead to slow parasite clearance times, with parasite death delayed until the next round of replication.

A third approach to drug discovery is to use the expertise on target families already shown to be successful in other therapeutic areas. Examples of this are signal transduction cascades, including kinases,[86] as well as oncology targets such as histone deacetylase.[87] Here, the challenge is to identify molecules that show sufficient discrimination against the host targets to ensure safety. Although lack of activity against human cell lines is an *in vitro* surrogate for safety, confirmation only comes in toxicology and clinical safety studies. Another challenge is presented by protease targets, such as the cysteine proteases, falcipain[88] and the serine protease PfSub1,[89] where our understanding of catalytic mechanism made it easy to find enzyme inhibitors, but which were hard to develop as drug candidates because of selectivity issues.

Most of these approaches target the erythrocyte stages of *P. falciparum*, although some of these drugs, if they can be delivered by intramuscular or intravenous route, rather than orally, and are fast acting, could be developed for severe malaria.[90] Developing compounds that are active against gametocytes, sporozoites, exo-erythrocytic liver schizonts and hypnozoites is much more challenging as a reliable cellular model is needed to be able to define whether key metabolic pathways, such as mitochondrial respiration or nucleoside biosynthesis, are crucial for the reactivation of the hypnozoïte. Our lack of information in this area underlines how neglected the research into *P. vivax* hypnozoïtes has been.

1.12 Conclusions

Spurred by the global spread of resistance to the current standard antimalarial drugs, such as chloroquine and sulphadoxine-pyramethamine, the last 10 years have seen a massive expansion in the research and development of new medicines to maintain and improve clinical control of malaria. Now, with the re-introduction of a malaria eradication agenda, the next 10 years are set to be even more demanding. In the short term, there should be a group of new ACTs available for use. The challenge will be to better understand how to deploy these for treatment and prophylaxis in the populations that are most at risk – small children and pregnant women. For the medium-term treatment of malaria, there is a strong clinical pipeline of new medicines, which could deliver additional Artemisinin Combination Therapies if these are required, a new, shorter-course co-packaged radical cure of *P. vivax* malaria, and the first of a series of innovative non-artemisinin-based combination therapies (NACTs) to deal with artemisinin resistance, should it start to spread outside of Cambodia. For the long term treatment and eventual eradication of malaria there are further challenges, in particular, the urgent need for more new medicines to target the dormant hypnozoite stage of *P. vivax*, and new products which will block the transmission of malaria by killing the gametocytes and inactivating sporozoites and exo-erythrocytic schizonts. At the same time we must maintain a watchful eye on the emergence of resistance and develop an understanding of its potential impact on priority needs for new medicines. With the successful registration of several new medicines, we at Medicines for Malaria Venture alongside our partners are confident that we can deliver what will be required of us.

Acknowledgements

Many thanks to the entire team at Medicines for Malaria Venture and the External Scientific Advisors over the years, for their help in defining the strategy for new antimalarial medicines. We apologise to those colleagues whose work was not cited for reasons of space. An enormous debt of gratitude has also to go to all the various governments, foundations and charities that have provided the finance required to operate our Product Development Partnership.

References

1. World Malaria Report 2010, World Health Organisation, Geneva www.who.int/world_malaria_report_2010/en/index.html
2. J. L. Gallup and J. D. Sachs, *Am. J. Trop. Med. Hyg.*, 2001, **64**, 85.
3. World Bank Booster Programme for Malaria Control in Africa (2007), www.worldbank.org.
4. J. T. Griffin, M. Cairns, A. C. Ghani, C. Roper, D. Schellenberg, I. Carneiro, R. D. Newman, M. P. Grobusch, B. Greenwood, D. Chandramohan and R. D. Gosling, *PLoS One.*, 2010, **9**, 2618.
5. R. M. Chico, R. Pittrof, B. Greenwood and D. Chandramohan, *Malaria Journal*, 2008, **7**, 255.
6. C. Kerah-Hinzoumbé, M. Péka, P. Nwane, I. Donan-Gouni, J. Etang, A. Samè-Ekobo and F. Simard, *Malaria Journal*, 2008, **7**, 192.

7. http://clinicaltrials.gov/ct2/show/NCT00866619 accessed 3 August 2011.
8. P. L. Alonso, J. Sacarlal, J. J. Aponte, A. Leach, E. Macete, P. Aide, B. Sigauque, J. Milman, I. Mandomando, Q. Bassat, C. Guinovart, M. Espasa, S. Corachan, M. Lievens, M. M. Navia, M. C. Dubois, C. Menendez, F. Dubovsky, J. Cohen, R. Thompson and W. R. Ballou, *Lancet*, 2005, **366**, 2012.
9. L. Roberts and M. Enserink, *Science*, 2007, **318**, 1544.
10. Roll Back Malaria, www.rollbackmalaria.org/
11. P. Das and R. Horton, *Lancet*, 2010, **2010**, 1515–17.
12. P. L. Alonso, G. Brown, M. Arevalo-Herrera, F. Binka, C. Chitnis, F. Collins, O. K. Doumbo, B. Greenwood, B. F. Hall, M. M. Levine, K. Mendis, R. D. Newman, C. V. Plowe, M. H. Rodríguez, R. Sinden, L. Slutsker and M. Tanner, *PLoS Med.*, 2011, **8**, e1000406.
13. R. N. Price, E. Tjiltra, C. A. Guerra, S. Yeung, N. J. White and N. M. Anstey, *Am. J. Trop. Med. Hyg.*, 2007, **77**, 79.
14. E. Tjitra, N. M. Anstey, P. Sugiarto, N. Warikar, E. Kenangalem, M. Karyana, D. A. Lampah and R. N. Price, *PLoS Med.*, 2008, **5**, 28.
15. M. Mayxay, S. Pukrittayakamee, P. N. Newton and N. J. White, *Trends Parasitol.*, 2004, **20**, 233.
16. W. A. Krotoski, D. M. Krotoski, P. C. Garnham, R. S. Bray, R. Killick-Kendrick, C. C. Draper, G. A. Targett and M. W. Guy, *Br. Med. J.*, 1980, **280**, 153.
17. W. Chin, P. G. Contacos, R. G. Coatney and H. R. Kimbal, *Science*, 1965, **149**, 865.
18. J. Cox-Singh, T. M. Davis, K. S. Lee, S. S. Shamsul, A. Matusop, S. Ratnam, H. A. Rahman, D. J. Conway and B. Singh, *Clinical Infection Diseases*, 2008, **46**, 165.
19. R. E. Sinden in *Malaria: Parasite Biology, Pathogenesis and Protection*, ed. I. W. Sherman, ASM Press, Washington, D. C. 1998, p. 25.
20. T. N. Wells, Natural products as starting points for future anti-malarial therapies: going back to our roots?, *Malaria J.*, 2011, **10**(Suppl 1), S3.
21. P. L. Olliaro and Y. Yuthavong, *Pharmacol. Ther.*, 1999, **81**, 91.
22. H. Y. Qu, H. Z. Gao, G. T. Hao, Y. Y. Li, H. Y. Li, J. C. Hu, X. F. Wang, W. L. Liu and Z. Y. Liu, *J. Clin. Pharmacol.*, 2010, **50**, 1310.
23. T. Tun, H. S. Tint, K. Lin, T. T. Kyaw, M. K. Myint, W. Khaing and Z. W. Tun, *Acta Trop.*, 2009, **3**, 275.
24. S. J. Burgess, J. X. Kelly, S. Shomloo, S. Wittlin, R. Brun, K. Liebmann and D. H. Peyton, *J. Med. Chem.*, 2010, **53**, 6477.
25. M. W. Dunne, N. Singh, M. Shukla, N. Valecha, P. C. Bhattacharyya, V. Dev, K. Patel, M. K. Mohapatra, J. Lakhani, R. Benner, C. Lele and K. Patki, *J. Infect. Dis.*, 2005, **191**, 1582.
26. L. H. Carvalho, E. M. Rocha, D. S. Raslan, A. B. Oliveira and A. U. Krettli, *Braz. J. Med. Biol. Res.*, 1988, **21**, 485.
27. L. F. Fieser and G. Fawaz, *J. Am. Chem. Soc.*, 1950, **72**, 996.
28. A. T. Hudson, *Parasitol. Today*, 1993, **9**, 66.
29. C. L. Yeates, J. F. Batchelor, E. C. Capon, N. J. Cheesman, M. Fry, A. T. Hudson, M. Pudney, H. Trimming, J. Woolven, J. M. Bueno, J. Chicharro,

E. Fernández, J. M. Fiandor, D. Gargallo-Viola, F. Gómez de las Heras, E. Herreros and M. L. León, *J. Med. Chem.*, 2008, **51**, 2845.
30. G. A. Biagini, N. Fisher, N. Berry, P. A. Stocks, B. Meunier, D. P. Williams, R. Bonar-Law, P. G. Bray, A. Owen, P. M. O'Neill and S. A. Ward, *Mol. Pharmacol.*, 2008, **73**, 1347.
31. X. D. Luo and C. C. Shen, *Med. Res. Rev.*, 1987, **7**, 29.
32. S. R. Meshnick, T. E. Taylor and S. Kamchonwongpaisan, *Microbiol. Rev.*, 1996, **60**, 301.
33. N. J. White, *Philos. Trans. R. Soc. Lond. B. Biol. Sci.*, 1999, **354**, 739.
34. World Health Organization. Guidelines for the Treatment of Malaria: whqlibdoc.who.int/publications/2010/9789241547925_eng.pdf
35. T. N. Wells and E. M. Poll, *Discov. Med.*, 2010, **9**, 389.
36. R. N. Price, J. Marfurt, F. Chalfein, E. Kenangalem, K. A. Piera, E. Tjitra, N. M. Anstey and B. Russell, *Antimicrob. Agents Chemother.*, 2010, **54**, 5146.
37. T. N. Wells, J. Burrows and J. K. Baird, *Trends in Parasitol.*, 2010, **26**, 1471.
38. P. Muehlens, *Arch. Shiffs-u Troppenhyg.*, 1926, **30**, 25.
39. E. Beutler and S. Duparc and G6PD Deficiency Working Group, *Am. J. Trop. Med. Hyg.*, 2007, **77**, 779.
40. R. P. Breukner, K. C. Lasseter, E. T. Lin and B. G. Schuster, *Am. J. Trop. Med. Hyg.*, 1998, **58**, 645.
41. A. M. Dondorp, F. Nosten, K. Stepniewska, N. Day and N. White, *Lancet*, 2005, **366**, 717.
42. A. M. Dondorp, C. I. Fanello, I. C. Hendriksen, E. Gomes, A. Seni, K. D. Chhaganlal, K. Bojang, R. Olaosebikan, N. Anunobi, K. Maitland, E. Kivaya, T. Agbenyega, S. B. Nguah, J. Evans, S. Gesase, C. Kahabuka, G. Mtove, B. Nadjm, J. Deen, J. Mwanga-Amumpaire, M. Nansumba, C. Karema, N. Umulisa, A. Uwimana, O. A. Mokuolu, O. T. Adedoyin, W. B. Johnson, A. K. Tshefu, M. A. Onyamboko, T. Sakulthaew, W. P. Ngum, K. Silamut, K. Stepniewska, C. J. Woodrow, D. Bethell, B. Wills, M. Oneko, T. E. Peto, L. von Seidlein, N. P. Day and N. J. White and AQUAMAT group, *Lancet*, 2010, **376**, 1647.
43. N. J. White, G. D. Turner, I. M. Medana, A. M. Dondorp and N. P. Day, *Trends Parasitol.*, 2010, **26**, 11.
44. M. F. Gomes, M. A. Faiz, J. O. Gyapong, M. Warsame, T. Agbenyega, A. Babiker, F. Baiden, E. B. Yunus, F. Binka, C. Clerk, P. Folb, R. Hassan, M. A. Hossain, O. Kimbute, A. Kitua, S. Krishna, C. Makasi, N. Mensah, Z. Mrango, P. Olliaro, R. Peto, T. J. Peto, M. R. Rahman, I Ribeiro, R. Samad and N. J. White and Study 13 Research Group, *Lancet*, 2009, **373**, 522.
45. D. Schellenberg, C. Menendez, J. J. Aponte, E. Kahigwa, M. Tanner, H. Mshinda and P. Alonso, *Lancet*, 2005, **365**, 1481.
46. H. Nakato, R. Vivancos and P. R. Hunter, *J. Antimicrob. Chemother.*, 2007, **60**, 929.
47. Stability data package for registration applications in climatic zones III and IV, www.ich.org/products/guidelines/quality/quality-single/article/Stability-data-package-for-registration-applications-in-climatic-zones-iii-and-iv.html

48. P. L. Alonso, A. Djimde, P. Kremsner, A. Magill, J. Milman, J. Nájera, C. V. Plowe, R. Rabinovich, T. Wells and S. Yeung. A research agenda for malaria eradication: drugs. malERA Consultative Group on Drugs *PLoS Med.*, 2011, **8**, e1000402.
49. www.mmv.org/research-development/essential-information-scientists/target-product-profiles, accessed 3 August 2011.
50. Z. Premji, R. E. Umeh, S. Owusu-Agyei, F. Esamai, E. U. Ezedinachi, S. Oguche S, S. Borrmann, A. Sowunmi, S. Duparc, P. L. Kirby, A. Pamba, L. Kellam, R. Guiguemdé, B. Greenwood, S. A. Ward and P. A. Winstanley, *PLoS One*, 2009, **4**, 6682.
51. H. Noedl, Y. Se, K. Schaecher K, B. L. Smith, D. Socheat and M. M. Fukuda and Artemisinin Resistance in Cambodia 1 (ARC1) Study Consortium, *New Engl. J. Med.*, 2008, **359**, 2619.
52. Y. Yuthavong, S. Kamchonwongpaisan, U. Leasrtsakulpanich and P. Chitnumsub, *Future Microbial*, 2006, **1**, 113.
53. M. J. Gardner, N. Hall, E. Fung, O. White, M. Berriman, R. W. Hyman, J. M. Carlton, A. Pain, K. E. Nelson, S. Bowman, I. T. Paulsen, K. James, J. A. Eisen, K. Rutherford, S. L. Salzberg, A. Craig, S. Kyes, M. S. Chan, V. Nene, S. J. Shallom, B. Suh, J. Peterson, S. Angiuoli, M. Pertea, J. Allen, J. Selengut, D. Haft, M. W. Mather, A. B. Vaidya, D. M. Martin, A. H. Fairlamb, M. J. Fraunholz, D. S. Roos, S. A. Ralph, G. I. McFadden, L. M. Cummings, G. M. Subramanian, C. Mungall, J. C. Venter, D. J. Carucci, S. L. Hoffman, C. Newbold, R. W. Davis, C. M. Fraser and B. Barrell, *Nature*, 2002, **419**, 498.
54. J. M. Carlton, S. V. Angiuoli, B. B. Suh, T. W. Kooij, M. Pertea, J. C. Silva, M. D. Ermolaeva, J. E. Allen, J. D. Selengut, H. L. Koo, J. D. Peterson, M. Pop, D. S Kosack, M. F. Shumway, S. L. Bidwell, S. J. Shallom, S. E. van Aken, S. B. Riedmuller, T. V. Feldblyum, J. K. Cho, J. Quackenbush, M. Sedegah, A. Shoaibi, L. M. Cummings, L. Florens, J. R. Yates, J. D. Raine, R. E. Sinden, M. A. Harris, D. A. Cunningham, P. R. Preiser, L. W. Bergman, A. B. Vaidya, L. H. van Lin, C. J. Janse, A. P. Waters, H. O. Smith, O. R. White, S. L. Salzberg, J. C. Venter, C. M. Fraser, S. L. Hoffman, M. J. Gardner and D. J. Carucci, *Nature*, 2008, **455**, 757.
55. Y. Zhou, V. Ramachandran, K. A. Kumar, S. Westenberger, P. Refour, B. Zhou, F. Li, J. A. Young, K. Chen, D. Plouffe, K. Henson, V. Nussenzweig, J. Carlton, J. M. Vinetz, M. T. Duraisingh and E. A. Winzeler, *PLoS ONE*, 2008, **3**, 570.
56. F. Aguero, B. Al-Lazikani, M. Aslett, M. Berriman, F. S. Buckner, R. K. Campbell, S. Carmona, I. M. Carruthers, A. W. Chan, F. Chen, G. J. Crowther, M. A. Doyle, C. Hertz-Fowler, A. L. Hopkins, G. McAllister, S. Nwaka, J. P. Overington, A. Pain, G. V. Paolini, U. Pieper, S. A. Ralph, A. Riechers, D. S. Roos, A. Sali, D. Shanmugam, T. Suzuki, W. C. Van Voorhis and C. L. Verlinde, *Nature Reviews. Drug Discovery*, 2008, **7**, 900.
57. C. Grundner, D. Perrin, R. Hooft van Huijsduijnen, D. Swinnen, J. Gonzalez, C. L. Gee, T. N. Wells and T. Alber, *Structure*, 2007, **15**, 499.

58. A. O. Talisuna and P. Bloland and D'Alessandro, *Clinical Microbiology Reviews*, 2004, **17**, 235.
59. D. J. Payne, M. N. Gwynn, D. J. Holmes and D. L. Pompliano, *Nature Rev. Drug Discov.*, 2007, **6**, 29.
60. D. Plouffe, A. Brinker, C. McNamara, K. Henson, N. Kato, K. Kuhen, A. Nagle, F. Adrián, J. T. Matzen, P. Anderson, T. G. Nam, N. S. Gray, A. Chatterjee, J. Janes, S. F. Yan, R. Trager, J. S. Caldwell, P. G. Schultz, Y. Zhou and E. A. Winzeler, *Proc. Natl. Acad. Sci. USA*, 2008, **105**, 9059.
61. F. J. Gamo, L. M. Sanz, J. Vidal, C. de Cozar, F. Alvarez, J. L. Lavandera, D. E. Vanderwall, D. V. Green, V. Kumar, S. Hasan, J. R. Brown, C. E. Peishoff, L. R. Cardon and J. F. Garcia-Bustos, *Nature*, 2010, **465**, 305.
62. W. A. Guiguemde, A. A. Shelat, D. Bouck, S. Duffy, G. J. Crowther, P. H. Davis, D. C. Smithson, M. Connelly, J. Clark, F. Zhu, M. B. Jiménez-Díaz, M. S. Martinez, E. B. Wilson, A. K. Tripathi, J. Gut, E. R. Sharlow, I. C. Bathurst, F. El Mazouni, J. W. Fowble, I. Forquer, P. L. McGinley, S. Castro, I. Angulo-Barturen, S. Ferrer, P. J. Rosenthal, J. L. Derisi, D. J. Sullivan, J. S. Lazo, D. S. Roos, M. K. Riscoe, M. A. Phillips, P. K. Rathod, W. C. Van Voorhis, V. M. Avery and R. K. Guy, *Nature*, 2010, **465**, 311.
63. T. N. Wells, *Science*, 2010, **329**, 1153.
64. www.ebi.ac.uk/chemblntd.
65. M. Rottmann, C. McNamara, B. K. Yeung, M. C. Lee, B. Zou, B. Russell, P. Seitz, D. M. Plouffe, N. V. Dharia, J. Tan, S. B. Cohen, K. R. Spencer, G. E. González- Páez, S. B. Lakshminarayana, A. Goh, R. Suwanarusk, T. Jegla, E. K. Schmitt, H. P. Beck, R. Brun, F. Nosten, L. Renia, V. Dartois, T. H. Keller, D. A. Fidock, E. A. Winzeler and T. T. Diagana, *Science*, 2010, **329**, 1175.
66. B. K. Yeung, B. Zou, M. Rottmann, S. B. Lakshminarayana, S. H. Ang, S. Y. Leong, J. Tan, J. Wong, S. Keller-Maerki, C. Fischli, A. Goh, E. K. Schmitt, P. Krastel, E. Francotte, K. Kuhen, D. Plouffe, K. Henson, T. Wagner, E. A. Winzeler, F. Petersen, R. Brun, V. Dartois, T. T. Diagana and T. H. Keller, *J. Med. Chem.*, 2010, **53**, 5155.
67. J. Bero, M. Frédérich and J. Quetin-Leclerq, *Jounal of Pharmacy and Pharmacology*, 2009, **61**, 1401.
68. K. Mesia, K Cimanga, L. Tona, M. M. Mampunza, N. Ntamabyaliro, T. Muanda, T. Muyembe, J. Totté, T. Mets, L. Pieters and A. Vlietinck, *Planta. Med.*, 2011, **77**, 111.
69. G. L. Tona Abstract MP02, www.edctpforum2009.org/fileadmin/documents/forum09/Forum2009_programme_book.pdf. Downloaded June 2010.
70. K. Mesia, R. K. Cimanga, L. Dhooghe, P. Cos, S. Apers, J. Totté, G. L. Tona, L. Pieters, A. J. Vlietinck and L. Maes, *Journal of Ethnopharmacology*, 2010, **131**, 0378.
71. B. Graz, M. L. Milcox, C. Diakite, J. Falquet, F. Dackuo, S. Giani and D. Siallo, *Transactions of the Royal Society of Tropical Medicine and Hygiene*, 2010, **104**, 33.

72. C. R. Chong, X. Chen, L. Shi and D. J. Sullivan, *Nature Chemical Biology*, 2006, **2**, 415.
73. T. C. Luke and S. L. Hoffman, *Journal of Experimental Biology*, 2003, **206**, 3803.
74. R. Udomsangpetch, O. Kaneko, K. Chotivanich and J. Sattabongkot, *Trends Parasitol.*, 2008, **24**, 85.
75. T. Dasgupta, P. Chitnumsub, S. Kamchonwongpaisan, C. Maneeruttanarungroj, S. E. Nichols, T. M. Lyons, J. Tirado-Rives, W. L. Jorgensen, Y. Yuthavong and K. S. Anderson, *ACS Chem. Biol.*, 2009, **4**, 29.
76. H. Xiang, J. McSurdy-Freed, G. S. Moorthy, E. Hugger, R. Bambal R, C. Han, S. Ferrer, D. Gargallo and C. B. Davis, *J. Pharm. Sci.*, 2006, **95**, 2657.
77. P. Oliaro and T. N. Wells, *Clinical Pharm. and Ther.*, 2009, **85**, 584.
78. J. X. Kelly, M. J. Smilkstein, R. Brun, S. Wittlin, R. A. Cooper, K. D. Lane, A. Janowsky, R. A. Johnson, R. A. Dodean, R. Winter, D. J. Hinrichs and M. K. Riscoe, *Nature*, 2009, **459**, 270.
79. J. Baldwin, C. H. Michnoff, N. A. Malmquist, J. White, M. G. Roth, P. K. Rathod and M. A. Phillips, *J. Biol. Chem.*, 2005, **280**, 21847.
80. R. Gujjar, A. Marwaha, F. El Mazouni, J. White, K. L. White, S. Creason, D. M. Shackleford, J. Baldwin, W. N. Charman, F. S. Buckner, S. Charman, P. K. Rathod and M. A. Phillips, *J. Med. Chem.*, 2009, **52**, 1864.
81. E. T. Larson, W. Deng, B. E. Krumm, A. Napuli, N. Mueller, W. C. Van Voorhis, F. S. Buckner, E. Fan, A. Lauricella, G. DeTitta, J. Luft, F. Zucker, W. G. Hol, C. L. Verlinde and E. A. Merritt, *J. Mol. Biol.*, 2008, **381**, 975.
82. E. A. Taylor Ringia and V. L. Schramm, *Current Topics in Medicinal Chemistry*, 2005, **5**, 1237.
83. D. C. Madrid, L. M. Ting, K. L. Waller, V. L. Schramm and K. J. Kim, *J. Biol. Chem.*, 2008, **283**, 35899.
84. R. F. Waller and G. I. McFadden, *Curr. Issues Mol. Biol.*, 2005, **7**, 57.
85. J. S. Freundlich, F. Wang, H. C. Tsai, M. Kuo, H. M. Shieh, J. W. Anderson, L. J. Nkrumah, J. C. Valderramos, M. Yu, T. R. Kumar, S. G. Valderramos, W. R. Jacobs, Jr. G. A. Schiehser, D. P. Jacobus, D. A. Fidock and J. C. Sacchettini, *J. Biol. Chem.*, 2007, **282**, 25436.
86. C. Doerig, O. Billker, T. Haystead, P. Sharma, A. B. Tobin and N. C. Waters, *Trends in Parasitology*, 2008, **24**, 570.
87. V. Patel, R. Mazitschek, B. Coleman, C. Nguyen, S. Urgaonkar, J. Cortese, R. H. Barker, E. Greenberg, W. Tang, J. E. Bradner, S. L. Schreiber, M. T. Duraisingh, D. F. Wirth and J. Clardy, *J. Med. Chem.*, 2009, **52**, 2185.
88. J. M. Coterón, D. Catterick, J. Castro, M. J. Chaparro, B. Díaz, E. Fernández, S. Ferrer, F. J. Gamo, M. Gordo, J. Gut, L. de las Heras, J. Legac, M. Marco, J. Miguel, V. Muñoz, E. Porras, J. C. de la Rosa, J. R. Ruiz, E. Sandoval, P. Ventosa, P. J. Rosenthal and J. M. Fiandor, *J. Med. Chem.*, 2010, **53**, 6129.

89. S. Yeoh, R. A. O'Donnell, K. Koussis, A. R. Dluzewski, K. H. Ansell, S. A. Osborne, F. Hackett, C. Withers-Martinez, G. H. Mitchell, L. H. Bannister, J. S. Bryans, C. A. Kettleborough and M. J. Blackman, *Cell*, 2007, **131**, 1072.
90. K. Wengelnik, V. Vidal, M. L. Ancelin, A. M. Cathiard, J. L. Morgat, C. H. Kocken, M. Calas, S. Herrera, A. W. Thomas and H. J. Vial, *Science*, 2002, **295**, 1311.
91. T. N. Wells, P. L. Alonso and W. E. Gutteridge, *Nat. Rev. Drug Discov.*, 2009, **8**, 879.

CHAPTER 2

Semisynthetic Artemisinin and Synthetic Peroxide Antimalarials

LEANN TILLEY,[a] SUSAN A. CHARMAN[b] AND JONATHAN L. VENNERSTROM*[c]

[a] Department of Biochemistry and Centre of Excellence for Coherent X-ray Science, La Trobe University, Melbourne, Victoria 3086, Australia; [b] Centre for Drug Candidate Optimisation, Monash Institute of Pharmaceutical Sciences, Monash University (Parkville Campus), 381 Royal Parade, Parkville, Victoria 3052, Australia; [c] College of Pharmacy, University of Nebraska Medical Center, 986025 Nebraska Medical Center, Omaha, NE, USA

2.1 Semisynthetic Artemisinins

2.1.1 Discovery of Artemisinin, Mechanism of Action, and SAR

In 1979, Chinese scientists[1] reported the discovery of qinghaosu or artemisinin (ART), an antimalarial endoperoxide sesquiterpene lactone isolated from *Artemisia annua* (Figure 2.1). A total of 2099 patients with *Plasmodium falciparum* and *P. vivax* malaria were cured with intramuscular or oral daily doses of ART administered for three consecutive days; total doses ranged from 0.5–1.2 g (intramuscular) and 2.5–3.2 g (oral). Parasite clearance times were less than 42 h for both plasmodia species although recrudescence rates at one month ranged from 13–85% for *P. falciparum* and from 9–31% for *P. vivax*. ART also proved useful in the treatment of cerebral malaria where 131/141 patients were cured. No obvious adverse reactions or side effects were recorded

RSC Drug Discovery Series No. 14
Neglected Diseases and Drug Discovery
Edited by Michael J. Palmer and Timothy N. C. Wells

Published by the Royal Society of Chemistry, www.rsc.org

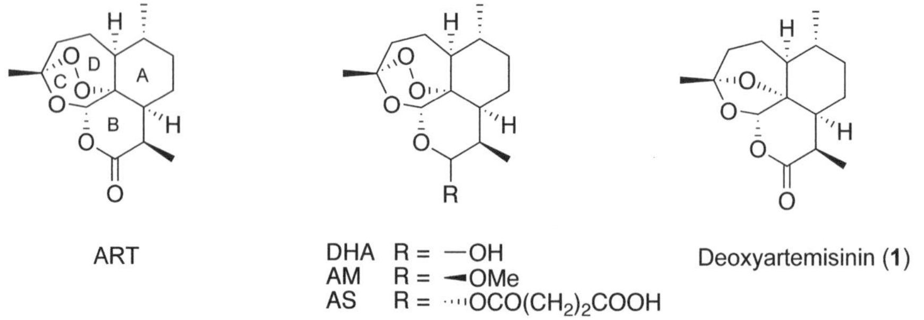

Figure 2.1 Artemisinin (ART) and its semisynthetic derivatives dihydroartemisinin (DHA), artemether (AM), and artesunate (AS) and the inactive deoxyartemisinin (**1**).

in these clinical studies. However, it was six years before the antimalarial properties of ART and its more effective semisynthetic derivatives dihydroartemisinin (DHA), artemether (AM), and artesunate (AS) (Figure 2.1) became more widely known by way of Klayman's thorough review[2] of the extensive body of ART data that had been generated by Chinese scientists.

The discovery of ART and its semisynthetic derivatives was the beginning of a new chapter in the chemotherapy of malaria. As briefly outlined below, data generated in the Avery, Jefford, Meshnick, Meunier, and Posner labs, summarized in a number of review articles,[3–12] form the foundation for much of our present understanding of the mechanism of action and structure-activity relationship (SAR) of the artemisinins. Other useful reviews include those by Li and Wu,[13] Haynes,[14] Haynes and Krishna,[15] and O'Neill *et al.*[16] As demonstrated by the complete lack of antimalarial activity of deoxyartemisinin (**1**) (Figure 2.1), the peroxide bond in ART is essential for activity, suggesting a chemistry-driven mechanism of action. Indeed, the lack of antagonism in drug combination experiments of ART and **1** show that the activity of ART does not derive from reversible interactions with parasite targets.[17]

Considerable evidence suggests that the peroxide bond in ART undergoes reductive activation by heme (or possibly by free ferrous iron derived from heme)[18–20] released by hemoglobin digestion in the parasite digestive vacuole. This irreversible redox reaction (Figure 2.2) produces carbon-centered radicals or carbocations that convey the parasiticidal effects and unique antimalarial specificity of the artemisinins by alkylation of heme or membrane-associated proteins. This is proposed to lead to perturbation of components of the parasite digestive vacuole[21–23] and other cellular targets. The ART-heme covalent adduct **4**[24] (Figure 2.3) has been identified in *P. falciparum* culture[7,8] and in *P. berghei*-infected mice[25] and accounts for as much as half of the parasite-associated radioactivity when *P. falciparum* is incubated with radiolabeled ART.[8] Two major pathways (Figure 2.2) for reductive cleavage of ART are initiated by delivery of an electron from ferrous iron to the antibonding σ^* orbital of the peroxide bond to form a pair of alkoxy radicals.

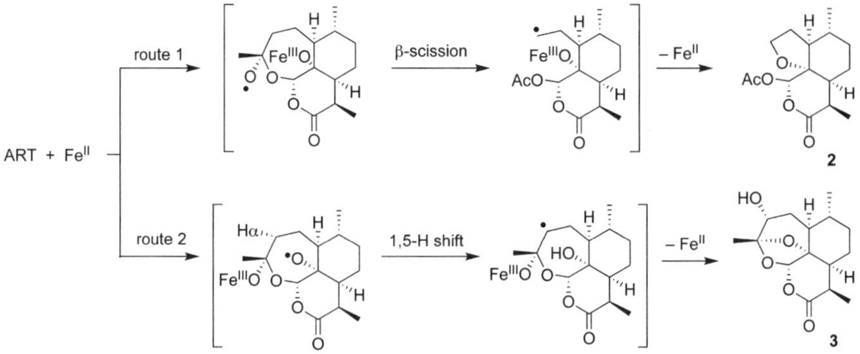

Figure 2.2 Reaction of ART with ferrous iron.

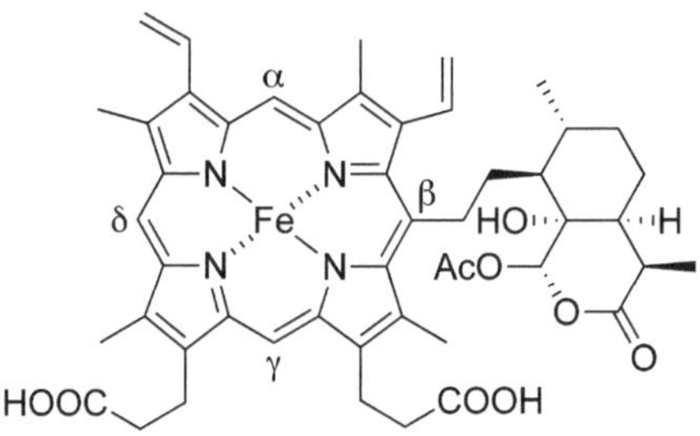

Figure 2.3 Structure of the ART-heme covalent adduct; only a single ART-heme covalent adduct (**1**) is shown. ART alkylates heme at each of the *meso* positions of the porphyrin macrocycle by an intramolecular reaction.

The first involves an attack of ferrous iron on O^1 of the peroxide bond, producing a short-lived O^2 alkoxy radical that rearranges via β-scission to a C4-centered primary radical, thermodynamically facilitated by concomitant formation of an ester; apparently, no methyl radical is produced by the alternate β-scission pathway. In the second, attack of ferrous iron on O^2 of the peroxide bond results in formation of an O^1 alkoxy radical, that rather than fragmenting by three possible β-scission pathways, undergoes a 1,5 H-shift giving rise to a C4-centered secondary radical. Both ART-derived carbon-centered radicals are irreversibly formed kinetic intermediates that, in the absence of external nucleophiles, self-quench in intramolecular reactions to produce **2** and **3** (Figure 2.2), both of which are devoid of antimalarial activity.

As noted above, hemoglobin digestion appears to be key to the mechanism of action of ART. First, ART is effective only against hemoglobin-degrading pathogens such as *Plasmodia* and *Schistosoma* species, and is orders of magnitude less potent against pathogens that do not degrade hemoglobin, such as other protozoa, bacteria, and fungi.[2,26–28] If, as is postulated by some,[29] the reaction of ART with heme in plasmodia is principally a degradation pathway that results in a loss of efficacy, the exquisite specificity of ART against hemoglobin-degrading pathogens is all the more remarkable. Second, the low activity of ART against non-hemoglobin-degrading pathogens is largely peroxide-bond independent.[26] Third, ART is effective against all plasmodial stages that actively digest hemoglobin (ring, schizont, trophozoite, early-stage gametocyte), but is much less active or inactive against plasmodial stages that do not digest hemoglobin (late stage gametocyte, sporozoite, liver stages).[30–34] Interestingly, early ring stage parasites tend to be somewhat less sensitive to ART,[33] consistent with the observation that hemoglobin digestion has only just begun in pre-digestive vacuole compartments;[35,36] consequently, less heme has been generated, although hemozoin can still be detected.[37–39] Fourth, disruption of the parasite digestive vacuole is an early event that occurs after exposure of *P. falciparum* to ART.[21,22,40,41]

Alternatively, it has been postulated that ART oxidizes parasite $FADH_2$ and parasite redox-active flavoenzymes[29] or undergoes reductive activation in the mitochondria of *P. falciparum*,[42,43] both of which are claimed to cause parasite death by an increase in reactive oxygen species (ROS). However, ART is an inefficient oxygen atom donor and does not act like a typical redox drug.[7,12] Further, if either redox-active flavoenzymes or mitochondria were important plasmodial targets or activators of ART, it is difficult to explain why all non-hemoglobin-degrading pathogens are so much less sensitive to ART. This is most clearly illustrated by the lack of activity of ART against the intraerythrocytic protozoa Babesia,[26] which unlike *Plasmodia*, does not catabolize hemoglobin.[44]

Turning our attention to the SAR of ART (Figure 2.4), we have already made the cardinal observation that ART contains an essential peroxide bond as part of its pharmacophoric 1,2,4-trioxane heterocycle (ring C). This is demonstrated by the complete lack of activity of **1** (Figure 2.1) and other non-peroxidic analogs. The high activity of deoxoartemsinin (**5**) reveals that lactone ring B is not required for activity; indeed, as reviewed by Tang *et al.*,[45] a large number of active synthetic tricyclic 1,2,4-trioxanes without the lactone ring B of ART have been prepared, largely in the Avery, Jefford, and Posner labs.

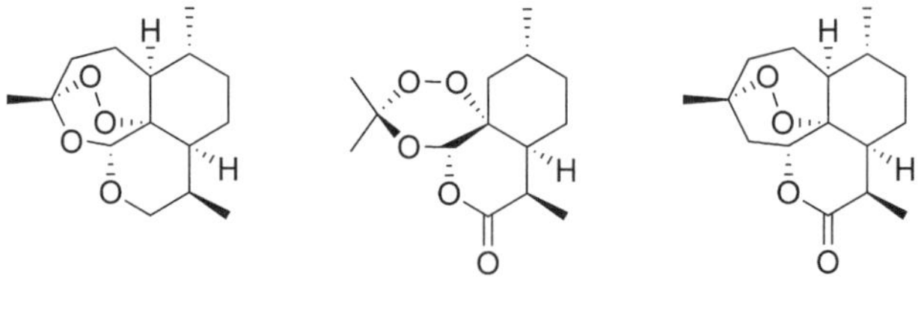

5 **6** **7**

Figure 2.4 ART structure-activity relationship (SAR): Representative compounds.

Similarly, tricyclic derivative **6** is as active as ART, indicating that ring D is also not required. As a consequence, the 1,2,4-trioxane in **6** is in a chair, not a twist-boat conformation, as is the case for ART. In addition, the methyl group on ring A is not required. The 1,2-dioxane isostere of ART (**7**), and similar 1,2-dioxane derivatives of **5**, are an order of magnitude less potent than the parent 1,2,4-trioxanes, indicating that the non-peroxide ketal oxygen atom at the 4 position of the 1,2,4-trioxane heterocycle is necessary for optimal activity. Finally, configuration and chirality play little, if any, role in the activity of ART derivatives. Clearly, the SAR of ART reveals that much of the complex structure of ART is not required[5] and more importantly, provides fertile ground for further optimization and structural simplification, as is illustrated below in our drug discovery case studies (Section 2.2).

2.1.2 Artemisinin Combination Therapy (ACT)

The semisynthetic artemisinins, DHA, AM, and AS are important antimalarial drugs (Figure 2.1) because they rapidly reduce parasite burden and have good therapeutic indices.[30,46] However, these drugs have short half-lives (1–3 h),[47,48] and as monotherapy, must be administered over a period of 5–7 days leading to noncompliance and recrudescence.[49] As a consequence of their short half-lives, ART and its derivatives are not suitable for prophylaxis. To ensure cure and encourage compliance, as well as minimize the potential for the development of resistance,[50] patients are usually treated with what has become known as artemisinin combination therapy (ACT), which are 3-day dose regimens comprising a semisynthetic artemisinin in combination with a second drug with a longer half-life.[46] These include the fixed dose combinations AM-lumefantrine, AS-mefloquine, and AS-amodiaquine. DHA-piperaquine and AS-pyronaridine will achieve registration soon.[48,51]

2.1.3 Pharmacokinetic Properties

The pharmacokinetic properties of artemisinin and its semisynthetic derivatives are not ideal, and as will be discussed later in this chapter, considerable work has been directed at synthesizing new derivatives that overcome these pharmacokinetic limitations. After oral administration to humans, ART absorption is rapid, but exhibits considerable intersubject variability.[52] The absolute bioavailability of ART is not known given the unavailability of an intravenous formulation, however, very high oral clearance (CL/F) suggests low oral absorption and high first pass metabolism.[52,53] The elimination half-life ($t_{1/2}$) ranges from 1–3 h, and there is evidence of autoinduction of cytochrome P450 enzymes leading to reduced concentrations upon repeat administration to both patients and volunteers.[54,55]

The clinical use of ART has essentially been replaced by the more effective semisynthetic derivatives, DHA, AS and AM.[56] Following oral administration of DHA to volunteers, absorption is variable[57] and somewhat delayed (Tmax

2–2.5 h).[58] The oral bioavailability of DHA measured relative to an intravenous dose of AS (which is rapidly and extensively metabolized to DHA via hydrolytic enzymes) is reported to be approximately 45% with a terminal $t_{1/2}$ of approximately 1 h.[58] Following administration of the more water-soluble AS, absorption is more rapid with peak DHA concentrations detected within approximately 1 h.[58–60] AS concentrations decline rapidly, and oral bioavailability (based on DHA) is approximately 80%.[58] In patients, DHA maximum concentrations (Cmax) and area under the curve (AUC) were found to increase relative to that in volunteers following either DHA or AS administration, a difference attributed to changes in hepatic clearance of DHA in the presence of malaria infection.[58] Following oral dosing, AM is also absorbed relatively rapidly (Tmax ~2 h) and is extensively converted to DHA.[61,62]

2.1.4 Toxicity

The semisynthetic artemisinins appear to be exceptionally safe for malaria treatment, with few if any serious adverse events.[63,64] However, these drugs are contraindicated in the first trimester of pregnancy since they are embryotoxic in experimental animals when administered over a narrow window of sensitivity corresponding to the clonal production of primitive erythroblasts.[65] While no developmental toxicity has yet been reported from use of these drugs in pregnant women, and even though the relevance of animal data to human pregnancy is uncertain and molecular mechanisms of toxicity are not completely understood,[63] this contraindication holds.[66] At high concentrations, the semisynthetic artemisinins are toxic in neuronal cell culture[67] and in animal models,[68,69] although neurotoxicity has not been shown to be a problem in humans.[63,64]

2.1.5 Potential Drug Resistance

Drug resistance is not yet a serious issue for the semisynthetic artemisinins; however, as reviewed by Dondorp *et al.*,[70] reports of higher recrudescence rates and longer parasite clearance times on the Thai–Cambodian border began to appear in about 2004. This was attributed to substandard products and the use of ART monotherapy in subtherapeutic doses.[70] The *P. falciparum* parasites in western Cambodia that show longer clearance times do not show decreased susceptibility to ART in standard *in vitro* growth inhibition assays. This resistance phenotype is not explained by genetic polymorphisms or overexpression of any of the proteins associated with the development of resistance *in vitro*. Nonetheless, studies of the heritability of ART resistance suggest that the underlying mechanism has a genetic basis and will spread through parasite populations unless the fitness cost of the associated mutations confers a significant growth disadvantage.[71] A quiescence or growth arrest mechanism for ART tolerance (or resistance) was recently postulated[72,73] on the basis of a sub-population of ring stage *P. falciparum* parasites that were driven into developmental arrest after exposure to ART or DHA. Microarray analysis of the ART-selected strain revealed no association with pfmdr1 or pfatp6.[72] Interestingly, ART dimers induced cell cycle arrest in prostate cancer cell lines,[74] providing additional support for a quiescence ART tolerance mechanism.

2.2 Investigational Semisynthetic Artemisinins and Synthetic Peroxides

2.2.1 Introduction

In the remainder of the chapter, using a case study approach, we will describe the discovery of two experimental semisynthetic artemisinins and five experimental synthetic peroxides. Even though ART is still isolated from *Artemisia annua* as its total synthesis is far too expensive for large-scale production,[75] progress continues in the effort to identify superior semisynthetic artemisinins as illustrated by the discovery of artelinic acid (**8**) and artemisone (**9**). Similarly, the SAR of ART suggested many avenues for profitable structural simplification that are illustrated by the discovery of the five antimalarial synthetic peroxides: arteflene (**10**), fenozan B07 (**11**), arterolane (**12**), PA1103/SAR116242 (**13**), and RKA182 (**14**).

2.2.2 Artelinic Acid (8)

In an effort to identify a hydrolytically stable derivative of AS, Lin and coworkers[76–79] described the synthesis of a number of carboxy ethers (acetals) of DHA prepared by boron trifluoride etherate-catalyzed etherification of DHA with various hydroxy esters followed by ester hydrolysis (Figure 2.5). Artelinic acid (**8**) is as potent as ART against *P. falciparum in vitro*, whereas derivatives without a phenyl substructure, such as **8a**, are 12- to 39-fold less potent; however, the corresponding methyl and ethyl esters are no less potent than ART.[76] When administered as 160 mg/kg subcutaneous doses on days 3, 4, and 5 post-infection to *P. berghei*-infected mice, the potassium salt of **8** cured

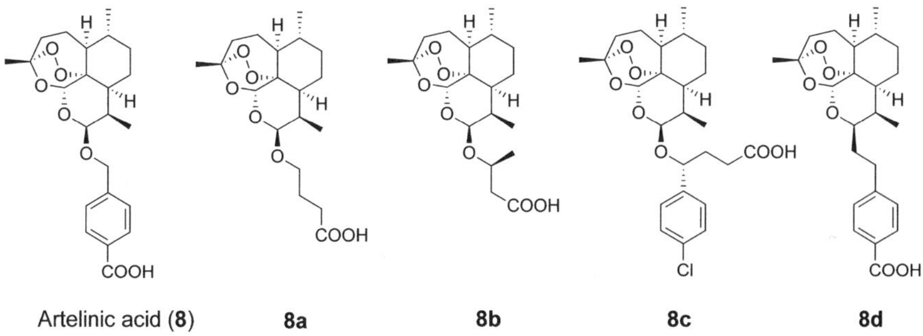

Figure 2.5 Artelinic acid (**8**) and analogs.

5/5 mice, whereas in the same experiment, AS cured 3/5 of the infected animals.[76] As illustrated by **8b**, branched derivatives of **8a** and homologous carboxy alkyl ethers of DHA were an order of magnitude less potent than **8** against *P. falciparum in vitro*.[77] Compound **8c** exemplifies efforts[78,79] to identify derivatives of **8** more stable to CYP3A4/5-mediated *O*-dealkylation to DHA[80] and to increase lipophilicity; it was the only derivative with *in vivo* activity superior to that of **8**. For example, when administered as 64 mg/kg oral doses on days 3, 4, and 5 post-infection to *P. berghei*-infected mice, **8c** cured 3/8 mice, whereas in the same experiment, **8** cured 1/8 of the infected animals. However, metabolism and pharmacokinetic data for **8c** were not disclosed.[78] In later work, Jung et al.[81] described the synthesis of deoxoartelinic acid **8d**, the non-acetal carbon isostere of **8**. Compound **8d** was as effective as **8** as an antimalarial, was 4-fold more water-soluble than **8**, and, not surprisingly, was 23-fold more stable than **8** in simulated stomach acid (pH 2.0, 37 °C). Recent work[82,83] demonstrates that the susceptibility of *P. falciparum* to **8** is, in part, a function of pfmdr1 copy number, although this resistance phenotype was unstable in the absence of drug pressure.

The pharmacokinetic profile of **8** in rats was similar to that of DHA following administration of AS. When **8** and AS were administered as 10 mg/kg intravenous doses,[84] the AUC for **8** and DHA was 5.5 and 3.2 µg.h/mL, Vd at steady state was 0.39 and 0.5 L, CL was 32 and 55 mL/min/kg, and elimination $t_{1/2}$ was 1.4 and 1.0 h. In this same experiment, the oral bioavailability for **8** and AS was equivalent (30%), although the percent of the total dose converted to DHA was much greater for AS (25–73%) than for **8** (1–4%). Artelinic acid (**8**) (10 mg/kg dose) also had a good pharmacokinetic profile[85] in dogs with a CL of 15 mL/min/kg, elimination $t_{1/2}$ of 2.6 h, and 80% oral bioavailability. In dogs, less than 0.5% of the total dose of **8** was converted to DHA.

Artelinic acid (**8**) was relatively well tolerated in rats; for example, 50 mg/kg intramuscular doses of **8** given for 7 consecutive days caused only mild anorectic toxicity.[84] In this same study, the LD_{50} of **8** in rats was 535 mg/kg. In another study,[86] single intravenous dose LD_{50} values for **8** and AS were 120 and 350 mg/kg. Given the neurotoxicity seen with DHA and AM,[67,68] **8** was also tested for neurotoxicity. In one study,[87] 36 mg/kg intramuscular doses of **8** were administered for 7 consecutive days to rats, and no behavioral abnormalities or brain damage were detected. However, in a more recent investigation,[88] neuronal injury was evident when 160 mg/kg **8** was administered orally to rats for 9 consecutive days.

2.2.3 Artemisone (9)

Given the neurotoxicity of DHA and the more lipophilic AM,[67,68] Haynes and coworkers set out to identify a more polar semisynthetic artemisinin that could not be metabolized to DHA.[89] The chosen class of compound was 10-alkylamino ART derivatives, prepared by conversion of DHA trimethylsilyl ether to the corresponding bromide followed by reactions with a number of aliphatic amines. Artemisone (**9**) (Figure 2.6) was chosen as the development candidate following assessment of the physicochemical properties, relative neurotoxicity and antimalarial efficacy for a series of analogs, exemplified by **9** and **9a–9c**.[89] Of these, **9** was marginally more polar and more soluble than DHA (log D of 2.5 vs. 2.9 and aqueous solubility of 89 *vs.* 63 μg/mL); **9a–9c** had Log D values ranging from 4.8 to 5.6 and aqueous solubilities from >2 to 8 μg/mL. Interestingly, the corresponding sulfoxide derivative (structure not shown) was an order of magnitude more water-soluble (1300 μg/mL), although it had Log D of 2.6, comparable to that of **9**. Relative *in vitro* neurotoxicity was assessed using primary neuronal brain stem culture where DHA, **9a–9c** and related derivatives were all neurotoxic to varying degrees, whereas **9** was uniquely devoid of toxicity; these data were confirmed in standard *in vivo* models of neurotoxicity.[89] In addition, **9** was considerably less anti-angiogenic than DHA[90] suggesting that it is also less embryotoxic.[65] Thus, although **9a–9c** were as active, or more active than **9** against *P. falciparum in vitro* and *P. berghei in vivo*, they were excluded from further consideration due to their less favorable physicochemical properties and higher toxicities.[89]

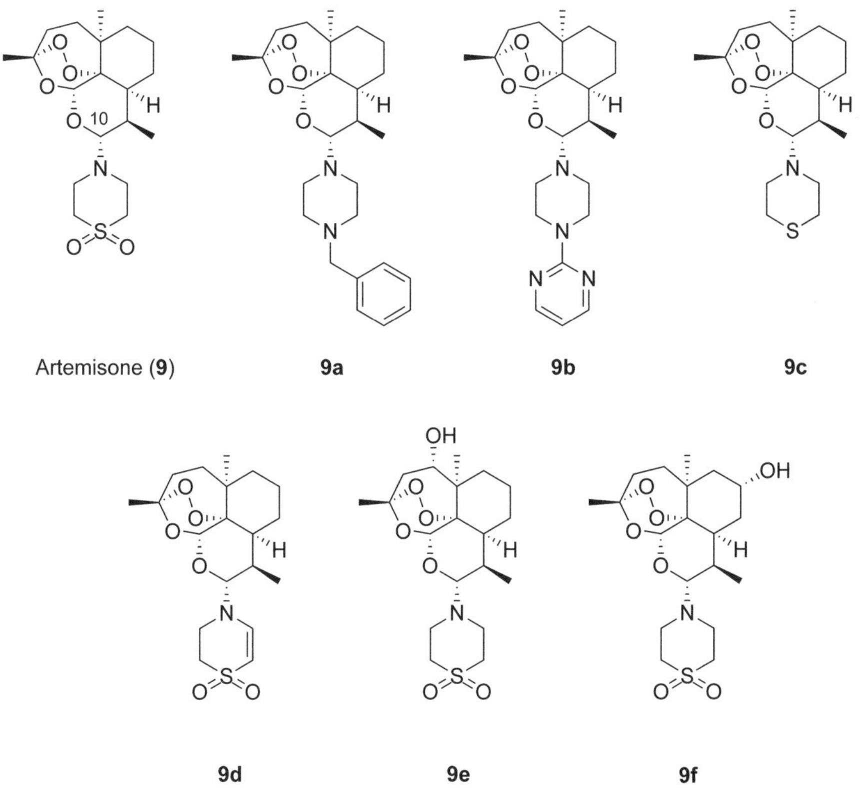

Artemisone (**9**) **9a** **9b** **9c**

9d **9e** **9f**

Figure 2.6 Artemisone (**9**), analogs, and major metabolites.

Artemisone (**9**) is 4 to 11-fold more potent than AS against *P. falciparum in vitro* and is an order of magnitude more effective than AS in the Peters 4-day suppressive test (oral administration).[91,92] In *P. falciparum* cultures, combinations of **9** with chloroquine, amodiaquine, atovaquone, tafenoquine, and pyrimethamine were slightly antagonistic, whereas combinations of **9** with mefloquine, lumefantrine, and quinine were slightly synergistic.[92] When assessed for curative efficacy in *P. berghei*-infected mice, **9** was considerably more effective than AS; after intraperitoneal administration of twice-daily doses of 5 mg/kg on each of days 7–9 post-infection (total dose 30 mg/kg), death was delayed by only one week in the AS treatment group, whereas all infected mice were cured in the **9** treatment group.[93] Artemisone (**9**) was also more effective than AS in the *P. falciparum* Aotus model where, after oral administration of 10 mg/kg daily doses of the two drugs for three consecutive days, **9** cleared parasites more rapidly than AS, parasite recrudescences tended to occur later for **9** than for AS, and only **9** was partially curative (1/4 monkeys).[89] In this same model, complete cures were obtained with single oral doses of 10 mg/kg **9** in combination with 5 mg/kg mefloquine, and with oral doses of 10 mg/kg/day **9** and 20 mg/kg/day amodiaquine given for three consecutive days.[94]

In mechanistic studies, treatment of **9** with ferrous acetate produces the C4 primary carbon-centered radical which reacts with the stable nitroxide radical 4-oxo-2,2,6,6-tetramethyl-1-piperidine-1-oxyl (4-oxo-TEMPO) to produce aminoxy adduct **9g** in 10% yield (Figure 2.7).[95] Similarly, exposure of **9**[96] and

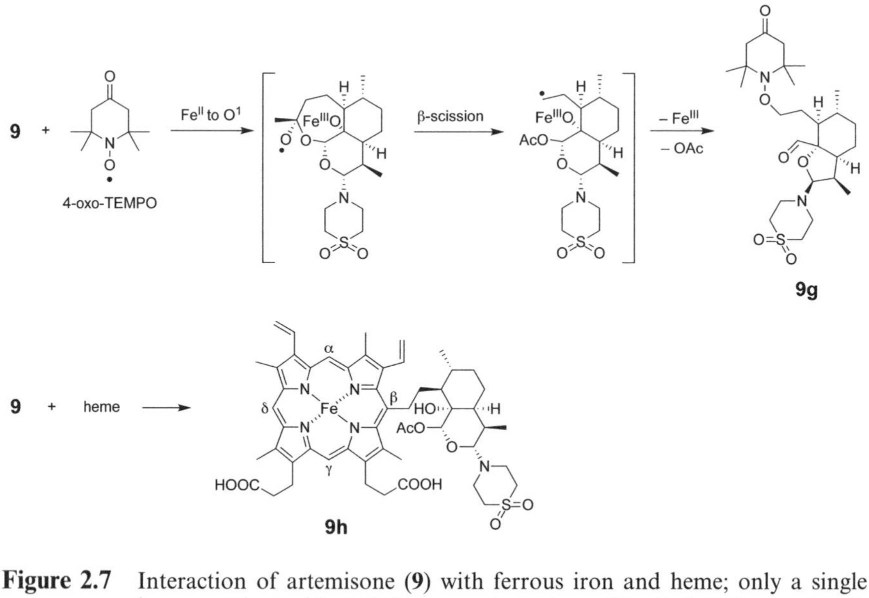

Figure 2.7 Interaction of artemisone (**9**) with ferrous iron and heme; only a single heme covalent adduct (**9h**) shown. Artemisone (**9**) alkylates heme at each of the *meso* positions of the porphyrin macrocycle by an intramolecular reaction.

9a[97] to heme produces C4 primary carbon-centered radicals that alkylate heme at the *meso* positions of the porphyrin macrocycle in high (70–90%) yields (**9h** = **9**-heme covalent adduct) (Figure 2.7). Thus, **9** and ART appear to react with ferrous iron in much the same way. An iron-dependent antiplasmodial mechanism of action for **9** is also suggested by its high potency against *P. falciparum* (IC_{50}s from 0.1 to 1 nM)[91,92] compared to its lower potencies against other non-hemoglobin degrading pathogens such as *Toxoplasma gondii* (IC_{50} = 30 nM), *Trypanosoma cruzi* and *Trypanosoma brucie rhod.* (IC_{50} = 2300 nM), and *Leishmania donovani* promastigotes (IC_{50} = 1300 nM).[27] As is the case for ART, the mechanism of action of **9** is still under investigation, and in this respect, it is interesting to note that **9** is a much more effective inhibitor of the putative ART target PfATP6[98] than ART; K_i values for **9** and ART are 1.7 and 170 nM, respectively.[99]

In Phase I studies, **9** was administered in single doses up to 80 mg and in multiple doses up to 80 mg/day for three consecutive days[100] and was well tolerated with no serious adverse events or clinically relevant alterations of laboratory parameters. For the single 80 mg dose, the Cmax was 140 ng/mL, the elimination $t_{1/2}$ was 2.8 h, the oral CL was 280 L/h, and the Vd was 15 L/kg. The oral clearance greatly exceeds hepatic blood flow, suggesting either extensive extrahepatic clearance processes or low oral bioavailability (or both). A dose proportional increase in AUC was observed for the five single-doses of **9** (10 to 80 mg). Artemisone (**9**) does not inhibit any of the CYP isoforms,[89] but it is rapidly oxidized by CYP3A4 to form three major active metabolites **9d**–**9f** (Figure 2.6). DHA is also formed following dosing of **9**, but only in low concentrations (Cmax of 10 ng/mL with an 80 mg dose of **9**); this was attributed to decomposition in the stomach.[100] Finally, preliminary data reveal that a combination of **9** and mefloquine is safe and effective in the treatment of uncomplicated *P. falciparum* malaria.[100]

2.2.4 Arteflene (10)

Yingzhaosu A (**10a**), a sesquiterpene 1,2-dioxane diol with weak antimalarial activity,[75,101] was isolated from the traditional Chinese herb, Yingzhao, *Artabotrys uncinatus L. Merr* (Figure 2.8). This peroxide natural product has *in vitro* IC_{50}s against *P. falciparum* ranging from 110–370 nM and a Peters 4-day suppressive test ED_{50} of 250 mg/kg/day in *P. berghei*-infected mice.[101] Due to the scarcity of **10a** and its lengthy 14-step synthesis from (*R*)-carvone,[75] or 8-step synthesis from (*S*)-limonene,[101] the chemistry of **10a** was little explored until a number of analogs containing its 2,3-dioxabicyclo[3.3.1]nonane core were prepared by the Roche group in the early 1990s.[102] The following synthetic sequence[102] furnished key aldehyde intermediate **10b** which provided convenient access to a variety of alkyl and arylvinyl derivatives: acid-catalysed epoxidation of (–)-carvone, ring-opening to the diol, oxidative cleavage to a ketone, Wittig olefination, singlet oxygenation to an allylic hydroperoxide, acid-catalysed ring-closure to an endoperoxide, and a final ozonolysis.

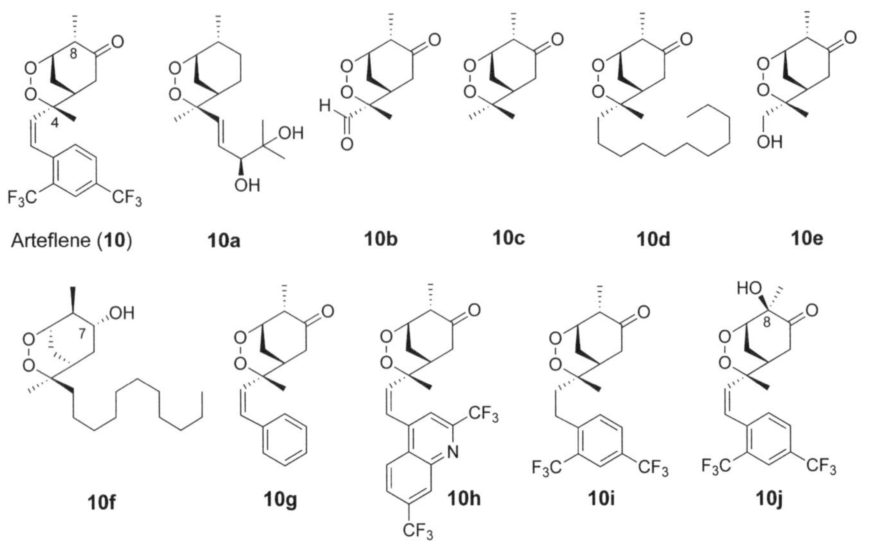

Figure 2.8 Arteflene (**10**) and analogs.

Although *in vitro* activity data for most of the analogs of **10a** were not disclosed, *in vivo* ED_{50} data (subcutaneous doses) obtained in the Peters 4-day suppressive test were reported (Figure 2.8).[102] Endoperoxide **10c**, the core structure of arteflene (Ro 42-1611, **10**), with an ED_{50} of 51 mg/kg/day, provides a benchmark for the series. Activity was insensitive to changes in absolute and relative stereochemistry. As illustrated by an ED_{50} of 5.9 mg/kg/day for **10d**, replacement of the methyl group at position 4 with *n*-alkyl chains of 9 to 11 carbon atoms led to nearly an order of magnitude increase in activity; analogs with shorter or longer chains were less active. As illustrated by **10b** and **10e**, compounds with polar functional groups such as alcohols, aldeyhdes, acids, esters, or amines at position 4 showed little or no activity, although as illustrated by **10f** with an ED_{50} of 5.9 mg/kg/day, sufficiently lipophilic alcohols were active. As shown by the inactive **10g**, replacement of the undecyl chain in **10d** with a styryl group abolished activity. However, several analogs of **10g**, including quinoline **10h** (ED_{50} of 5.4 mg/kg/day) and most notably, 2,4-di(trifluoromethyl)styryl derivative **10** (ED_{50} of 2.3 mg/kg/day), had very good efficacies. Compound **10i**, the dihydro derivative of **10**, was only 2- to 3-fold less potent against *P. falciparum* than **10**;[103] no *in vivo* activity data for **10i** were reported.

With IC_{50} values against *P. falciparum* ranging from 51 to 170 nM compared to 6.4 to 34 nM for ART, **10** is an order of magnitude less potent than ART, but its activity against *P. berghei in vivo* is comparable to that of ART.[104] For example, in the Peters 4-day suppressive test, **10** and ART had oral ED_{50}s of 10.4 and 5.0 mg/kg/day, and when administered as single subcutaneous 320 mg/kg doses in the Rane test, **10** and ART cured 9/10 and 4/5 infected mice.[104] In addition, **10** had prophylactic activity comparable to that of chloroquine.

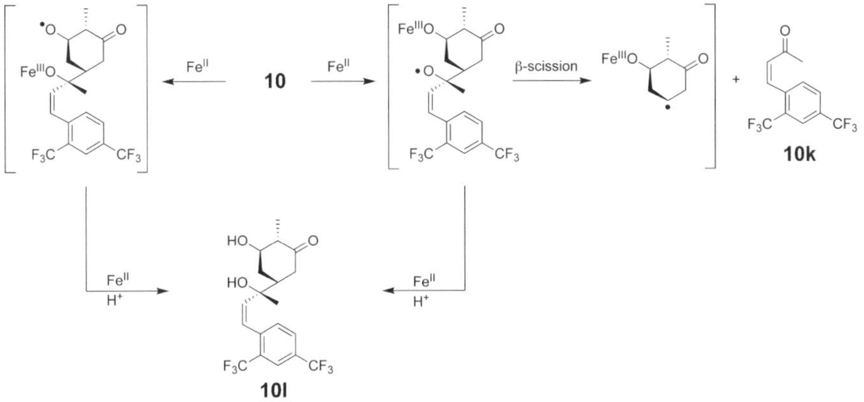

Figure 2.9 Reaction of arteflene (**10**) with ferrous iron.

Exposure of **10** to ferrous chloride produces a secondary carbon-centered radical[105] and enone **10k** via β-scission,[103] but the predominant reaction product was inactive diol **10l**[106] resulting from two-electron reduction (Figure 2.9). This reaction pathway was confirmed by ESR spectra of the radicals formed by reaction of **10** with the spin traps sodium 3,5-dibromo-4-nitrosobenzenesulfonate (DBNBS) and 5,5-dimethyl-1-pyrrolidine *N*-oxide (DMPO)[103] and by the chemically stable adduct formed by reaction of **10** with the stable nitroxide radical 2,2,6,6-tetramethylpiperidine 1-oxyl (TEMPO).[105] The same secondary carbon-centered radical was also formed by treatment of **10i** with ferrous chloride, revealing that enone **10k** does not contribute to the antimalarial activity of **10**.[103] Like ART, **10** alkylates heme and parasite proteins in *P. falciparum*-infected erythrocytes.[107]

The metabolism and pharmacokinetic properties of selected clinical candidate **10**[102] were assessed in mice, rats, dogs, marmosets, and monkeys.[108] In the various species, plasma $t_{1/2}$ values ranged from 1.3 to 4.7 h, Vd ranged from 4.1 to 38 L/kg and clearance was greater than or equal to hepatic blood flow in all species except the dog. Oral bioavailability was low and variable in all species. The high clearance of the relatively hydrophobic **10** (cLog P 5.1) was attributed to CYP450 metabolism to form hydroxylated metabolite **10j** (Figure 2.8), which was 4-fold less potent against *P. falciparum* than the parent. Preclinical toxicology studies[102] revealed that doses of 400 and 700 mg/kg/day **10** administered over a period of 4 weeks to rats and dogs, respectively, were well tolerated; minor effects on the liver, kidney, ovaries, uterus and clinical chemistry were reversible post-treatment. Arteflene (**10**) was not mutagenic, but it was embryotoxic (blood vessel abnormalities and fetal deaths) in rats at a dose of 20 mg/kg, but less so in rabbits, where only a slight increase in fetal deaths was observed at a dose of 40 mg/kg.

In a Phase I study,[109] single oral doses of 100–3600 mg **10** were administered and no serious adverse events or clinically relevant abnormalities were observed. Mean Cmax and AUC values increased proportionally for doses up to 1800 mg, at which Cmax was 580 ng/mL. Elimination $t_{1/2}$ values were between 2 and 4 h. Cmax values for the 8-hydroxy metabolite **10j** were 3-fold greater than those of **10**; elimination $t_{1/2}$ values for **10j** were similar to those of **10**. Arteflene (**10**) then progressed to Phase II clinical trials in semi-immune African patients with mild *P. falciparum* malaria.[110,111] In these trials, the drug was given as a single 25 mg/kg oral dose in a lipid suspension. In the first trial in Nigeria and Burkina Faso, **10** cleared parasites in 10/19 patients at 48 h; in the second trial in Cameroon, **10** cleared parasites in 24/30 patients at 48 h. In a later trial in Gabon, the same dose of **10** cured only 1/20 patients.[112] The inconsistent results were attributed to physicochemical problems with the suspension formulation of **10**.[110] Other factors contributing to the failure of **10** may have been its relatively weak potency, difficulty in scaling up its rather long synthesis, and its poor oral bioavailability that, in part, is a consequence of extensive first-pass metabolism to the 4-fold less active C8-hydroxy metabolite **10j**.[108]

2.2.5 Fenozan B07 (11)

In early work[113] towards elucidating the SAR of ART, a spiro ring-fused trioxane **11a** without ring D of the tetracyclic ART was synthesized in which the trioxane ring adopts a chair rather than the twist-boat conformation present in ART; **11a** was only slightly less potent than ART and was the entry into this compound series (Figure 2.10). Reactions of bicyclic 1,2-dioxolanes (formed from photooxygenation of 1,4-diaryl-1,3-cyclopentadienes) with various ketones and aldehydes catalyzed by Me_3SiOTf produced a large series of *cis*-fused cyclopenteno-1,2,4-trioxanes, exemplified by Fenozan B07 (**11**).[114]

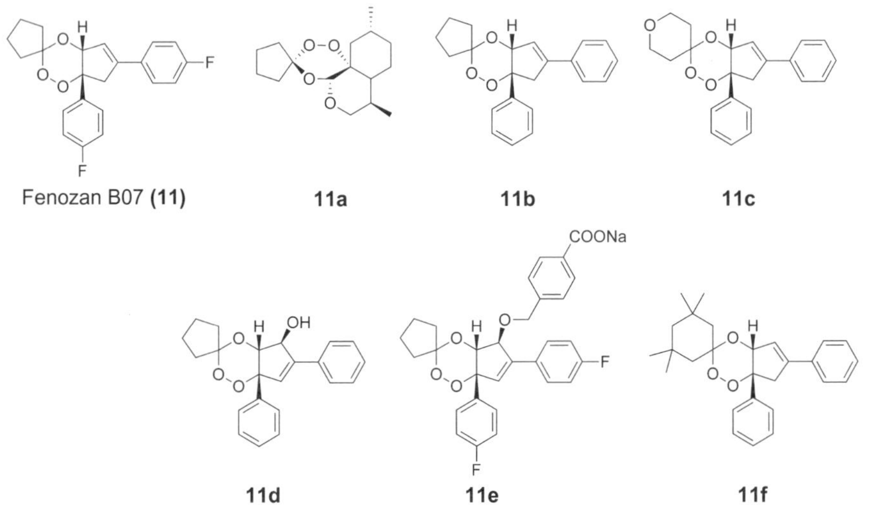

Figure 2.10 Fenozan B07 (**11**) and analogs.

A detailed SAR of this trioxane family was described by Jefford and his coworkers[5,115–119] and is briefly summarized above (Figure 2.10). First, antimalarial assays of several sets of racemates and their corresponding enantiomers showed that activity was configuration independent. Second, as illustrated by **11b** and **11c**, most spirocyclic analogs at the 3-position were as potent, or slightly more potent than **11** (IC_{50} 6.5-12 nM) against *P. falciparum in vitro* and these were much more active than the corresponding 3-monoalkyl or 3,3-dialkyl derivatives. Third, as measured in the Peters 4-day suppressive test, *in vivo* efficacy was uniquely enhanced by *p*-fluoro substitution; for example, **11** (ED_{50} 2.5 mg/kg/day) was an order of magnitude more effective than **11b** (ED_{50} 25 mg/kg/day); other *p*-substituted analogs were less active than either **11** or **11b**. Fourth, introduction of polar functional groups such as ethers (**11c**), alcohols (**11d**), or carboxylic acids (**11e**), decreased overall antimalarial efficacy. Although **11c** and **11d** are as potent as **11** against *P. falciparum in vitro*, they are 3- and 9-fold less effective than **11** in *P. berghei*-infected mice, and **11e** is inactive. Fifth, analogs such as **11f** with sterically hindered peroxide bonds are two orders of magnitude less potent than **11**. As shown in Figure 2.11, reaction of **11** with ferrous iron produces an alkoxy radical that undergoes β-scission to form a primary carbon-centered radical[5,118] that can readily alkylate heme[11] or parasite proteins. It was suggested[5] that the low activity of **11f** could be accounted for either by an inability to react with iron, or should such a reaction take place, by formation of a much less reactive neopentyl carbon-centered radical.

Among these *cis*-fused cyclopenteno-1,2,4-trioxanes, **11** (cLog P 5.1) had the most promising activity profile and was chosen for further development.[119–123] Notably, **11** is equally active whether administered subcutaneously or orally, and has an antimalarial profile equal to that of AM. Fenozan B07 (**11**) is also quite effective in mice infected with a multidrug-resistant strain of *P. yoelii nigeriensis*; in this model, five consecutive daily oral doses of **11** cured 15/16 mice.[124] Like the semisynthetic artemisinins, the potency of **11** varies less than 4-fold against a wide spectrum of *P. falciparum* strains. When **11** was administered to malaria-infected mice, marked ultrastructural changes in the membranes and ribosomes of trophozoites, young schizonts and immature gametocytes were noted, but few ultrastructural changes in mature gametocytes were seen. Its toxicity in mice appears to be low when it was administered in single oral 3000 mg/kg or subcutaneous 600 mg/kg doses.

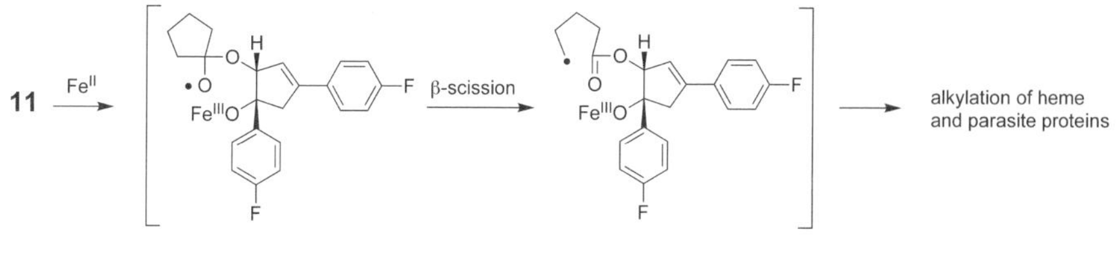

Figure 2.11 Reaction of fenozan B07 (**11**) with ferrous iron.

2.2.6 Arterolane (12)

Two fortuitous events led to the discovery of arterolane (**12**), also known as OZ277 (tosylate) or RBx11160 (hydrogen maleate) (Figure 2.12).[125] The first was the rediscovery of symmetrical dispiro 1,2,4,5-tetraoxane antimalarials in the early 1990s (see **14a**; Figure 2.16),[126] and the second was the invention of a new type of cross-ozonolysis reaction[127] in which *O*-alkyl ketone oximes (oxime ethers) are ozonized in the presence of ketones to give cross-ozonides, providing for the first time a widely applicable synthesis of symmetrical and unsymmetrical tetrasubstituted secondary ozonides (1,2,4-trioxolanes). Applying this new reaction, achiral dispiro ozonides **12a-12c** were synthesized.[128] Ozonides **12b** and **12c** were completely inactive against *P. falciparum in vitro* (IC_{50}s > 1,500 nM), whereas **12a** had IC_{50}s ranging from 4–6 nM.[128] Moreover, **12a** was no less active than AS in the *P. berghei* mouse model,[128] and this ozonide turned out to be the core structure of the compound series.

These initial data for **12a**–**12c** suggest that antimalarial activity decreases when the peroxide bond is too exposed (**12b**) or is sterically inaccessible (**12c**) to ferrous iron.[125,128] For ozonide **12a**, a middle ground seems to have been met as one side of the ozonide heterocycle is sterically hindered, but the other allows for an attack of ferrous iron on a relatively sterically unhindered peroxide oxygen atom to form an alkoxy radical which undergoes β-scission to form a secondary carbon-centered free radical. This radical was trapped by the stable nitroxide radical 4-oxo-TEMPO to form aminoxy acid **12l** in 56% yield (Figure 2.13).[129] This hypothesis is also supported by the relative reaction rates of **12a**–**12c** with ferrous sulfate where **12b** reacts at double the rate of **12a**, whereas **12c** is completely unreactive.[130] The lack of activity of the

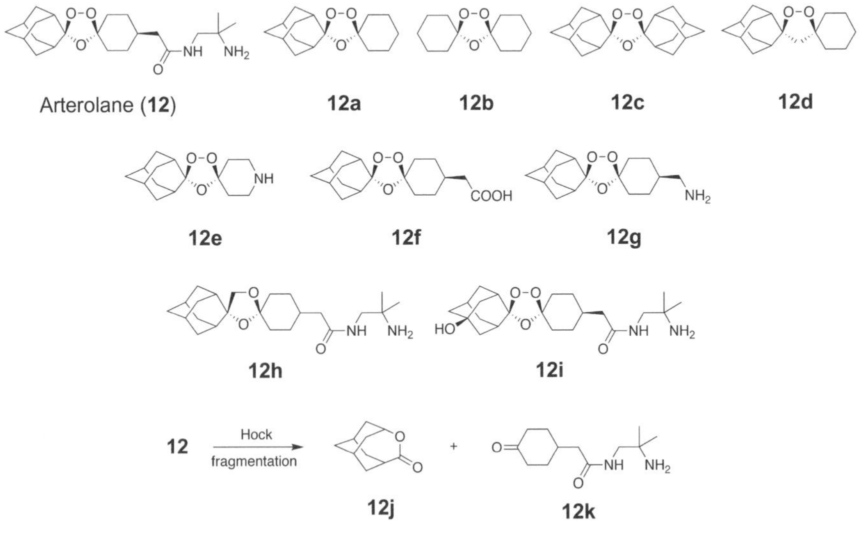

Figure 2.12 Arterolane (**12**), analogs, and major metabolites.

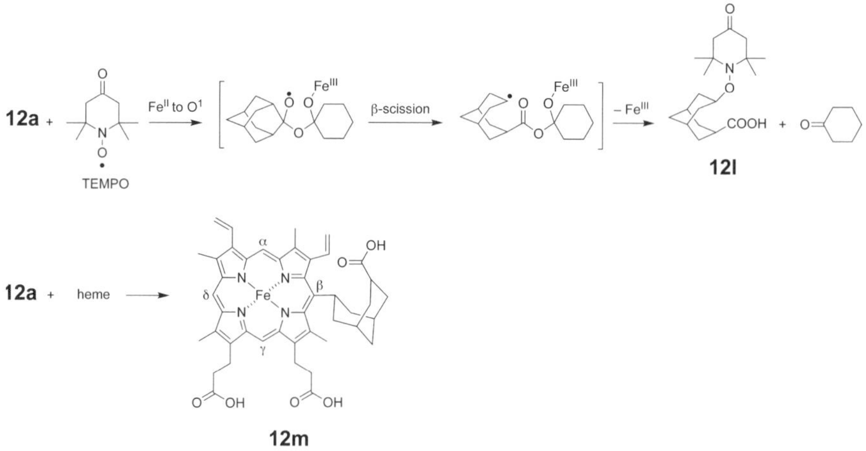

Figure 2.13 Reaction of ozonide **12a** with ferrous iron and heme; only a single heme covalent adduct (**12m**) is shown. Ozonides (**12**) alkylate heme at each of the *meso* positions of the porphyrin macrocycle by an intramolecular reaction.

non-peroxidic 1,3-dioxolane isosteres of **12a** confirmed that the peroxide bond is key to the antimalarial activity of **12a** and its analogs. However, the peroxidic 1,2-dioxolane isostere **12d** was also inactive ($IC_{50} > 2{,}500$ nM) against *P. falciparum in vitro*, an outcome explained by the very different reaction pathways of **12a** and **12d** with ferrous iron.[131] Although **12d** reacts with ferrous iron at a rate only 3-fold less than that of **12a, 12d** and similar 1,2-dioxolanes react with ferrous iron primarily by two-electron *vs.* one-electron reduction to form inactive diol reaction products rather than carbon-centered radicals,[131] the latter of which are formed by β-scission reactions of the initially formed alkoxy radicals (Figure 2.13). These β-scission reactions are accelerated by the adjacent oxygen atom[132] present in ozonides (1,2,4-trioxolanes), but absent in 1,2-dioxolanes.

The driving force for compound optimization was to improve upon the physicochemical and ADME properties of the sparsely functionalized and highly lipophilic lead **12a** (Log P 6.1) by the synthesis of more polar ozonides. Indeed, **12a** had an aqueous solubility of less than 1 μg/mL and a 1% oral bioavailability in rats.[125] To avoid the synthesis of chiral analogs of **12a**, we chose to use 4-substituted piperidones or cyclohexanones. Due to mirror plane symmetry, only two *cis* and *trans* achiral reaction products are formed in the Griesbaum coozonolysis reaction[127] between the symmetrical *O*-methyl 2-adamantanone oxime and 4-substituted cyclohexanones. Fortunately, we discovered that the coozonolysis reactions proceeded to afford primarily the *cis* ozonides, the configurations of which were established by X-ray crystallographic data.[133] As illustrated by ozonides **12** (Log D 3.2), **12e** (Log D 0.25), **12f** (Log D 2.8), and **12g** (Log D 2.6), it was possible to decrease lipophilicity considerably. However, more lipophilic ozonides tended to have better oral activities than their more polar counterparts (Dong *et al.*, 2006),[134] an outcome consistent with that seen for other classes of synthetic peroxides[45] and a formidable obstacle in compound optimization.

Building upon prototype **12a**, one can illustrate the SAR of **12** by considering ozonides **12e-12g**. Piperidine ozonide **12e** was a potent inhibitor of *P. falciparum* growth with IC_{50} values ranging from 0.5 to 1.1 nM, but **12e** and its derivatives had unexpectedly low activities against *P. berghei.*[135] This was attributed to their poor pharmacokinetic profiles that were postulated to result from their 10-fold higher reactivity with ferrous iron compared to that of protototype **12a.**[130,135] With IC_{50} values > 100 nM against *P. falciparum*, carboxy ozonide **12f** was two orders of magnitude less potent than **12, 12e**, and **12g**, but it did have high (74%) oral bioavailability.[125] The high *in vitro* and *in vivo* antimalarial efficacy of amino ozonide **12g**[125,136] combined with its acceptable (31%) oral bioavailability led to the synthesis of a wide range of carboxamide, sulfonamide, and urea derivatives of **12g**[136,137] and weak base amides of **12f.**[138] Arterolane (**12**), an example of the latter, has IC_{50} values ranging from 0.8 to 2.8 nM against *P. falciparum* and 1.1 nM against *P. vivax* and cures 3/5 *P. berghei*-infected mice with three consecutive daily oral doses of 10 mg/kg.[125,139] In the same experiment, AS increased survival to 11 days post-infection compared to 5 days for control animals. In summary, ozonides with a wide range of neutral and basic but not acidic functional groups differed little in their potencies against *P. falciparum* in vitro, but weak-base ozonides had the highest efficacies in the *P. berghei* mouse model, likely a function of their superior pharmacokinetic profiles.[134,136–138]

Further mechanistic characterization of the antimalarial ozonides, most particularly **12**, included experiments with heme, infected red blood cells, PfATP6, and the non-peroxidic isostere **12h** (Figure 2.12). Like ART, this class of ozonides does not react with ferric hematin, but does undergo rapid electron-transfer with heme followed by an intramolecular alkylation to form heme covalent adduct **12m** (Figure 2.13).[140] For 22 representative ozonides, there was a strong correlation between the extent of heme alkylation and antimalarial activity against *P. falciparum in vitro.*[140] Data from this and the following experiments support the hypothesis that interaction with heme is key to the mechanism of action of **12**. First, a sample of **12** with a radiolabel on its adamantane substructure accumulated in infected erythrocytes up to 270-fold, whereas it distributed to uninfected erythrocytes only 1.5-fold.[33] A specific LC/MS assay[33] suggested that this accumulation is not due to intact **12**, but rather to accumulation of the radiolabeled adamantyl substructure subsequent to radical formation and alkylation of heme and parasite proteins. Second, a fluorescent derivative of **12** accumulated within the parasite digestive vacuole.[18] Third, using **12h**, the non-peroxidic isostere of **12**, the antiplasmodial activity of **12** was shown to be peroxide-bond dependent, whereas the very weak activities of **12** against a range of non-hemoglobin degrading protozoa was peroxide-bond independent.[26] Finally, although PfATP6 may be a target of ART,[98] the very different K_i values of 79 and 7700 nM for ART and **12**[141] suggest that this transporter is not an important target of the latter.

In pharmacokinetic experiments in rats, **12** had a Vd of 4 L/kg and a CL of 61 mL/min/kg after intravenous administration (based on blood concentrations), and an oral $t_{1/2}$ of 1.4 h and oral bioavailability of 35% (17.4 mg/kg oral dose).[125] The clearance of **12** is accounted for by contributions of CYP450-catalyzed oxidation at the distal bridgehead carbon atoms of the spiroadamantane substructure to form **12i** (major CYP metabolite),[142] and Hock fragmentation[133,143] to form adamantane lactone **12j** and ketone **12k**; none of these has antimalarial activity (Figure 2.12). With the exception of 2C19 (IC_{50} 4.9 μM), **12** did not inhibit the five standard CYP isoforms at concentrations up to 50 μM. In standard *in vitro* assays, **12** was not mutagenic, but it had a hERG IC_{50} of 2.4 μM and was weakly embryotoxic in rat whole embryo culture, but much less so than DHA with respective NOEL values of 175 and 10 ng/mL.[66] When mice were treated with daily doses of 300 mg/kg **12** for five consecutive days, the minor clinical findings were completely reversible after a 1-week recovery period. After progressing through preclinical toxicology studies and a successful Phase I trial, **12** underwent a Phase II trial with 230 patients with once daily doses of 50–200 mg for 7 consecutive days.[144] In patients, **12** had a 3-fold lower exposure compared to that observed in volunteers, ascribed to an increase in the rate of degradation of **12** in infected *vs.* uninfected blood.[145,146] The 100 and 200 mg doses cleared 90% of parasitemias within 24 hours with recrudescence rates of 28–37% after PCR adjustment. Arterolane (**12**) is now in Phase III clinical trials in combination with piperaquine phosphate.[144,146]

2.2.7 PA1103/SAR116242 (13)

Meunier and colleagues applied the hybrid drug design concept in their discovery of drug development candidate **13**, one of the so-called trioxaquines, designed to have a dual mode of action: Heme alkylation by a 1,2,4-trioxane substructure, and inhibition of hemozoin formation by a 4-aminoquinoline substructure, and thereby embody in a single molecule the beneficial attribute of combination chemotherapy to overcome parasite drug resistance. The prototype in this work was **13a** (DU1102),[147] a construct of a fenozan-type 1,2,4-trioxane[118] and a 4-aminoquinoline (Figure 2.14), tested as both the free base and dicitrate. Diastereomeric trioxaquine **13a** was obtained in high yield by a reductive amination reaction between trioxane ketone **13c**[118] and quinoline **13b**. Initial data for **13a** was encouraging as it had an IC_{50} of 2 nM against a chloroquine-sensitive *P. falciparum* isolate compared to IC_{50}s of 45 and 28 nM for the component substructures **13b** and **13c**.[147] Analogs of **13a** with longer *n*-alkyl linkers were less active. Subsequent evaluation of **13a** revealed that it was somewhat less potent than was observed in the initial investigation;[147] for example, against a panel of 32 Cameroonian *P. falciparum* isolates, **13a** had mean IC_{50}s of 43 and 40 nM, respectively, against chloroquine-sensitive and -resistant isolates, and the IC_{50}s for **13a** did not correlate with those of chloroquine.[148]

Exploration of the trioxaquine 1,2,4-trioxane substructure led to the synthesis of **13d** (DU1302) obtained by a reductive amination reaction between a

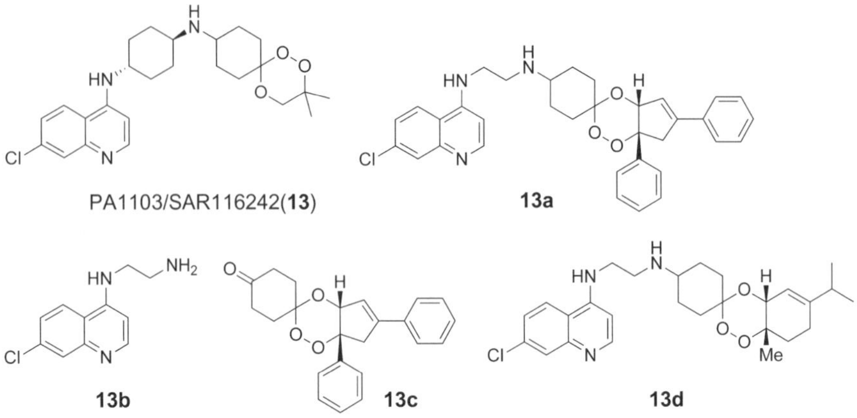

Figure 2.14 Trioxaquine PA1103/SAR116242 (**13**) and analogs.

trioxane ketone derived from α-terpinene and quinoline **13b.**[149,150] Trioxaquine **13d** (tested as both the free base and dicitrate) had IC_{50}s ranging from 6–17 nM against *P. falciparum in vitro* and oral ED_{50}s of 15 and 18 mg/kg/day against *P. yeolii* and *P. vinckei* in the Peters 4-day suppressive test. The individual diastereomers of **13d** were equally active against *P. falciparum in vitro*. With respective IC_{50}s of 67 and 110 nM, trioxaquine **13d** was also more active than AS against stage IV and V *P. falciparum* gametocytes. Trioxaquine **13d** was not genotoxic, and there was no apparent toxicity when mice were treated with daily doses of 400 mg/kg **13d** for four consecutive days.

Mechanistic studies with **13d** confirmed that trioxaquines have a potential dual mode of action. First, **13d** is a 5-fold more effective inhibitor of hemozoin formation than chloroquine.[151] Second, **13d** efficiently alkylates heme (Figure 2.15) in both model studies[152] and in *P. berghei*-infected mice.[153] As depicted in Figure 2.15, heme covalent adduct **13e** is formed in 50–60% yield by the reaction of **13d** with heme.[152] Heme also attacks the other peroxide oxygen atom to form the regioisomeric alkoxy radical that can fragment by three β-scission pathways, but heme covalent adducts formed by these reactions were produced in less than 10% yield.[152]

As the stereochemical complexity of **13a** and **13d** was deemed to exclude them from further consideration, 120 more symmetric trioxaquine analogs containing 4-aminoquinoline and 1,2,4-trioxane or 1,2,4-trioxolane (ozonide) substructures were synthesized.[154] For this compound library, *in vitro* activity against *P. falciparum* ranged from 5 to 74 nM. Seventy-two of the most potent analogs were selected for efficacy testing against *P. vinckei*-infected mice using the Peters 4-day suppressive test. On the basis of the *in vivo* efficacy data, 25 compounds underwent ADME profiling which led to the selection of **13** as the drug development candidate.[154] Compounds with *trans*-1,4-cyclohexyl substructure linkers had better metabolic stabilities than did their *n*-alkyl counterparts, as did compounds with 1,2,4-trioxane *vs.* 1,2,4-trioxolane peroxide heterocycles.

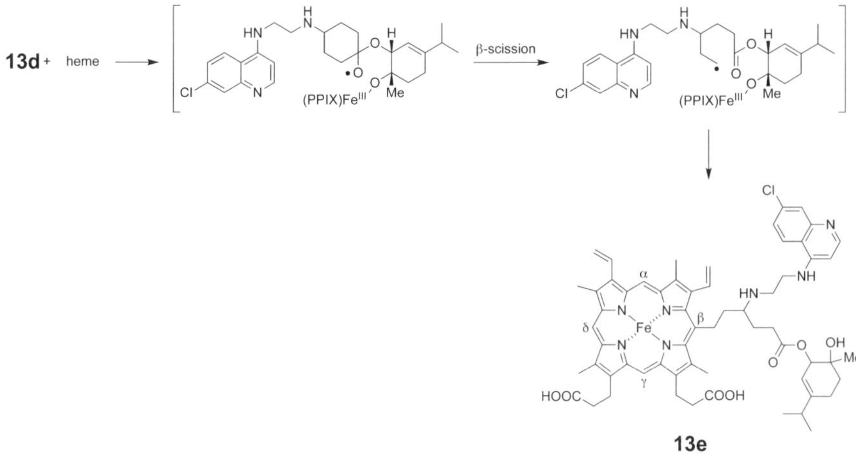

Figure 2.15 Reaction of trioxaquine **13d** with heme; only a single heme covalent adduct (**13e**) is shown. Trioxaquine **13d** alkylates heme at each of the *meso* positions of the porphyrin macrocycle by an intramolecular reaction.

Trioxaquine **13** is a mixture of two achiral *cis* and *trans* epimers, and since both epimers have the same *in vitro* potencies against *P. falciparum in vitro* and *P. vinckei in vivo* and very similar ADME profiles, **13** is undergoing development as the 1:1 mixture of epimers. Trioxaquine **13** appears to share an iron (heme)-dependent mechanism of action similar to that of **13d**. Against seven *P. falciparum* clones, *in vitro* IC_{50}s for **13** and ART ranged from 7–24 and 7–10 nM, respectively. To mice infected with *P. vinckei vinckei, P. vinckei petteri*, or *P. falciparum* (humanized mice), the curative doses of **13** and AS administered orally once daily over a period of four days were 30, 32, 63 and 100, 32, 54 mg/kg/day, respectively. When *P. vinckei vinckei*-infected mice were treated with single equimolar doses of **13** (25 mg/kg) and AS (22 mg/kg) on day 4 post-infection (20–40% parasitemia), **13** and AS reduced parasitemia by 58 and 13% at day 1 post-treatment.

ADME and toxicology data for **13** were broadly summarized by Cosledan et al.[154] In experiments with human liver microsomes and hepatocytes, **13** is 5 to 10-fold more stable than AS; however, results were not compared to those for DHA which is the more relevant species *in vivo* following dosing with AS. Trioxaquine **13** also did not inhibit or induce any of the major human CYP450 isoforms. Data from the Ames test and micronucleus assay indicate that **13** was neither mutagenic or clastogenic. Trioxaquine **13** and chloroquine had similar hERG IC_{50} values of 1.5 and 3.0 μM. Experiments in rats, however, demonstrated that a single 100 mg/kg oral dose of **13** had no effect on blood pressure, heart rate, QT, or QTc intervals. Finally, when mice were treated with daily doses of 300 mg/kg **13** for four consecutive days, there was no apparent toxicity.

2.2.8 RKA182 (14)

In 1992, the symmetrical *meso* dispiro 1,2,4,5-tetraoxane **14a** (WR 148999), readily obtained in one step by reaction of 2-methylcyclohexanone with acidified hydrogen peroxide, was demonstrated to have relatively high antimalarial activity with IC_{50}s against *P. falciparum* ranging from 30 to 60 nM and efficacy in the *P. berghei* mouse model equal or superior to that of ART.[126,155] This set in motion efforts to discover analogs of **14a** with superior antimalarial activities (Figure 2.16). The culmination of this work was the discovery of drug development candidate RKA182 (**14**) by the O'Neill group.[145]

Early efforts to improve the activity and ADME profile of **14a** by incorporation of neutral and acidic functional groups in this symmetrical tetraoxane skeleton were unsuccessful.[155] However, incorporation of weak base functional groups in these simple tetraoxanes was considerably more promising.[156] As illustrated by prototype RKA216 (**14b**), a major step forward was achieved by incorporation of polar functional groups in an achiral unsymmetrical spiroadamantane tetraoxane skeleton.[157] Tetraoxane **14b** had IC_{50} values ranging from 0.8 to 5.2 nM against *P. falciparum in vitro* and an oral ED_{50} of 3.2 mg/kg/day against *P. berghei* in the Peters 4-day suppressive test; in contrast, the spirocyclohexyl analog **14c** was 4-fold less potent against *P. falciparum* and 3-fold less effective against *P. berghei*.

With a goal to identify a metabolically stable polar side chain to counterbalance the lipophilic spiroadamantane substructure, optimization efforts proceeded on two fronts: amino amide analogs of **14b** and piperidine sulfonamides **14d**; the former were found to have superior antimalarial efficacy and ADME profiles. Representative compounds from this series include **14, 14f** and **14g**. Each of these had against IC_{50}s of less than 1.5 nM against *P. falciparum in vitro* and respective oral ED_{50}s of 1.3, 3.5, and 0.99 mg/kg/day against *P. berghei* in the Peters 4-day suppressive test. Low oral bioavailabilities of

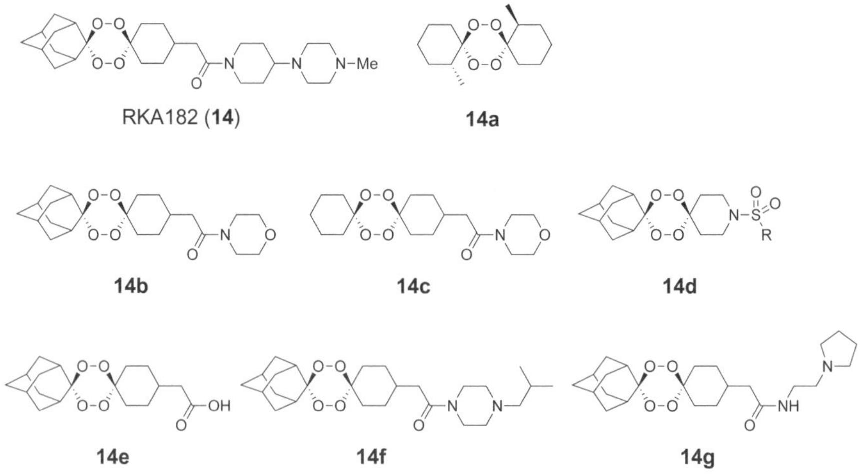

Figure 2.16 RKA182 (**14**) and analogs.

approximately 9% and complex metabolic profiles disqualified **14f** and **14g** from further consideration. In addition to **14f** and **14g**, several other tetraoxanes were also considered, but on the basis of toxicity and/or ADME data, **14** was chosen for further development.

Tetraoxane **14** (RKA182) was synthesized on a kilo scale by *in situ* formation of the *gem*-dihydroperoxide of ethyl 4-oxocylohexyl acetate with 30% H_2O_2/formic acid followed by Re_2O_7 catalyzed condensation[158] of the *gem*-dihydroperoxide with 2-adamantanone to form the tetraoxane ester which was hydrolyzed to yield **14e**.[145] Mixed anhydride coupling of **14e** with the corresponding amine afforded RKA182 (**14**). Formulation and salt selection studies led to the preferred form of **14** as its ditosylate salt.

In an assessment of efficacy in *P. berghei*-infected mice, administration of 3 consecutive 10 mg/kg oral doses of **14** and AS increased survived to 22 and 9 days post-infection compared to 4 days for control animals.[145] Even though no specific metabolites of **14** were identified, Phase I metabolic pathways for the compound class were identified as adamantane/cyclohexane hydroxylation, *N*-oxide formation and *N*-dealkylation. With the exception of weakly inhibiting CYP2C19 (IC_{50} of 59 μM), **14** did not inhibit any of the major human CYP450 isoforms at concentrations up to 100 μM. In pharmacokinetic experiments in rats conducted over 6 h, **14** (1 mg/kg intravenous and 10 mg/kg oral doses) had an oral Cmax of 90 ng/mL, intravenous and oral half-lives of 0.80 and 2.4 h, CL of 136 mL/min/kg, Vd of 9.5 L/kg, and an oral bioavailability of 40%. In the same pharmacokinetic experiment, the Cmax and oral bioavailability for **12** were reported to be only 2.8 ng/mL and 5%, respectively; however, these values differ considerably from previous reports.[125] RKA182 (**14**) was also more stable than **12** upon exposure to infected red blood cells; the latter was reported to undergo complete degradation in 35 minutes whereas the former degraded by only 21% in 4 hours under the conditions used for the study. In toxicity experiments, the maximum tolerated dose of **14** in rats was 400 mg/kg.

In mechanistic studies, treatment of **14f** with ferrous bromide produces primary and secondary carbon-centered radicals by β-scission of the initially formed regioisomeric alkoxy radicals (Figure 2.17). These were trapped by

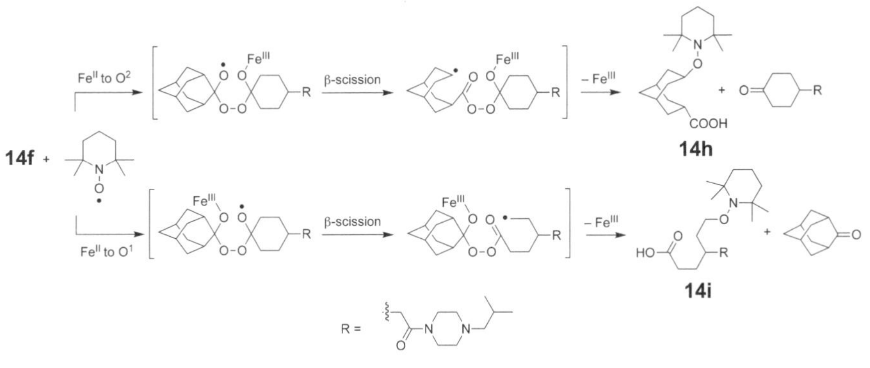

Figure 2.17 Reaction of tetraoxane **14f** with ferrous iron.

TEMPO to form the corresponding aminoxy adducts **14h** and **14i**.[145] In addition, **14f** reacts with heme to form covalent adduct **12k** (Figure 2.13), the same heme adduct produced by the reaction of arterolane (**12**) with heme.[140] Single-cell confocal imaging experiments with a fluorescent derivative of **14d** revealed accumulation only in infected erythrocytes, largely within the digestive vacuole.[159] Thus, it is evident that **14** reacts with ferrous iron in much the same way as does ART and **12** and appears to possess an iron-dependent mechanism of action.

2.3 Conclusions

The semisynthetic artemisinins are the most important class of antimalarial drugs. Their high antimalarial specificity and unique mechanism of action derive from activation of the peroxide pharmacophore during parasite hemoglobin digestion. Since their discovery, numerous second-generation semisynthetic artemisinins and synthetic peroxides have been prepared and tested for their antimalarial properties. Some of these, including the semisynthetic artemisinins artelinic acid (**8**) and artemisone (**9**), and the structurally diverse synthetic peroxides arteflene (**10**), fenozan B07 (**11**), arterolane (**12**), PA1103/SAR116242 (**13**), and RKA182 (**14**) have undergone assessment as drug development candidates. Of these, **8** may undergo NDA filing as an injectable for treatment of severe and complicated malaria, **9** is in Phase II trials, **10** progressed to Phase II trials before its development was discontinued, **11** underwent extensive and largely favorable preclinical development before being dropped, **12** is in Phase III trials, and **13** and **14** are in preclinical development. Thus, despite all of the many challenges in drug discovery, some unique to organic peroxides, it is clear that substantial progress has been made in the discovery and development of a new generation of antimalarial semisynthetic artemisinins and synthetic peroxides.

2.4 Abbreviations

ACT = artemisinin combination therapy; ADME = adsorption-distribution-metabolism-excretion; ART = artemisinin; AM = artemether; AS = artesunate; CL = clearance; DHA = dihydroartemisinin; hERG = human ether-a-go-go-related gene; NOEL = no observed effect level; PfMDR = *Plasmodium falciparum* multidrug resistance transporter; PfATP6 = *Plasmodium falciparum* sarco/endoplasmic reticulum calcium ATPase; ROS = reactive oxygen species; SAR = structure-activity relationship; $t_{1/2}$, half-life; Vd = volume of distribution.

Acknowledgements

The authors acknowledge the support of Medicines for Malaria Venture, the Nebraska Research Initiative, the Australian Research Council, and the Australian National Health and Medical Research Council.

References

1. Qinghaosu Antimalaria Coordinating Research Group, *Chin. Med. J.*, 1979, **92**, 811.
2. D. L. Klayman, *Science*, 1985, **228**, 1049.
3. K. M. Muraleedharan and M. A. Avery, *Drug Discovery Today*, 2009, **14**, 793.
4. J. A. Vroman, M. Alvim-Gaston and M. A. Avery, *Curr. Pharmaceutical Design*, 1999, **5**, 101.
5. C. W. Jefford, *Curr. Med. Chem.*, 2001, **8**, 1803.
6. C. W. Jefford, *Adv. Drug Res.*, 1997, **29**, 271.
7. S. R. Meshnick, *Int. J. Parasitol.*, 2002, **32**, 1655.
8. S. R. Meshnick, T. E. Taylor and S. Kamchonwongpaisan, *Microbiol. Rev.*, 1996, **60**, 301.
9. P. M. O'Neill and G. H. Posner, *J. Med. Chem.*, 2004, **47**, 2945.
10. G. H. Posner and P. M. O'Neill, *Acc. Chem. Res.*, 2004, **37**, 397.
11. A. Robert and B. Meunier, *Chem. Soc. Rev.*, 1998, **27**, 273.
12. A. Robert, O. Dechy-Cabaret, J. Cazelles and B. Meunier, *Acc. Chem. Res.*, 2002, **35**, 167.
13. Y. Li and Y.-L. Wu, *Curr. Med. Chem.*, 2003, **10**, 2197.
14. R. K. Haynes, *Curr. Topics Med. Chem.*, 2006, **6**, 509.
15. R. K. Haynes and S. Krishna, *Microb. Infect.*, 2004, **6**, 1339.
16. P. M. O'Neill, J. Chadwick and S. L. Rawe in *Chemistry of Peroxides 2 (Pt. 2)*, ed. Z. Rappoport, Wiley, Chichester, 2006, p. 1279.
17. M. A. Fügi, S. Wittlin, Y. Dong and J. L. Vennerstrom, *Antimicrob. Agents Chemother.*, 2010, **54**, 1042.
18. P. A. Stocks, P. G. Bray, V. E. Barton, M. Al-Helal, M. Jones, N. C. Araujo, P. Gibbons, S. A. Ward, R. H. Hughes, G. A. Biagini, J. Davies, R. Amewu, A. E. Mercer, G. Ellis and P. M. O'Neill, *Angew. Chem. Int. Ed.*, 2007, **46**, 6278.
19. H. Ginsburg, O. Famin, J. Zhang and M. Krugliak, *Biochem. Pharmacol.*, 1998, **56**, 1305.
20. P. Loria, S. Miller, M. Foley and L. Tilley, *Biochem. J.*, 1999, **339**, 363.
21. M. del Pilar Crespo, T. D. Avery, E. Hanssen, E. Fox, T. V. Robinson, P. Valente, D. K. Taylor and L. Tilley, *Antimicrob. Agents Chemother.*, 2008, **52**, 98.
22. C. L. Hartwig, A. S. Rosenthal, J. D'Angelo, C. E. Griffin, G. H. Posner and R. A. Cooper, *Biochem. Pharmacol.*, 2009, **77**, 322.
23. P. A. Berman and P. A. Adams, *Free Radic. Biol. Med.*, 1997, **22**, 1283.
24. A. Robert, Y. Coppel and B. Meunier, *Chem. Commun.*, 2002, 414.
25. A. Robert, F. Benoit-Vical, C. Claparols and B. Meunier, *Proc. Natl. Acad. Sci. USA*, 2005, **102**, 13676.
26. M. Kaiser, S. Wittlin, A. Nehrbass-Stuedli, Y. Dong, X. Wang, A. Hemphill, H. Matile, R. Brun and J. L. Vennerstrom, *Antimicrob. Agents Chemother.*, 2007, **51**, 2991.
27. S. Krishna, L. Bustamante, R. K. Haynes and H. M. Staines, *Trends Pharmacol. Sci.*, 2008, **29**, 520.

28. J. Keiser and J. Utzinger, *Curr. Opin. Infect. Dis.*, 2007, **20**, 605.
29. R. K. Haynes, W. C. Chan, H. N. Wong, K. Y. Li, W. K. Wu, K. M. Fan, H. H. Sung, I. D. Williams, D. Prosperi, S. Melato, P. Coghi and D. Monti, *ChemMedChem*, 2010, **5**, 1282.
30. F. ter Kuile, N. J. White, P. Holloway, G. Pasvol and S. Krishna, *Exp. Parasitol.*, 1993, **76**, 85.
31. A. U. Orjih, *Br. J. Haematol.*, 1996, **92**, 324.
32. T. S. Skinner, L. S. Manning, W. A. Johnston and T. M. Davis, *Int. J. Parasitol.*, 1996, **26**, 519.
33. S. Maerki, R. Brun, S. A. Charman, A. Dorn, H. Matile and S. Wittlin, *J. Antimicrob. Chemother.*, 2006, **58**, 52.
34. D. A. Elliott, M. T. McIntosh, H. D. Hosgood III, S. Chen, G. Zhang, P. Baevova and K. A. Joiner, *Proc. Natl. Acad. Sci. USA*, 2008, **105**, 2463.
35. N. A. Abu Bakar, N. Klonis, E. Hanssen, C. Chan and L. Tilley, *J. Cell Sci.*, 2010, **123**, 441.
36. A. R. Dluzewski, I. T. Ling, J. M. Hopkins, M. Grainger, G. Margos, G. H. Mitchell, A. A. Holder and L. H. Bannister, *PLoS One*, 2008, **3**, e3085.
37. S. E. Francis, D. J. Jr. Sullivan and D. E. Goldberg, *Annu. Rev. Microbiol.*, 1997, **51**, 97.
38. A. U. Orjih and C. D. Fitch, *Biochim. Biophys. Acta*, 1993, **1157**, 270.
39. W. Asawamahasakda, I. Ittarat, C.-C. Chang, P. McElroy and S. R. Meshnick, *Mol. Biochem. Parasitol.*, 1994, **67**, 183.
40. Y. Maeno, T. Toyoshima, H. Fujioka, Y. Ito, S. R. Meshnick, A. Benakis, W. K. Milhous and M. Aikawa, *Am. J. Trop. Med. Hyg.*, 1993, **49**, 485.
41. A. V. Pandey, B. L. Tekwani, R. L. Singh and V. S. Chauhan, *J. Biol. Chem.*, 1999, **274**, 19383.
42. W. Li, W. Mo, D. Shen, L. Sun, J. Wang, S. Lu, J. M. Gitschier and B. Zhou, *PLoS Genet.*, 2005, **1**, e36.
43. J. Wang, L. Huang, J. Li, Q. Fan, Y. Long, Y. Li and B. Zhou, *PLoS One*, 2010, **5**, e9582.
44. E. Richier, G. A. Biagini, S. Wein, F. Boudou, P. G. Bray, S. A. Ward, E. Precigout, M. Calas, J.-F. Dubremetz and H. J. Vial, *Antimicrob. Agents Chemother.*, 2006, **50**, 3381.
45. Y. Tang, Y. Dong and J. L. Vennerstrom, *Med. Res. Rev.*, 2004, **24**, 425.
46. N. White, *Phil. Trans. R. Soc. Lond. B Biol. Sci.*, 1999, **354**, 739.
47. P. I. German and F. T. Aweeka, *Clin. Pharmacokinet.*, 2008, **47**, 91.
48. R. T. Eastman and D. A. Fidock, *Nat. Rev. Microbiol.*, 2009, **7**, 864.
49. R. Price, M. van Vugt, F. Nosten, C. Luxemburger, A. Brockman, L. Phaipun, T. Chongsuphajaisiddhi and N. White, *Am. J. Trop. Med. Hyg.*, 1998, **59**, 883.
50. N. J. White, *J. Clin. Inves.*, 2004, **113**, 1084..
51. N. J. White, *Science*, 2008, **320**, 330.
52. M. Ashton, T. Gordi, T. N. Hai, N. V. Huong, N. D. Sy, N. T. Niêu, D. X. Huong, M. Johansson and L. D. Công, *Biopharm. Drug Dispos.*, 1998, **19**, 245.

53. D. D. Duc, P. J. deVries, N. X. Khanh, L. N. Binh, P. A. Kager and C. J. Van Boxtel, *Am. J. Trop. Med. Hyg.*, 1994, **51**, 785.
54. M. Hassan Alin, M. Ashton, C. M. Kihamia, G. J. B. Mtey and A. Björkman, *Trans. Roy. Soc. Trop. Med. Hyg.*, 1996, **90**, 61.
55. M. Ashton, T. N. Hai, N. D. Sy, D. X. Huong, N. V. Huong, N. T. Niêu and L. D. Công, *Drug Metab. Dispos*, 1998, **26**, 25.
56. F. Nosten and N. J. White, *Am. J. Trop. Med. Hyg.*, 2007, **77**(Suppl 6), 181.
57. K. Na-Bangchang, S. Krudsood, U. Silachamroon, P. Molunto, O. Tasanor, K. Chalermrut, O. Matangkasombut, S. Kano and S. Looareesuwan, *Southeast Asian J. Trop. Med. Public Health*, 2004, **35**, 575.
58. T. Q. Binh, K. F. Ilett, K. T. Batty, T. M. E. Davis, N. C. Hung, S. M. Powell, L. T. A. Thu, H. V. Thien, H. L. Phuöng and V. D. B. Phuong, *Br. J. Clin. Pharmacol.*, 2001, **51**, 541.
59. P. Teja-Isavadharm, G. Watt, C. Eamsila, K. Jongsakul, Q. Li, D. Keeratithakul, N. Sirisopana, L. Luesutthiviboon, T. G. Brewer and D. E. Kyle, *Am. J. Trop. Med. Hyg.*, 2001, **65**, 717.
60. C. Orrell, F. Little, P. Smith, P. Folb, W. Taylor, P. Olliaro and K. I. Barnes, *Eur. J. Clin. Pharmacol.*, 2008, **64**, 683.
61. P. Teja-Isavadharm, F. Nosten, D. E. Kyle, C. Luxemburger, F. Ter Kuile, J. O. Peggins, T. G. Brewer and N. J. White, *Br. J. Clin. Pharmacol.*, 1996, **42**, 599.
62. J. Karbwang, K. Na-Bangchang, K. Congpuong, P. Molunto and A. Thanavibul, *Eur. J. Clin. Pharmacol.*, 1997, **52**, 307.
63. B. K. Park, P. M. O'Neill, J. L. Maggs and M. Pirmohamed, *Br. J. Clin. Pharmacol.*, 1998, **46**, 521.
64. T. Gordi and E.-I. Lepist, *Toxicol. Lett.*, 2004, **147**, 99.
65. R. L. Clark, *Repro. Toxicol.*, 2009, **28**, 285.
66. M. Longo, S. Zanoncelli, M. Brughera, P. Colombo, S. Wittlin, J. L. Vennerstrom, J. Moehrle and J. C. Craft, *Repro. Toxicol.*, 2010, **30**, 583.
67. D. L. Wesche, M. A. DeCoster, F. C. Tortella and T. G. Brewer, *Antimicrob. Agents Chemother.*, 1994, **38**, 1813.
68. T. G. Brewer, S. J. Grate, J. O. Peggins, P. J. Weina, J. M. Petras, B. S. Levine, M. H. Heiffer and B. G. Schuster, *Am. J. Trop. Med. Hyg.*, 1994, **51**, 251.
69. A. Nontprasert, S. Pukrittayakamee, S. Prakongpan, W. Supanaranond, S. Looareesuwan and N. J. White, *Trans. R. Soc. Trop. Med. Hyg.*, 2002, **96**, 99.
70. A. M. Dondorp, S. Yeung, L. White, C. Nguon, N. P. Day, D. Socheat and L. von Seidlein, *Nat. Rev. Microbiol.*, 2010, **8**, 272.
71. T. J. Anderson, S. Nair, S. Nkhoma, J. T. Williams, M. Imwong, P. Yi, D. Socheat, D. Das, K. Chotivanich, N. P. Day, N. J. White and A. M. Dondorp, *J. Infect. Dis.*, 2010, **201**, 1326.
72. B. Witkowski, J. Lelievre, M. J. Barragan, V. Laurent, X. Z. Su, A. Berry and F. Benoit-Vical, *Antimicrob. Agents Chemother.*, 2010, **54**, 1872.
73. F. Teuscher, M. L. Gatton, N. Chen, J. Peters, D. E. Kyle and Q. Cheng, *J. Infect. Dis.*, 2010, **202**, 1362.

74. A. A. Alagbala, A. J. McRiner, K. Borstnik, T. Labonte, W. Chang, J. G. D'Angelo, G. H. Posner and B. A. Foster, *J. Med. Chem.*, 2006, **49**, 7386.
75. W.-S. Zhou and X.-X. Xu, *Acc. Chem. Res.*, 1994, **27**, 211.
76. A. J. Lin, D. L. Klayman and W. K. Milhous, *J. Med. Chem.*, 1987, **30**, 2147.
77. A. J. Lin, M. Lee and D. L. Klayman, *J. Med. Chem.*, 1989, **32**, 1249.
78. A. J. Lin, A. B. Zikry and D. E. Kyle, *J. Med. Chem.*, 1997, **40**, 1396.
79. A. J. Lin and R. E. Miller, *J. Med. Chem.*, 1995, **38**, 764.
80. J. M. Grace, D. J. Skanchy and A. J. Aguilar, *Xenobiotica*, 1999, **29**, 703.
81. M. Jung, K. Lee, H. Kendrick, B. L. Robinson and S. L. Croft, *J. Med. Chem.*, 2002, **45**, 4940.
82. M. Chavchich, L. Gerena, J. Peters, N. Chen, Q. Cheng and D. E. Kyle, *Antimicrob. Agents Chemother.*, 2010, **54**, 2455.
83. N. Chen, M. Chavchich, J. M. Peters, D. E. Kyle, M. L. Gatton and Q. Cheng, *Antimicrob. Agents Chemother.*, 2010, **54**, 3395.
84. Q.-G. Li, J. O. Peggins, L. L. Fleckenstein, K. Masonic, M . H. Heiffer and T. G. Brewer, *J. Pharm. Pharmacol.*, 1998, **50**, 173.
85. Q.-G. Li, J. O. Peggins, A. J. Lin, K. J. Masonic, K. M. Trotman and T. G. Brewer, *Trans. R. Soc. Trop. Med. Hyg.*, 1998, **92**, 332.
86. Q. Li, L. H. Xie, T. O. Johnson, Y. Si, A. S. Haeberle and P. J. Weina, *Trans. R. Soc. Trop. Med. Hyg.*, 2007, **101**, 104.
87. R. F. Genovese, D. B. Newman and T. G. Brewer, *Pharmacol. Biochem. Behav.*, 2000, **67**, 37.
88. Y. Si, Q.-G. Li, L. Xie, K. Bennett, P. J. Weina, S. Mog and T. O. Johnson, *Int. J. Toxicol.*, 2007, **26**, 401.
89. R. K. Haynes, B. Fugmann, J. Stetter, K. Rieckmann, H.-D. Heilmann, H.-W. Chan, M.-K. Cheung, W.-L. Lam, H.-N. Wong, S. L. Croft, L. Vivas, L. Rattray, L. Stewart, W. Peters, B. L. Robinson, M. D. Edstein, B. Kotecka, D. E. Kyle, B. Beckermann, M. Gerisch, M. Radtke, G. Schmuck, W. Steinke, U. Wollborn, K. Schmeer and A. Romer, *Angew. Chem. Int. Ed.*, 2006, **45**, 2082.
90. S. D'Alessandro, M. Gelati, N. Basilico, A. Eugenio, R. K. Haynes and D. Taramelli, *Toxicol.*, 2007, **241**, 66.
91. M. Ramharter, D. Burkhardt, J. Nemeth, A. A. Adegnika and P. G. Kremsner, *Am. J. Trop. Med. Hyg.*, 2006, **75**, 637.
92. L. Vivas, L. Rattray, L. B. Stewart, B. L. Robinson, B. Fugmann, R. K. Haynes, W. Peters and S. L. Croft, *J. Antimicrob. Chemother.*, 2007, **59**, 658.
93. J. H. Waknine-Grinberg, N. Hunt, A. Bentura-Marciano, J. A McQuillan, H.-W. Chan, W.-C. Chan, Y. Barenholz, R. K. Haynes and J. Golenser, *Malaria J.*, 2010, **9**, 227.
94. N. Obaldia III, B. M. Kotecka, M. D. Edstein, R. K. Haynes, B. Fugmann, D. E. Kyle and K. H. Rieckmann, *Antimicrob. Agents Chemother.*, 2009, **53**, 3592.
95. R. K. Haynes, W. C. Chan, C.-M. Lung, A.-C. Uhlemann, U. Eckstein, D. Taramelli, S. Parapini, D. Monti and S. Krishna, *ChemMedChem*, 2007, **2**, 1480.

96. F. Bousejra-El Garah, B. Meunier and A. Robert, *Eur. J. Inorg. Chem.*, 2008, 2133.
97. S. A.-L. Laurent, A. Robert and B. Meunier, *Angew. Chem. Int. Ed.*, 2005, **44**, 2060.
98. U. Eckstein-Ludwig, R. J. Webb, I. D. Van Goethem, J. M. East, A. G. Lee, M. Kimura, P. M. O'Neill, P. G. Bray, S. A. Ward and S. Krishna, *Nature*, 2003, **424**, 957.
99. A.-C. Uhlemann, A. Cameron, U. Eckstein-Ludwig, J. Fischbarg, P. Iserovich, F. A. Zuniga, M. East, A. Lee, L. Brady, R. K. Haynes and S. Krishna, *Nature Struct. Mol. Biol.*, 2005, **12**, 628.
100. J. Nagelschmitz, B. Voith, G. Wensing, A. Roemer, B. Fugmann, R. K. Haynes, B. M. Kotecka, K. H. Rieckmann and M. D. Edstein, *Antimicrob. Agents Chemother.*, 2008, **52**, 3085.
101. A. M. Szpilman, E. E. Korshin, H. Rozenberg and M. D. Bachi, *J. Org. Chem.*, 2005, **70**, 3618.
102. W. Hofheinz, H. Bürgin, E. Gocke, C. Jaquet, R. Masciadri, G. Schmid, H. Stohler and H. Urwyler, *Trop. Med. Parasitol.*, 1994, **45**, 261.
103. P. M. O'Neill, L. P. D. Bishop, N. L. Searle, J. L. Maggs, R. C. Storr, S. A. Ward, B. K. Park and F. Mabbs, *J. Org. Chem.* 2000, **65**, 1578.
104. C. Jaquet, H. R. Stohler, J. Chollet and W. Peters, *Trop. Med. Parasitol.*, 1994, **45**, 266.
105. J. Cazelles, A. Robert and B. Meunier, *J. Org. Chem.* 1999, **64**, 6776.
106. P. M. O'Neill, L. P. D. Bishop, N. L. Searle, J. L. Maggs, S. A. Ward, P. G. Bray, R. C. Storr and B. K. Park, *Tetrahedron Lett.*, 1997, **38**, 4263.
107. W. Asawamahasakda, I. Ittarat, Y.-M. Pu, H. Ziffer and S. R. Meshnick, *Antimicrob. Agents Chemother.*, 1994, **38**, 1854.
108. M. A. Girometta, R. Jauch, C. Ponelle, A. Guenzi and R. C. Wiegand-Chou, *Trop. Med. Parasitol.*, 1994, **45**, 272.
109. E. Weidekamm, E. Dumont and C. Jaquet, *Trop. Med. Parasitol.*, 1994, **45**, 278.
110. L. A. Salako, R. Guiguemde, M.-L. Mittelholzer, L. Haller, F. Sorenson and D. Stürchler, *Trop. Med. Parasitol.*, 1994, **45**, 284.
111. R. Somo-Moyou, M.-L. Mittelholzer, F. Sorenson, L. Haller and D. Stürchler, *Trop. Med. Parasitol.*, 1994, **45**, 288.
112. P. D. Radloff, J. Philipps, M. Nkeyi, D. Stürchler, M.-L. Mittelholzer and P. G. Kremsner, *Amer. J. Trop. Med. Hyg.*, 1996, **55**, 259.
113. C. W. Jefford, J. A. Velarde, G. Bernardinelli, D. H. Bray, D. C. Warhurst and W. K. Milhous, *Helv. Chim. Acta*, 1993, **76**, 2775.
114. C. W. Jefford, J. Boukouvalas and S. Kohmoto, *J. Chem. Soc. Chem. Commun.*, 1984, 523.
115. C. W. Jefford, D. Misra, J. C. Rossier, P. Kamalaprija, U. Burger, J. Mareda, G. Bernardinelli, W. Peters, B. L. Robinson, W. K. Milhous, F. Zhang, D. K. Gosser, Jr. and S. R. Meshnick, *Perspectives in Medicinal Chemistry* ed. B. Testa E. Kyburz W. Fuhrer R. Gifer, VCH, Basel, 1993, p. 459.
116. C. W. Jefford, S. Kohmoto, D. Jaggi, G. Timari, J. C. Rossier, M. Rudaz, O. Barbuzzi, D. Gerard and U. Burger, *Helv. Chim. Acta*, 1995, **78**, 647.

117. C. W. Jefford, D. Jaggi, S. Kohmoto, G. Timari, G. Bernardinelli, C. J. Canfield and W. K. Milhous, *Heterocycles*, 1998, **49**, 375.
118. C. W. Jefford, J. C. Rossier and W. K. Milhous, *Heterocycles*, 2000, **52**, 1345.
119. W. Peters, B. L. Robinson, J. C. Rossier, D. Misra and C. W. Jefford, *Ann. Trop. Med. Parasitol.*, 1993, **87**, 9.
120. W. Peters, B. L. Robinson, J. C. Rossier and C. W. Jefford, *Ann. Trop. Med. Parasitol.*, 1993, **87**, 1.
121. W. Peters, B. L. Robinson, G. Tovey, J. C. Rossier and C. W. Jefford, *Ann. Trop. Med. Parasitol.*, 1993, **87**, 111.
122. S. L. Fleck, B. L. Robinson, W. Peters, F. Thevin, Y. Boulard, C. Glenat, V. Caillard and I. Landau, *Ann. Trop. Med. Parasitol.*, 1997, **91**, 25.
123. S. L. Fleck, B. L. Robinson and W. Peters, *Ann. Trop. Med. Parasitol.*, 1997, **91**, 33.
124. R. Tripathi, C. W. Jefford and G. P. Dutta, *Parasitol.*, 2006, **133**, 1.
125. J. L. Vennerstrom, S. Arbe-Barnes, R. Brun, S. A. Charman, F. C. Chiu, J. Chollet, Y. Dong, A. Dorn, D. Hunziker, H. Matile, K. McIntosh, M. Padmanilayam, J. Santo Tomas, C. Scheurer, B. Scorneaux, Y. Tang, H. Urwyler, S. Wittlin and W. N. Charman, *Nature*, 2004, **430**, 900.
126. J. L. Vennerstrom, H. N. Fu, W. Y. Ellis, A. L. Ager Jr, J. K. Wood, S. L. Andersen, L. Gerena and W. K. Milhous, *J. Med. Chem.*, 1992, **35**, 3023.
127. K. Griesbaum, X. Liu, A. Kassiaris and M. Scherer, *Liebigs Ann./Recueil*, 1997, 1381.
128. Y. Dong, J. Chollet, H. Matile, S. A. Charman, F. C. K. Chiu, W. N. Charman, B. Scorneaux, H. Urwyler, J. Santo Tomas, C. Scheurer, C. Snyder, A. Dorn, X. Wang, J. M. Karle, Y. Tang, S. Wittlin, R. Brun and J. L. Vennerstrom, *J. Med. Chem.*, 2005, **48**, 4953.
129. Y. Tang, Y. Dong, X. Wang, K. Sriraghavan, J. K. Wood and J. L. Vennerstrom, *J. Org. Chem.*, 2005, **70**, 5103.
130. D. J. Creek, W. N. Charman, F. C. K. Chiu, R. J. Prankerd, K. J. McCullough, Y. Dong, J. L. Vennerstrom and S. A. Charman, *J. Pharm. Sci.*, 2007, **96**, 2945.
131. X. Wang, Y. Dong, S. Wittlin, D. Creek, J. Chollet, S. A. Charman, J. Santo Tomas, C. Scheurer, C. Snyder and J. L. Vennerstrom, *J. Med. Chem.*, 2007, **50**, 5840.
132. S. Erhardt, S. A. Macgregor, K. J. McCullough, K. Savill and B. J. Taylor, *Org. Lett.*, 2007, **9**, 5569.
133. Y. Tang, Y. Dong, J. M. Karle, C. A. DiTusa and J. L. Vennerstrom, *J. Org. Chem.*, 2004, **69**, 6470.
134. Y. Dong, Y. Tang, J. Chollet, H. Matile, S. Wittlin, S. A. Charman, W. N. Charman, J. Santo Tomas, C. Scheurer, C. Snyder, B. Scorneaux, S. Bajpai, S. A. Alexander, X. Wang, M. Padmanilayam, C. S. Rao, R. Brun and J. L. Vennerstrom, *Bioorg. Med. Chem.*, 2006, **14**, 6368.
135. M. Padmanilayam, B. Scorneaux, Y. Dong, J. Chollet, H. Matile, S. A. Charman, D. J. Creek, W. N. Charman, J. Santo Tomas, C. Scheurer, S. Wittlin, R. Brun and J. L. Vennerstrom, *Bioorg. Med. Chem. Lett.*, 2006, **16**, 5542.

136. Y. Tang, Y. Dong, S. Wittlin, S. A. Charman, J. Chollet, F. C. K. Chiu, W. N. Charman, H. Matile, H. Urwyler, A. Dorn, S. Bajpai, X. Wang, M. Padmanilayam, J. M. Karle, R. Brun and J. L. Vennerstrom, *J. L. Bioorg. Med. Chem. Lett.*, 2007, **17**, 1260.
137. Y. Tang, S. Wittlin, S. A. Charman, J. Chollet, F. C. K. Chiu, J. Morizzi, L. M. Johnson, J. Santo Tomas, C. Scheurer, C. Snyder, L. Zhou, Y. Dong, W. N. Charman, H. Matile, U. Urwyler, A. Dorn and J. L. Vennerstrom, *Bioorg. Med. Chem. Lett.*, 2010, **20**, 563.
138. Y. Dong, S. Wittlin, K. Sriraghavan, J. Chollet, S. A. Charman, W. N. Charman, C. Scheurer, U. Urwyler, J. Santo Tomas, C. Snyder, D. J. Creek, J. Morizzi, M. Koltun, H. Matile, X. Wang, M. Padmanilayam, Y. Tang, A. Dorn, R. Brun and J. L. Vennerstrom, *J. Med. Chem.*, 2010, **53**, 481.
139. C. H. M. Kocken, A. van der Wel, S. Arbe-Barnes, R. Brun, H. Matile, C. Scheurer, S. Wittlin and A. W. Thomas, *Exp. Parasitol.*, 2006, **113**, 197.
140. D. J. Creek, W. N. Charman, F. C. K. Chiu, R. J. Prankerd, Y. Dong, J. L. Vennerstrom and S. A. Charman, *Antimicrob. Agents Chemother.*, 2008, **52**, 1291.
141. A. C. Uhlemann, S. Wittlin, H. Matile, L. Y. Bustamante and S. Krishna, *Antimicrob. Agents Chemother.*, 2007, **51**, 667.
142. L. Zhou, A. Alker, A. Ruf, X. Wang, F. C. K. Chiu, J. Morizzi, S. A. Charman, W. N. Charman, C. Scheurer, S. Wittlin, Y. Dong, D. Hunziker and J. L. Vennerstrom, *Bioorg. Med. Chem. Lett.*, 2008, **18**, 1555.
143. C. S. Perry, S. A. Charman, R. J. Prankerd, F. C. K. Chiu, Y. Dong, J. L. Vennerstrom and W. N. Charman, *J. Pharm. Sci.*, 2006, **95**, 737.
144. N. Valecha, S. Looareesuwan, A. Martensson, S. M. Abdulla, S. Krudsood, N. Tangpukdee, S. Mohanty, S. K. Mishra, P. K. Tyagi, S. K. Sharma, J. Moehrle, A. Gautam, A. Roy, J. K. Paliwal, M. Kothari, N. Saha, A. P. Dash and Anders Björkman, *Clin. Infect. Dis.*, 2010, **51**, 684.
145. P. M. O'Neill, R. K. Amewu, G. L. Nixon, F. Bousejra ElGarah, M. Mungthin, J. Chadwick, A. E. Shone, L. Vivas, H. Lander, V. Barton, S. Muangnoicharoen, P. G. Bray, J. Davies, B. K. Park, S. Wittlin, R. Brun, M. Preschel, K. Zhang and S. A. Ward, *Angew. Chem. Int. Ed.*, 2010, **49**, 5693.
146. P. Olliaro and T. N Wells, *Clin. Pharmacol. Ther.*, 2009, **85**, 584.
147. O. Dechy-Cabaret, F. Benoit-Vical, A. Robert and B. Meunier, *ChemBioChem*, 2000, 281.
148. L. K. Basco, O. Dechy-Cabaret, M. Ndounga, F. S. Meche, A. Robert and B. Meunier, *Antimicrob. Agents Chemother.*, 2001, **45**, 1886.
149. O. Dechy-Cabaret, F. Benoit-Vical, C. Loup, A. Robert, H. Gornitzka, A. Bonhoure, H. Vial, J.-F. Magnaval, J.-P. Séguéla and B. Meunier, *Chem. Eur. J.*, 2004, **10**, 1625.
150. F. Benoit-Vical, J. Lelievre, A. Berry, C. Deymier, O. Dechy-Cabaret, J. Cazelles, C. Loup, A. Robert, J.-F. Magnaval and B. Meunier, *Antimicrob. Agents Chemother.*, 2007, **51**, 1463.

151. C. Loup, J. Lelievre, F. Benoit-Vical and B. Meunier, *Antimicrob. Agents Chemother.*, 2007, **51**, 3768.
152. S. A.-L. Laurent, C. Loup, S. Mourgues, A. Robert and B. Meunier, *ChemBioChem*, 2005, **6**, 653.
153. F. Bousejra-El Garah, C. Claparols, F. Benoit-Vical, B. Meunier and A. Robert, *Antimicrob. Agents Chemother.*, 2008, **52**, 2966.
154. F. Cosledan, L. Fraisse, A. Pellet, F. Guillou, B. Mordmüller, P. G. Kremsner, A. Moreno, D. Mazier, J.-P. Maffrand and B. Meunier, *Proc. Natl. Acad. Sci. USA*, 2008, **105**, 17579.
155. Y. Dong, H. Matile, J. Chollet, R. Kaminsky, J. K. Wood and J. L. Vennerstrom, *J. Med. Chem.*, 1999, **42**, 1477.
156. I. Opsenica, D. Opsenica, K. S. Smith, W. K. Milhous and B. A. Solaja, *J. Med. Chem.*, 2008, **51**, 2261.
157. R. Amewu, A. V. Stachulski, S. A. Ward, N. G. Berry, P. G. Bray, J. Davies, G. Labat, L. Vivas and P. M. O'Neill, *Org. Biomol. Chem.*, 2006, **4**, 4431.
158. P. Ghorai and P. H. Dussault, *Org. Lett.*, 2009, **11**, 213.
159. G. L. Ellis, R. Amewu, S. Sabbani, P. A. Stocks, A. Shone, D. Stanford, P. Gibbons, J. Davies, L. Vivas, S. Charnaud, E. Bongard, C. Hall, K. Rimmer, S. Lozanom, M. Jesús, D. Gargallo, S. A. Ward and P. M. O'Neill, *J. Med. Chem.*, 2008, **51**, 2170.

CHAPTER 3

Antimalarial Agents Targeting Nucleotide Synthesis and Electron Transport: Insight from Structural Biology

MARGARET A. PHILLIPS

Department of Pharmacology, University of Texas Southwestern Medical Center, Dallas, Texas, 75390-9041, US

3.1 Introduction

Considerable effort is ongoing to identify new antimalarial compounds. Both target based approaches and phenotypic[1,2] screens have contributed to the identification of new chemical entities with activity against the parasite. Target-based drug discovery provides several advantages that highlight the appeal of this approach. Targets can be selected based on genetic studies demonstrating essentiality and target selection can be refined by considering data from existing chemical databases to prioritize those targets most likely to bind to drug-like molecules.[3] In comparison to whole organism screening, isolated protein targets, such as purified enzymes, are typically more amenable to a wider range of lead discovery techniques. Target proteins can be obtained by heterologous expression and purified through the use of recombinant proteins containing purification tags. Enzyme assays that are amenable to high throughput screening technology can be developed for many potential target enzymes. This

RSC Drug Discovery Series No. 14
Neglected Diseases and Drug Discovery
Edited by Michael J. Palmer and Timothy N. C. Wells

Published by the Royal Society of Chemistry, www.rsc.org

facilitates the discovery of novel chemical species with inhibitory activity against the target. Furthermore, a target-based approach allows for the potential to utilize protein structure data to identify inhibitors and to inform the lead optimization program. Finally, knowledge of the target helps provide insight into the potential modes of toxicity that may be observed as a candidate molecule advances towards the clinics.

Despite these key advantages, few successful target-based drug discovery programs have been described for microbial pathogens. Review of bacterial target-based drug discovery programs suggests too great a focus on establishing the biological relevance of potential targets and insufficient attention to their chemical attributes, e.g. the probability that the targets will bind to drug-like molecules (druggability) and the chemical space of likely ligands (diversity).[4] The task is made more difficult in the malaria parasite, *Plasmodium* species, because few target enzymes have been both genetically and chemically validated.[5] Of the marketed antimalarials only 3 enzymatic targets have been identified: dihydrofolate reductase (DHFR), the target of pyrimethamine and cycloguanil; dihydropteroate synthase (DHPS), the target of sulfonamides and sulfones; and the bc1 complex, the target of atovoquone (Figure 3.1).[5,6] Interestingly, all three of these targets are involved in pyrimidine biosynthesis and metabolism (Figure 3.2), underscoring the importance of this pathway to parasite growth, and suggesting it may be mined for the identification of additional chemotherapeutic agents against *Plasmodium* parasites.

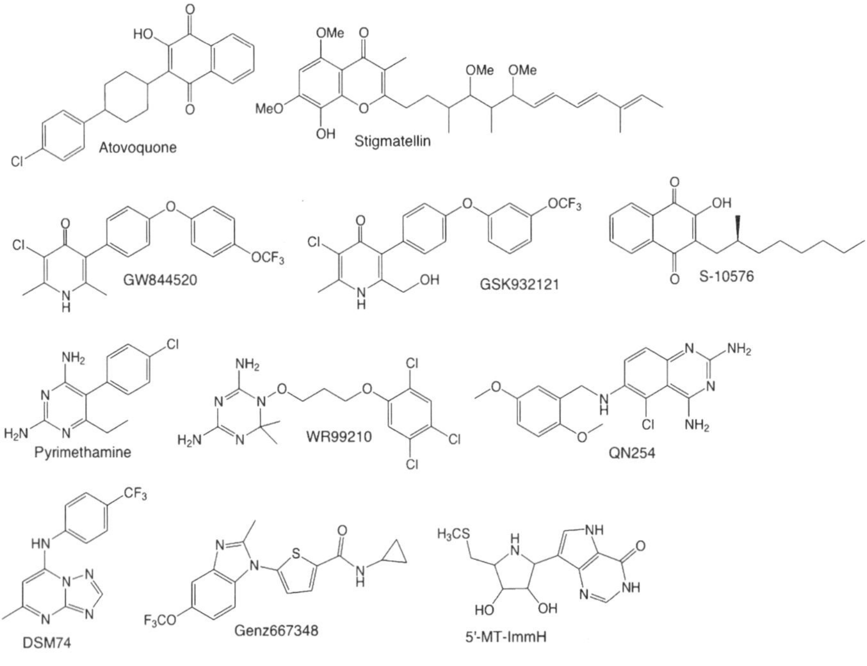

Figure 3.1 Selected inhibitors of pyrimidine biosynthetic enzymes.

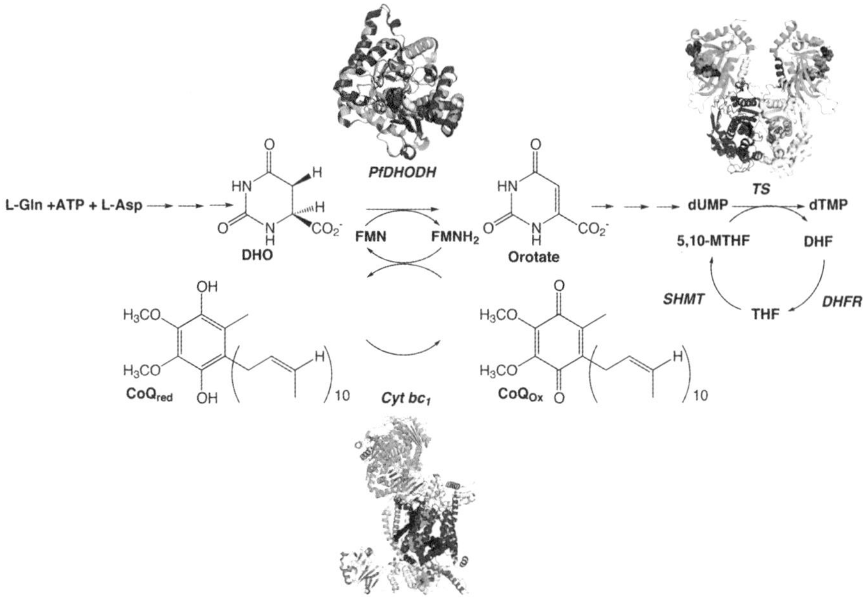

Figure 3.2 Pyrimidine metabolism and mitochondrial electron transport in *Plasmodium falciparum*. *De novo* pyrimidine metabolism requires six enzymes to produce UMP; (bifunctional glutamine amidotransferase (GAT) and carbamoyl-phosphate synthetase (CPS), aspartate carbamoyltransferase (ACT), dihydroorotase (DHOtase), dihydroorotate dehydrogenase (DHODH), orotate phosphoribosyltransferase (OPRT) and orotidine-5'-monophosphate decarboxylase (OMPDC)). UMP is then metabolized to dUMP. Only the reaction catalyzed by DHODH is shown. Three enzyme activities are required to complete the cycle for the conversion of dUMP to dTMP (bifunctional TS/DHFR thymidylate synthetase/dihydrofolate reductase and SHMT, serine hydroxymethyltransferase), where dihydrofolate, DHF, tetrahydrofolate, THF and 5, 10 methenyltetrahydrofolate, 5, 10-MTHF are shown. The Cyt bc_1 complex is required to replenish the oxidized ubiquinone (CoQ_{ox}) pools. Structure figure insets: *pf*DHODH – overlay of pdb 3I6R (turquoise) with pdb 3O8A (pink) showing *pf*DHODH (ribbon) bound to DSM74 and Genz667348[63,65], respectively; *P. falciparum* TS/DHFR, pdb 1J3I [41] showing the TS domains in pink and yellow and the DHFR domains in green and turquoise; yeast bc1 complex pdb 1EZV[83], key described in Figure 3.3. Structure figures were generated with PyMol (DeLano, W. L. The PyMOL Molecular Graphics System (2002) at: www.pymol.org).

This chapter will focus on three avenues of research being pursued to discover novel antimalarials using target-based approaches that impact on the synthesis of precursors for DNA and RNA biosynthesis. These include: 1) efforts to find second-generation compounds against existing targets that overcome issues of resistance; 2) programs targeting alternative pyrimidine biosynthetic enzymes, with the highest impact studies focused on dihydroorotate dehydrogenase (DHODH); and 3) efforts directed at targeting purine salvage enzymes. A common theme in all of these approaches is the integration of structural biology to identify and/or optimize novel inhibitors of these essential parasite enzymes, or to understand the mechanism of drug resistance and how it might be overcome.

3.2 Electron Transport – the bc1 Complex

The ubiquinol-cytochrome c oxidoreductase (bc1 complex) (Complex III) is responsible for generating an electrochemical proton gradient across the inner mitochondrial membrane in cells.[7,8] The electrochemical gradient is used for the generation of ATP and for other energy requirements such as the transport of solutes. In *Plasmodim falciparum* the primary role of the bc1 complex appears to be the generation of oxidized unbiquinone (CoQ) for use in pyrimidine biosynthesis.[9]

Extensive structural and mechanistic data have been described for mammalian, yeast, plant and bacterial bc1 complexes and these data have led to considerable insight into the reaction mechanism (reviewed in several papers).[7,8,10,11] The bc1 complex in eukaryotes contains up to 11 subunits per monomer, which are organized as a dimer in the inner membrane of the mitochondria. The catalytic machinery is housed in three subunits (cytochrome b (Cyt b), cytochrome c (Cyt c), and the Rieske iron-sulfur protein) that function in concert to catalyze electron transfer from ubiquinol to Cyt c1, generating the transfer of 1 proton/electron across the membrane (Figure 3.3A). Cyt b contains 2 separate heme binding-sites (b_L and b_H) situated within the membrane spanning helical region of the protein, and each is associated with an independent ubiquinol binding site (quinol oxidation (Qo) and quinone reduction (Qi)). The iron-sulfur protein containing a 2Fe2S Rieske center, packs against Cyt b proximal to the intermembrane space. Cyt c interacts with Cyt b within the intermembrane space and contains the third heme (H3) center required for the catalytic cycle. According to the hypothesized Q cycle mechanism, ubiquinol generated by mitochondrial dehydrogenases (e.g. dihydrorotate dehydrogenase) is oxidized at the Qo site, and the 2 electrons generated from this reaction are routed along separate pathways. One electron is transfered to the 2Fe2S Rieske center, which after a conformational change transfers an electron to Cyc c, and the other is transferred to the b_L heme in closest proximity to the Q_0 site. From the b_L heme the electron travels to the b_h heme where it is utilized at the Q_i site to regenerate reduced ubiquinol, near the matrix side of the inner mitochondrial membrane.

3.2.1 Atovaquone and Mechanism of Resistance to bc1 Inhibitors

Atovaquone is a hydroxynapthoquinone that was developed in the Wellcome Research Laboratories in the 1980s (reviewed in several papers).[6,12,13] It has

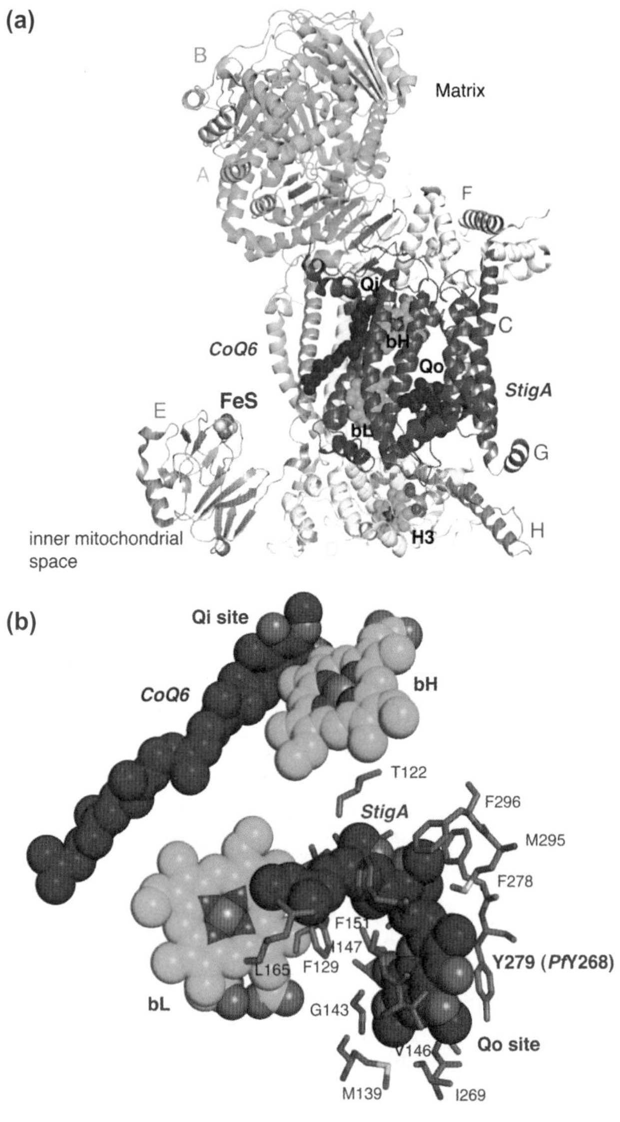

Figure 3.3 Structure of the yeast bc1 complex. (a) Ribbon diagram showing subunits as follows: (A) ubiquinol-cytochrome C reductase complex core protein I, (B) ubiquinol-cytochrome C reductase complex core protein II, (C) cytochrome B, (D) Cytochrome C1, (E) ubiquinol-cytochrome C reductase iron-sulfur subunit, (F) ubiquinol-cytochrome C reductase complex 14 KD protein (H) ubiquinol-cytochrome C reductase complex 17 KD protein (G) ubiquinol-cytochrome C reductase complex unbiquinone-binding protein QP-C. (b) Stigmatellin A (StigA) Qo binding-site. The coordinates pdb 1EZV were used to generate the displayed figures. CoQ and Stigmatellin are displayed as space filling balls with carbon in dark blue and nitrogen in red. Heme is displayed with carbon in green, nitrogen in blue and oxygen in red, and the FeS cluster is displayed in yellow. The heme (bH, bL and H3) and CoQ (Qi and Qo) binding sites are labeled.

broad-spectrum activity and is useful for the clinical treatment of *P. falciparum* malaria, *Toxoplasma gondii* and *Pneumocystis* pneumonia. Early on it was recognized that the cellular action of this class of compounds was the inhibition of the parasite mitochondrial respiration chain, targeting the Q_0 site of Cyt b.[14,15] Despite the good activity of atovaquone against the parasite, resistance developed readily both in the lab and in patients during clinical trials. The mechanism of resistance was mapped to point mutations in Cyt b. This observation led to the decision that atovaquone would not be useful as a monotherapy. It was subsequently combined with proguanil, which was found to be synergistic with atovaquone in the malaria parasite. The atovaquone/proguanil combination is marketed as Malarone™ and, because of its expense, is used almost exclusively for prophylaxis in western travelers to the infected regions.

The atovaquone binding-site on Cyt b has been elucidated by mapping point mutations observed in resistant parasites onto the X-ray structures of the mammalian and yeast bc1 complex[12,16–19] (Figure 3.3B). Cocrystal structures of the yeast and mammalian enzymes bound to a number of different Qo site (e.g. stigmatellin, myxothiazol, famoxadone) and Qi site inhibitors (e.g. antimycin A) have elucidated the two quinone binding-sites. At the Qo site a number of potential binding orientations for inhibitors have been observed, and they have been correlated with conformational changes in the orientation of the Qo site relative to the Rieske iron-sulfur protein and to the b_L heme.[11,20–22] Atovaquone resistance in patients strongly correlates to mutation of Tyr268 in *P. falciparum* Cyt b. This mutation leads to a decreased catalytic efficiency (by up to 30%) but this apparently is not enough to lead to a significant reduction in fitness. While the moderate level of the decreased catalytic efficiency associated with the mutation is not consistent with a major role in catalysis, the mutations caused up to a 730-fold increase in ED_{50} for atovaquone versus the parental strain, suggesting a major role for these residues in stabilizing the inhibitor enzyme complex. Site-directed mutagenesis on the yeast enzyme, and EPR spectroscopy were used to show that atovaquone binds to the Qo site of Cyt b. Atovaquone is a competitive inhibitor of quinone and its binding leads to a locked conformation of the Rieske iron-sulfur complex in an orientation that does not allow electron transfer. This collapses the mitochondrial membrane potential, leading to parasite death.[23] The mechanism of proguanil synergy is not fully understood, however it has been shown to potentiate the effects of atovaquone on the membrane potential[24] and more recent data suggest that the parasites contain a proguanil sensitive pathway that does not require the electron transport chain to generate mitochondrial membrane potential.[9]

3.2.2 Next-generation bc1 Complex Inhibitors

Considerable effort is underway to identify novel inhibitors of *P. falciparum* bc1 complex with better bioavailability than atovaquone, while also focused on designing inhibitors that will avoid the development of rapid resistance. Modeling of the atovaquone binding-site is an important tool in understanding how to avoid inhibitors that will yield to ready resistance. These efforts are hampered by the lack of an X-ray structure for the *P. falciparum* bc1 complex. In lieu of these data, the yeast enzyme has been used to build an energy-minimized structure of the atovaquone binding-site.[12,25] Several classes of newer-generation bc1 complex inhibitors have been reported, including the pyridones, acridones, acridinediones, quinolones and some newer-generation hydroxy-napthoquinones.[13]

Pyridones. The pyridones were developed by GSK based on the observation that clopidol, a known anticoccidial drug, likely acts by inhibiting mitochondrial respiration. A series of derivatives were synthesized by replacing the 5-Cl of clopidol with lipophilic side chains.[26] Analogs with a biaryl or 4'-phenoxyaryl side chain were potent inhibitors of *P. falciparum* growth both *in vitro* and *in vivo*, and against both drug sensitive and resistant strains, including atovaquone resistant strains.[25–27] This series yielded two preclinical candidates (GW844520 and GSK932121; Figure 3.1) that showed sustained plasma exposure and good oral bioavailability,[28] but progression of these compounds has apparently been stopped due to toxicity.[13] A series of modified pyridones incorporating an aromatic moiety at the imine nitrogen have also been reported, however in all cases these compounds showed lower activity than the series generated by GSK.[25] The molecular modeling suggests that the pyridones bind to the same site as atovaquone, yet they retain activity against the atovaquone resistant strains of *P. falciparum* that contain point mutations in the bc1 complex. The published modeled structures have not provided a clear structural basis for the lack of cross-resistance, making it difficult to extend these results for the development of other bc1 inhibitors that lack cross-resistance to Atovaquone. The unexpected toxicity of the two preclinical candidates has been speculated to arise from inhibition of mammalian bc1 complex.

Acridones. While working on the synthesis of tricyclic heme complexing compounds, the Riscoe group discovered that an acridone intermediate in their synthesis had potent antimalarial activity on its own.[29] A dual function acridone that also includes a heme-targeting moiety has also been reported to have potent antimalarial activity *in vitro* and *in vivo*.[30] Based on structural similarity to known bc1 complex inhibitors it was proposed that the mechanism of action of the acridones was similar to atovaquone. This hypothesis was supported by the generation of a highly acridone resistant cell line (SBI-A6) that showed cross-resistance to atovaquone and other bc1 inhibitors.[31] Interestingly no point mutations were found in the bc1 complex, thus the molecular basis of the drug resistance is unclear. Further adding to the mystery, it has recently been shown that the acridone resistant cell line is also resistant to inhibitors of dihydroorotate dehydrogenase (DHODH),[2] thus this cell line appears to have bypassed the need to utilize the electron transport chain to synthesize pyrimidines. Our current understanding of the parasite biology cannot explain the behavior of the acridone resistant cell line, though it does suggest that clinical use of acridone analogs could endanger more than one class of potentially clinically useful antimalarial agents.

Hydroxy-napthoquinones. Chemistry to improve the properties of the hydroxy-napthoquinones has also been reported. A series of additional analogs with substitutions at the 3-hydroxy group were made, with the goal of improving bioavailability over that observed for atovoquone.[32] A number of potent analogs were identified with good antimalarial activity, however in silico experiments suggested they will not have improved bioavailability. A second group synthesized a series of analogs with a branched aliphatic chain (e.g. S-10576; Figure 3.1), and they tested the effects of the addition of a CF_3 group at the end of this chain with the goal of improving metabolic stability.[33] Molecular modeling was used to develop correlations between predicted and observed binding energy within the series, based on the yeast enzymes structure. However, they did not observe a strong correlation between the observed and predicted values, demonstrating the difficulties of this type of modeling approach. In addition, the study stopped short of demonstrating improved metabolic stability, and the studies performed to test species selectivity were based on comparison between the yeast and bovine enzymes, when comparison of the *P. falciparum* and human enzyme would clearly be the more useful indicator of selectivity. In summary, it remains unclear whether the goal of identifying new clinically useful bc1 inhibitors can be realized, however the efforts to date suggest that further development of new inhibitors against the bc1 target should include building good selectivity between the parasite and human enzymes as a key design feature.

3.3 Pyrimidine Nucleoside and Nucleotide Metabolism

3.3.1 Dihydrofoloate Reductase (DHFR) – Therapeutically used Inhibitors and Structural Basis of Resistance

DHFR and dihydropteroate synthase (DHPS) were among the first molecular targets to be identified for the treatment of malaria and inhibitors of these enzymes formed the corner stone of treatment after the development of resistance to chloroquine.[34–36] The DHFR inhibitor pyrimethamine, discovered in the mid 1940s, has been used in combination with sulfadoxine (marketed as Fansidar) since the early 1960s, proving to be a highly effective antimalarial therapy. Sulfadoxine is an inhibitor of DHPS and is synergistic with pyrimethamine as both agents interfere with folate metabolism. Other clinically utilized DHFR inhibitors include proguanil and chlorproguanil, both of which have been used in combination with dapsone

DHFR catalyzes the NADPH dependent reduction of dihydrofolate to tetrahydrofolate, a key step in the regeneration of 5,10-methenyltetrahydrafolate, which is required by thymidylate synthetase (TS)[34–36] (Figure 3.2). In *Plasmodium* species DHFR and TS are expressed as a bifunctional enzyme on a single polypeptide chain, which is in contrast to the situation in human cells where the two enzymes are encoded by separate gene products. The selective toxicity of pyrimethamine has been linked to differences in the binding affinity of the inhibitors to DHFR from *Plasmodium* species versus the human enzyme, and to differences in regulation of the enzyme under drug pressure. DHFR has been shown to bind its own messenger RNA (mRNA) leading to translational repression.[37] While human DHFR interacts with RNA at the active site, the malarial enzyme binds mRNA at a non-active site location and, therefore, translational inhibition is not relieved in the presence of an active site inhibitor. Additionally, it has been shown by microarray analysis that the parasite does not up-regulate the mRNA levels of DHFR or other enzymes in the pathway under the pressure of DHFR inhibition.[38] The inability of the parasite to regulate mRNA or protein levels in response to drug pressure reduces the ability of the parasite to overcome inhibition of DHFR relative to the host cell, which is able to increase the amount of DHFR that it translates when inhibitors are present.

Despite the early success of DHFR inhibitors for the treatment of malaria, resistance to sulfadoxine/pyrimethamine (SP) began to develop soon after introduction, and is now widespread throughout Africa, South America and Asia.[34–36,39] Furthermore, the recent addition of artesunate to the SP combination has not slowed resistance development to either compound, suggesting that this triple drug combination will have a short life-span.[40] The molecular basis for resistance to pyrimethamine has been tracked to point mutations in DHFR. The single point mutation of S108N leads to initial resistance, and then is often followed by mutations of N51I and C59R, leading to heightened resistance over the single point mutation. A quadruple mutation that includes I164L in addition to the other three mutations has been isolated that leads to greater than 1000-fold resistance to pyrimethamine.

The crystal structures of DHFR-TS have been solved for the wild-type enzyme and for the double (C59R/S108N) and quadruple mutant (C59R/S108N/N51I/I164L) enzymes, which are involved in clinical resistance to pyrimethamine and cycloquanil[41,42] (Figure 3.4). These mutant enzymes retain sensitivity to WR99210 an experimental anti-folate that is also a potent inhibitor of DHFR, but which did not have the oral bioavailability to be developed as an antimalarial agent.[43] The analysis includes a comparison between the binding modes of pyrimethamine and WR99210, leading to a good understanding of the structural basis for resistance in one case, and for sensitivity in the other. The substitution of S108N leads to a close contact between the Asn108 side chain and the p-chlorophenyl portion of pyrimethamine. The flexibility of the WR99210 side chain allows it to rotate to avoid the steric clash, thus providing a structural rationale for the resistance phenotypes observed for this mutation. In the structure of the quadruple mutant it appears that the N51I change leads to a substantial main chain displacement of residues 48–51, which in turn leads to a more open active site structure. The enlarged pocket is postulated to have a greater affect on the binding affinity of smaller inhibitors like pyrimethamine in contrast to WR99210, which can still fill the mutant binding-pocket effectively for inhibition.

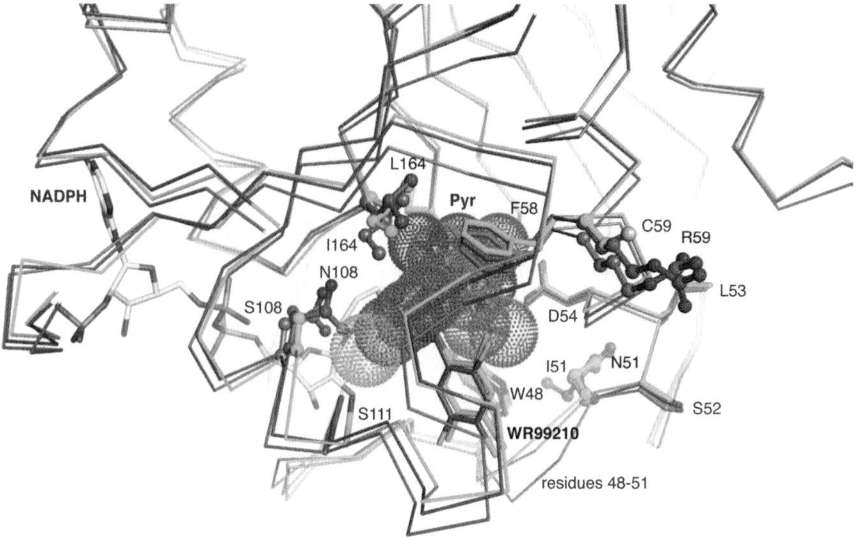

Figure 3.4 Structural basis of resistance to DHFR inhibitors. Overlay of the X-ray structures of wild-type (1J3I) (green) in complex with WR99210, the double mutant (1J3J) (purple) complexed with pyrimethamine and the quadruple mutant (1J3K) (pink) in complex with WR99210. Inhibitor binding site residues are labeled. NADPH (yellow) and WR99210 are displayed as sticks, amino acids with mutations are displayed as ball and sticks and pyrimethamine is displayed using space filling balls. Non-carbon atoms are colored as follows: Nitrogen (blue), oxygen (red), sulfur (yellow), chlorine (green).

3.3.2 Structure-based Design of Next-generation DHFR Inhibitors

The successful solution of the *P. falciparum* DHFR-TS X-ray structure provided the potential to develop redesigned DHFR inhibitors that would be able to inhibit the wild-type and mutant DHFRs, including the potential to develop inhibitors with broad enough spectrum to inhibit the quadruple mutant.

Computational active-site inhibitor design. Molecular field analysis and quantum chemical calculations have been performed for a series of both cycloguanil and pyrimethamine derivatives in a comparative analysis between the wild-type and quadruple mutant.[44–46] These studies provide additional support for the hypothesis that the S108N mutation leads to steric clash with pyrimethamine and cycloguanil, reducing binding affinity. A good correlation was observed between predicted binding constants to both the wild-type and mutant enzymes for the inhibitor sets that were studied, suggesting this method provides a computational prediction to guide new inhibitor design with the goal of finding analogs that show good activity against the quadruple mutant enzyme. QSAR analysis has also been used to evaluate the relative predicted activity of a series of trisubstituted-s-triazine derivatives and the method gave a good overall correlation to the observed potency against the parasite suggesting that these methods provide reasonable predictive power to generate new analogs within a series.[47] Computational methods have also been used to screen virtual libraries to identify potential inhibitors that are predicted to bind to the quadruple mutant DHFR active site, though these leads have not been biochemically validated.[48,49]

Non-active site computational approaches. Structure-based computational methods have also been used to identify inhibitors of DHFR-TS that would potentially interact with a linker region outside of the active site that is parasite-specific.[50] Glide SP was used for a virtual screen of 16 000 compounds from the Maybridge-HitFinder library and three bis-guanide derivatives with IC_{50}s in the 20 uM range for both the wild-type and quadruple mutant were identified. However, while the initial computational screen was conducted against a binding region in the linker domain subsequent crystallography showed that these compounds bind the DHFR active site. The relatively weak binding affinity of these compounds, however, suggests that they are not ideal leads for a drug discovery program.

Quinazolines. Quinazolines are known inhibitors of DHFR and are used for the treatment of cancer and other human proliferative diseases. A quinazoline derivative (QN254) with low nanomolar affinity against both the wild-type and quadruple mutant *Pf*DHFR was identified and taken through a full preclinical evaluation.[43] QN254 showed excellent potency against a range of pyrimethamine resistant field isolates, it was orally bioavailable, had a long half-life in rodents and showed good efficacy against the *P. berghei* mouse model of the disease (ED_{90} = 12 mg/kg). X-ray structure analysis was performed, providing insight into the high-affinity binding. Unfortunately, QN254 was toxic in the 7-day rat toxicity study that was performed with marked gastrointestinal and bone marrow toxicity being observed that was consistent with inhibition of host DHFR. The selectivity of QN254 inhibition between the *P. falciparum* and human DHFR was only 15-fold, about 4-fold less than observed for pyrimethamine, likely contributing to the lack of selectivity when dosed in animals. These studies show the importance of identifying compounds with better levels of selectivity against human DHFR before progressing compounds into preclinical evaluation.

Other. While none of the published compounds appear to be well-developed leads for a second-generation DHFR inhibitor for the treatment of malaria, the potential to develop new inhibitors that are able to inhibit both the wild-type and mutant enzyme remains. The Medicines for Malaria Venture's website (www.mmv.org) currently lists a DHFR inhibitor (P218) as a preclinical candidate suggesting that a compound with sufficient efficacy in rodent models of the disease, good drug like properties and acceptable toxicity in exploratory rat toxicology studies has been identified.

3.3.3 Other Targets in Pyrimidine and Folate Metabolism

While DHFR is the only enzyme in the pathway to have been targeted for treatment of malaria in the clinics, work towards targeting other enzymes in the pathway has also been published. 5-fluoroorotate has been shown to have potent activity against parasites both *in vitro* and *in vivo* and its mechanism of action is thought to be via conversion to 5-fluoro-2'deoxyuridylate, a potent inhibitor of TS.[51–55] 5-Fluoroororotate however showed mechanism-based toxicity in mice that could be rescued by the co-administration of uridine. At present there are insufficient published data to determine if *Plasmodium* TS can be targeted with a sufficient therapeutic window for the development of novel antimalarials.

Two enzymes are required for the replenishment of the 5, 10 methenyltetrahydrofolate (MTHF) to complete the thymidylate synthesis cycle, DHFR and serine hydroxymethyl transferase (SHMT). The enzymatic properties of SHMT have been recently published, providing an avenue to study this enzyme further for potential drug discovery.[56]

3.4 *De novo* Pyrimidine Biosynthesis

In the malaria parasite the pyrimidine building blocks for DNA and RNA biosynthesis can only be acquired through the *de novo* biosynthetic pathway.[57,58] The parasite lacks the salvage enzymes found in most other cells, including mammals, which are required for the incorporation of preformed nucleosides and bases. In the *de novo* pathway, six enzymes are required to generate uridine monophosphate (UMP) from the starting precursors of L-Gln and bicarbonate (Figure 3.2). UMP then serves as the precursor for the remaining pyrimidine nucleotides including dTMP. Enzymes in the *de novo* pyrimidine biosynthetic pathway are essential for parasite growth, as demonstrated using a genetic rescue strategy for the fourth enzyme in the pathway, DHODH.[9]

3.4.1 Dihydroorotate Dehydrogenase (DHODH) as a New Drug Target

DHODH catalyzes the flavin-dependent oxidation of dihydroorotate.[59] DHODH from different species differ in both localization and mechanism with some species utilizing a mitochondrial bound enzyme that requires CoQ for the reoxidation of flavin, while other species have a cytoplasmic enzyme that functions with fumarate or NADH for this final oxidation step. Both the malarial parasite and the human host have mitochondrial DHODH. Inhibitors of human DHODH have been characterized for their potential as immunosuppressive agents and indeed a human DHODH inhibitor (A77 1726 the active metabolite of leflunomide) is on the market for the treatment of rheumatoid arthritis. Inhibitors of human DHODH are not effective inhibitors of the malarial enzyme,[60] and X-ray structure analysis has shown that the inhibitor binding-site contains many amino acid substitutions between the host and parasite enzymes that lead to selective inhibitor binding.[61–65] Thus, the findings that DHODH is essential to *Plasmodium* species, that it is a known drug target, and that species selective inhibition was feasible, has led to an extensive effort to target DHODH for the development of new antimalarials. The utilization of target-based high throughput screening (HTS) led to the identification of a number of different structural classes of *P. falciparum* DHODH inhibitors, and two of these compound series are currently undergoing lead optimization.[63,64] The availability of X-ray structures of DHODH in complex with both lead series has provided important insight to advance these programs.[63,65]

3.4.2 Identification of Novel Inhibitors: Triazolopyrimidines

A series of triazolopyrimidines were identified to be both potent and selective inhibitors of *P. falciparum* DHODH (*Pf*DHODH) by my group, as part of a collaborative effort between University of Texas Southwestern Medical Center, University of Washington and Monash University with support from Medicines for Malaria Venture and the NIH.[64,66] The initial lead compound (DSM1) was identified by HTS and inhibited both *Pf*DHODH and *P. falciparum* in a whole cell *in vitro* assay with an IC_{50} in the range of 50 nM, while not inhibiting the human enzyme.[66] This compound however was rapidly metabolized, and, furthermore, the data suggested that it was a likely metabolic inducer. This finding explained the lack of activity of DSM1 against the mouse model of malaria infection (*P. berghei*). The subsequent lead optimization program identified a triazolopyrimidine analog DSM74 (Figure 3.1), that showed prolonged plasma exposure after oral dosing in mice or rats, and that was able to significantly suppress a *P. berghei* infection in mice at a dose of 50 mg/kg bid for 4 days.[64] These studies provided the first proof of concept that DHODH inhibitors could have *in vivo* activity against *Plasmodium* species. However, while DSM74 was less rapidly metabolized than DSM1, it is also about 6-fold less potent, and subsequent lead optimization has been directed at improving the potency of the compound series.

Thiophenecarboxamides. Genzyme in collaboration with Harvard and Broad identified a second series of potent *Pf*DHODH inhibitors by HTS that also showed good cell based activity against *P. falciparum in vitro* with activity in the 0.5–1 μM range.[67] Modification of the bulky aromatic group that projects from the thiophene ring during the subsequent lead optimization program identified several compounds (Genz-667348, Genz-668857 and Genz-669178) that inhibited *Pf*DHODH with IC_{50}s ranging from 20–50 nM, and that showed improved potency against the parasite with activity in the range required of a drug candidate (*P. falciparum* 3D7 cells; $EC_{50} = 7–20$ nM).[63] These compounds show good plasma exposure after oral dosing and were able to suppress parasitemia in the *P. berghei* mouse model ($ED_{50} = 13–21$ mg/kg with twice daily oral dosing). Genz-667348 was shown to provide a sterile cure at a dose of

100 mg/kg/day b.i.d. These compounds have drug-like physical properties, were shown not to have any significant Cyp inhibition and some in the series were also clean in a hERG analysis, setting the stage for potential preclinical development of a compound from the series.

3.4.3 Insights from X-ray Structural Analysis of DHODH Bound to Inhibitors

The X-ray structure of *Pf*DHODH in complex with the triazolopyrimidines (including DSM1 and DSM74), led to insight into the binding mode of the inhibitor class and has provided insight to direct the lead optimization program[65] (Figure 3.5). The triazolopyrimidines bind in a pocket adjacent to the flavin cofactor, in a site that overlaps the position of A77 1726 in structures of human DHODH. The triazolopyrimidine ring binds in a pocket close to the flavin cofactor, which contains the only two residues that are capable of H-bond interactions within the inhibitor binding-site. The pyridine nitrogen (N5) forms an H-bond with Arg-265, while the bridging nitrogen (N1) forms an H-bond with His-185. Both of these residues are conserved in human DHODH. The aromatic moiety of the inhibitor attached through N1 occupies a completely hydrophobic pocket that extends towards the membrane spanning

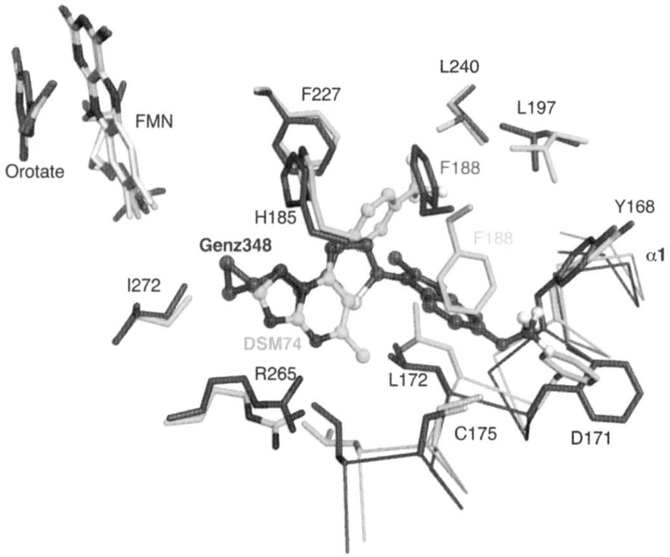

Figure 3.5 Inhibitor binding-site of *P. falciparum* DHODH. Overlay of DHODH bound to DSM74 (3I6R) (turquoise) and Genz348 (3O8A) (pink). Flavin is displayed in yellow. The N-terminal α-helix (α1) is labeled. Non-carbon atoms are colored as follows: Nitrogen (blue), oxygen (red), sulfur (yellow), fluorine (grey).

helices. This pocket is created by a rotation of Phe-188 relative to the position observed in the complex of A77 1726 to *Pf*DHODH.

X-ray structure analysis of Genz-667348 bound to *Pf*DHODH was also reported.[63] For this inhibitor class the thiophenecarboxamide binds in the H-bond pocket overlapping the position of the triazolopyrimidine ring of DSM1/DSM74. H-bond interactions are formed between Arg-265 and the carbonyl oxygen of the inhibitor and between His-185 and the amide NH. The cyclopropyl extends towards the flavin cofactor and the benzimidazole ring binds in a hydrophobic pocket towards the N-terminal helices. Phe-188 is found in the configuration similar to that observed in A77 1726, thus the benzimidazole ring occupies a different hydrophobic pocket than that observed for the 4-CF_3-aniline ring in DSM74. The finding of significant structural flexibility within the inhibitor binding-pocket of *Pf*DHODH, explains the ability of the enzyme to bind multiple inhibitor classes from different chemical space, increasing its value as a potential target in the quest for new antimalarial agents.

Structural basis for species selectivity. A number of key amino acid differences are observed in the inhibitor binding sites that explain the species selectivity of both the triazolopyrimdines and the thiophenecarboxamides.[63,65] Key differences for the triazolopyrimidines include the substitution of *Pf*DHODH Leu-240 for Met-111 in human DHODH, and the replacement of Ala-59 in human DHODH with Phe-188 in *Pf*DHODH. For the Genzyme series, again the hydrophobic pocket appears to be the key to species selective binding. The substitution of Leu-172 and Cys-175 in *Pf*DHODH for Met-43 and Leu-46 in human DHODH appears to open up the benzimidazole binding-pocket that allows Genz-667348 to bind *Pf*DHODH.

In summary, DHODH represents a promising new potential target for the treatment of malaria. To date, two inhibitor scaffolds with advanced lead optimization programs have been described, providing optimism that a clinically useful DHODH inhibitor will be discovered. However, final validation of the target awaits the identification of a clinical candidate and the demonstration that a well-tolerated compound with good efficacy in humans can be identified that meets the standards required for registration.

3.5 Purine Salvage Enzymes

Protozoan parasites are not capable of *de novo* purine biosynthesis and instead utilize salvage pathways to obtain the base and nucleoside precursors for the synthesis of purine nucleotides, DNA and RNA.[58,68,69] In *Plasmodium* species, the most active salvage enzymes found in parasite lysates are hypoxanthine-xanthine-guanine-phosphoribosyltransferase (HGXPRT), adenonsine deaminase (ADA) and purine phosphonucleoside phosphorylase (PNP).[57] In addition to their roles in purine salvage, PNP and ADA have a dual function in *Plasmodium* species and both are involved in the recycling of methylthioadenosine (MTA) generated during polyamine biosynthesis.[70] While mammalian cells utilize MTA phosphorylase followed by adenosine phosphoribosyltransferase to convert MTA back to AMP and methionine, *Plasmodium* species lack both enzymes and instead the dual function ADA is able to convert MTA into methylthioinosine, which is then converted to hypoxanthine by the dual function PNP. The parasites thus lack the ability to recycle methionine after polyamine biosynthesis. A series of elegant studies have been carried out by the Schramm laboratory to identify potent transition stage analogs of PNP, ADA and HGXPRT using kinetic isotope effect (KIE) studies to define the transition state for each enzyme as the prelude to designing and synthesizing inhibitors. These inhibitors have been used to probe the suitability of targeting these purine salvage enzymes for the treatment of malaria.[71]

3.5.1 Purine Nucleoside Phosphorylase

Immucillin H was designed as an inhibitor of human PNP based on transition state analogy using a TS model derived from KIEs and computational chemistry, and was later demonstrated to have antimalarial activity *in vitro* (ED_{50} = 63 nM).[72,73] The compound is a potent inhibitor of both the human and *P. falciparum* enzymes though it is 15-fold more potent on the human enzyme (IC_{50} = 86 *vs.* 550 pM). Immucillin H was postulated to inhibit both the host and parasite PNP from producing hypoxanthine, thus decreasing both the parasites ability to synthesize hypoxanthine from inosine and its ability to obtain hypoxanthine by uptake from the host red cell. The X-ray structure of immucillin H bound to *P. falciparum* PNP was solved uncovering a cavity near the 5'carbon of the inhibitor that provides the structural basis for the enzymes ability to also catalyze the conversion of methylthioinosine into hypoxanthine (Figure 3.6).[74] Based on both the structure and the biochemical demonstration of the enzymes dual function 5'methylthio-immucillin-H (MT-ImmH) was synthesized and found to be a selective inhibitor of the malarial enzyme (ED_{50} = 2.7 nM *vs.* 303 nM for the human enzyme) that showed good activity against cultured *P. falciparum* parasites (ED_{50} = 50 nM).[70] Despite inhibiting parasite growth *in vitro*, there are no reports demonstrating that inhibitors of PNP have efficacy against the parasite *in vivo* raising questions about the essentiality of the target and the ability of inhibitors of this enzyme to be developed into drugs. A series of genetic studies have been published that shed light on this issue.

The essentiality of PNP for parasite growth was studied using gene knockout (KO) strategies in both *P. falciparum* parasites grown in culture and in *P. yoelli* parasites within the mouse host.[75,76] In *P. falciparum* the PNP KO line was generated by performing the transfection in the presence of high exogenous hypoxanthine to provide rescue (500 uM). The doubling time of the selected KO line was 2-fold lower than for wild-type at 10 uM hypoxanthine but it grew normally at 100 uM. Ethanolamine incorporation as a measure of cell growth was reduced in the KO by 5-fold in the complete absence of hypoxanthine but

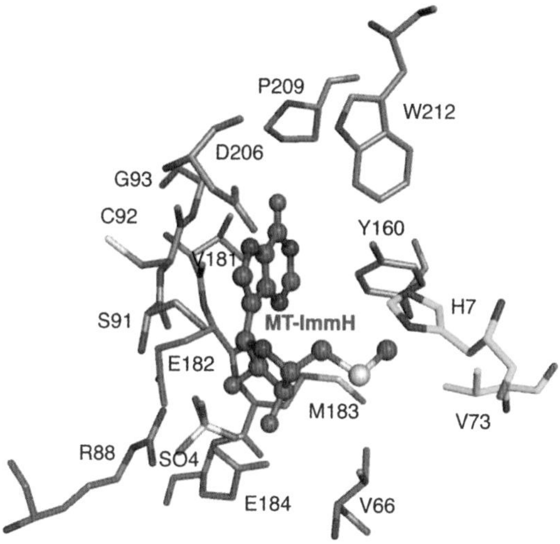

Figure 3.6 Inhibitor binding-site of *P. falciparum* purine nucleoside phosphorylase. *Pf*PNP bound to 5-methylthio-immucillin-H (MT-ImmH) is displayed (1QIG).[74] Residues from subunit A are displayed in purple and residues from subunit B are displayed in turquoise. MT-ImmH is displayed in pink using ball and stick. A bound SO_4^{-2} ion is also displayed. Nitrogen (blue), oxygen (red).

at 6 uM incorporation was similar to wild-type. Hypoxanthine levels in human serum were reported to be 0.4–6 uM, suggesting that the cells may not have impaired growth *in vivo*. In addition to hypoxanthine, adenosine and inosine were also able to rescue the cell growth defect of the KO.

A knockout of PNP was also generated in *P. yoelli* and growth of the KO parasites in mice relative to wild-type was substantially slower.[75] Wild-type parasites killed mice in 10 days, while in contrast mice inoculated with the PNP KO strain eventually cleared infection, leading to the conclusion that PNP is important for optimal growth of the erythrocyte stage. Mosquitoes fed on KO strain did not develop sporozoites. This study provides the strongest published data supporting the target as important. However, while growth was attenuated it was not completely inhibited and, given the fact that a small molecule will not yield 100% inhibition as observed in the KO, the effects of inhibitors would be expected to be less pronounced.

Studies to understand how the parasites are able to survive in the absence of PNP led to the discovery that *P. falciparum* parasites are able to salvage AMP from the host erythrocyte and convert it to IMP using AMP deaminase.[77] This pathway provides a clear mechanism for the parasites to avoid the PNP block and also provides an explanation as to how endogenous adenosine and inosine are able to rescue cell growth in the presence of PNP inhibitors (i.e. via conversion to AMP).

In summary, PNP essentiality is predicated on the idea that the parasite will not be able to salvage sufficient hypoxanthine from the host if the parasite PNP is selectively inhibited. Reported serum levels of hypoxanthine appear close to sufficient to rescue full growth levels of the PNP knockout (KO) *P. falciparum* parasites. Additionally adenosine and inosine can also sustain growth of the KO parasites and parasites can salvage AMP and convert this into hypoxanthine. Thus redundant salvage pathways are present that can get around the block and operating together sufficient pools are likely to exist to bypass the block, particularly given the fact that an inhibitor can never be as effective as the KO. Thus the available data do not support PNP as a target in *P. falciparum*.

3.5.2 Other Purine Salvage Enzymes

P. falciparum adenosine deaminase (ADA) is a dual function enzyme that also catalyzes the deamination of methylthioadenosine.[70] The structural basis for the dual specificity has been elucidated by X-ray crystallography.[78] 5'methylthiolcoformycin, a specific transition state analogue of *P. falciparum* ADA, is a potent low nM inhibitor of the enzyme, highly selective against the human enzyme, and is able to suppress the synthesis of parasite DNA *in vitro* in the same concentration range. No *in vivo* studies have been reported nor have genetic studies addressing the essentiality of this enzyme been undertaken. Since ADA catalyzes the reaction that generates inosine (substrate for PNP), the story is likely to be the same as for PNP and inhibitors of this enzyme are unlikely to have good activity *in vivo* because the block is likely to be rescued by exogenous hypoxanthine.

Hypoxanthine-guanine phosphoribosyltransferase (HGPRT) is the third purine salvage enzyme in the pathway and is likely to be the most essential of the three. Hyopxanthine would not be able to rescue growth effects caused by the inhibition of HGPRT, though it is unknown to what extent AMP salvage would be able to overcome a block in this enzyme. A recent study suggests that adenine salvage can overcome a block of HGPRT.[79] Transition state inhibitors of human HGPRT have been reported providing the tools to address the importance of this enzyme to the parasite life cycle.[80] Acyclic nucleoside phosphonates with activity against parasites *in vitro* have also been reported.[81,82]

3.6 Conclusions

Target-based drug discovery to identify novel antimalarials through inhibition of pyrimidine or purine biosynthesis has led to both successes and failures. Utilization of structural information to guide lead optimization programs has proven a successful route to identify inhibitors of target enzymes with the appropriate potency. Translating this effort into clinically useful drugs has proven more difficult. Work to identify new chemical species that overcome drug resistance against previously established targets such as DHFR and the bc1 complex has yielded some new chemical classes that meet potency criteria against both wild-type and resistant parasites, but as of yet none of these compounds has progressed to clinical evaluation. Efforts to target new enzymes in the pyrimidine biosynthetic pathway have led to the identification of DHODH as a promising new target, and the publication of two lead series with potent antimalarial activity and good plasma exposure after oral dosing. These data provide hope that a DHODH inhibitor will progress to clinical evaluation. Studies to target purine biosynthesis have provided an elegant biochemical evaluation of the role of the pathway in parasite growth and have uncovered novel aspects of metabolism in the parasite. These studies have provided good evidence for the existence of redundant pathways making targeting of purine biosynthesis for the treatment of malaria unlikely to be a fruitful approach.

Acknowledgements

This work was supported by the United States National Institutes of Health grant, U01AI075594. Margeret A. Phillips holds the Carolyn R. Bacon Professorship in Medical Science.

References

1. F. J. Gamo, L. M. Sanz, J. Vidal, C. de Cozar, E. Alvarez, J. L. Lavandera, D. E. Vanderwall, D. V. Green, V. Kumar, S. Hasan, J. R. Brown, C. E. Peishoff, L. R. Cardon and J. F. Garcia-Bustos, *Nature*, 2010, **465**, 305.
2. W. A. Guiguemde, A. A. Shelat, D. Bouck, S. Duffy, G. J. Crowther, P. H. Davis, D. C. Smithson, M. Connelly, J. Clark, F. Zhu, M. B. Jimenez-Diaz, M. S. Martinez, E. B. Wilson, A. K. Tripathi, J. Gut, E. R. Sharlow, I. Bathurst, F. El Mazouni, J. W. Fowble, I. Forquer, P. L. McGinley, S. Castro, I. Angulo-Barturen, S. Ferrer, P. J. Rosenthal, J. L. Derisi, D. J. Sullivan, J. S. Lazo, D. S. Roos, M. K. Riscoe, M. A. Phillips, P. K. Rathod, W. C. Van Voorhis, V. M. Avery and R. K. Guy, *Nature*, 2010, **465**, 311.
3. F. Aguero, B. Al-Lazikani, M. Aslett, M. Berriman, F. S. Buckner, R. K. Campbell, S. Carmona, I. M. Carruthers, A. W. Chan, F. Chen, G. J. Crowther, M. A. Doyle, C. Hertz-Fowler, A. L. Hopkins, G. McAllister, S. Nwaka, J. P. Overington, A. Pain, G. V. Paolini, U. Pieper, S. A. Ralph, A. Riechers, D. S. Roos, A. Sali, D. Shanmugam, T. Suzuki, W. C. Van Voorhis and C. L. Verlinde, *Nat. Rev. Drug Discov.*, 2008, **7**, 900.
4. D. J. Payne, M. N. Gwynn, D. J. Holmes and D. L. Pompliano, *Nat. Rev Drug Discov.*, 2007, **6**, 29.
5. B. M. Greenwood, D. A. Fidock, D. E. Kyle, S. H. Kappe, P. L. Alonso, F. H. Collins and P. E. Duffy, *J. Clin. Invest*, 2008, **118**, 1266.
6. A. B. Vaidya and M. W. Mather, *Annu. Rev. Microbiol.*, 2009, **63**, 249.

7. A. R. Crofts, *Annu. Rev. Physiol.*, 2004, **66**, 689.
8. C. Hunte, H. Palsdottir and B. L. Trumpower, *FEBS Lett.*, 2003, **545**, 39.
9. H. J. Painter, J. M. Morrisey, M. W. Mather and A. B. Vaidya, *Nature*, 2007, **446**, 88.
10. E. A. Berry, M. Guergova-Kuras, L. S. Huang and A. R. Crofts, *Annu. Rev. Biochem.*, 2000, **69**, 1005.
11. A. R. Crofts, J. T. Holland, D. Victoria, D. R. Kolling, S. A. Dikanov, R. Gilbreth, S. Lhee, R. Kuras and M. G. Kuras, *Biochim. Biophys. Acta*, 2008, **1777**, 1001.
12. J. J. Kessl, S. R. Meshnick and B. L. Trumpower, *Trends Parasitol.*, 2007, **23**, 494.
13. V. Barton, N. Fisher, G. A. Biagini, S. A. Ward and P. M. O'Neill, *Curr. Opin. Chem. Biol.*, 2010, **14**, 440.
14. A. T. Hudson, A. W. Randall, M. Fry, C. D. Ginger, B. Hill, V. S. Latter, N. McHardy and R. B. Williams, *Parasitology*, 1985, **90** (Pt 1), 45.
15. M. Fry, A. T. Hudson, A. W. Randall and R. B. Williams, *Biochem. Pharmacol.*, 1984, **33**, 2115.
16. J. J. Kessl, B. B. Lange, T. Merbitz-Zahradnik, K. Zwicker, P. Hill, B. Meunier, H. Palsdottir, C. Hunte, S. Meshnick and B. L. Trumpower, *J. Biol. Chem.*, 2003, **278**, 31312.
17. M. W. Mather, E. Darrouzet, M. Valkova-Valchanova, J. W. Cooley, M. T. McIntosh, F. Daldal and A. B. Vaidya, *J. Biol. Chem.*, 2005, **280**, 27458.
18. I. K. Srivastava, J. M. Morrisey, E. Darrouzet, F. Daldal and A. B. Vaidya, *Mol. Microbiol.*, 1999, **33**, 704.
19. N. Fisher and B. Meunier, *FEMS Yeast Res.*, 2008, **8**, 183.
20. H. Kim, D. Xia, C. A. Yu, J. Z. Xia, A. M. Kachurin, L. Zhang, L. Yu and J. Deisenhofer, *Proc. Natl. Acad. Sci. U S A*, 1998, **95**, 8026.
21. Z. Zhang, L. Huang, V. M. Shulmeister, Y. I. Chi, K. K. Kim, L. W. Hung, A. R. Crofts, E. A. Berry and S. H. Kim, *Nature*, 1998, **392**, 677.
22. C. R. Lancaster, C. Hunte, J. Kelley, B. L. Trumpower and R. Ditchfield, *J. Mol. Biol.*, 2007, **368**, 197.
23. I. K. Srivastava, H. Rottenberg and A. B. Vaidya, *J. Biol. Chem.*, 1997, **272**, 3961.
24. I. K. Srivastava and A. B. Vaidya, *Antimicrob. Agents Chemother.*, 1999, **43**, 1334.
25. T. Rodrigues, R. C. Guedes, D. J. dos Santos, M. Carrasco, J. Gut, P. J. Rosenthal, R. Moreira and F. Lopes, *Bioorg. Med. Chem. Lett.*, 2009, **19**, 3476.
26. C. L. Yeates, J. F. Batchelor, E. C. Capon, N. J. Cheesman, M. Fry, A. T. Hudson, M. Pudney, H. Trimming, J. Woolven, J. M. Bueno, J. Chicharro, E. Fernandez, J. M. Fiandor, D. Gargallo-Viola, F. Gomez de las Heras, E. Herreros and M. L. Leon, *J. Med. Chem.*, 2008, **51**, 2845.
27. M. B. Jimenez-Diaz, T. Mulet, S. Viera, V. Gomez, H. Garuti, J. Ibanez, A. Alvarez-Doval, L. D. Shultz, A. Martinez, D. Gargallo-Viola and I. Angulo-Barturen, *Antimicrob. Agents Chemother.*, 2009, **53**, 4533.

28. H. Xiang, J. McSurdy-Freed, G. S. Moorthy, E. Hugger, R. Bambal, C. Han, S. Ferrer, D. Gargallo and C. B. Davis, *J. Pharm. Sci.*, 2006, **95**, 2657.
29. R. W. Winter, J. X. Kelly, M. J. Smilkstein, R. Dodean, G. C. Bagby, R. K. Rathbun, J. I. Levin, D. Hinrichs and M. K. Riscoe, *Exp. Parasitol.*, 2006, **114**, 4.
30. J. X. Kelly, M. J. Smilkstein, R. Brun, S. Wittlin, R. A. Cooper, K. D. Lane, A. Janowsky, R. A. Johnson, R. A. Dodean, R. Winter, D. J. Hinrichs and M. K. Riscoe, *Nature*, 2009, **459**, 270.
31. M. J. Smilkstein, I. Forquer, A. Kanazawa, J. X. Kelly, R. W. Winter, D. J. Hinrichs, D. M. Kramer and M. K. Riscoe, *Mol. Biochem. Parasitol.*, 2008, **159**, 64.
32. S. El Hage, M. Ane, J. L. Stigliani, M. Marjorie, H. Vial, G. Baziard-Mouysset and M. Payard, *Eur. J. Med. Chem.*, 2009, **44**, 4778.
33. L. M. Hughes, R. Covian, G. W. Gribble and B. L. Trumpower, *Biochim. Biophys. Acta.*, 2010, **1797**, 38.
34. A. Nzila, *J. Antimicrob. Chemother.*, 2006, **57**, 1043.
35. J. E. Hyde, *Acta Trop.*, 2005, **94**, 191.
36. A. Gregson and C. V. Plowe, *Pharmacol. Rev.*, 2005, **57**, 117.
37. K. Zhang and P. K. Rathod, *Science*, 2002, **296**, 545.
38. K. Ganesan, N. Ponmee, L. Jiang, J. W. Fowble, J. White, S. Kamchonwongpaisan, Y. Yuthavong, P. Wilairat and P. K. Rathod, *PLoS Pathog.*, 2008, **4**, e1000214.
39. D. J. Bacon, D. Tang, C. Salas, N. Roncal, C. Lucas, L. Gerena, L. Tapia, A. A. Llanos-Cuentas, C. Garcia, L. Solari, D. Kyle and A. J. Magill, *PLoS One*, 2009, **4**, e6762.
40. J. Raman, F. Little, C. Roper, I. Kleinschmidt, Y. Cassam, R. Maharaj and K. I. Barnes, *Am. J. Trop. Med. Hyg.*, 2010, **82**, 788.
41. J. Yuvaniyama, P. Chitnumsub, S. Kamchonwongpaisan, J. Vanichtanankul, W. Sirawaraporn, P. Taylor, M. D. Walkinshaw and Y. Yuthavong, *Nat. Struct. Biol.*, 2003, **10**, 357.
42. Y. Yuthavong, J. Yuvaniyama, P. Chitnumsub, J. Vanichtananankul, S. Chusacultanachai, B. Tarnchompoo, T. Vilaivan and S. Kamchonwongpaisan, *Parasitology*, 2005, **130**, 249.
43. A. Nzila, M. Rottmann, P. Chitnumsub, S. M. Kiara, S. Kamchonwongpaisan, C. Maneeruttanarungroj, S. Taweechai, B. K. Yeung, A. Goh, S. B. Lakshminarayana, B. Zou, J. Wong, N. L. Ma, M. Weaver, T. H. Keller, V. Dartois, S. Wittlin, R. Brun, Y. Yuthavong and T. T. Diagana, *Antimicrob. Agents Chemother.*, 2010, **54**, 2603.
44. P. Maitarad, S. Kamchonwongpaisan, J. Vanichtananankul, T. Vilaivan, Y. Yuthavong and S. Hannongbua, *J. Comput. Aided Mol. Des.*, 2009, **23**, 241.
45. P. Maitarad, P. Saparpakorn, S. Hannongbua, S. Kamchonwongpaisan, B. Tarnchompoo and Y. Yuthavong, *J. Enzyme Inhib. Med. Chem.*, 2009, **24**, 471.
46. K. Choowongkomon, S. Theppabutr, N. Songtawee, N. P. Day, N. J. White, C. J. Woodrow and M. Imwong, *Malar. J.*, 2010, **9**, 65.

47. H. Ojha, P. Gahlot, A. K. Tiwari, M. Pathak and R. Kakkar, *Chem. Biol. Drug Des.*, 2010, **77**, 57.
48. L. Adane, D. S. Patel and P. V. Bharatam, *Chem. Biol. Drug Des.*, 2010, **75**, 115.
49. L. Adane, P. V. Bharatam and V. Sharma, *J. Enzyme Inhib. Med. Chem.*, 2010, **25**, 635.
50. T. Dasgupta, P. Chitnumsub, S. Kamchonwongpaisan, C. Maneeruttanarungroj, S. E. Nichols, T. M. Lyons, J. Tirado-Rives, W. L. Jorgensen, Y. Yuthavong and K. S. Anderson, *ACS Chem. Biol.*, 2009, **4**, 29.
51. M. Hekmat-Nejad and P. K. Rathod, *Antimicrob. Agents Chemother.*, 1996, **40**, 1628.
52. S. Gassis and P. K. Rathod, *Antimicrob. Agents Chemother.*, 1996, **40**, 914.
53. P. K. Rathod, M. Khosla, S. Gassis, R. D. Young and C. Lutz, *Antimicrob. Agents Chemother.*, 1994, **38**, 2871.
54. P. K. Rathod, N. P. Leffers and R. D. Young, *Antimicrob. Agents Chemother.*, 1992, **36**, 704.
55. F. W. Muregi, S. Kano, H. Kino and A. Ishih, *Exp. Parasitol.*, 2009, **121**, 376.
56. C. K. Pang, J. H. Hunter, R. Gujjar, R. Podutoori, J. Bowman, D. G. Mudeppa and P. K. Rathod, *Mol. Biochem. Parasitol.*, 2009, **168**, 74.
57. P. Reyes, P. K. Rathod, D. J. Sanchez, J. E. Mrema, K. H. Rieckmann and H. G. Heidrich, *Mol. Biochem. Parasitol.*, 1982, **5**, 275.
58. W. E. Gutteridge and P. I. Trigg, *J. Protozool.*, 1970, **17**, 89.
59. M. A. Phillips and P. K. Rathod, *Infect. Disord. Drug Targets*, 2010, **10**, 226.
60. J. Baldwin, A. M. Farajallah, N. A. Malmquist, P. K. Rathod and M. A. Phillips, *J. Biol. Chem.*, 2002, **277**, 41827.
61. D. E. Hurt, J. Widom and J. Clardy, *Acta Crystallogr. D Biol. Crystallogr.*, 2006, **62**, 312.
62. S. Liu, E. A. Neidhardt, T. H. Grossman, T. Ocain and J. Clardy, *Structure*, 2000, **8**, 25.
63. M. L. Booker, C. M. Bastos, M. L. Kramer, R. H. Barker, Jr. R. Skerlj, A. Bir Sidhu, X. Deng, C. Celatka, J. F. Cortese, J. E. Guerrero Bravo, K. N. Krespo Llado, A. E. Serrano, I. Angulo-Barturen, M. B. Jimenez-Diaz, S. Viera, H. Garuti, S. Wittlin, P. Papastogiannidis, J. W. Lin, C. J. Janse, S. M. Khan, M. Duraisingh, B. Coleman, E. J. Goldsmith, M. A. Phillips, B. Munoz, D. F. Wirth, J. D. Klinger, R. Wiegand and E. Sybertz, *J. Biol. Chem.*, 2010, **285**, 33054.
64. R. Gujjar, A. Marwaha, F. El Mazouni, J. White, K. L. White, S. Creason, D. M. Shackleford, J. Baldwin, W. N. Charman, F. S. Buckner, S. Charman, P. K. Rathod and M. A. Phillips, *J. Med. Chem.*, 2009, **52**, 1864.
65. X. Deng, R. Gujjar, F. El Mazouni, W. Kaminsky, N. A. Malmquist, E. J. Goldsmith, P. K. Rathod and M. A. Phillips, *J. Biol. Chem.*, 2009, **284**, 26999.
66. M. A. Phillips, R. Gujjar, N. A. Malmquist, J. White, F. El Mazouni, J. Baldwin and P. K. Rathod, *J. Med. Chem.*, 2008, **51**, 3649.

67. V. Patel, M. Booker, M. Kramer, L. Ross, C. A. Celatka, L. M. Kennedy, J. D. Dvorin, M. T. Duraisingh, P. Sliz, D. F. Wirth and J. Clardy, *J. Biol. Chem.*, 2008, **283**, 35078.
68. D. J. Hammond and W. E. Gutteridge, *Mol. Biochem. Parasitol.*, 1984, **13**, 243.
69. H. F. Hassan and G. H. Coombs, *FEMS Microbiol. Rev.*, 1988, **4**, 47.
70. L. M. Ting, W. Shi, A. Lewandowicz, V. Singh, A. Mwakingwe, M. R. Birck, E. A. Ringia, G. Bench, D. C. Madrid, P. C. Tyler, G. B. Evans, R. H. Furneaux, V. L. Schramm and K. Kim, *J. Biol. Chem.*, 2005, **280**, 9547.
71. T. Donaldson and K. Kim, *Infect. Disord. Drug Targets*, 2010, **10**, 191.
72. G. A. Kicska, P. C. Tyler, G. B. Evans, R. H. Furneaux, K. Kim and V. L. Schramm, *J. Biol. Chem.*, 2002, **277**, 3219.
73. G. A. Kicska, P. C. Tyler, G. B. Evans, R. H. Furneaux, V. L. Schramm and K. Kim, *J. Biol. Chem.*, 2002, **277**, 3226.
74. W. Shi, L. M. Ting, G. A. Kicska, A. Lewandowicz, P. C. Tyler, G. B. Evans, R. H. Furneaux, K. Kim, S. C. Almo and V. L. Schramm, *J. Biol. Chem.*, 2004, **279**, 18103.
75. D. C. Madrid, L. M. Ting, K. L. Waller, V. L. Schramm and K. Kim, *J. Biol. Chem.*, 2008, **283**, 35899.
76. L. M. Ting, M. Gissot, A. Coppi, P. Sinnis and K. Kim, *Nat. Med.*, 2008, **14**, 954.
77. M. B. Cassera, K. Z. Hazleton, P. M. Riegelhaupt, E. F. Merino, M. Luo, M. H. Akabas and V. L. Schramm, *J. Biol. Chem.*, 2008, **283**, 32889.
78. M. C. Ho, M. B. Cassera, D. C. Madrid, L. M. Ting, P. C. Tyler, K. Kim, S. C. Almo and V. L. Schramm, *Biochemistry*, 2009, **48**, 9618.
79. S. Mehrotra, M. P. Bopanna, V. Bulusu and H. Balaram, *Exp. Parasitol.*, 2010, **125**, 147.
80. H. Deng, R. Callender, V. L. Schramm and C. Grubmeyer, *Biochemistry*, 2010, **49**, 2705.
81. D. Hockova, A. Holy, M. Masojidkova, D. T. Keough, J. de Jersey and L. W. Guddat, *Bioorg. Med. Chem.*, 2009, **17**, 6218.
82. D. T. Keough, D. Hockova, A. Holy, L. M. Naesens, T. S. Skinner-Adams, J. Jersey and L. W. Guddat, *J. Med. Chem.*, 2009, **52**, 4391.
83. C. Hunte, J. Koepke, C. Lange, T. Rossmanith and H. Michel, *Structure Fold. Des.*, 2000, **8**, 669.

CHAPTER 4

Human Targets Repositioning and Cell-based Approaches for Antimalarial Discovery

ARNAB K. CHATTERJEE* AND
ELIZABETH A. WINZELER

Genomics Institute of the Novartis Research Foundation, 10675 John J. Hopkins Dr., Room C226, San Diego, CA 92121, US

4.1 Introduction

Many drugs used to treat neglected diseases were developed decades ago. The widely used, synthetic antimalarial drugs including chloroquine and primaquine were developed by mass-testing the products of the German dye industry for their ability to cure experimentally infected canaries, nonhuman primates and ultimately humans, including inmates of the Dusseldorf psychiatric hospital and United States prisons. Needless to say such approaches would be considered unacceptable in today's research climate thus newer approaches to developing drugs for neglected diseases have been sought. Because few parasite "targets" are known, recent drug discovery efforts for neglected diseases have primarily been driven by "piggybacking" approaches that have tried to reposition compounds from other indications, such as oncology and inflammatory diseases. The first part of this chapter will outline some of these efforts and best practices in relation to five target classes with analogous proteins in plasmodium: farnesyltransferase, histone deacetylase (HDAC), protein kinases, proteases and

RSC Drug Discovery Series No. 14
Neglected Diseases and Drug Discovery
Edited by Michael J. Palmer and Timothy N. C. Wells

Published by the Royal Society of Chemistry, www.rsc.org

dihydrofolate reductase (DHFR). The key issues here are to understand the needs for selectivity and improving overall pharmacokinetic and toxicity profiles. The second part of the chapter will outline how modern cell-based screening can provide small-molecule pharmacologically validated targets and methods used for optimization of these hits. In addition, the chemical biology tools available for determination of mechanism of action will also be discussed.

4.2 Human Targets Classes as a Source for Antimalarials

4.2.1 Farnesyltransferase Inhibitors

Protein farnesyltransferase (PFT) is a promising target for oncology with several compounds in late-stage clinical trials.[1] PFT catalyzes the transfer of a farnesyl group from farnesylpyrophosphate (FPP) to the thiol of a cysteine side chain of proteins that carry a CAAX-sequence. PFT inhibitors have long been studied by the pharmaceutical community for their potential as anticancer agents.[2] Lacking protein geranylgeranyltransferase-1 (PGGT-1), *P. falciparum* parasites seem to be highly reliant on PFT for many essential activities that are carried out by both PFT and PGGT-1 in mammalian cells. As short-term inhibition of PFT alone is relatively non-toxic to mammalian cells, inhibition of PFT could be highly toxic to parasites while not affecting the host to a similar degree.[3]

The Gelb group at University of Washington started their search of *P. falciparum* PFT inhibitors by looking at known PFT inhibitors that were optimized based on mammalian cells.[3b,4] Among many of the scaffolds, the tetrahydroquinoline (THQ) series reported by Bristol-Myers Squibb (BMS) stood out for its *in vitro* enzymatic activity against *P. falciparum* PFT and whole cell inhibition of *P. falciparum* growth.

A representative synthetic route towards tetrahydroquinolines is shown in Reaction Scheme 4.1. This route starts with amino-tetrahydroquinoline and

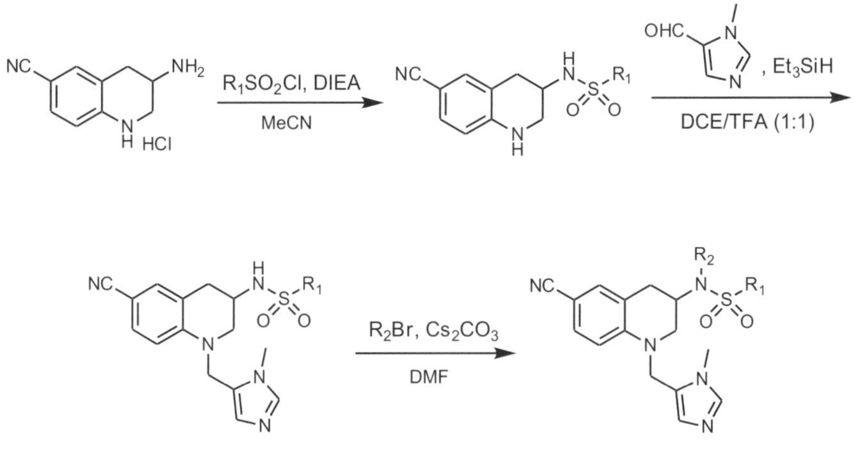

Scheme 4.1 Synthesis of Farnesyltransferase Inhibitors.

Table 4.1 Enzymatic PFT inhibition and whole cell activity of representative tetrahydroquinolines.

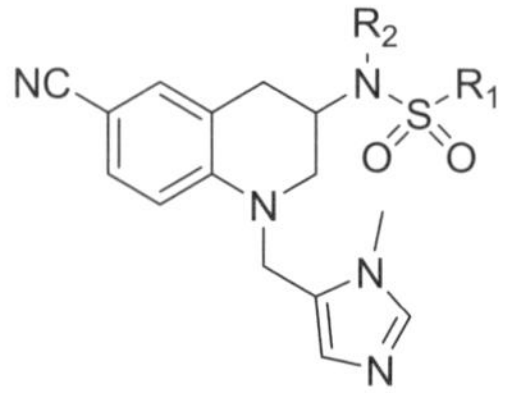

Compd	R_1	R_2	*Rat-PFT* IC_{50}(nM)	*Pf-PFT* IC_{50}(nM)	*3D7* EC_{50}(nM)	*W2* EC_{50}(nM)
1	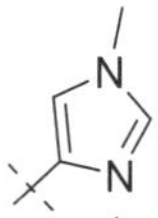	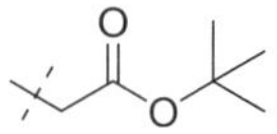	1.2	0.9	5	–
2	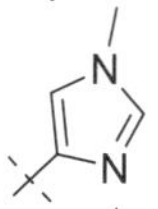	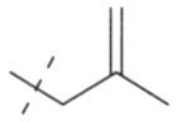	0.7	0.6	7	21
3	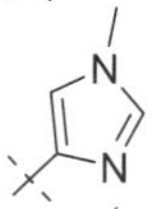	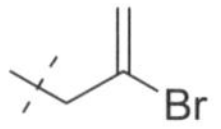	0.8	0.4	6	14
4	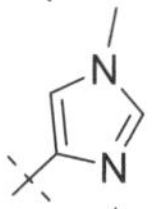	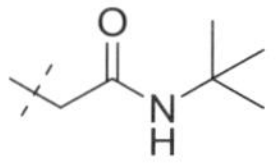	1.5	1.2	5	8
5	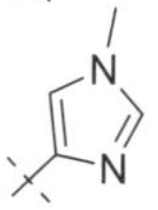	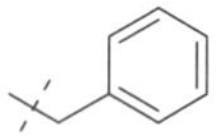	1.2	0.6	5	14

the efficient synthesis comprises three straight-forward steps that can readily provide analogs with different R_1 and R_2 groups.

Initial SAR study (Table 4.1) showed that an *N*-methylimidazole group is optimal for R_1. Modification of the R2 position indicated that several groups can provide compounds that inhibit Rat-PFT and Pf-PFT at sub- to low-nanomolar concentration. These compounds also have low nanomolar EC_{50} values against 3D7, W2 and three other parasite strains.

With these potent PFT inhibitors in hand, the group carried out a series of experiments to understand the mechanism of parasite growth inhibition and to determine whether *Pf*PFT is actually the target for the THQs. First, they showed that compounds with activity against the enzyme were also able to inhibit parasite growth. When **2** was incubated with synchronized cultures at the ring stage of *P. falciparum* the maturation was affected at 5 and 10 nM concentration.

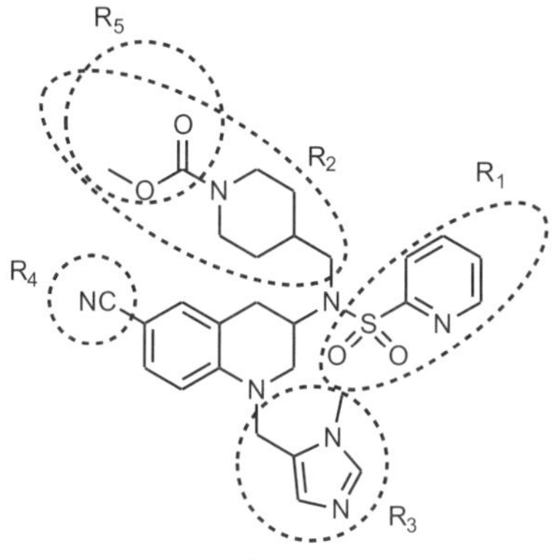

6

Figure 4.1 Substituent Definition of Tetrahydroquinoline (THQs).

The Gelb group conducted extensive SAR study on this series to optimize on potency, *in vitro* ADMET properties and *in vivo* pharmacokinetic properties.[3b] More than 260 analogs were prepared through modular synthesis to evaluate the effect of different substituents on the THQ core with the lead compound highlighted, **6** in Figure 4.1. With few exceptions, the PfPFT IC_{50}s generally agree with Pf EC_{50}s (3D7 and K1).

Testing results showed that 2-pyridylsulfonyl or N1-methyl-4-imidazolylulfonyl is optimal for R_1. When R_1 is 2-pyridylsulfonyl or N1-methyl-4-imidazolylulfonyl, a number of R_2 groups bring the enzymatic potency down to low nanomolar range. Structural studies show that the imidazole appended to N-1 of the THQ ring directly coordinates the Zn^{2+} ion at the active site of mammalian PFT. Removal of the methyl group from the imidazole ring has little effect on PfPFT IC_{50}s but greatly increases Pf EC_{50}s. Replacement and placing substitution on the methylene group on R_3 both adversely affect enzymatic and cellular potencies. Experience with mammalian PFT indicates that the R_4 6-cyano group is important in conferring tight enzyme binding. This seems to also be the case for PfPFT. Many attempted replacements of the 6-cyano functionality, such as phenyl, all failed to improve potency.

Due to the poor pharmacokinetic profile of early leads like **1**, implantable osmotic pumps were employed to deliver a high dose of 200 mg/kg/day in the *in vivo* efficacy model. Despite this dosing regimen, only moderate parasitemia eradication was achieved. These data illustrate the challenges of discovering and developing a malaria drug that is potent and orally bioavailable even when the target is known to be essential for parasite growth. It will be imperative to obtain potent THQs with improved potency and bioavailability, if the target product profile of three days of treatment is to be achieved.

Liver microsomal stability and Caco-2 assays were employed to help improve oral exposure. Firstly, the Caco-2 cell permeability assay was used to improve the GI absorption of the THQs. A reasonably good correlation was found between *in vitro* Caco-2 cell permeability and *in vivo* C_{max} values after oral dosing. Compounds with N_1-methyl-4-imidazolylulfonyl as R_1 turned out to have low Caco-2 A-B values and high B-A/A-B ratios, which indicated that these compounds may be actively pumped out, resulting in low absorption. However, compounds with a 2-pyridylsulfonyl R_1 group had improved Caco-2 A-B values and reduced B-A values. This is reflected by the significant improvement of *in vivo* C_{max} values in mice. Second, the group found that *in vitro* liver microsome data correlated well with *in vivo* clearance of this series of compounds. Combined with *in vitro* metabolite identification results, compounds with methyl carbamate as R_5 were found to be more metabolically stable and gave better oral exposure. Having improved the balance of potency and PK profile, the lead compound **6** was selected for testing in the *P. berghei* model. Compound **6** was tested for efficacy in rats due to its better oral exposure in rats than in mice. Rats were infected with *P. berghei* at about 2% parasitemia level and then treated with the same compound at 50 mg per kg every 8 h for 3 days. No parasites were detected in the compound treated group, while the level of parasitemia persisted at 1.4% in the untreated controls. While the *in vivo* efficacy of these compounds needs to be improved, the use of *in vitro* ADME assays was able to significantly improve the oral exposure profile.

In summary, THQ compounds were proved to be potent inhibitors of *Pf*PFT as well as agents that efficiently suppress parasite proliferation *in vitro*. In regards to toxicity, the lead compound is negative in mutagenicity and genotoxicity tests and shows minimal potential for off-target effects as demonstrated in receptor binding panel studies. However, this most advanced compound **6** still needed to be dosed at least every 8 h to show efficacy. This is likely due to its less than optimal metabolic stability. In a following study,[5] a series of 2-oxotetrahydroquinolines are investigated. There compounds are more metabolically stable but potency will need further improvement.

4.2.2 HDAC Inhibitors

Histone deacetylases (HDACs, E.C. 3.5.1) and Histone acetyl transferases (HATs) are enzymes that are crucial for modulating cellular chromatin structure. These enzymes are involved in the removal (HDACs) or the addition (HATs) of acetyl groups at the histones' lysine residues. The lysines contain the positively charged amines that interact with negatively charged phosphate units in DNA and thus their acetylation status can activate or suppress transcription. As transcriptional regulators, HDACs modulate the processes of cell-cycle arrest, angiogenesis (blood supply to tumors) and apoptosis (programmed cell-death). These enzymes have been widely investigated as therapeutic targets, especially in oncology. They are also known as lysine deacetylases, since their enzyme activity is not only limited to histones.[6] Five putative HDAC family members have been identified in the *P. falciparum* genome. Furthermore, the parasite appears to use histone modification to regulate gene expression.[7]

Three of the five *falciparum* HDACs have homology to mammalian classes I and II, and the other two are analogous to class III and related to the sirtuins. Only two of the plasmodium HDACs have been investigated in detail; pfHDAC1 and PfSir2 have demonstrated expression in the symptomatic blood-stages of the malaria. The cellular distribution of these two HDACs has not been investigated in detail, although pfHDAC1 does appear to be localized to the nucleus.[8] PfSir2 is not a target, since it can be genetically disrupted in *P. falciparum* parasites.[7b] Initial results support Pf HDAC to be a target of some potential – this is the thrust of the Fairlie paper and others.[9] The precise roles the class I/II HDACs play in transcriptional regulation and their essentiality and their specificity is not known, which limits the ability to fully investigate them as malaria drug discovery targets.

Homology models have been used to obtain an initial assessment of potential selectivity on pfHDAC1. With only 40% sequence identity to human HDAC8, these models may not be very predictive, however, they can provide some insights into selective inhibitors.[10] For example, the C-terminus region, which is further extended beyond the catalytic domain, could be exploited for further hydrophobic interactions. Docking simulations based on these models showed whole-cell parasite activity did correlate loosely with docking scores. These methods would be greatly enhanced with biochemical data as well as SAR on a wider variety of chemotypes.

The majority of work in the pfHDAC field has been related to testing known human HDAC inhibitors for parasite-killing activity. The general binding motifs mimic suberoylanilide hydroxamic acid (SAHA – compound **7**), which contains a hydroxamic acid group for zinc binding as well as an aliphatic linker to the hydrocarbon head group (Figure 4.2). The key issue of maintaining good metabolic stability and oral bioavailability in a series of inhibitors with flexible hydrocarbon linkers, has been investigated by Kozikowski and co-workers.[11] In their compounds, rigidification through the introduction of an aminothiazole moiety allowed for compounds with good metabolic stability. *In vitro* demonstration of antagonism with quinine-based compounds has been one significant drawback for compounds from this series and was not predicted in the target-validation studies described above. In addition, curative doses in the mouse were quite high, >50 mg/kg/day for compounds like WR301801, even in combination of with a subtherapeutic dose of chloroquine. This is surprising, given potent *in vitro* potencies on the parasite (0.5 nM EC_{50}) for WR301801 (compound **8**) (Figure 4.2). The primary liability for these compounds is hydrolysis of the hydroxamate to the carboxylic acid, presumably by a non-microsomal mechanism, since it was formed in the absence of cofactors and was not a chemical decomposition product. Despite these findings, a moderate $t_{1/2}$ was achieved (3.5 h) upon oral dosing of the compound. This work was able to demonstrate hyper acetylation of histones by Western blotting. The compound displays high functional potency on the human HDACs and although it demonstrated good cellular selectivity, it would still be useful to gain biochemical selectivity to mitigate toxicity liabilities associated with these types of inhibitors.

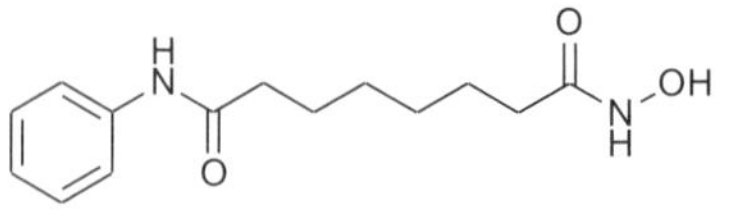

7 (SAHA)

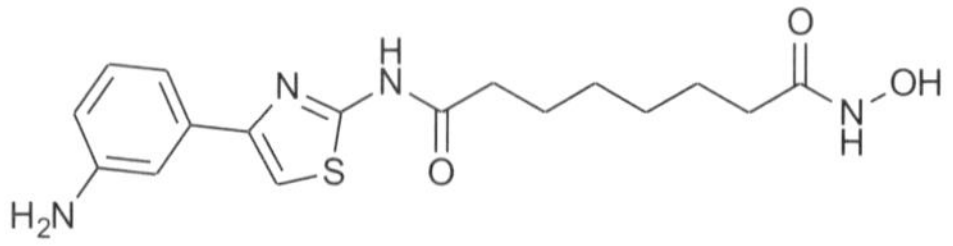

8 (WR301801)

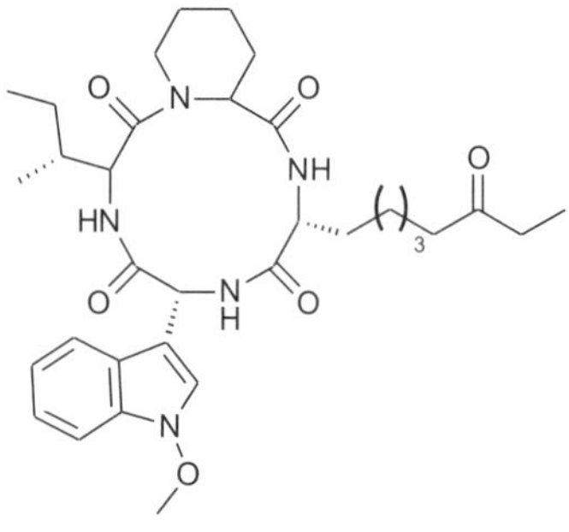

9 (Apicidin)

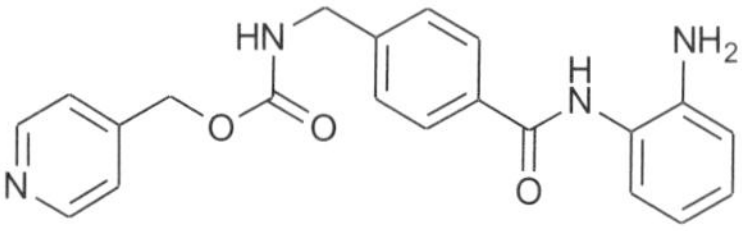

10 (MS-275)

Figure 4.2 Human HDAC Inhibitor Motifs.

In addition, natural product cyclic tetrapeptides, including apicidin (compound **9**), were the first human HDAC inhibitors found to inhibit *P. falciparum* growth *in vitro* with an EC50 of 125 nM. The apicidins are not selective, and have poor cellular toxicity, and exhibit poor oral bioavailability (as expected for a natural product with many amide bonds and saturated aliphatic groups). Their mechanism of action does not involve zinc binding. Instead, the aliphatic ketone bearing sidechain acts as a substrate mimic of acylated lysine.[12] An improvement in potency is achieved (~2 fold) with modification to a diketone or α-hydroxy ketone. In addition, modification of the ketone functionality to a hydroxamate leads to a ~200-fold improvement in potency. Incorporation of other zinc binding groups (such as α-hydroxyepoxides) also retains activity. Recent reports of non-hydroxamic acid HDAC inhibitors such as MS-275 (compound **10**) for oncology indications may also suggest new scaffolds with antimalarial activity (Figure 4.2).[13] These compounds are far from optimal, having a 400-fold decrease in potency and the mutagenicity concerns associated with anilines, but they do serve to illustrate that other metal-binding mimics could be used. The discovery of more toxicologically benign zinc binding motifs will be discussed at the end of this chapter.

4.2.3 Kinase Inhibitors

Protein kinases are important targets for human diseases due to their importance in regulating signal transduction. It is estimated that about one-third of current human drug discovery programs are centered around kinase inhibitors.[14] *P. falciparum* protein kinases are also extensively studied for their potential as drug targets. The Doerig group in the University of Glasgow has published several excellent reviews on this topic.[15]

Shortly after the Plasmodium genome was sequenced, it became clear there are two important avenues available for malaria drug discovery. The first approach involved cloning clear orthologues of human kinases, such as CK1 and pfMAP1, with the idea that compounds from human-targeted drug discovery efforts could be used against these. The other method involved finding kinases with unique activities that are structurally distinct from mammalian kinases. For example, the "orphan" or "semi-orphan" *P. falciparum* protein kinases (such as FIKK kinases) and the plant-like protein kinases (such as calcium-dependent kinases) have been proposed as interesting drug targets because they lack human homologs. There are five calcium-dependent protein kinases (CDPKs) that have been identified. In many cases, reverse genetic approaches have been employed (coupled with pharmacological activity with selective inhibitors) to determine essentiality of kinase targets. More recently, cdpk5 has been identified as a key regulator or parasite egress from infected erythrocytes at the invasive merozoite stage.[16] While these methods do provide some insight into kinase prioritization, it does not necessarily translate in the potential profile of an optimized drug candidate.

One of the first examples of developing malaria kinase inhibitors has been in the case of pfCDPK1. Inhibitors of serine/threonine calcium-dependant protein kinase 1 (pfCDPK1) have been reported from two groups and are highlighted in Figure 4.3.[17] The work by Kato and co-workers identified pfCDPK1 as expressed in the erythrocytic and sporozoite stages, with a role in parasite invasion and motility. A family of trisubstituted purines was found by screening in-house kinase libraries, which was followed by optimization of enzymatic and cellular potencies. Affinity pull down studies followed by mass-spectrometry based proteomics analysis were completed to demonstrate direct binding to the enzyme. Purfalcamine (compound **11**) was found to be a 17 nM

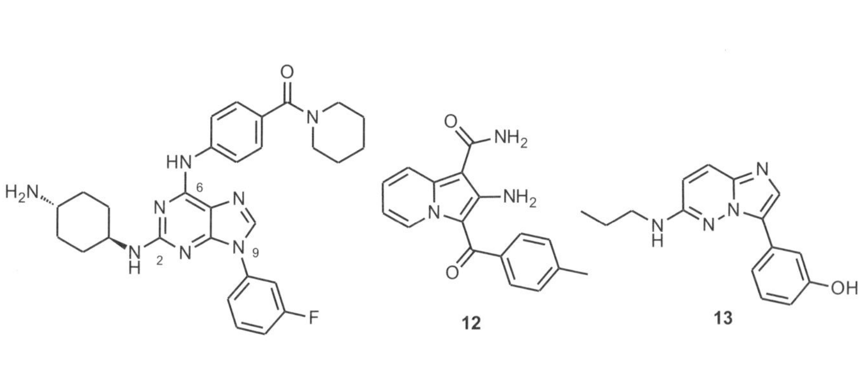

11 (Purfalcamine)

12

13

Lemercier and co-workers

Figure 4.3 Plasmodium Falciparum CDPK1 Inhibitors.

inhibitor of the enzyme (IC_{50}) and have a 230 nM EC_{50} on the parasite. SAR studies on the purine template demonstrated a tight pattern with regards to steric bulk on the N9 phenyl ring, a cycloaliphatic group at C2 and a simple aniline benzamide at C6. However, analogues of compound **11** showed differing correlations between kinase inhibition and inhibition of parasite, suggesting that parasite inhibition also was due to off-target activity. The inhibitors from Lemercier and co-workers (compound **12** and **13**), underwent a variety of studies to determine the mode of inhibition and human kinase selectivity. A Millipore panel of 46 human kinases did not show any significant activity on either the pyrrolopyridines (**12**) or imidazolopyridazine (**13**) scaffolds, demonstrating good initial selectivity. However, whole-cell activity of these compounds was not reported, making it unclear whether improvements in biochemical inhibition would lead to parasite killing.

Plasmodium PK7, which is distantly related to the human mitogen-activiated protein kinases (MAPK) is expressed in the sexual and asexual stages of the parasite life cycle.[18] Several kinase scaffolds with activity against MAPK were discovered by high-throughput screening and have been disclosed (Figure 4.4). A scaffold similar to the imidazolopyridazine reported for pfCDPK1 has also been described by Merckx and co-workers along with inhibitor bound structures.[19] In addition, another purine-like pyrazolopyrimidine scaffold was also disclosed by the same group and biochemical optimization was disclosed on this series. The initial hit (compound **14**) was optimized by 100-fold in biochemical potency to aminocyclohexyl derivative compound **15** and while there is a good correlation between enzyme and whole cell activity for the compound series, the overall cellular potency was marginable (1 μM). All of these templates are still commonly used in optimization for oncology indications, suggesting that it may indeed be possible to overcome the issues of cellular penetration and improve the parasite growth inhibitory properties.

There have been other non-kinase templates that have been reported, including chalcones for pfMRK (a cyclin-dependant protein kinase) by Geyer and co-workers.[20] These compounds were potent on the enzyme (9 nM), but demonstrated only low potency in the parasite-killing assay (~3 uM). A related series of coumarins were also disclosed by the same group[21] and more recently a series of thiophene sulfonamides.[22] Some interesting chemical starting

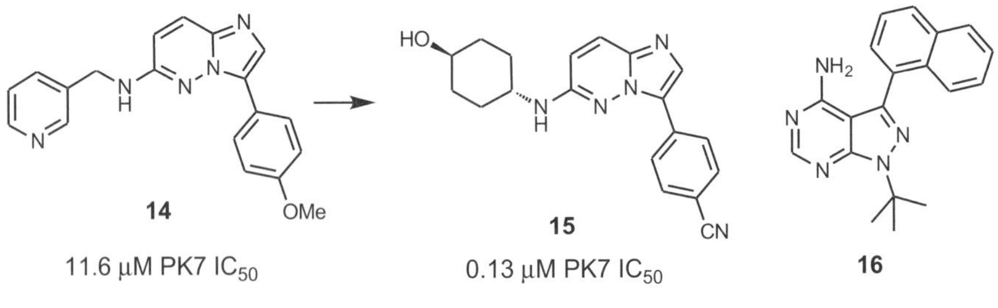

14 11.6 μM PK7 IC_{50} **15** 0.13 μM PK7 IC_{50} **16**

Figure 4.4 Plasmodial Kinase PK7 Inhibitors.

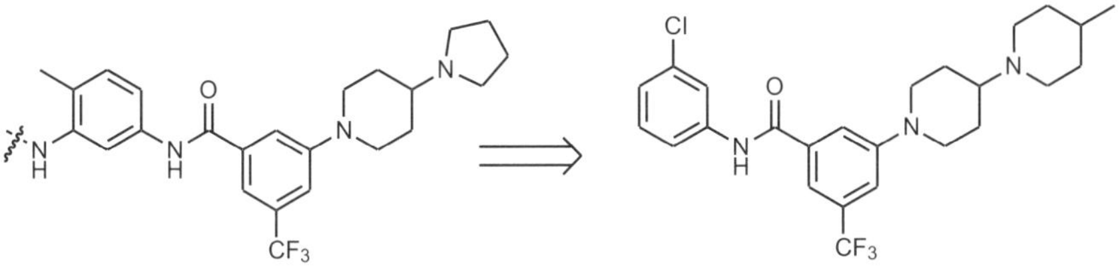

Human kinase template (**17**) Wu and co-workers (**18**)

Figure 4.5 Optimization of Malaria Whole-cell Potency from a Kinase Inhibitor Template.

points have also been found for pfNEK1, where natural product extracts have had substantial activity.[23] These approaches highlight the potential for finding chemically diverse and selective starting points for drug discovery. At this point, optimization of these hits to viable preclinical candidates has not been reported.

It is clear that the screening of malaria kinases has generated a wide variety of starting points where chemical space overlaps with human kinase inhibitors. The field has been limited by poor activity against the whole-parasite and a lack of correlation between biochemical and whole-cell activity. Key questions are whether activity against the target limits cellular activity and whether limited cell-penetration of the compound is most responsible for poor whole-cell activity. To determine the minimal pharmacophore from a human-kinase inhibitor needed for cellular activity, Wu and co-workers [24] started with a non-selective human kinase inhibitor template starting point and removed all the human kinase binding elements, but were still able to improve activity of compounds in a *P. falciparum* growth assay to around 50 nM (Figure 4.5). This demonstrated that the kinase-binding portion of the inhibitors had little to do with potent cellular activity. It is clear that cellular activity of many kinase inhibitors described in the studies above are unlikely to correlate with one biochemical target alone.

In addition, this suggests (as similarly observed in the human kinase field) that cellular potency may be driven by multi-kinase activity (and in some cases even non-kinase activities) Any medicinal chemistry optimization will require parallel interrogation of whole-parasite activity as well as a broad biochemical study of potential human and parasite biochemical targets. The target profile is therefore quite complex: compounds are required which show selectivity against human kinases (to avoid toxicity issues) but are still able to inhibit multiple parasite targets.

4.2.4 Protease Inhibitors

Proteases are essential enzymes involved in polypeptide cleavage and they represent a major class of drug targets for human diseases. The nucleophiles can be amino acids such as aspartate, serine, cysteine, or threonine residues. In general, cysteine and serine proteases are the most common drug targets. In addition, metal-assisted metalloproteases, in which zinc (or cobalt) atoms are present as cofactors in both exopeptidases and endopeptidases (including the matrix metalloproteases) have also been recently exploited as drug targets. Examples of early inhibitors include peptide substrate mimics that contain an electrophilic "warhead", which covalently (reversibly or irreversibly) interacts with the enzymatic nucleophile (cysteine, for example) to provide the primary thermodynamic force for inhibition. Therefore, key challenges for medicinal chemistry include engineering appropriate selectivity through modifying the peptidomimetic portion or by removal/modification of the warhead. Selectivity over other nucleophiles in the body, such as detoxifying glutathiones and free thiols in the tissues makes the development of non-covalent inhibitors particularly important. In addition, the introduction of non-peptidic starting points is likely to improve the drug-like properties for oral bioavailability, metabolic stability, cell-permeability and selectivity.

In regards to protozoan drug discovery, two classes of proteases have been widely investigated, the papain-like cysteine proteases falcipains and the aspartyl "plasmepsin" proteases.[25] In *P. falciparum*, the falcipains are comprised of four different cysteine proteases and are involved in erthyrocytic invasion, hemoglobin degradation and erythrocytic rupture. Their biological role is described in detail elsewhere.[26] A variety of irreversible fluoromethylketones have been described by the Rosenthal group as inhibitors or falcipain 2 (FP-2) and are outlined in Figure 4.6 (compound **19**). All these compounds have good *in vitro* and *in vivo* efficacy, but significant toxicity due to their reactive functionality.[27] As vinylsulfone based warheads were being investigated for the human cathepsins and illustrated by compound **20**[28], several related analogs with different amino acids in the inhibitor design led to potent falcipain inhibitors (IC_{50} = 5 nM) with potent cellular activities (EC_{50} = 3 nM) (compound **21**). Furthermore, these compounds showed activity *in vivo* but do have the drawbacks of irreversible alkylation of the catalytic cysteine as well as poor selectivity over human cysteine proteases.[29] Development of reversible α-ketoamides and aldehydes addresses some of these issues, but these compounds can suffer from racemization of the adjacent stereocenter that arise from chemical stability issues from hemi-ketals/hemi-acetals and enol tautomers.[30]

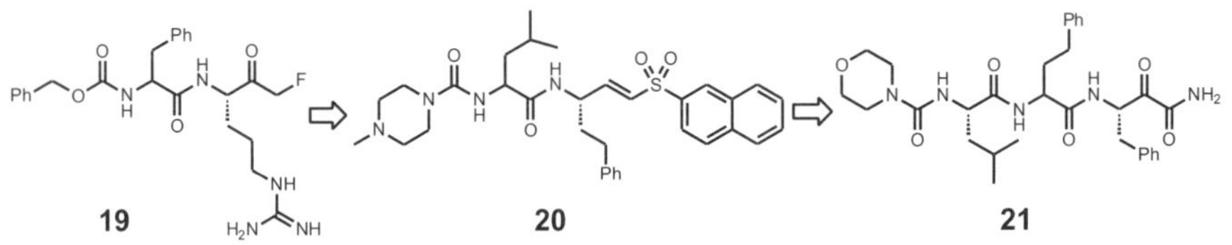

Figure 4.6 Progression of Peptidic Falcipain Inhibitors.

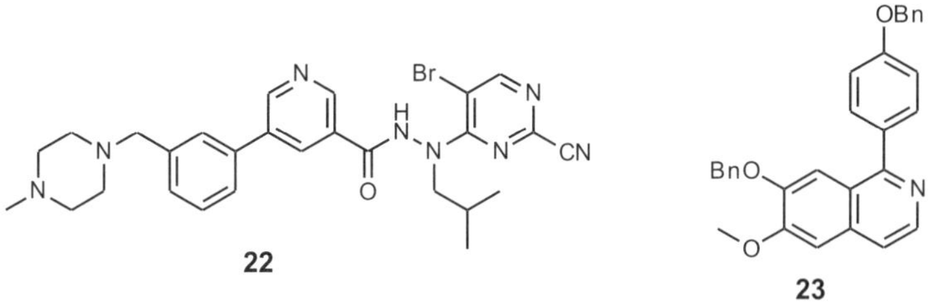

Figure 4.7 Non-peptidic Falcipain Inhibitors.

For the discovery of non-peptidic inhibitors, empirical human data on alkyl and arylnitriles serve as starting points for optimization. In a recent publication by Coteron and co-workers (Figure 4.7), a wide variety of pyrimidine nitriles (compound **22**) were optimized for FP-2 and FP-3 activity with subnanomolar enzymatic potencies and 2 nM whole parasite EC_{50}s.[31] These compounds were not developed further because of their ability to modify free thiols in the tissues. The challenge here is clearly to have an electrophilic group reactive enough to inhibit the target enzyme without extensive non-specific covalent modification *in vivo* and thus avoid the poor toxicity profile of vinylsulfone compounds. Structure-guided design has been used to find new hit series. Homology models have been used for both falcipain-2 and falcipain-3 with and without bound vinylsulfone ligands.[32] and a set of isoquinolines were docked, with follow-up of scored hits. Figure 4.7 (compound **23**) details a hit found from these methodologies. The best compounds are still of modest potency compared with other hits (~3–10 uM IC_{50}), but could serve as a viable hit to lead starting point. One further cysteine protease, dipeptidyl peptidase 3 (DPAP3), has been identified as a regulator of erythrocytic rupture and virulence and may be an important target. While these cysteine proteases have been investigated extensively, there needs to be development of novel chemical matter that enables compounds with a better toxicity profile and selectivity profile over human proteases.

Several recent reports have also suggested aspartic acid proteases, the plasmepsins, as important regulators of parasite survival. Aspartate protease inhibitors developed against HIV are active against *Plasmodium falciparum in vitro,* although at relatively high concentration. Plasmepsin-4 is expressed on the apical surface and secreted during ookinete formation, suggesting plasmepsin 4 inhibitors could block transmission.[33] In addition, work by Boddey and co-workers describe plasmepsin 5 as an ER-bound protein that is key in modifying effector proteins that are exported from infected erythrocytes.[34] In addition, the subtilisin-family serine protease, PfSUB1, has also been identified as a regulator of erythrocytic rupture and virulence of the parasite.[35] As for inhibitors of the more recent proteases, some peptidic inhibitors have been described, but cellular potencies need to be optimized.[36] To discover interesting new drugs, coupling these interesting biological observations with novel hit-finding methods (as described below) will be needed.

4.2.5 Folate Biosynthesis

The pyrimethamine/sulfadoxine combination (Fansidar) has provided a very useful drug combination targeting dihydrofolate reductase (DHFR) and dihydropteroate synthase (DHPS), respectively. These drugs inhibit the folate pathway synergistically and, along with thymidylate synthase are responsible for thymidylate monophosphate biosynthesis and ultimately DNA biosynthesis.[37] DHPS has not been pursued extensively as a drug target (largely due to toxicity risk of sulfur-based compounds like sulfadoxine), but DHFR has been thoroughly investigated.[38] Pyrimethamine is in a class of "non-classical" DHFR inhibitors that are lipophilic, neutral, non-glutamate compounds (unlike anticancer compounds such as methotrexate) and crosses membranes by passive diffusion. It should be noted that the exact mechanism of parasite killing by pyrimethamine was not known at the time of first disclosure.[39]

Development of resistance has led to additional drug discovery efforts to find novel compounds with improved activity on resistant parasites, such as WR-99210 (**24**)[40] and its related orally available bis-guanidino prodrug PS-15 (**25**)[41] (Figure 4.8).

More recently, even more potent antifolates have been disclosed by several groups, including Yuthavong (**26**)[42] and Nzila (**27**)[43]. These compounds are active against the DHFR VS-4 quadruple mutant, but while compounds like **26** are in the 5-7 μM range, compound **27** is low nanomolar. Unfortunately, compound **27** has a very poor therapeutic index in toxicity studies,[44] likely due to the quinone pharmacophore present. While this field has certainly provided a useful area for bringing new drugs like pyrimethamine to patients, it has been disappointing to see no new drugs in this target class in the last 60 years. New techniques for hit finding, especially given the depth of structural data that is in this field, may provide new pharmacophores for this important target and some insight into these approaches have been recently disclosed.[45]

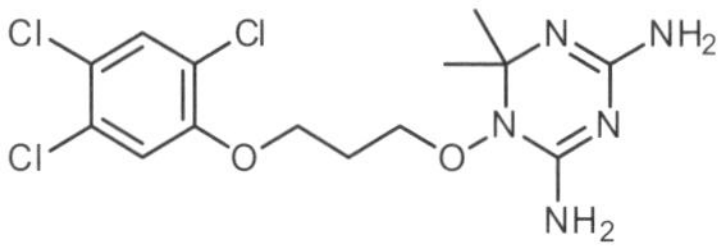

WR-99210 (**24**)

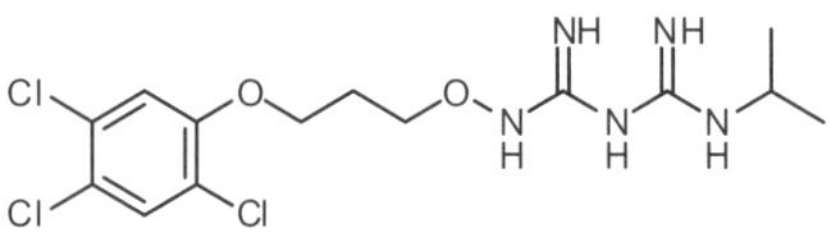

PS-15 (**25**)

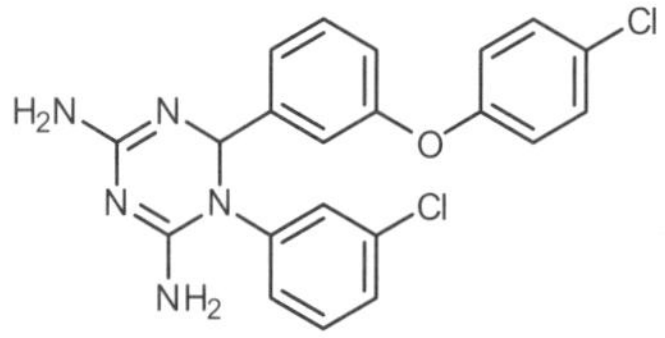

26

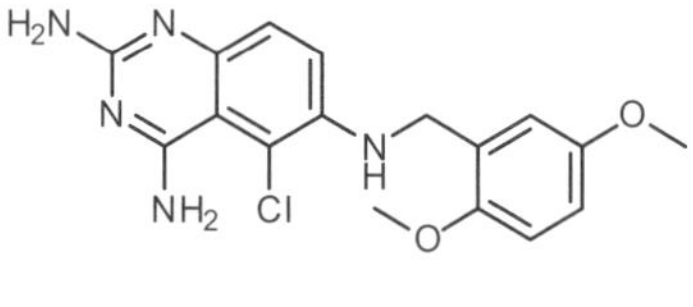

27

Figure 4.8 Progression of DHFR Inhibitors.

4.2.6 Future Perspectives on Target-based Discovery using Novel Hit-finding Methods

For HDAC and protease inhibitors, novel pharmacophores that can bind zinc in the case of HDACs are needed and novel serine and cysteine interacting groups are required in the case of proteases. Toxicity related to HDAC inhibitors has recently been reviewed[46] and further testing of hydroxamate (e.g. Phase III Novartis compound LBH259) and the aniline-based MS-275 (Phase II) will likely determine their potential usefulness in malaria. The development of Odanacatib, a peptidic covalent protease inhibitor of cathepsin K from Merck is in Phase III testing for osteoporosis. Since the toxicity from bisphosphonates is quite low, it will be useful to observe the side-effect profile for these compounds. Related non-covalent and covalent inhibitors for Dipeptidyl peptidase-4 (DPP4) are on the market or are in late-stage testing and these have also demonstrated the importance of proteases as a target class. Work on both of these target classes for malaria will need to consider cell penetration[47] with particular scrutiny on molecular weight and ligand efficiency. As for hydrophobic compounds that are weak enzymatic inhibitors, it will be important to ensure that non-specific aggregation is not a key factor for inhibition.[48]

Recent reports of non-covalent inhibitors for cysteine proteases (such as the cathepsins)[49,50] as well as the serine protease HCV NS3[51] illustrate the feasibility of creating potent inhibitors without the need for an electrophilic center. Serine protease inhibitors, such as for Factor Xa[52] and the inhibitor of the 20S proteasome[53] also highlight the ability to generate novel non-warhead containing compounds. In fact, the literature around DDP-IV has provided several examples of the success of non-covalent protease inhibitors in the clinic.[54] These mammalian system examples demonstrate how generation of related novel pharmacophores could be applied to anti-parasitics.

Fragment-based screening is certainly one new approach for hit finding and has not been broadly applied in the area of malaria biochemical inhibitors. Several success stories from the kinase and protease fields make this an attractive method if enough protein is available to allow structure determination (NMR and crystallography).[55] In addition to fragment-based approaches, alternative computational hit-finding methods can be employed. For example, using a non-Bayesian protocol, incomplete kinase experimental profiling data can be completed by individual homology models. When coupled to available databases with kinome-wide homology models based on sequence (such as the Eidogen database), one can imagine using these methods to discover novel chemical matter.[56] In fact, if coupled to the large commercial source of known pharmacophores with whole-cell malaria activity (as recently disclosed by GSK and Novartis among others) then scaffold hopping/scaffold morphing may be possible to achieve cellular and biochemical activity.

It is important not to abandon tractable scaffolds if a good correlation between target and cell activity does not exist, given the poly-pharmacological nature of many ligands. In fact, one could argue that compounds with excellent cellular activity devoid of mammalian or parasitic activities are still valuable starting points since off-target toxicities might be less likely. This approach does require broad-scale profiling of compounds in a variety of mammalian target assays, which has become much more commonplace with outsourced profiling centers (Life Technologies, Caliper, Nanosyn, etc.). Broad-scale profiling of compounds in parasitic target assays is relatively still in its infancy, but can allow for characterization of many compounds within a given chemical scaffold to determine the level of polypharmacology. Compounds with excellent cellular activity from a scaffold template that is related to a known target-based drug discovery effort may serve as excellent starting points from which targets can be determined in parallel with cellular optimization. Particularly exciting would be to use pharamacophores with structural similarities to known human inhibitors, such as Tanimoto similarity coefficients >0.8, but where no human inhibition is known but whole-cell activity has been demonstrated. These compounds would serve as excellent chemical biological tools to understand novel parasite biology, and could provide chemical validation of novel druggable targets.

4.3 Phenotypic Drug Discovery

4.3.1 Overview of Cell-based Assays and Drug Discovery

Most of the current antimalarials were originated from a phenotypic drug discovery approach. Quinine, chloroquine, mefloquine, and artemisinin derivatives were either natural products or derived analogs that did not have a well-defined mechanism of action at their time of discovery and clinical approval. Recent target-based antimalarial drug discovery effort has yielded many impressive candidates. However, a good balance between target-based and cell-based optimization is important to ensure a diversified chemical portfolio of antimalarial agents. In addition to our efforts,[24] there are 15 more cell-based projects that are supported by MMV. Not surprisingly, many of these cell-based programs are from pharmaceutical companies or are academia–pharma collaborations. Optimization of benzamides from human kinase inhibitors as described above, plus the recent breakthrough from Yeung and co-workers[57] on a family of spiroindolone compounds culminating in NITD609 (compound **28**) (Figure 4.9), are examples that demonstrate the value of cell-based optimization of tractable starting points. These examples also demonstrate the

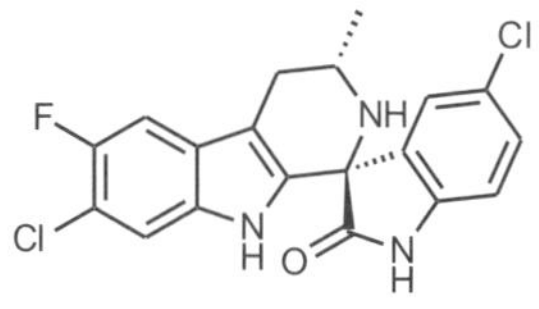

28

Figure 4.9 Cell-based Optimization Clinical Candidate NITD609.

benefit of determining *in vivo* pharmacokinetics and efficacy early in the hit optimization process.

Furthermore, advances in genomics have greatly speeded the process by which targets can be discovered and will be discussed in a later section. Here, we highlight some of the methods used to conduct high-throughput cell-based screens that have provided a large small-molecule collection of drug discovery starting points.

The human malaria parasite, *P. falciparum*, replicates in mature human red blood cells that have lost their nuclei. When whole blood is depleted of white cells using a leukocyte depletion filter prior to a parasite culture being added, increases in total DNA content over a 72-hour period can be attributed to parasite growth. If compounds are added that prevent the parasite from replicating, this can be detected using a simple staining system in which a DNA binding dye is added to treated and untreated cultures. Parasite growth is then monitored with a fluorometric assay. In theory, this method is similar to the traditional method for measuring antimalarial activity in which radiolabeled hypoxanthine or ethanolamine is added to drug-treated cultures, but is more compatible with modern high-throughput screening where the use of radioactivity is generally avoided. Several adaptations of this basic method have been published, including the Cyquant assay, GR dye [58], SYBR green[59] and DAPI.[60] While most were developed for use in 96 well-plates, a few modifications, such as increasing the cell lysis time and the use of precision liquid dispensers make them suitable for use in 1536 well-plates and modern screening systems.[61] In addition, enzymatic assays that rely on the activity of parasite lactate dehydrogenase have also been developed and adapted to high throughput screening.

In addition to the work by Plouffe and co-workers, there have been several recent reports of cell-based screening campaigns from both GSK[62] and the Guy group.[61b] These hit sets serve as a valuable starting point for medicinal chemistry programs. With such a large number of structures available in the public domain, hit triaging becomes particularly important. Several methods are available for hit triaging including a standard activity based cut-off, but the ontology-based pattern identification (OPI) method described by Yan and co-workers pays particular attention to structure-activity relationships derived from the screen.[63] Integration of these relationships into the analysis of screening data prioritizes interesting singletons as well as lower activity compound series where there is a good structure-activity relationship. With series where there is good structure-activity relationship, the rate of reconfirmation from powder stocks is significantly higher and also provides confidence that potency can be readily optimized. As demonstrated with the spiroindolones, having reconfirmed hits from powder stock, early assessment of animal efficacy is then extremely valuable in generating viable lead compounds. The identification of the molecular target can be done in parallel, as the leads are being optimized. One advantage of this approach is the speed: the time from screening of NITD609 to the first administration to human volunteers was less than four years.

While most effort has been focused on cell-based screens of the blood stages, high-throughput screens for hepatic stage activity are also feasible. Such screens are considerably more challenging because of the difficulty associated with obtaining infectious sporozoites as well as receptive hepatic cell lines. *P. falciparum* sporozoites can be challenging to obtain, and do not readily infect most hepatic cell lines, although they do infect primary hepatocytes. Reports that they can grow in the HC04 cell line[64] have been difficult to repeat because of the limitations in cell line availability and *P. falciparum* is not known to invade other cell lines. There are no ready sources of *P. vivax* sporozoites, although they do infect readily available HepG2 cells. Nevertheless, the rodent models of malaria *P. berghei* and *P. yoelii*, have sporozoites that are able to invade several cell lines, included HepG2-cd81. The *P. yoelii* HepG2/*P. yoelii* system has been used to discover compounds with antimalarial activity.[65]

4.3.2 Lab-evolved Resistance and Genome-scanning for Target Discovery

Once compounds are selected from cell-based screens and optimized to have the appropriate pharmacokinetic and potency properties, there is generally an interest in determining the protein target and/or mechanism of action. Whilst not absolutely essential for licensing and development, knowing the *Plasmodium* target of a bioactive small molecule can help interpret toxicity and offer a starting point for further rounds of structure-based drug design. Various strategies have been tested, including affinity chromatography,[17a] but the method that has yielded the most success involves evolving parasites to low levels of resistance and then examining their genomes for newly acquired lesions, which are often found in the target. Several methods can be used for evolving to resistance. In one case, 10^8–10^9 parasites are simply placed in a 25 ml flask containing the compound of interest at various concentrations (e.g. five times to ten times the IC_{50}). These are incubated, and fresh erythrocytes, medium and drug are replaced every week. The alternative is to expose parasites to ever-increasing amounts of compound over a period of months. In the former method, the experiment is ended when parasites emerge, while in the latter, dose response measurements are performed periodically and the experiment is terminated once parasites have become 2- to 10-fold resistant. Afterwards, the parasites are cloned. Because one is looking for single base changes in the genome it is imperative that the investigator is rigorous in parasite handling and cloning the parasites before and after selection. Under such circumstances typically a handful of single-nucleotide and copy number variants will emerge in the resistant clones, often in, or encompassing, the same gene.

Different methods are available for detecting these changes. One method that has been used with much success is a full genome tiling array. In this system overlapping 25mers that probe a large proportion of the organism's genome are placed on the array. Not all regions can be probed because of redundancies in the genome but coverage of up to 85% of the coding regions are feasible on a single microarray for *P. falciparum*'s 26M base pair genome. DNA from the starting clone and the ending clone is obtained and digested to fragments of 25–50 bases. Then DNA from both clones is labeled, denatured and hybridized to the array. Because hybridization will be disrupted between a 25mer probe and its target under the temperature and salt conditions used, if there is a base pair change, the genomic locations where changes have occurred can be detected. Furthermore, increases in the copy number can also be detected. When malaria parasites grow in red blood cells they are haploid, which means heterozygosity is not a problem. An advantage of this method is the relatively low cost compared to sequencing, the fact that the evolved strain is compared directly to a reference, and the fact that appropriate software exists[66] to compute the probability that a change has occurred at a given location using knowledge of the genome and the microarray design. Next generation sequencing should in principle also work and will be essential for diploid organisms, but the computational infrastructure needed to determine whether a given change has a low probability of being a false positive, has the capacity to detect copy number variants, and to determine how even coverage is for each sequencing run (e.g. the false negative rate) has not yet been developed. For microarrays there is also an initial investment in design, which limits the extent to which the method can be applied to non-falciparum malaria species, such as *P. berghei*, but the fact that copy number variants can be readily visualized makes them the method of choice. With drug-evolved parasites, the detected mutation may be in a gene involved in resistance or in the actual target.

Methods for performing allelic replacement are relatively laborious in *P. falciparum*, often requiring 3 months or more of labor. Thus it is imperative that strong statistical support for a given allele or target be obtained before performing genetic engineering. For both whole genome sequencing and tiling array analysis, random events and changes in the predicted target or resistance gene can be distinguished by the use of three to six independently evolved lines. If five to ten plausible mutations are obtained per line and most lines contain mutations in the same gene, then it is likely that this gene will be the one of interest. However, compiled data from other strains and evolved lines may also be useful. Certain gene types, such as *var* genes, accumulate mutations at high rates in the absence of drug selection and thus can likely be excluded from consideration. This method was used with great success to determine the likely target of NITD609. Here, genome-scanning revealed that six independently derived strains which had acquired resistance to the spiroindolone, NITD609, all bore SNPs or copy number variants in PfATP4.[57a] The importance of the gene to the process of resistance was confirmed by Rottmann and co-workers by transferring the mutations into a clean genetic background. Genome-scanning has also been used to determine the mechanism of resistance to fosmidomycin.[66]

Genome-scanning and *in vitro* selection is not without limitations. For example, only genes involved in resistance may be revealed. Thus, biochemical approaches may also be attempted. As mentioned previously, Kato *et al.* were able to attach a linker to the kinase inhibitor, purfalcamine, and show that resins bearing an active form were able to retain the predicted target, PfCDPK1. Likewise, Knockaert *et al.* used affinity chromatography to identify Casein Kinase as a target of purvalanol.[67] The disadvantage of affinity chromatography is that there can be a rather large background and it is not unusual to obtain scores or even hundreds of possible targets. Furthermore, abundant proteins are found much more often than low abundance ones. An advantage of genetic selection is that every gene is generally equally represented in the genome while for protein based methods; abundant proteins are much more likely to be positives than rare ones. Ironically, abundant proteins might be less likely to be targets of cell-active compound because of the stoichiometry of inhibition – within a cell's volume it may take 100-times more compound to knock down the activity of an abundant protein relative to a rare one.

Another approach is to screen cell active compounds against known targets as a method to understand mechanism of action for these cell-based hits. For example, Crowther and co-workers screened 9 predicted *P. falciparum* targets against a set of 5655 low molecular weight cell-active compounds, predicting that the target of some of these might be identified.[68] These efforts have had limited success in identification of putative targets for the more chemically tractable cell-based hits and assessment of whether or not the inhibition was specific for a given target was difficult. However, the screening of cell-actives against a target may be a useful method to assess chemical validation of new targets before large-scale screening campaigns of millions of compounds are undertaken. On the other hand, pathway-based screens have in some cases led to a possible mechanism of action or have been used to confirm the interaction between a small molecule and a predicted target. For example, transgenic malaria parasites expressing the *S. cerevisiae* dihydroorotate dehydrogenase are resistant to *P. falciparum* DHOD inhibitors and this has been used to support the argument that compounds derived from biochemical screens against DHOD exert their mechanism of action in this way.[69]

4.4 Conclusions

The current leads from cell-based optimization projects represent the largest portion of the pre-clinical pipeline for novel antimalarials. The chemical diversity and early successes on optimization of leads from these approaches opens new avenues for understanding parasite biology and novel therapeutics. Targeted approaches based on validated molecular targets still represent an important way forwards. What is key here is that the existing work on inhibitors of the human orthologs or members of related families means that a large amount of medicinal chemistry experience is available to speed the design and synthesis of inhibitors. Ultimately, the test of activity and validation of the target is in man, and so the key to success is always to build on the experience of clinical testing of these compound classes.

References

1. A. Wittinghofer and H. Waldmann, *Agnew. Chem., Int. Ed. Engl.*, 2000, **39**, 4193.
2. (*a*) A. D. Cox and C. J. Der, *Curr. Opin. Pharmacol.*, 2002, **2**, 388; (*b*) W. T. Purcell and R. C. Donehower, *Curr. Oncol. Rep.*, 2002, **4**, 29.
3. (*a*) F. S. Buckner, R. T. Eastman, K. Yokoyama, M. H. Gelb and W. C. Van Voorhis, *Curr. Opin. Investig. Drugs*, 2005, **6**, 791; (*b*) P. Bendale, S. Olepu, P. K. Suryadevara, V. Bulbule, K. Rivas, L. Nallan, B. Smart, K. Yokoyama, S. Ankala, P. R. Pendyala, D. Floyd, L. J. Lombardo, D. K. Williams, F. S. Buckner, D. Chakrabarti, C. L. Verlinde, W. C. Van Voorhis and M. H. Gelb, *J. Med. Chem.*, 2007, **50**, 4585.
4. (*a*) L. Nallan, K. D. Bauer, P. Bendale, K. Rivas, K. Yokoyama, C. P. Horney, P. R. Pendyala, D. Floyd, L. J. Lombardo, D. K. Williams, A. Hamilton, S. Sebti, W. T. Windsor, P. C. Weber, F. S. Buckner, D. Chakrabarti, M. H. Gelb and W. C. Van Voorhis, *J. Med. Chem.*, 2005, **48**, 3704; (*b*) W. C. Van Voorhis, K. L. Rivas, P. Bendale, L. Nallan, C. Horney, L. K. Barrett, K. D. Bauer, B. P. Smart, S. Ankala, O. Hucke, C. L. Verlinde, D. Chakrabarti, C. Strickland, K. Yokoyama, F. S. Buckner, A. D. Hamilton, D. K. Williams, L. J. Lombardo, D. Floyd and M. H. Gelb, *Antimicrob. Agents Chemother.*, 2007, **51**, 3659.
5. V. J. Bulbule, K. Rivas, C. L. Verlinde, W. C. Van Voorhis and M. H. Gelb, *J. Med. Chem.*, 2008, **51**, 384.
6. C. Choudhary, C. Kumar, F. Gnad, M. L. Nielsen, M. Rehman, T. C. Walther, J. V. Olsen and M. Mann, *Science*, 2009, **325**, 834.
7. (*a*) L. H. Freitas-Junior, R. Hernandez-Rivas, S. A. Ralph, D. Montiel-Condado, O. K. Ruvalcaba-Salazar, A. P. Rojas-Meza, L. Mancio-Silva, R. J. Leal-Silvestre, A. M. Gontijo, S. Shorte and A. Scherf, *Cell*, 2005, **121**, 25; (*b*) M. T. Duraisingh, T. S. Voss, A. J. Marty, M. F. Duffy, R. T. Good, J. K. Thompson, L. H. Freitas-Junior, A. Scherf, B. S. Crabb and A. F. Cowman, *Cell*, 2005, **121**, 13.
8. M. B. Joshi, D. T. Lin, P. H. Chiang, N. D. Goldman, H. Fujioka, M. Aikawa and C. Syin, *Mol. Biochem. Parasitol.*, 1999, **99**, 11.
9. K. T. Andrews, T. N. Tran, A. J. Lucke, P. Kahnberg, G. T. Le, G. M. Boyle, D. L. Gardiner, T. S. Skinner-Adams and D. P. Fairlie, *Antimicrob. Agents Chemother.*, 2008, **52**, 1454.
10. P. Mukherjee, A. Pradhan, F. Shah, B. L. Tekwani and M. A. Avery, *Bioorg. Med. Chem.*, 2008, **16**, 5254.
11. G. S. Dow, Y. Chen, K. T. Andrews, D. Caridha, L. Gerena, M. Gettayacamin, J. Johnson, Q. Li, V. Melendez, N. Obaldia, T. N. Tran and A. P. Kozikowski, *Antimicrob. Agents Chemother.*, 2008, **52**, 3467.
12. S. L. Colletti, R. W. Myers, S. J. Darkin-Rattray, A. M. Gurnett, P. M. Dulski, S. Galuska, J. J. Allocco, M. B. Ayer, C. Li, J. Lim, T. M. Crumley, C. Cannova, D. M. Schmatz, M. J. Wyvratt, M. H. Fisher and P. T. Meinke, *Bioorg. Med. Chem. Lett.*, 2001, **11**, 113.
13. T. M. Suzuki and Naoki, *Mini-Reviews in Medicinal Chemistry*, 2006, **6**, 515.

14. H. Weinmann and R. Metternich, *Chembiochem*, 2005, **6**, 455.
15. (*a*) D. Leroy and C. Doerig, *Trends in Pharmacological Sciences*, 2008, **29**, 241; (*b*) C. Doerig, A. Abdi, N. Bland, S. Eschenlauer, D. Dorin-Semblat, C. Fennell, J. Halbert, Z. Holland, M. P. Nivez, J. P. Semblat, A. Sicard and L. Reininger, *Biochimica Et Biophysica Acta-Proteins and Proteomics*, 2010, **1804**, 604; (*c*) C. Doerig, O. Billker, T. Haystead, P. Sharma, A. B. Tobin and N. C. Waters, *Trends in Parasitology*, 2008, **24**, 570.
16. J. D. Dvorin, D. C. Martyn, S. D. Patel, J. S. Grimley, C. R. Collins, C. S. Hopp, A. T. Bright, S. Westenberger, E. Winzeler, M. J. Blackman, D. A. Baker, T. J. Wandless and M. T. Duraisingh, *Science*, 2010, **328**, 910.
17. (*a*) N. Kato, T. Sakata, G. Breton, K. G. Le Roch, A. Nagle, C. Andersen, B. Bursulaya, K. Henson, J. Johnson, K. A. Kumar, F. Marr, D. Mason, C. McNamara, D. Plouffe, V. Ramachandran, M. Spooner, T. Tuntland, Y. Zhou, E. C. Peters, A. Chatterjee, P. G. Schultz, G. E. Ward, N. Gray, J. Harper and E. A. Winzeler, *Nat. Chem. Biol.*, 2008, **4**, 347; (*b*) G. Lemercier, A. Fernandez-Montalvan, J. P. Shaw, D. Kugelstadt, J. Bomke, M. Domostoj, M. K. Schwarz, A. Scheer, B. Kappes and D. Leroy, *Biochemistry*, 2009, **48**, 6379.
18. N. Bouloc, J. M. Large, E. Smiljanic, D. Whalley, K. H. Ansell, C. D. Edlin and J. S. Bryans, *Bioorg. Med. Chem. Lett.*, 2008, **18**, 5294.
19. A. Merckx, A. Echalier, K. Langford, A. Sicard, G. Langsley, J. Joore, C. Doerig, M. Noble and J. Endicott, *Structure*, 2008, **16**, 228.
20. J. A. Geyer, S. M. Keenan, C. L. Woodard, P. A. Thompson, L. Gerena, D. A. Nichols, C. E. Gutteridge and N. C. Waters, *Bioorg. Med. Chem. Lett.*, 2009, **19**, 1982.
21. Z. Xiao, N. C. Waters, C. L. Woodard, Z. Li and P.-K. Li, *Bioorg. Med. Chem. Lett.*, 2001, **11**, 2875.
22. D. Caridha, A. K. Kathcart, D. Jirage and N. C. Waters, *Bioorg. Med. Chem. Lett.*, 2010, **20**, 3863.
23. (*a*) D. Laurent, V. Jullian, A. Parenty, M. Knibiehler, D. Dorin, S. Schmitt, O. Lozach, N. Lebouvier, M. Frostin, F. Alby, S. Maurel, C. Doerig, L. Meijer and M. Sauvain, *Bioorg. Med. Chem.*, 2006, **14**, 4477; (*b*) D. Desoubzdanne, L. Marcourt, R. Raux, S. V. Chevalley, D. Dorin, C. Doerig, A. Valentin, F. D. R. Ausseil and C. C. Debitus, *Journal of Natural Products*, 2008, **71**, 1189.
24. T. Wu, A. Nagle, T. Sakata, K. Henson, R. Borboa, Z. Chen, K. Kuhen, D. Plouffe, E. Winzeler, F. Adrian, T. Tuntland, J. Chang, S. Simerson, S. Howard, J. Ek, J. Isbell, X. Deng, N. S. Gray, D. C. Tully and A. K. Chatterjee, *Bioorg. Med. Chem. Lett.*, 2009, **19**, 6970.
25. F. Shah, P. Mukherjee, P. Desai and M. Avery, *Curr. Comput. Aided Drug Des.*, 2010, **6**, 1.
26. P. S. Sijwali, B. R. Shenai and P. J. Rosenthal, *J. Biol. Chem.*, 2002, **277**, 14910.
27. P. J. Rosenthal, W. S. Wollish, J. T. Palmer and D. Rasnick, *J. Clin. Invest.*, 1991, **88**, 1467.
28. J. T. Palmer, D. Rasnick, J. L. Klaus and D. Bromme, *J. Med. Chem.*, 1995, **38**, 3193.

29. J. E. Olson, G. K. Lee, A. Semenov and P. J. Rosenthal, *Bioorg. Med. Chem.*, 1999, **7**, 633.
30. B. J. Lee, A. Singh, P. Chiang, S. J. Kemp, E. A. Goldman, M. I. Weinhouse, G. P. Vlasuk and P. J. Rosenthal, *Antimicrob. Agents Chemother.*, 2003, **47**, 3810.
31. J. M. Coterón, D. Catterick, J. Castro, M. a. J. Chaparro, B. Díaz, E. Fernández, S. Ferrer, F. J. Gamo, M. Gordo, J. Gut, L. de las Heras, J. Legac, M. Marco, J. Miguel, V. Muñoz, E. Porras, J. C. de la Rosa, J. R. Ruiz, E. Sandoval, P. Ventosa, P. J. Rosenthal and J. M. Fiandor, *J. Med. Chem.*, 2010, **53**, 6129.
32. Y. A. Sabnis, P. V. Desai, P. J. Rosenthal and M. A. Avery, *Protein Sci.*, 2003, **12**, 501.
33. F. Li, K. P. Patra, C. A. Yowell, J. B. Dame, K. Chin and J. M. Vinetz, *J. Biol. Chem.*, 2010, **285**, 8076.
34. J. A. Boddey, A. N. Hodder, S. Gunther, P. R. Gilson, H. Patsiouras, E. A. Kapp, J. A. Pearce, T. F. de Koning-Ward, R. J. Simpson, B. S. Crabb and A. F. Cowman, *Nature*, 2010, **463**, 627.
35. S. Arastu-Kapur, E. L. Ponder, U. P. Fonovic, S. Yeoh, F. Yuan, M. Fonovic, M. Grainger, C. I. Phillips, J. C. Powers and M. Bogyo, *Nat. Chem. Biol.*, 2008, **4**, 203.
36. (*a*) D. Gupta, R. S. Yedidi, S. Varghese, L. C. Kovari and P. M. Woster, *J. Med. Chem.*, 2010, **53**, 4234; (*b*) C. Fäh, R. Mathys, L. A. Hardegger, S. Meyer, D. Bur and F. Diederich, *European Journal of Organic Chemistry*, 2010, **2010**, 4617.
37. A. Gangjee, S. Kurup and O. Namjoshi, *Curr. Pharm. Des.*, 2007, **13**, 609.
38. A. Nzila, *Drug Discov. Today*, 2006, **11**, 939.
39. M. E. Balis, G. B. Brown, G. B. Elion, G. H. Hitchings and H. Vanderwerff, *J. Biol. Chem.*, 1951, **188**, 217.
40. G. E. Childs and C. Lambros, *Ann. Trop. Med, Parasitol.*, 1986, **80**, 177.
41. C. J. Canfield, W. K. Milhous, A. L. Ager, R. N. Rossan, T. R. Sweeney, N. J. Lewis and D. P. Jacobus, *Am. J. Trop. Med. Hyg.*, 1993, **49**, 121.
42. S. Kamchonwongpaisan, R. Quarrell, N. Charoensetakul, R. Ponsinet, T. Vilaivan, J. Vanichtanankul, B. Tarnchompoo, W. Sirawaraporn, G. Lowe and Y. Yuthavong, *J. Med. Chem.*, 2004, **47**, 673.
43. S. Ommeh, E. Nduati, E. Mberu, G. Kokwaro, K. Marsh, A. Rosowsky and A. Nzila, *Antimicrob. Agents Chemother.*, 2004, **48**, 3711.
44. A. Nzila, M. Rottmann, P. Chitnumsub, S. M. Kiara, S. Kamchonwongpaisan, C. Maneeruttanarungroj, S. Taweechai, B. K. Yeung, A. Goh, S. B. Lakshminarayana, B. Zou, J. Wong, N. L. Ma, M. Weaver, T. H. Keller, V. Dartois, S. Wittlin, R. Brun, Y. Yuthavong and T. T. Diagana, *Antimicrob. Agents Chemother.*, 2010, **54**, 2603.
45. T. Dasgupta, P. Chitnumsub, S. Kamchonwongpaisan, C. Maneeruttanarungroj, S. E. Nichols, T. M. Lyons, J. Tirado-Rives, W. L. Jorgensen, Y. Yuthavong and K. S. Anderson, *ACS Chem. Biol.*, 2009, **4**, 29.
46. S. Subramanian, S. E. Bates, J. J. Wright, I. Espinoza-Delgado and R. L. Piekarz, *Pharmaceuticals*, 2010, **3**, 2751.

47. J. K. DiStefano and R. M. Watanabe, *Pharmaceuticals*, 2010, **3**, 2610.
48. K. E. Coan and B. K. Shoichet, *J. Am. Chem. Soc.*, 2008, **130**, 9606.
49. A. K. Chatterjee, H. Liu, D. C. Tully, J. Guo, R. Epple, R. Russo, J. Williams, M. Roberts, T. Tuntland, J. Chang, P. Gordon, T. Hollenbeck, C. Tumanut, J. Li and J. L. Harris, *Bioorg. Med. Chem. Lett.*, 2007, **17**, 2899.
50. J. J. Wiener, S. Sun and R. L. Thurmond, *Curr. Top. Med. Chem.*, 2010, **10**, 717.
51. F. Narjes, K. F. Koehler, U. Koch, B. Gerlach, S. Colarusso, C. Steinkuhler, M. Brunetti, S. Altamura, R. De Francesco and V. G. Matassa, *Bioorg. Med. Chem. Lett.*, 2002, **12**, 701.
52. V. J. Klimkowski, B. M. Watson, M. R. Wiley, J. Liebeschuetz, J. B. Franciskovich, J. Marimuthu, J. A. Bastian, D. J. Sall, J. K. Smallwood, N. Y. Chirgadze, G. F. Smith, R. S. Foster, T. Craft, P. Sipes, M. Chastain and S. M. Sheehan, *Bioorg. Med. Chem. Lett.*, 2007, **17**, 5801.
53. C. Garcia-Echeverria, P. Imbach, J. Roesel, P. Fuerst, M. Lang, V. Guagnano, M. Noorani, J. Zimmermann and P. Furet, *Chimia*, 2003, **57**, 179.
54. (*a*) S. M. Sheehan, H. J. Mest, B. M. Watson, V. J. Klimkowski, D. E. Timm, A. Cauvin, S. H. Parsons, Q. Shi, E. J. Canada, M. R. Wiley, G. Ruehter, B. Evers, S. Petersen, L. C. Blaszczak, S. R. Pulley, B. J. Margolis, G. N. Wishart, B. Renson, D. Hankotius, M. Mohr, J. C. Zechel, J. Michael Kalbfleisch, E. A. Dingess-Hammond, A. Boelke and A. G. Weichert, *Bioorg. Med. Chem. Lett.*, 2007, **17**, 1765; (*b*) N. A. Thornberry and A. E. Weber, *Curr. Top. Med. Chem.*, 2007, **7**, 557.
55. R. C. Gozalbes, R. J., Pineda-Lucena and A., *Current Medicinal Chemistry*, 2010, **17**, 1769.
56. E. J. Martin and D. C. Sullivan, *Journal of Chemical Information and Modeling*, 2008, **48**, 873.
57. (*a*) M. Rottmann, C. McNamara, B. K. Yeung, M. C. Lee, B. Zou, B. Russell, P. Seitz, D. M. Plouffe, N. V. Dharia, J. Tan, S. B. Cohen, K. R. Spencer, G. E. Gonzalez-Paez, S. B. Lakshminarayana, A. Goh, R. Suwanarusk, T. Jegla, E. K. Schmitt, H. P. Beck, R. Brun, F. Nosten, L. Renia, V. Dartois, T. H. Keller, D. A. Fidock, E. A. Winzeler and T. T. Diagana, *Science*, 2010, **329**, 1175; (*b*) B. K. S. Yeung, B. Zou, M. Rottmann, S. B. Lakshminarayana, S. H. Ang, S. Y. Leong, J. Tan, J. Wong, S. Keller-Maerki, C. Fischli, A. Goh, E. K. Schmitt, P. Krastel, E. Francotte, K. Kuhen, D. Plouffe, K. Henson, T. Wagner, E. A. Winzeler, F. Petersen, R. Brun, V. Dartois, T. T. Diagana and T. H. Keller, *J. Med. Chem.*, 2010, **53**, 5155.
58. N. Sriwilaijaroen, J. X. Kelly, M. Riscoe and P. Wilairat, *Southeast Asian J. Trop. Med. Public Health*, 2004, **35**, 840.
59. M. Smilkstein, N. Sriwilaijaroen, J. X. Kelly, P. Wilairat and M. Riscoe, *Antimicrob. Agents Chemother.*, 2004, **48**, 1803.
60. M. L. Baniecki, D. F. Wirth and J. Clardy, *Antimicrob. Agents Chemother.*, 2007, **51**, 716.

61. (*a*) D. Plouffe, A. Brinker, C. McNamara, K. Henson, N. Kato, K. Kuhen, A. Nagle, F. Adrian, J. T. Matzen, P. Anderson, T. G. Nam, N. S. Gray, A. Chatterjee, J. Janes, S. F. Yan, R. Trager, J. S. Caldwell, P. G. Schultz, Y. Zhou and E. A. Winzeler, *Proc. Natl. Acad. Sci. U.S.A.* 2008, **105**, 9059; (*b*) W. A. Guiguemde, A. A. Shelat, D. Bouck, S. Duffy, G. J. Crowther, P. H. Davis, D. C. Smithson, M. Connelly, J. Clark, F. Zhu, M. B. Jimenez-Diaz, M. S. Martinez, E. B. Wilson, A. K. Tripathi, J. Gut, E. R. Sharlow, I. Bathurst, F. El Mazouni, J. W. Fowble, I. Forquer, P. L. McGinley, S. Castro, I. Angulo-Barturen, S. Ferrer, P. J. Rosenthal, J. L. Derisi, D. J. Sullivan, J. S. Lazo, D. S. Roos, M. K. Riscoe, M. A. Phillips, P. K. Rathod, W. C. Van Voorhis, V. M. Avery and R. K. Guy, *Nature*, 2010, **465**, 311.
62. F. J. Gamo, L. M. Sanz, J. Vidal, C. de Cozar, E. Alvarez, J. L. Lavandera, D. E. Vanderwall, D. V. Green, V. Kumar, S. Hasan, J. R. Brown, C. E. Peishoff, L. R. Cardon and J. F. Garcia-Bustos, *Nature*, 2010, **465**, 305.
63. S. F. Yan, H. Asatryan, J. Li and Y. Zhou, *J. Chem. Inf. Model.*, 2005, **45**, 1784.
64. J. Sattabongkot, N. Yimamnuaychoke, S. Leelaudomlipi, M. Rasameesoraj, R. Jenwithisuk, R. E. Coleman, R. Udomsangpetch, L. Cui and T. G. Brewer, *Am. J. Trop. Med. Hyg.*, 2006, **74**, 708.
65. (*a*) M. Carraz, A. Jossang, P. Rasoanaivo, D. Mazier and F. Frappier, *Bioorg. Med. Chem.*, 2008, **16**, 6186; (*b*) N. Mahmoudi, R. Garcia-Domenech, J. Galvez, K. Farhati, J. F. Franetich, R. Sauerwein, L. Hannoun, F. Derouin, M. Danis and D. Mazier, *Antimicrob. Agents Chemother.*, 2008, **52**, 1215; (*c*) M. Carraz, A. Jossang, J. F. Franetich, A. Siau, L. Ciceron, L. Hannoun, R. Sauerwein, F. Frappier, P. Rasoanaivo, G. Snounou and D. Mazier, *PLoS Med.*, 2006, **3**, e513.
66. N. V. Dharia, A. B. Sidhu, M. B. Cassera, S. J. Westenberger, S. E. Bopp, R. T. Eastman, D. Plouffe, S. Batalov, D. J. Park, S. K. Volkman, D. F. Wirth, Y. Zhou, D. A. Fidock and E. A. Winzeler, *Genome Biol.*, 2009, **10**, R21.
67. M. Knockaert, N. Gray, E. Damiens, Y. T. Chang, P. Grellier, K. Grant, D. Fergusson, J. Mottram, M. Soete, J. F. Dubremetz, K. Le Roch, C. Doerig, P. Schultz and L. Meijer, *Chem. Biol.*, 2000, **7**, 411.
68. G. J. Crowther, A. J. Napuli, J. H. Gilligan, K. Gagaring, R. Borboa, C. Francek, Z. Chen, E. F. Dagostino, J. B. Stockmyer, Y. Wang, P. P. Rodenbough, L. J. Castaneda, D. J. Leibly, J. Bhandari, M. H. Gelb, A. Brinker, I. H. Engels, J. Taylor, A. K. Chatterjee, P. Fantauzzi, R. J. Glynne, W. C. Van Voorhis and K. L. Kuhen, *Mol. Biochem. Parasitol.*, 2010.
69. V. Patel, M. Booker, M. Kramer, L. Ross, C. A. Celatka, L. M. Kennedy, J. D. Dvorin, M. T. Duraisingh, P. Sliz, D. F. Wirth and J. Clardy, *J. Biol. Chem.*, 2008, **283**, 35078.

CHAPTER 5

The Medicinal Chemistry of Eradication: Hitting the Lifecycle where it Hurts.

Approaches to Blocking Transmission

JEREMY NICHOLAS BURROWS[a] AND
ROBERT EDWARD SINDEN[b]

[a] Medicines for Malaria Venture, Route de Pré-Bois 20, Geneva 1215, Switzerland; [b] Department of Life Sciences, Imperial College London, South Kensington, London SW7 2AZ, UK

5.1 Introduction

Since the time when malaria and its modes of transmission were first understood, attempts to treat the disease have, perhaps unwittingly, identified two effective modes of intervention. First, to prevent and treat the disease. Second, to prevent the transmission of the parasite from person to person via the mosquito vector. One of the first recorded treatments for malaria, in the 17th century, was quinine, the active ingredient in 'Jesuit's bark'. This was followed by chloroquine in 1934. Whilst quinine is still in use today, widespread resistance to chloroquine has severely restricted its effectiveness. Transmission of the parasite through communities has principally been managed by

RSC Drug Discovery Series No. 14
Neglected Diseases and Drug Discovery
Edited by Michael J. Palmer and Timothy N. C. Wells

Published by the Royal Society of Chemistry, www.rsc.org

controlling the mosquito vector with bed nets and insecticides such as DDT. Such was the success of these measures that, in 1957, the scientific community felt confident enough to use them as the central theme of a global malaria eradication campaign. Following significant early successes, this campaign was discontinued fourteen years later for many reasons including resistance of the parasite to chloroquine, and of the mosquito to DDT. Future eradication campaigns will need new classes of medicines and insecticides to overcome resistance. Some of these could be specifically targeted towards preventing transmission of the parasite.

5.2 Features of *Plasmodium* Biology Relevant to Drug Design

First, we must appreciate that five species of parasite are known to infect man, *Plasmodium falciparum, P. vivax, P. malariae, P. ovale* and *P. knowlesi* and at least two others that have done so in the laboratory (*P. cynomolgi* and *P. inui*). Of these it is essential to recognise that the biology of *P. falciparum* (which is in the sub-genus *Laverania*) differs significantly with respect to other malaria species (sub-genus *Plasmodium*).

This raises a number of key points. First, differences in parasite biology mean data gained on other species of parasite need to be reconfirmed on *P. falciparum*, especially those obtained from species which infect rodents. Second, it may be that there is no medicine for targeting all clinical malaria transmission.

Any therapy attacking transmission of the parasite will need components that are designed to stop the appearance of the sexual stages of parasite. These are normally detected in the blood of the patient some 8–10 days after the main wave of disease in a *P. falciparum*-infected patient. They are most likely present in small numbers during the main wave of infection, but their presence is masked by the sheer biomass of the normal asexual parasites. In *P. vivax* the situation is different, and the sexual stages are already present at significant levels when patients first come and seek treatment.

There are two further twists to the problem. For *P. vivax, P. ovale* and the primate infecting strain *P. cynomolgi* there are dormant forms in the liver (known as hypnozoites) which can reactivate, giving the potential of a relapse in the absence of a mosquito bite. Second, there are significant reservoirs of *P. knowlesi* parasites in macaque monkeys and other great apes of Africa.[1] These species can also be transmitted to humans and the medicines need to take this into account.

The lifecycles of all malaria parasites exhibit common properties (See Figure 5.1; Table 1), which help us design new medicines to block transmission.

1. The parasite population undergoes exponential expansion in the asexual blood stages. At the peak of infection an adult patient may host 200 000 parasites per microlitre. This total of 10^{11} parasites is a formidable biomass.
2. Two population bottlenecks occur. The first in the pre-erythrocytic stage in the liver where, following bites from infected mosquitoes, a hundred

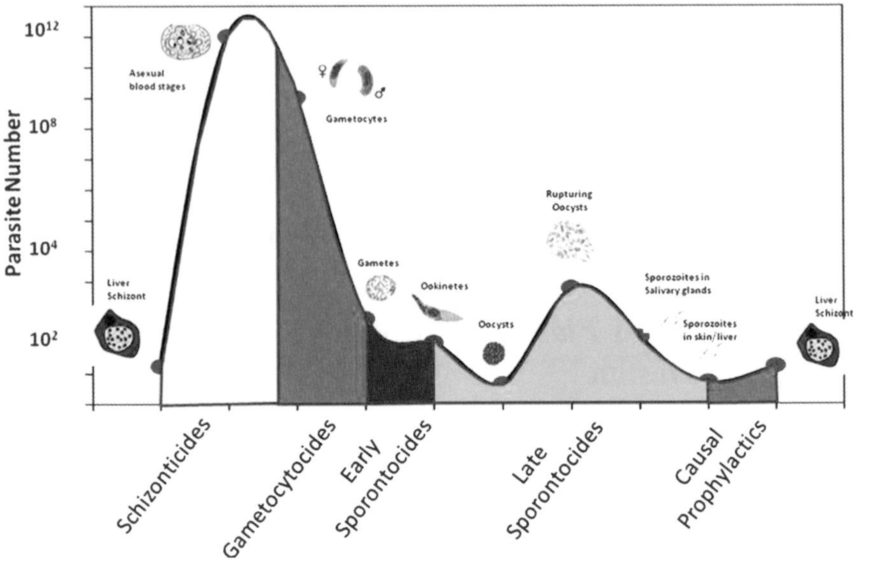

Figure 5.1 The replication cycle of *Plasmodium*. The numbers of parasites at each stage varies considerably. The critical bottlenecks are in the early liver stages in humans, and in the oocysts in the mosquito.

Table 5.1 Key metabolic activities and cellular properties of the different life stages that might influence the design and implementation of drugs targeting each stage.

	Vegetative growth	*DNA replication*	*Energy metabolism*	*Motility*	*Host cell barrier?*
Liver schizont	+++	+++	+++	-	Hepatocyte
Asexual blood stages	++	++	++	+merozoite	RBC
Gametocyte	±(immature)	?	+(all ages)	-	RBC
Gamete	-	+♂	++♂	++♂	None
Ookinete	±	+(meiosis)	++ (mitochondria)	++	None
Oocyst	+++	+++	++	-	None/capsule
Sporozoite	-	-	++	++	Salivary gland Kupffer Cell Hepatocyte

parasites can be found in the liver of the infected human host. The second, in the mosquito, where following the very inefficient processes leading to fertilisation, commonly less than five individual parasites survive to form oocysts in the infected mosquito.

3. The parasites are intracellular for most of their lifecycle. Pre-erythrocytic schizonts are resident in liver cells for several days. Blood-stage schizonts reside in anucleate red blood cells. The mosquito stages of parasite

development are extracellular, but the oocyst is enveloped by a capsule that is highly permeable to amino acids;[2] nucleotides[3] and drugs.[4] Conversely, all gametic and invasive stages are extracellular.

4. Mature gametocytes, salivary gland sporozoites and hypnozoites are all terminally differentiated cells that require specific external cues to progress through the cell cycle. As such they may share certain regulatory and metabolic similarities. It is therefore possible that drugs which kill the mature stage V gametocytes will have an impact on the hypnozoites. Indeed this is true for the 8-aminoquinoline family including primaquine and tafenoquine.

Prevention of disease requires that we reduce the large asexual blood-stage population, in which we might anticipate rapid selection of resistant mutants. However, if we want to reduce transmission we might profitably attack the small populations of parasites responsible for the transit of the parasite between human and mosquito. Here we might hypothesise that the selection of resistant genotypes will be substantially delayed (in inverse proportion to the population size).

5.3 Status of Current Biological Assays and Future Needs

For a thorough review of the early work on assays used we direct the reader to the excellent two-volume thesis of Peters.[5]

5.3.1 Pre-erythrocytic (Liver-stage) Assays

***In vivo*:** Although *in vivo* assays have very low throughput, they have been the mainstay of drug development for compounds addressing the liver stages. These could be for causal prophylaxis (preventing the parasite leaving the liver prevents malaria blood stage infections) or for radical cure (wiping out the dormant reservoirs of *P. vivax, P. ovale* or *P. cynomolgi* in the liver). Until recently, the impact of drugs was primarily determined by the delay or cessation of blood patent parasitemia. The major development in recent years is using PCR to detect the appearance of parasites in the blood. This can be used in human studies since detection precedes clinical symptoms by a safe margin.[6] The sensitivity of PCR-based methodologies has *inter alia* permitted quantisation of pre-erythrocytic parasites directly in liver biopsies from model species, notably the rodent malarial parasites. The recent transgenic parasites (commonly in the rodent malaria species) expressing reporter molecules such as green fluorescent protein (GFP) or luciferase at selected points in the lifecycle permits their widespread application to time-series observations of parasites *in situ*,[7] and might provide a route to *in vivo* observations of the hypnozoite stages of *P. vivax* and *P. ovale* (or the model *P. cynomolgi*); or gametocyte sequestration in *P. falciparum*.

In vitro: The successful culture of rodent and then human parasite species in hepatoma cells,[8,9] and subsequently primary hepatocytes[10,11] has raised the importance of such techniques in drug discovery, especially now the assays can be conducted in 96-well formats using automated microscopic analysis,[12] thereby permitting the screening of large libraries of compounds.[13] Questions still remain as to the metabolic relevance of the hepatocyte monoculture and, to a lesser extent, the absence of 3D organisation of the host cells. Attempts to address this by the use of prefabricated 3D matrices are at an early stage.

Attempts have been made to investigate the hypnozoites of *P. vivax in vitro*,[14] but these are still experimental. The problem is how to prevent the host cell from totally outgrowing the static parasite population in culture. Recognising that the hypnozoite is an arrested parasite, identification of markers to distinguish hypnozoites from young primary schizonts is problematic. In the literature there are clear descriptions of strains of *P. vivax* (Nickolaiev strain) that exclusively produce relapsing forms.[15] Might it be feasible to re-discover these parasites and adapt them to culture?

5.3.2 Asexual Blood-stage (Schizonticide) Assays

Since the majority of drug discovery over the last seventy years has focused on the asexual blood stages responsible for the symptoms of malaria, there is a wide range of assays available.

In vivo: The main development in the field has been that of transgenic reporter parasites. However, they have yet to receive widespread application in understanding the impact of drugs on the dynamics of infection *in vivo*.

In vitro: The revolution in drug screening was brought about by the culture of the asexual blood stages of the most pathogenic of human parasites: *P. falciparum*.[16] The subsequent use of radio-labelling methods ([^{3}H] –hypoxanthine uptake), and more recently of non-radioactive (SYBR-green) approaches to measure parasite replication will remain a mainstay of antimalarial discovery. However, despite these successes there is still no routine procedure to support the growth of the blood-stages of *P. vivax* and the other human pathogens, and this is an area which requires more attention.

5.3.3 Mature Gametocyte (Gametocytocide) Assays

Development of gametocytocidal assays have been heavily influenced by the atypical biology of sexual development in *P. falciparum*.

In vivo: The specific search for gametocytocidal compounds was pioneered by Terzian,[17] but has received scarce attention since. Studies on the rodent malaria parasites in the 1980s, at the Department of Entomology at the Walter Reed Army Institute of Research, however, remain notable.[18] Studies on *P. falciparum* gametocytes are complicated by their biology. *In vivo* immature gametocytes are sequestered in the deep tissue for the first six to eight days and so are not seen in the bloodstream. The mature crescent-shaped gametocytes appear nine to twelve days after the waves of asexual parasites. If the infection is treated with a pure schizonticide (classically this was chloroquine), all the blood stages and the immature gametocytes are killed up to stage III.[19,20] Thus, there is a short period where no parasites can be detected in the blood. This is followed by the emergence in the peripheral circulation of the late-stage gametocytes (which are not affected by 4-aminoquinolines such as chloroquine). Some have interpreted such data[21] to suggest that drugs with such selective killing activity induce gametocytogenesis, but have not recognised they may simply be observing the inevitable consequence of stage-specific lethality. Methods to distinguish gametocyte-selection from induction require careful design.[22] Concepts on the use of drug combinations to avoid enhanced transmission[23] require a more secure understanding of gametocyte biology. Studies on all of the other species apart from *P. falciparum* are free of this confounder.

In vitro: Methods to grow and monitor (by microscopy) mature gametocytes of *P. falciparum in vitro*,[24] in 96-well format[25] and by flow cytometry[26] have advanced our understanding of the metabolic organisation and drug sensitivities of these stages.[27] However, we need to carefully evaluate these assays in the light of *Plasmodium falciparum* as explained above. The biology of the first three stages of gametocyte development is sensitive to drugs which kill the blood-stage asexual parasites, and so there is less need for drugs specifically targeting these stages. Where assays are really needed are the metabolically down-regulated stage IV and V gametocytes. Currently, the problem is that most markers such as Pfs16 are for early gametocytes. The search for later markers is ongoing. Current assays therefore focus on whether the infected cell is dead or can progress to gametes. Alternatives would be to look at the ability of stage IV and V gametocytes to develop into metabolically active cells.

Molecular assays: Whilst we currently understand little of the molecular pathways regulating gametocytogenesis, markers of sexual development are being described e.g. Pfs16;[28] telomerase,[29] and proteosomes.[30] These are a first step to describing molecular targets.

5.3.4 Mosquito-stage Assays (Gametogenesis; Ookinete and Oocyst Formation)

In vivo: The ability to transmit many species of malarial parasite through the mosquito in the laboratory provides a powerful (but expensive) *in vivo* assay for transmission-blocking agents – widely exploited in the vaccine field as the 'gold' standard membrane feeding assay. Clearly the most amenable (safe) systems are the rodent parasites when transmitted by laboratory vectors such as *A. stephensi*. The success of transmission of the parasites from a drug-treated/infected host may be measured at many stages in development, including the ookinete, oocyst and the sporozoite. Clearly monitoring gametocyte-to-sporozoite development embraces the widest biology. All endpoints can now be measured conveniently using genetically tagged reporter parasites.[31,32] Data have in the past been commonly represented as the average reduction in parasite number/intensity, or more rarely as a reduction in the prevalence of infected mosquitoes. Notwithstanding that these two parameters are mathematically related,[33] it is easier for the non-specialist to understand the impact of the intervention using the latter format.

Use of infected rodent hosts in standard membrane feeding assays benefits from the inclusion of host–drug PK/PD, but may be less appropriate for compounds targeted at *P. falciparum*. It is therefore convenient that *P. falciparum* gametocytes can be cultured throughout their development and then offered in membrane feeders to mosquitoes – a highly relevant assay.

Antivector drugs (smart insecticides) to disturb parasite–mosquito interactions delivered through the human host are a theoretical, but currently sidelined possibility, as demonstrated by studies using insect-immune peptides and their analogues SHIVA,[34] magainin[35] and cercropin.[36] *In vivo* studies on vivax, ovale, malariae and knowlesi malaria, such as those on *P. falciparum,* remain feasible using non-human primate hosts, and therefore have an important position in the development pipeline.

In vitro: The culture of the entire mosquito phase of parasite development is now technically possible, but currently impractical for drug screening.[37,38] Nevertheless, useful assays have been developed to screen for compounds that block early gametic and sporogonic development.[39] Exflagellation of mature cultured drug-treated male gametocytes of *P. falciparum* is a glycolysis-powered event.[40] It represents perhaps the most relevant assay to monitor drugs that kill mature gametocytes. It could also be applied to all species of parasite.

The importance of finding new compounds has led to testing known compounds in these assays, with some success. Compounds already demonstrated to possess particular activity in this assay include atovaquone and thiostrepton. On published evidence this assay could readily be adapted to *P. vivax*.[41,42] Assays on cultured ookinetes of *P. falciparum* are still not at the stage where they can be used for routine testing.[43,44]

Molecular assays: Molecular pathways, targets and inhibitors that may be appropriate to monitor in the search for inhibitors of microgametogenesis include: Cysteine proteases,[45] PfPK7,[46] Pfnek-2,[47] Pfnek-4,[48] cGMP-dep protein kinase,[49] pbmap-2,[50] pbKch1,[51] cdpk3,[52,53] PMSF protease inhibitors,[54] PLA2,[55] antimicrobial peptides,[56] protein synthesis,[57] *Azadirachta indica*,[58] and NADPH-dependant dual oxidases.[59]

Compounds known to categorically impact transmission *via* either the gametocytes or mosquito stages are listed in Figure 5.2.

5.4 Clinical Aspects of Transmission-blocking Approaches

The differences between the timing of gametocyte appearance for *P. falciparum* and the other species of malaria mean they must be considered separately.

In all malaria species a small fraction of asexual blood stages are committed at each round of cell division in the red blood cell to become sexual stages.

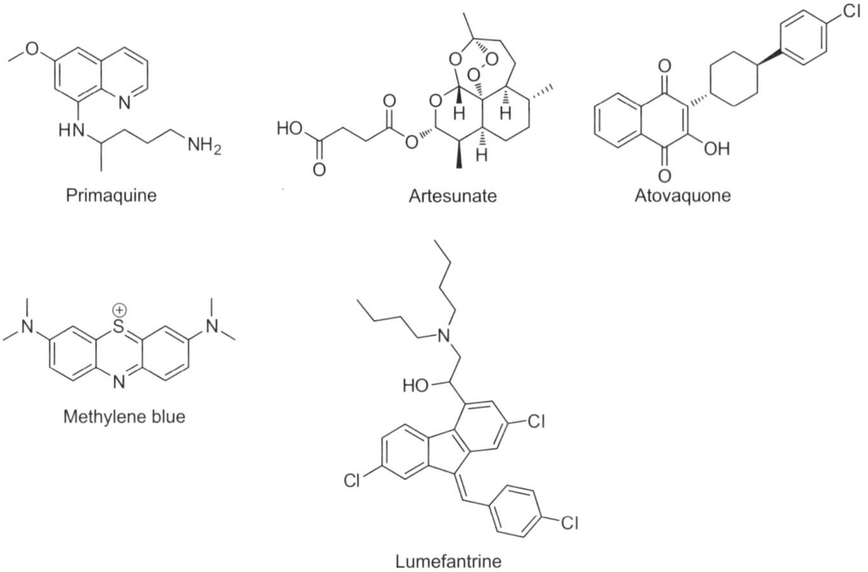

Figure 5.2 Drugs known to impact transmission in *P. berghei* or *P. falciparum*.[60,61]

Where the gametocyte and asexual blood stage maturation are of similar duration, the infected human must be considered to be infectious to the mosquito from the moment the patient has schizonts present in the blood. The waves of synchronous rupture of schizonts, which mark the clinical course of disease, have been shown to temporarily suppress this infectivity.[62] The short half-life of the mature gametocyte emphasises the notion that co-administration of a schizonticide and a (mature) gametocytocide is a very desirable strategy, but is this stratagem applicable to *P. falciparum? P. falciparum* gametocyte maturation is significantly slower than that of the asexual blood stage, onset of theoretical infectivity is delayed by 8–12 days and the population persists for 22 days (half life of 4–7 days). Periods of high gametocytaemia and infectivity follow the waves of asexual burden for extended periods of infection.[63] Does this temporal 'dislocation' influence the strategy of drug delivery? Do gametocytocidal compounds need to be delivered 8–12 days after a *P. falciparum*-infected patient has visited the clinic, and then must it persist in the bloodstream for a further 22 days? We suggest that the delivery of an effective schizonticide (blood-stage killer) combined with a fast-acting compound targeted to kill the late-stage gametocytes should also be effective here. The former will kill all input to the gametocyte population and potentially the standing population of immature gametocytes (stages I-III); and the latter will kill all the standing population of mature (stages IV-V) gametocytes. There will be no ensuing infectious population. The key conclusion from this hypothesis is that the half-life of a gametocytocidal drug needs to be sufficient to ensure that the gametocyte population is killed, though it need not be inordinately long.

5.4.1 Development of Transmission-blocking Drugs

One of the key questions raised by the concept of transmission blocking is to clearly explain the benefit of such a medicine to the patient. A gametocytocidal or sporozonticidal (killing the mosquito stages) molecule will benefit the recipient because it lowers the probability of being reinfected. Mosquitoes do not travel far and often feed on many people within the same room. Treating infectious patients to prevent transmission is not an altruism; it is sound public health policy.

One important question is how to test the efficacy of a transmission-blocking medicine clinically. Here the best model is another transmission-blocking agent, namely insecticide-treated nets (ITNs), and use randomised cluster approaches. What endpoint should we use? Whilst in the laboratory the screens have correctly measured the reduction in number of parasites (exflagellating microgametocytes; ookinetes; oocysts; sporozoites, or the prevalence of infected mosquitoes), we venture to suggest that this is not the endpoint that should be used in field trials other than early proof-of-concept studies. Here the key parameter to be measured is the reduction in the number of new infections in man following the introduction of the intervention. Clearly the issue of identifying new infections is key, but if impractical, the reduction in the total number of cases may be an indirect but relevant endpoint as has been used in bed net trials.[64]

5.5 Medicinal Chemistry Perspectives on Transmission Blocking

5.5.1 Liver-stage Parasites

Plasmodium vivax

The substantial worldwide burden of vivax malaria and its relapsing dormant liver-stage (hypnozoite) coupled with inadequate first-line therapy has raised the drug discovery priority for this indication. Consequently, it is important to have new drugs which kill liver-stage vivax hypnozoites.[65] Primaquine, an 8-aminoquinoline is the only marketed vivax malaria anti-relapse agent, though is contraindicated in patients with a glucose-6-phosphate dehydrogenase (G6PD) deficient phenotype and in expectant mothers, where the G6PD classification of the foetus is unknown.

From a medicinal chemistry perspective, a range of potential issues need to be considered for successful prosecution of a drug discovery project and these fall broadly into two categories. First, those dependent on the limitations of available assays and thus the nature of the test cascade. Second, what combination of parameters are necessary to be clinically relevant. These areas can be summarised for a *P. vivax* liver-stage drug as follows.

1. Test cascade: A *P. vivax in vitro* hepatocyte assay exhibiting both liver schizonts and hypnozoites is still not available. There are many reasons for this and these have been discussed at length elsewhere.[68] Furthermore, although an

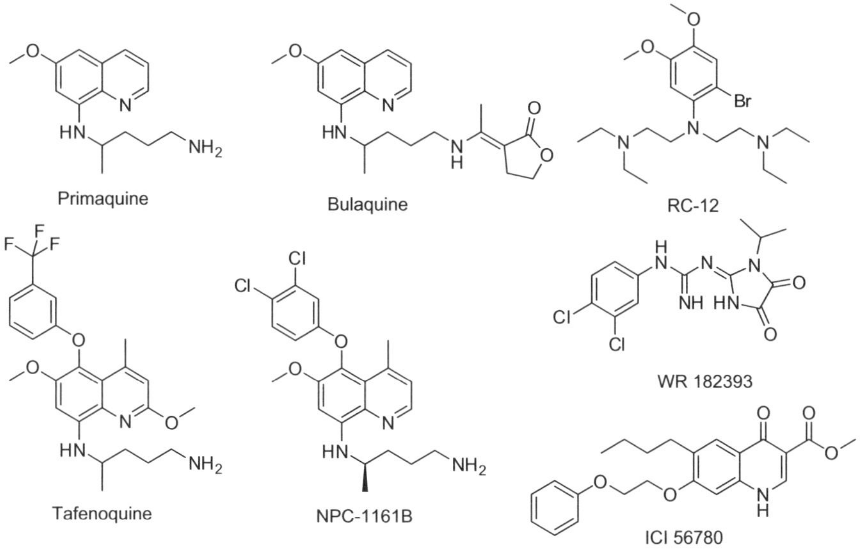

Figure 5.3 Structures of compounds known to block relapse in either *P. vivax, P. ovale* or *P. cynomolgi.*[66,67]

in vivo model utilising a *P. vivax* surrogate, *P. cynomolgi,* to infect rhesus monkeys is available with a reproducible relapse frequency,[69] a direct, simple validated *P. vivax* translational relapse model is not available.

Several models with validation underway include the following *in vitro* assays: *P. cynomolgi*-infected primary rhesus hepatocytes[70] and *P. vivax* infected HepG2 cells.[71] A validated *P. cynomolgi* assay will have the advantage that it should, assuming acceptable pharmacokinetics, translate well to the *in vivo* standard model (same host and parasite), though is not the human relevant parasite.

Current test cascades are pragmatic and can be influenced by hypotheses regarding the higher likelihood of finding anti-hypnozoiticidal agents from amongst liver schizonticides, rather than from non-liver acting molecules. In such scenarios, high-throughput rodent liver-stage malaria assays are performed and used as a filter for the lower throughput, more demanding vivax or cynomolgi malaria liver-stage assays. Alternatively, screening against vivax or cynomolgi liver-stage malaria can be performed with advanced compounds optimised for erythrocyte activity.

Such an approach is a profiling, rather than an optimisation strategy, and can be applied to any exo-erythrocytic stage, as shown in Figure 5.4. Should any activity be observed, follow-up according to the more detailed test cascades is possible.

The ideal test cascade for liver-stage vivax would have a validated and high-throughput vivax hypnozoite assay as the primary assay, in a metabolically functional human cell line. This is followed by *in vitro* metabolism, physical

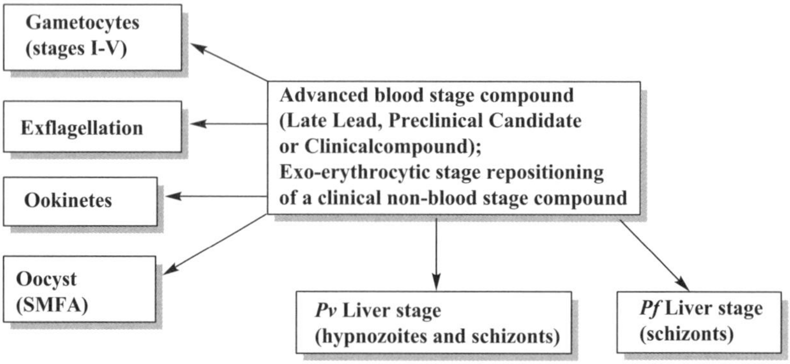

Figure 5.4 Profiling to determine extra-erythrocytic activity of advanced compounds. *Pv* = *Plasmodium vivax; Pf* = *Plasmodium falciparum*; SMFA = standard membrane feeding assay.

property measurements, and *in vivo* pharmacokinetics. Ultimately, the impact on relapse in the *in vivo P. cynomolgi* rhesus efficacy model, including chloroquine to eliminate blood-stage parasites, is the pre-clinical *in vivo* model. Given the issues of haemolysis with 8-aminoquinolines in G6PD deficient patients, a further requirement for a new molecule would be safety in an *in vivo* G6PD deficiency model.[72] One such idealised test cascade for a vivax liver drug discovery project is shown in Figure 5.5. It is to be noted that currently some of the specific parasitological assays do not exist, whilst other non-pharmacological assays are part of the standard drug discovery battery.

2. Parameters required for clinical utility: The lifetime of a hypnozoite is highly variable, from 26 days to a year. For a single exposure radical cure, exposure over a time period equal to three- to four-times (or less) the human half-life, after which the drug is likely to be below efficacious levels, must be sufficient to completely sterilize the liver. If not, multiple doses will be necessary. Clinically, we are searching for something as safe for everyone as primaquine is in G6PD normal patients, but with a single administration sufficient to kill all hypnozoites.

Plasmodium falciparum

Plasmodium falciparum does not form hypnozoites. A compound such as atovaquone, which kills the liver schizonts can be used as a causal prophylactic. Other examples of compounds with this activity are shown in Figure 5.6.

In those instances where, for pragmatic reasons, the parasite used in the assay is different from that relevant for human disease there is always a question relating to species differences. For full confidence, it is always important to demonstrate activity in the most physiologically relevant setting: the human parasite, in the human cell in the right tissue context.

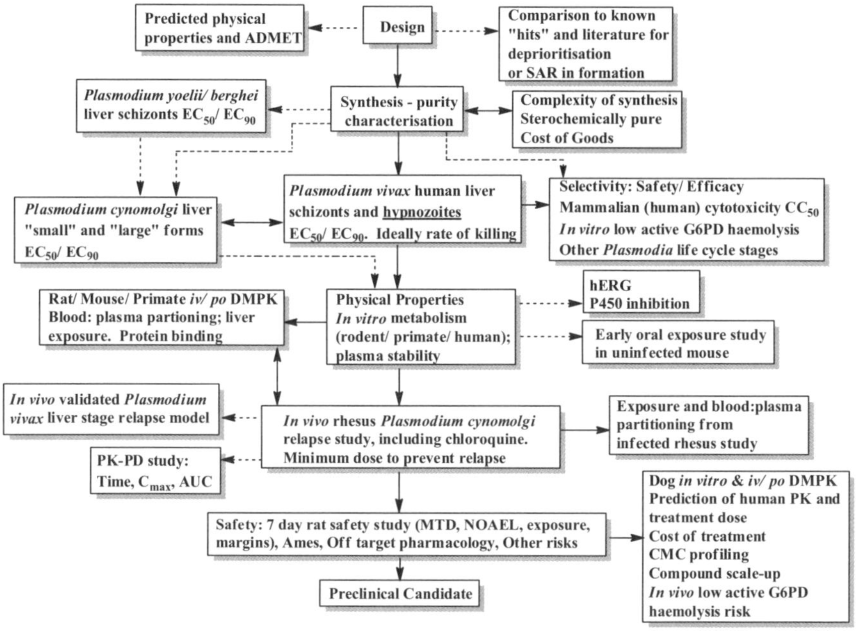

Figure 5.5 Idealised test cascade for *Plasmodium vivax* liver stage. ADMET = Absorption, Distribution, Metabolism, Excretion, Toxicity; SAR = Structure Activity Relationship; EC_{50}/EC_{90} = concentration of compound necessary to inhibit growth of the parasite by 50%/90% respectively; CC_{50} = concentration of compound resulting in 50% inhibition of an endpoint linked with cytotoxicity; DMPK = Drug Metabolism and Pharmacokinetics; hERG = human ether-a-go-go related gene (an ion channel implicated in possible cardiovascular risk); P450 = Cytochrome enzymes (those liver enzymes effecting metabolism of a compound); *iv/po* = *intra venous/per oral* (dosing routes directly into the vein or by mouth; *in vitro/in vivo* within glass/within living organisms); PK = Pharmacokinetics; PK-PD = Pharmacokinetic-Pharmacodynamic relationship; Tmax, Cmax, AUC = Time associated with maximal concentration, maximum concentration, area under the curve; MTD = maximum tolerated dose; NOAEL = No adverse effect level; Ames = a bacterial genotoxicity assay; CMC = chemistry, manufacturing and control; G6PD = glucose-6-phosphate dehydrogenase.

Consequently, primary cultures give more physiological readouts.[76] It is critical to understand whether activity is driven by parent molecule or a metabolite and to be aware to what extent turnover is an issue in the primary assay. This can be explored in several ways; comparing with metabolically inert cell lines, monitoring metabolic stability in the *in vitro* assay conditions or pre-treatment with pan-P450 inhibitors.[77] Usually, it is better for activity to be driven by the parent compound and to optimize the structure to reduce metabolism and thus extend half-life and duration of action.

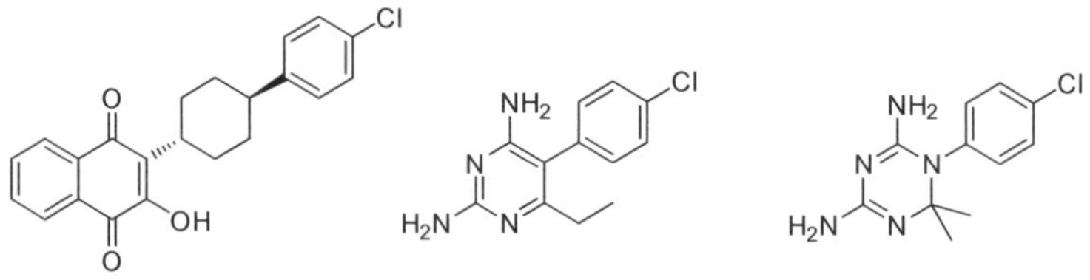

Figure 5.6 Compounds affecting pre-erythrocytic parasites.[73–75]

Liver targeting may also be an advantage and comparisons of liver *versus* blood exposure can be instructive. Strategies to selectively increase higher free concentrations in the liver usually involve exploiting active transport mechanisms and designing in characteristics that make the compounds better substrates for active uptake, such as has been demonstrated with the statins. One such transporter is OATP, the oxyanion transporter protein that is particularly beneficial for anionic species.[78] To the authors' knowledge, no substantial examples of such active transport strategies have been adopted in the malaria community, although many compounds are highly concentrated in the liver such as primaquine.

5.5.2 Gametocyte-stage Parasites

Disrupting transmission to the mosquito vector to render the mosquito non-infective to humans can be achieved by either eliminating the sexual gametocyte pool in the patient (gametocytocidal) or by eliminating the ookinete, oocyst and sporozoite load within the mosquito (sporozonticidal).[79–81]

Whilst the existing and developing host gametocyte assays measure various endpoints and stages, it is clear that the preferred phenotype would be rapid killing of all five stages. Clearly, killing of stage IV and V gametocytes, especially the latter, is of most value given that they are the infective parasite species. Furthermore, although in falciparum malaria there is an 8-day peak delay between asexual and sexual parasite load, at the point when a patient experiences symptoms, committed gametocytes will be present in the blood. For vivax malaria the gametocyte differentiation and production is coincident with the emergence of the blood-stage infection.[82] Consequently, having a drug combination treatment that kills stage V gametocytes is critical for full gametocytocidal efficacy. It is important to note that, particularly for vivax malaria, there is a window prior to symptoms, in which a patient may be infective.

1. Test cascade: The medicinal chemistry strategy and test-cascade adopted for such a transmission-blocking programme could depend on whether gametocytocidal activity was the chief optimisation parameter or not. Most pragmatically, advanced falciparum asexual acting compounds can be profiled *in vitro* against falciparum gametocyte stages I–V and for the differential stage activity to be assessed. The most relevant subsequent transmission-blocking assay would be to incubate the compound in *P. falciparum* culture, for a duration and at a concentration relevant to the predicted *in vivo* duration and trough level following three daily doses in humans. Membrane feeding with the female *Anopheles gambiae* vector followed by dissections and oocyst counting then gives an idea of transmission-blocking potential.

Naturally the clinical endpoint (measuring oocyst counts in the female *Anopheles gambiae* vector) is the same, whether performed in a non-clinical or clinical setting, providing a useful translational link. However, in mosquitoes ingesting a blood meal through the skin of an infected human in which gametocyte load has been impacted by the drug, the metabolites and the immune response is the most relevant measure of response. Some compounds are only active via their metabolites, such as primaquine.

A medicinal chemistry strategy to optimise potency against gametocytes clearly requires a primary assay to examine killing of the required stages *in vitro*. *In vivo* assays looking at impact on gametocytes are feasible though will commonly involve both rodent malaria parasites and rodent liver drug metabolism. Gametocyte carriage in the *P. falciparum* severe combined immunodeficient (SCID) mouse model can be assessed,[83,84] though again rodent drug metabolism is in operation and the immunological component absent. The more complex *in vivo* transmission-blocking assays involve mosquito feeding of *Anopheles stephensi* or *Anopheles gambiae* (for the *Plasmodium falciparum* SCID mouse) on an infected mouse previously treated with the oral drug. Oocysts are then counted in the mosquito vector as a measurement of the transmission blocking potential. As described earlier, this assay can be substituted by the membrane feeding study with human parasites and human blood, with the metabolic caveats already mentioned.[85]

An idealised test cascade for this approach is shown in Figure 5.7.

2. Parameters required for clinical utility: Understanding the full gametocyte stage profile of a compound is key. For a compound acting on multiple lifecycle stages then it can be suggested that a similar mechanism of action is responsible for killing in the different phases. Since resistance generation is related to parasite numbers, any transmission-blocking agent with asexual-stage activity could potentially be rendered ineffective over time due to asexual resistance induction. From this specific perspective a pure gametocytocidal-only agent would be preferable. In practice, primaquine, which has only weak blood-stage efficacy, kills mature gametocytes and has shown impact on gametocytes in clinical trials. Consequently, it illustrates the target product profile and is the compound to compare new molecules against to ensure superiority.

It is critical that the half-life in humans of a gametocytocidal drug is appropriate relative to the rate of killing of gametocytes. In other words, that three once-daily doses of the agent results in exposure that will eliminate gametocytes below an infective concentration.[86] Rapid action would be preferred to truncate the transmission window.

Since gametocyte cultures are non-proliferating there is no opportunity for resistance generation through errors occurring via clonal expansion. Resistant

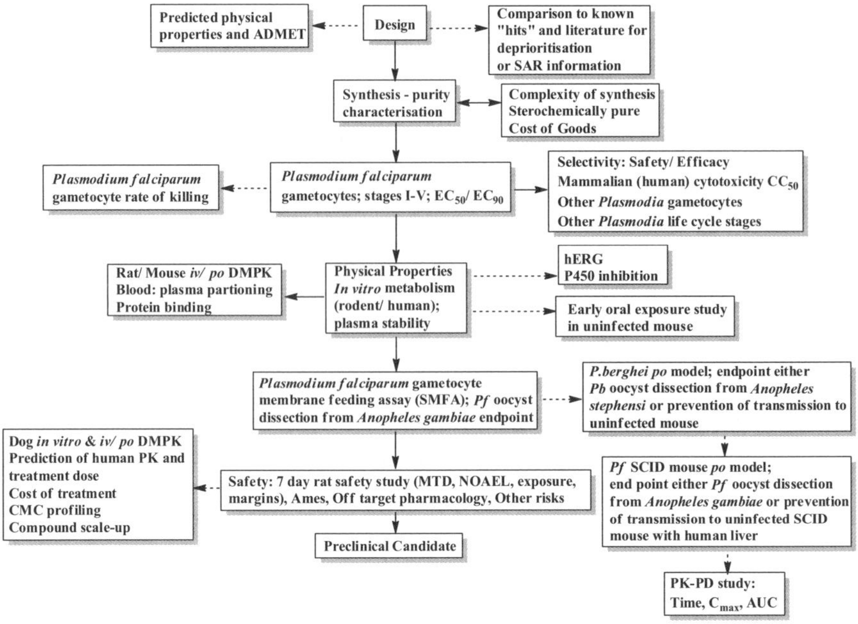

Figure 5.7 Idealised test cascade for gametocyte stages. ADMET = Absorption, Distribution, Metabolism, Excretion, Toxicity; SAR = Structure Activity Relationship; EC_{50}/EC_{90} = concentration of compound necessary to inhibit growth of the parasite by 50%/90% respectively; CC_{50} = concentration of compound resulting in 50% inhibition of an endpoint linked with cytotoxicity; DMPK = Drug Metabolism and Pharmacokinetics; hERG = human ether-a-go-go related gene (an ion channel implicated in possible cardiovascular risk); P450 = Cytochrome enzymes (those liver enzymes effecting metabolism of a compound); *iv/po* = *intra venous/per oral* (dosing routes directly into the vein or by mouth; *in vitro/in vivo* within glass/within living organisms); PK = Pharmacokinetics; PK-PD = Pharmacokinetic-Pharmacodynamic relationship; Tmax, Cmax, AUC = Time associated with maximal concentration, maximum concentration, area under the curve; MTD = maximum tolerated dose; NOAEL = No adverse effect level; Ames = a bacterial genotoxicity assay; CMC = chemistry, manufacturing and control.

gametocytes presumably thus occur as an end product of the asexual stage or via direct chemo-mutagenic events once committed. The ideal combination would thus include compounds with orthogonal gametocytocidal mechanisms to eliminate the transmission of the resistant phenotype.

5.5.3 Vector-stage Parasites

A transmission-blocking target product profile is also achievable *via* killing the parasites in the mosquito (sporozonticidal), in addition to or independent of host-stage parasiticidal action. Due to parasite numbers, the ookinete and oocyst are extremely attractive choke-point targets for intervention. The idealised test cascade is covered in Figure 5.8.

1. Test cascade: The medicinal chemistry strategy and test cascade adopted for such a transmission-blocking programme would depend on whether vector-stage activity was seen as a selective property or as an additional value to a blood-stage acting compound. Most pragmatically, compounds acting against asexual stages can be profiled in the dedicated falciparum or rodent vector-stage assays. These include gametocyte exflagellation, ookinete and oocyst

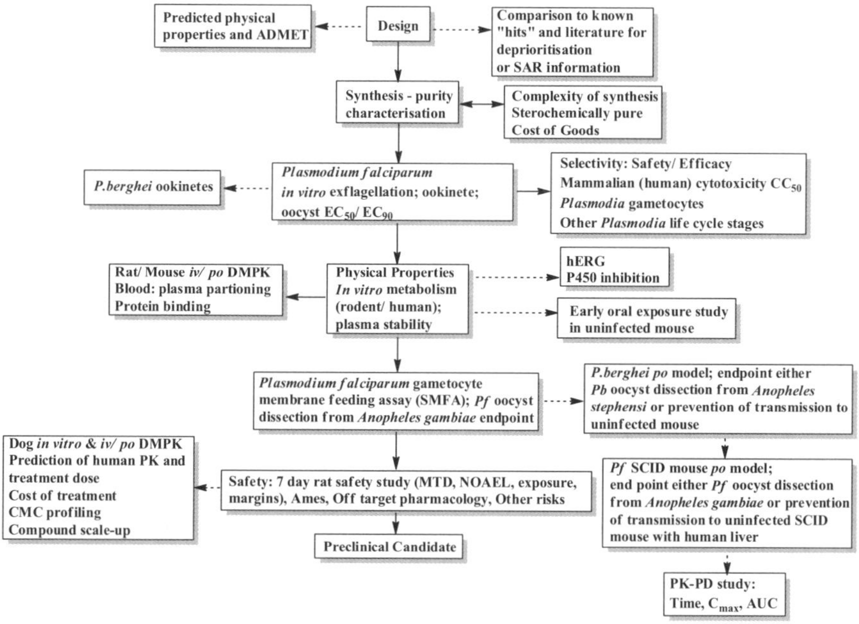

Figure 5.8 Idealised test cascade for mosquito stages. ADMET = Absorption, Distribution, Metabolism, Excretion, Toxicity; SAR = Structure Activity Relationship; EC_{50}/EC_{90} = concentration of compound necessary to inhibit growth of the parasite by 50%/90% respectively; CC_{50} = concentration of compound resulting in 50% inhibition of an endpoint linked with cytotoxicity; DMPK = Drug Metabolism and Pharmacokinetics; hERG = human ether-a-go-go related gene (an ion channel implicated in possible cardiovascular risk); P450 = Cytochrome enzymes (those liver enzymes effecting metabolism of a compound); *iv/po* = *intra venous/per oral* (dosing routes directly into the vein or by mouth; *in vitro/in vivo* within glass/within living organisms); PK = Pharmacokinetics; PK-PD = Pharmacokinetic-Pharmacodynamic relationship; Tmax, Cmax, AUC = Time associated with maximal concentration, maximum concentration, area under the curve; MTD = maximum tolerated dose; NOAEL = No adverse effect level; Ames = a bacterial genotoxicity assay; CMC = chemistry, manufacturing and control.

inhibition (Figure 5.4). As with gametocytes, the most relevant transmission-blocking assay involves counting oocysts in the vector following a membrane feed of an infective culture and drug. There is clearly a trade-off between throughput and relevance of the assay.

Alternatively, if the objective is a mosquito-stage-only compound, then the approach would be to screen compounds in the mosquito-based parasite assays. The results can be confirmed through periodic testing of optimised compounds in the membrane-feeding assay. *In vivo* rodent studies could be performed to determine the dose and duration of transmission blocking.

Pharmacokinetics and metabolism in the vector could be considered a potential issue for such a strategy, but the pragmatic way forward is simply to only pursue series that present correlations between a primary phenotypic assay and the oocyst count because, by definition their properties must have been sufficiently good in the vector to elicit the desired response.

2. Parameters required for clinical utility: For a compound to be clinically useful on mosquito-stage parasites, the drug must be present in blood, at concentrations commensurate with transmission blocking, for as long as infective gametocytes are in circulation. Given the data showing sub-microscopic gametocyte concentrations out to beyond 50 days,[87,88] it is clear that the major challenge with this approach is ensuring sufficient compound cover (i.e. long half-life). Therefore, it is likely to be viable only for a vector-stage-acting compound to be dosed as part of a combination that also kills gametocytes, and thus inherently reduces circulating mature gametocytes.

From a translational perspective, a concentration response curve in the standard membrane feeding assay with falciparum gametocytes can be obtained (i.e. defining the concentration at which oocyst counts are inhibited). This minimum effective sporozonticidal concentration can then be used, along with the desired duration of cover, to model the dose necessary in man to achieve blocking of transmission. The caveats of this are the lack of metabolism, immunological factors and the feeding *via* a non-skin model.

5.6 Conclusions

Antimalarial drug discovery, historically, has focused on treating the asexual blood stages of the infection. With increased understanding of the parasite lifecycle and an availability of assays to measure infectivity, researchers are finally starting to have the tools that will allow a new era of antimalarial drug discovery: focused for the first time on exoerythrocytic stages. In particular, hitting the lifecycle where it hurts – at the key choke points – remains a compelling strategy. Only by addressing transmission between the infected human and the vector (and *vice versa*) can the moral imperative of eradication ever be realised.

Acknowledgements

The authors wish to thank colleagues at Medicines for Malaria Venture for their proof-reading and helpful discussions. In particular Dr Tim Wells, Dr David Waterson, Dr Xavier Ding, Dr Didier Leroy and Elizabeth Poll.

References

1. For example, M. Kaiser, D. H. Bray, A. Zommers, E. Couacy-Hymann, T. R. Gillespie, H. Ellerbrok and F. B. Leendertz, *Malaria Journal*, 2010, **16**(Suppl 2), O21.
2. G. H. Ball and J. Chao, *Experimental Parasitology*, 1976, **39**, 115.
3. E. E. Davies and R. E. Howells, *Trans. R. Soc. Trop. Med. Hyg.*, 1973, **67**, 20.
4. E. J. Gerberg, *Trans. R. Soc. Trop. Med. Hyg.*, 1971, **65**, 358.
5. W. Peters, *Chemotherapy and drug resistance in malaria*, Academic Press, London, 2nd edn, 1987, vols 1 & 2.
6. D. J. Pombo, G. Lawrence, C. Hirunpetcharat, C. Rzepczyk, M. Bryden, N. Cloonan, K. Anderson, Y. Mahakunkijcharoen, L. B. Martin, D. Wilson, S. Elliott, S. Elliott, D. P. Eisen, J. B. Weinberg, A. Saul and M. F. Good, *Lancet*, 2002, **360**, 610.
7. B. Franke-Fayard, D. Djokovic, M. W. Dooren, J. Ramesar, A. P. Waters, M. O. Falade, M. Kranendonk, A. Martinelli, P. Cravo and C. J. Janse, *Int. J. Parasitol.*, 2008, **38**, 1651.
8. C. P. A. Strome, P. L. DeSantis and R. L. Beaudoin, *In Vitro*, 1979, **15**, 531.
9. R. E. Sinden and J. E. Smith, *Trans. R. Soc. Trop. Med. Hyg.*, 1980, **74**, 134.
10. D. Mazier, I. Landau, P. Druihle, F. Miltgen, C. Guguen-guillouzo, D. Baccam, J. Baxter, J. P. Chigot and M. Gentilini, *Nature*, 1984, **307**, 367.
11. J. E. Smith, J. F. G. M. Meiss, J. P. Verhave, T. Ponnudurai and H. J. Moshage, *Lancet*, 1984, **2**, 757.
12. I. H. J. Ploemen, M. Prudêncio, B. G. Douradinha, J. Ramesar, J. Fonager, G.-J. van Gemert, A. J. F. Luty, C. C. Hermsen, R. W. Sauerwein, F. G. Baptista, M. M. Mota, A. P. Waters, I. Que, C. W. G. M. Lowik, S. M. Khan, C. J. Janse and B. M. D. Franke-Fayard, *PLoS One*, 2009, **4**, e7881.
13. D. Plouffe, A. Brinker, C. McNamara, K. Henson, N. Kato, K. Kuhen, A. Nagle, F. Adriín, J. T. Matzen, P. Anderson, T. Nam, N. S. Gray, A. Chatterjee, J. Janes, S. F. Yan, R. Trager, J. S. Caldwell, P. G. Schultz, Y. Zhou and E. A. Winzeler, *PNAS*, 2008, **105**, 9059.
14. M. R. Hollingdale, W. E. Collins, C. C. Campbell and A. L. Schwartz, *Am. J. Trop. Med. Hyg.*, 1985, **34**, 216.

15. R. S. Bray, *Studies on the exo-erythrocytic cycle in genus Plasmodium*, H. K. Lewis, London University, 1957.
16. W. Trager and J. B. Jensen, *Science*, 1976, **193**, 673.
17. L. A. Terzian, *Science*, 1947, **106**, 449.
18. R. E. Coleman, A. M. Clavin and W. K. Milhous, *Am. J. Trop. Med. Hyg.*, 1992, **46**, 169.
19. M. E. Smalley, *Trans. R. Soc. Trop. Med. Hyg.*, 1977, **71**, 526.
20. R. E. Sinden, *Ann. Trop. Med. Parasitol.*, 1982, **76**, 15.
21. J. W. Field, P. G. Shute, *The microscopic diagnosis of human malaria*, The Institute for Medical Research, Malaya, Kuala Lumpur, 1955, Study No. 24.
22. B. Hogh, R. Thompson, C. Hetzel, S. L. Fleck, N. A. A. Kruse, I. Jones, M. Dgedge, J. Barreto and R. E. Sinden, *Am. J. Trop. Med. Hyg.*, 1995, **52**, 50.
23. R. L. Hallett, C. J. Sutherland, N. Alexander, R. Ord, M. Jawara, C. J. Drakeley, M. Pinder, G. Walraven, G. A. T. Targett and A. Alloueche, *Antimicrob. Agents. Chemother.*, 2004, **48**, 3940.
24. M. E. Smalley, *Nature*, 1976, **264**, 271.
25. R. E. Sinden and M. E. Smalley, *Parasitology*, 1979, **79**, 277.
26. S. Chevalley, A. Coste, A. Lopez, B. Pipy and A. Valentin, *Malaria Journal,*, 2010, **9**, 49.
27. B. Coulibaly, A. Zoungrana, F. P. Mockenhaupt, R. H. Schirmer, C. Klose, U. Mansmann, P. E. Meissner and O. Muller, *PLoS One*, 2009, **4**, e5318.
28. I. I. M. D. Moelans, C. H. W. Klaassen, D. C. Kaslow, R. N. H. Konings and J. G. G. Schoenmakers, *Mol. Biochem. Parasitol.*, 1991, **46**, 311.
29. D. K. Raj, B. R. Das, A. P. Dash and P. C. Supakar, *Biochem. Biophys. Res. Commun.*, 2003, **309**, 685.
30. B. Czesny, S. Goshu, J. L. Cook and K. C. Williamson, *Antimicrob. Agents. Chemother.*, 2009, **53**, 4080.
31. M. Delves and R. E. Sinden, *Malaria Journal*, 2010, **9**, 35.
32. A. M. Talman, A. M. Blagborough and R. E. Sinden, *PLoS One*, 2010, **5**, e9156.
33. G. F. Medley, R. E. Sinden, S. Fleck, P. F. Billingsley, N. Tirawanchai and M. H. Rodriguez, *Parasitology*, 1993, **106**, 441.
34. M. D. Rodriguez, F. Zamudio, J. A. Torres, L. Gonzalezceron, L. D. Possani and M. H. Rodriguez, *Exp. Parasitol.*, 1995, **80**, 596.
35. R. W. Gwadz, D. Kaslow, J.-Y. Lee, W. L. Maloy, M. Zasloff and L. H. Miller, *Infection and Immunity*, 1989, **57**, 2628.
36. V. Kokoza, A. Ahmed, S. Woon Shin, M. Okafor, Z. Zou and A. S. Raikhel, *PNAS*, 2010, **107**, 8111.
37. G. H. Ball and J. Chao, *J. Parasitol.*, 1957, **43**, 409.
38. E. B. Al-Olayan, A. L. Beetsma, G. A. Butcher, R. E. Sinden and H. Hurd, *Science* 2002, **295**, 677.
39. M. Delves, D. Plouffe, C. Scheurer, S. Meister, S. Wittlin, E. A. Winzeler, R. E. Sinden and D. Leroy, 2011 submitted.

40. R. E. Sinden, A. Talman, S. R. Marques, M. N. Wass and M. J. E. Sternberg, *Curr. Opin. Microbiol.*, 2010, **13**, 491.
41. L. Gonzalez-Ceron, M. H. Rodriguez, F. Santillan, B. Chavez, J. A. Nettel, J. E. Hernandez-Avila and K. C. Kain, *Exp. Parasitol.*, 2001, **98**, 152.
42. C. M. McClean, H. G. Alvarado, V. Neyra, A. Llanos-Cuentas and J. M. Vinetz, *Am. J. Trop. Med. Hyg.*, 2010, **83**, 1183.
43. A. Ghosh, R. Dinglasan, H. Ikadai and M. Jacobs-Lorena, *Malaria Journal*, 2010, **9**, 194.
44. V. Bounkeua, F. Li and J. M. Vinetz, *Am. J. Trop. Med. Hyg.*, 2010, **83**, 1187.
45. S. Eksi, B. Czesny, G.-J. van Gemert, R. W. Sauerwein, W. Eling and K. C. Williamson, *Antimicrob. Agents. Chemother.*, 2007, **51**, 1064.
46. D. Dorin-Semblat, A. Sicard, C. Doerig, L. Ranford-Cartwright and C. Doerig, *Eukaryotic Cell*, 2008, **7**, 279.
47. L. Reininger, R. Tewari, C. Fennell, Z. Holland, D. Goldring, L. Ranford-Cartwright, O. Billker and C. Doerig, *J. Biol. Chem.*, 2009, **284**, 20858.
48. L. Reininger, O. Billker, R. Tewari, A. Mukhopadhyay, C. Fennell, D. Dorin-Semblat, C. Doerig, D. Goldring, L. Harmse, L. Ranford-Cartwritght, J. Packer and C. Doerig, *J. Biol. Chem.*, 2005, **280**, 31957.
49. L. McRobert, C. J. Taylor, W. Deng, Q. L. Fivelman, R. M. Cummings, S. D. Polley, O. Billker and D. A. Baker, *PLoS Biology*, 2008, **6**, e139.
50. R. R. Tewari, D. Dorin, R. Moon, C. Doerig and O. Billker, *Mol. Microbiol.*, 2005, **58**, 1253.
51. P. Ellekvist, J. Maciel, G. Mlambo, C. H. Ricke, H. Colding, D. A. Klaerke and N. Kumar, *PNAS*, 2008, **105**, 6398.
52. T. Ishino, Y. Oriton, Y. Chinzei and M. Yuda, *Mol. Microbiol.*, 2006, **59**, 1175.
53. I. Siden-Kiamos, A. Ecker, S. Nyback, K. Louis, R. E. Sinden and O. Billker, *Mol. Microbiol.*, 2006, **60**, 1355.
54. I. I. Rupp, R. Bosse, T. Schirmeister and G. Pradel, *Mol. Biochem. Parasitol.*, 2008, **158**, 208.
55. H. Zieler, D. B. Keiste, J. A. Dvorak and J. M. C. Ribeiro, *J. Exp. Biol.*, 2001, **204**, 4157.
56. V. Carter and H. Hurd, *Trends Parasitol.*, 2010, **26**, 582.
57. S. Shimizu, Y. Osada, T. Kanazawa, Y. Tanaka and M. Arai, *Malaria Journal*, 2010, **9**, 73.
58. L. Lucantoni, R. Yerbanga, G. Lupidi, L. Pasqualini, F. Esposito and A. Habluetzel, *Malaria Journal*, 2010, **9**, 66.
59. S. Kumar, A. Molina-Cruz, L. Gupta, J. Rodrigues and C. Barillas-Mury, *Science*, 2010, **327**, 1644.
60. S. Enosse, G. A. Butcher, G. Margos, J. Mendoza, R. E. Sinden and B. Hogh, *Trans. R. Soc. Trop. Med. Hyg.*, 2000, **94**, 77.
61. R. E. Fowler, R. E. Sinden and M. Pudney, *J. Parasitol.*, 1995, **81**, 452.
62. P. Gautret and A. Motard, *Parasite*, 1999, **6**, 103.
63. W. E. Collins, M. C. W. Warren, J. C. Skinner, J. B. Richardson and T. S. Kearse, *J. Parasitol.*, 1977, **63**, 57.

64. U. D'Alessandro, B. O. Olaleye, W. McGuire, P. Langerock, S. Bennett, M. K. Aikins, M. C. Thomson, M. K. Cham, B. A. Cham and B. M. Greenwood, *Lancet*, 1995, **345**, 479.
65. http://www.mmv.org/research-development/essential-information-scientists/target-product-profiles
66. J. Guan, X. Wang, K. Smith, A. Ager, M. Gettayacamin, D. E. Kyle, W. K. Milhous, M. P. Kozar, A. J. Magill and A. J. Lin, *J. Med. Chem.*, 2007, **50**, 6226.
67. S. K. Puri and G. P. Dutta, *Trans. R. Soc. Trop. Med. Hyg.*, 1990, **84**, 759.
68. T. N. Wells, J. N. Burrows and J. K. Baird, *Trends Parasitol.*, 2010, **26**, 145.
69. L. H. Schmidt, *Antimicrob. Agents Chemother.*, 1983, 615.
70. L. Dembele, A. Gego, A. M. Zeeman, J. F. Franetich, O. Silvie, A. Rametti, R. Le Grand, N. Dereuddre-Bosquet, R. Sauerwein, G. J. van Gemert, J. C. Vaillant, A. W. Thomas, G. Snounou, C. H. Kocken and D. Mazier, *PLoS One*, 2011, **6**, e18162.
71. R. Chattopadhyay, S. Velmurugan, C. Chakiath, D. L. Andrews, W. Milhous, J. W. Barnwell, W. E. Collins and S. L. Hoffman, *PLoS One*, 2010, **5**, e14275.
72. E. Beutler and S. Duparc, *Am. J. Trop. Med. Hyg.*, 2007, **77**, 779.
73. A. Mwakingwe, L.-M. Ting, S. Hochman, J. Chen, P. Sinnis and K. Kim, *J. Inf. Diseases*, 2009, **200**, 1470.
74. F. Castelli, S. Odolini, B. Autino, E. Foca and R. Russo, *Pharmaceuticals*, 2010, **3**, 3212.
75. A. M. Croft, F. A. Jacquerioz and K. L. Jones, *Human Parasitic Diseases*, 2010, **2**, 1.
76. J. Sahi, S. Grepper and C. Smith, *Curr. Drug Disc. Technol.*, 2010, **7**, 188.
77. C. Emoto, S. Murase, Y. Sawada, B. C. Jones and K. Iwasaki, *Drug Metabolism and Pharmacokinetics*, 2003, **18**, 287.
78. H. Glaeser, M. F. Fromm and J. Koenig, *Methods and Principles in Medicinal Chemistry*, 2008, **38**, 341.
79. G. G. A. Butcher, *Int. J. Parasitol.*, 1997, **27**, 975.
80. A. E. Kiszewski, *Pharmaceuticals*, 2011, **4**, 44.
81. A. Kuehn and G. Pradelle, *J. Biomed. Biotechnol.*, 2010, **2010**, 975827.
82. D. A. Baker, *Mol. Biochem. Parasitol.*, 2010, **172**, 57.
83. J. M. Moore, N. Kumar, L. D. Schultz and T. V. Rajan, *J. Exp. Med.*, 1995, **181**, 2265.
84. I. Angulo-Barturen, M. B. Jimenez-Diaz, T. Mulet, J. Rullas, E. Herreros, S. Ferrer, E. Jimenez, A. Mendoza, J. Regadera, P. J. Rosenthal, I. Bathurst, D. L. Pompliano, F. Gomez de las Heras and D. Gargallo-Viola, *PloS One*, 2008, **3**, e2252.
85. M. Diallo, A. M. Toure, S. F. Traore, O. Niare, L. Kassambara, A. Konare, M. Coulibaly, M. Bagayogo, J. C. Beier, R. K. Sakai, Y. T. Toure and O. K. Doumbo, *Malaria Journal*, 2008, **7**, 248.

86. For example see the example of the timing of primaquine dosing: S. Lawpoolsri, E. Y. Klein, P. Singhasivanon, S. Yimsamran, N. Thanyavanich, W. Maneeboonyang, L. L. Hungerford, J. H. Maguire and D. L. Smith, *Malaria Journal*, 2009, **8**, 159.
87. T. Bousema, L. Okell, S. Shekalaghe, J. T. Griffin, S. Omar, P. Sawa, C. Sutherland, R. Sauerwein, A. C. Ghani and C. Drakeley, *Malaria Journal*, 2010, **9**, 136.
88. A. L. Ouedraogo, T. Bousema, P. Schneider, S. J. de Vlas, E. Ilboudo-Sanogo, N. Cuzin-Ouattara, I. Nebie, W. Roeffen, J. P. Verhave, A. J. F. Luty and R. Sauerwein, *PloS One*, 2009, **4**, e8410.

CHAPTER 6

Drugs for Kinetoplastid Diseases – Current Situation and Challenges

SIMON L. CROFT

Faculty of Infectious and Tropical Diseases, London School of Hygiene & Tropical Medicine, London WC1E 7HT, UK

6.1 Introduction

The human diseases leishmaniasis, human African trypanosomiasis and Chagas disease are caused by related protozoan parasites, often referred to as kinetoplastids or trypanosomatids. The protozoa which cause the diseases are closely related. They share structural and biochemical features, for example, a single mitochondrion with a discrete structured DNA body – the kinetoplast, specific organelles for glycolysis – the glycosomes, sub-pellicular microtubular corset, and a specific thiol metabolism. Their biological relationship has recently been given a new focus by the publication of the genome sequences and subsequent analysis of metabolic pathways.[1] These important studies have provided the basis for new understanding about the pathogenesis of disease[2] and with relevance to this series of papers, to further description of biochemical/molecular pathways that might serve as potential drug targets.[1] These three diseases also fall under the label "neglected diseases" and continue to have limited funding for discovery and development, delivery of new tools, and a low profile on the agenda.[3,4]

RSC Drug Discovery Series No. 14
Neglected Diseases and Drug Discovery
Edited by Michael J. Palmer and Timothy N. C. Wells

Published by the Royal Society of Chemistry, www.rsc.org

Most of the drugs used for treatment of the leishmaniases, Chagas disease and HAT are fraught with problems of toxicity, variable efficacy, often needing to be injected, and need a long period of treatment. For HAT, Chagas disease and cutaneous leishmaniasis, there are only a few drugs or treatments in clinical development, though several are in pre-clinical development (see Chapter 7). In contrast, for visceral leishmaniasis there has been some progress, with liposomal amphotericin B, miltefosine and paromomycin all becoming available in India. Importantly, single-dose liposomal amphotericin B and the co-administration of the above three drugs show potential for both treatment and control of visceral leishmaniasis in the Indian subcontinent. Although considerable advances in the identification, validation and characterization of drug targets have come with the completion of the genomes for *Trypanosoma cruzi, Trypanosoma brucei* and *Leishmania major,* and new tools such as RNAi[5,6] have been developed, this is only one early part of the long and complex process of drug discovery and development. New and established pharmacophores, based upon synthetic and natural product chemistry, have also to be put into place along with improved screening technologies to identify hits.[7] Fortunately, large screening programmes in industry and in institutes and academia are now in place (see www.novartis.com; www.pasteur.or.kr), as is described by Jacobs in Chapter 7). Appropriate predictive models of infection and pharmacokinetic studies to evaluate leads also need to be integrated into the process. The support of public–private partnerships, for example, the Drugs for Neglected Diseases Initiative (www.dndi.org) over the past decade and improved interaction with and commitment of the private sector in the past 5 years offers hope for the future. Development will depend upon the expertise in lead optimization, toxicology and pharmacology available in the pharmaceutical and biotech sectors and contract research organizations, with the associated requirement for high levels of funding. Clinical trials capacity is also required and is being addressed in India, Africa (see www.stphi.ch; www.dndi.org) and South America. The issues around drug production, delivery and sustainability need to be addressed, especially if control programmes are successful and patient numbers decrease.

6.2 Leishmaniasis

Leishmaniasis as a disease is endemic in 98 countries across Asia, Africa, South and Central America and southern Europe. It is a disease complex with two major manifestations, visceral leishmaniasis (VL) and cutaneous leishmaniasis (CL), as well as a number of other mucocutaneous and diffuse forms. The *Leishmania* parasite is transmitted by female phlebotomine sandflies, in which the flagellated promastigote form divides and develops into an infective metacyclic form (Figure 6.1). In the mammalian host the amastigote form survives and multiplies in the phagolysosomal compartment of various macrophage populations. There are around half a million cases annually of visceral leishmaniasis and 1.5 million cases of cutaneous disease.

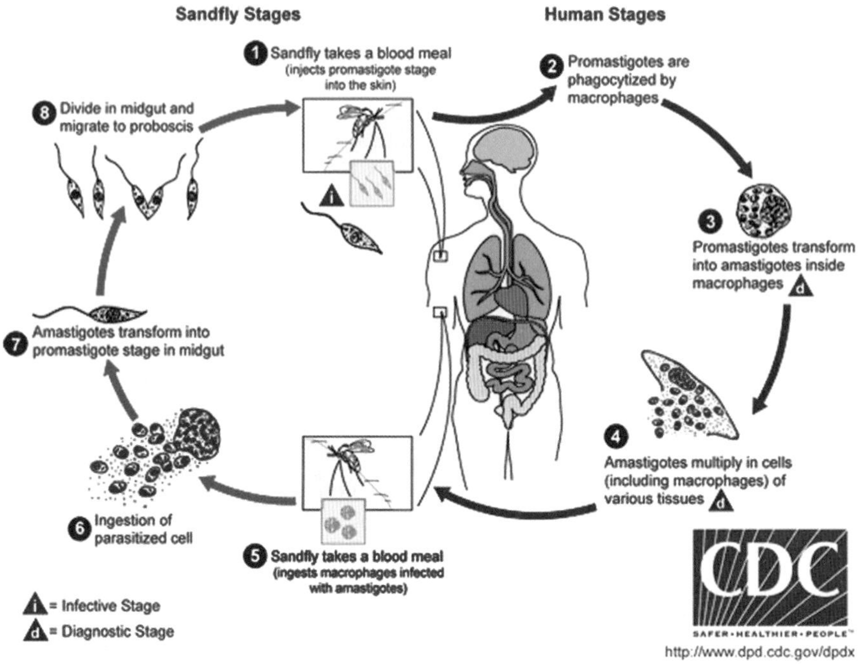

Figure 6.1 Life cycle for leishmaniasis (reproduced from the CDC website).

There have been significant differences in progress and approaches to drug development for VL and CL and these manifestations will be discussed separately. For drug discovery and development, there are four key issues about the parasite to consider. First, the amastigote form of the parasite is located in the low pH phagolysosomal compartment of different macrophage populations. Second, different pharmacokinetic properties are required for the target organs of the visceral disease (liver, spleen, bone marrow) compared to the skin in the cutaneous disease. Third, there are seventeen species of *Leishmania* which cause disease in humans that can be distinguished by molecular markers but also have significantly different drug sensitivities.[8] Finally, leishmaniasis may occur where there is severe immune suppression and where the drugs are less effective.

6.2.1 Visceral Leishmaniasis

Visceral leishmaniasis is caused by different species of *Leishmania*: *L. donovani* and *L. infantum* (equivalent to *L. chagasi* in South America), that can be clearly differentiated.[9] It is potentially fatal, with different pathologies associated with each species. *L. donovani* even shows different pathologies in India and Sudan.[10] This is all further complicated by the observations that some strains of *L. infantum* can cause cutaneous leishmaniasis and that post-treatment, some

L. donovani infected patients develop a diffuse cutaneous form termed post kala-azar leishmaniasis (PKDL).

Pentavalent antimonials, sodium stibogluconate (Pentostam, GlaxoSmithKline) and meglumine antimoniate (Glucantime, Sanofi-Aventis)[11] have been the standard of care for over 60 years (Table 6.1). Their mechanism of action is poorly understood, but they have been reported to inhibit a wide variety of cellular processes, including DNA replication (toposiomerase) and cell signalling (tyrosine phosphatases). They have many drawbacks: sodium stibogluconate needs to be administered by slow intravenous infusion, which is problematic, since the treatment is given at 20 mg/kg/day over 30 days. In addition, these medicines are almost useless in the key endemic area in the Indian state of Bihar due to drug resistance.[12] The introduction of a generic brand of sodium stibogluconate has reduced costs,[13] but these are still expensive treatments. Médicins sans Frontiers estimates a treatment of a 35 kg patient with Glucantime costs US$49. The same treatment with Pentostam costs US$155, and US$27 with generic sodium stibogluconate (www.essentialdrugs.org/edrug/archive/200705/msg00062.php).

Amphotericin B is an antifungal medicine, originally identified in extracts of *Streptomyces nodosus*. Its mechanism of action is similar to an anti-fungal. It forms complexes with the cell membrane sterol ergosterol, leading to the production of channels, which cause monovalent ion leakage and cell death. It is normally considered a second-line drug, except in Bihar State, India, because of the loss of effectiveness of antimonial drugs. Although a number of amphotericin B lipid formulations, developed during the 1980s for treatment of systemic mycoses in immunocompromised patients, have proved effective in the treatment of visceral leishmaniasis only one of these, the liposomal formulation AmBisome, has become a standard treatment. It is registered in various countries and its use prescribed by a WHO working group.[14] Recently, a single course therapy of 10 mg/kg has been shown to cure 95% of patients in India.[15] Again, price is an issue: the WHO has negotiated a price with the producers (Gilead Sciences) of $18 per 50 mg ampoule compared with the commercial price of around $150. However this still leads to a treatment cost of over $250 for a 35 kg patient treated at 20 mg/kg or $125 for the lower dose.[16] In addition there are still a large number of reported adverse events,[15] and temperature stability above 25 °C remains an issue. Research to identify other amphotericin B formulations that might overcome these problems is in progress.

A recurring issue around all the visceral leishmaniasis treatments is regional differences in response rates. A study over 10 years ago suggested that AmBisome was most effective in treatment of visceral leishmaniasis patients in India, less so in East Africa, and even less so against *L. infantum* (*L. chagasi*) in South America.[17] Clinical studies to confirm the efficacy of AmBisome are underway in East Africa (www.dndi.org).

The aminoglycoside antibiotic paromomycin (aminosidine, monomycin, identified from *Streptomyces krestomuceticus*) acts as an antibacterial by binding to the 16S ribosomal RNA, and preventing protein synthesis. It has

Table 6.1 Available treatments and compounds in trials for visceral leishmaniasis.

Structure	*Name*	*Properties*	*Comment*
Available treatments			
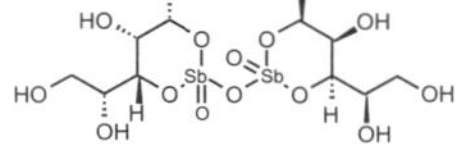 (representative structure) 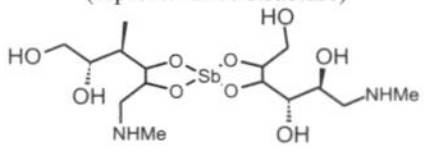 (representative structure)	Sodium stibogluconate (Pentostam and SSG) Meglumine antimoniate (Glucantime)	Organo-metallic complexes, polymeric forms. Pentostam contains around 33% pentavalent antimony, Glucantime around 28% pentavalent antimony. Administered by intravenous or intramuscular routes.	Generic sodium stibogluconate (often referred to as SSG in the literature) from Albert David (India) has made treatment cheaper. Sanofi-Aventis have reduced their price for Glucantime
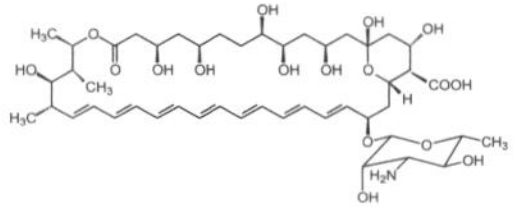	Amphotericin B (Fungizone) Liposomal amphotericin B (AmBisome)	Polyene antibiotic, fermentation product of Streptomyces nodus, intravenous Unilamellar liposome, intravenous	Other polyenes, for example nystatin, also have proven antileishmanial activity. Proved to be most effective lipid formulation for VL and available at \$18/50 mg ampoule via WHO.
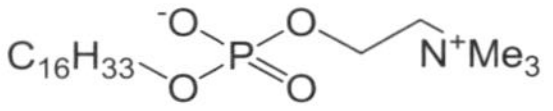	Miltefosine	Hexadecylphosphocholine, oral	First oral drug for VL. Contraindicated in pregnancy as found to be teratogenic in rats.

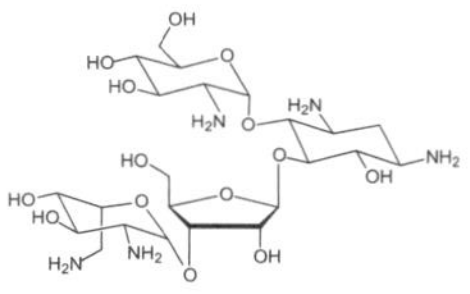	Paromomycin	Aminoglycoside (also known as aminosidine and monomycin in the literature), fermentation product of *Streptomyces rimosus*. Supplied as sulphate. Intramuscular administration.	Completed phase IV studies in India as monotherapy. Completed phase III studies in East Africa as monotherapy and in co-administration with SSG.
Compounds in clinical trials			
(see above)	Amphotericin B	Lipid formulations	Other lipid formulations, including Abelcet, Amphocil and multi-lamellar liposomes have been in clinical studies. Studies on a lipid emulsion (ABLE) reported in 2009
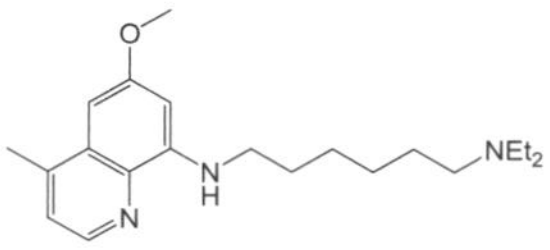	Sitamaquine	Other members of the 8-aminoquinolines are currently oral, and are currently being developed to prevent relapse of the dormant liver stage of the malarial parasite *Plasmodium vivax*	Completed phase II trials, but only shows 85% efficacy.

been known to possess anti-leishmanial activity for over 50 years; other aminoglycosides have markedly less activity. Since the 1980s, when an injectable formulation was first shown to have a curative effect in visceral leishmaniasis, paromomycin has moved slowly through clinical trials with WHO/TDR (the WHO, World Bank and UN Development Special Programme for Research and Training in Tropical Diseases) in the 1990s and Institute of One World Health in the 2000s. An extensive study by the Institute of One World Health in India showed 94% efficacy after intramuscular injection of 15 mg/kg every day for 21 days, in phase III clinical trials in India.[18] Following this, paromomycin was registered for visceral leishmaniasis in India in 2006. Despite the low cost (estimated to be $10 per course) the drug has not yet become part of the treatment options for visceral leishmaniasis patients in the subcontinent, although a phase IV trial in 2008–09 indicated a cure rate of 94.2% at 6 months post-treatment (P. Desjeux, personal communication). Trials conducted by DNDi in East Africa, showed reduced efficacy, with 15 mg/kg giving a $>50\%$ cure rate, and the increased dose of 20 mg/kg for 21 days gave only a 85% cure rate. This is insufficient for consideration as a monotherapy.[19] The reason for this geographical difference in sensitivity is not understood.

The first oral treatment of leishmaniasis was the phospholipid derivative, miltefosine (1-*O*-hexadecylphosphocholine, Table 6.1) that was first identified at the Wellcome Laboratories, UK[20,21] in the 1980s. It is a membrane-active synthetic lipid analogue developed separately for the treatment of cutaneous metastasis from mammary carcinomas. Its mechanism of action against *Leishmania* is not clear. It showed 94% efficacy in adults and children in clinical trials supported by WHO/TDR and Zentaris.[22] It was registered for treatment of visceral leishmaniasis in India in 2002. Miltefosine was the first anti-leishmanial to undergo phase IV studies,[23] and was incorporated into the visceral leishmaniasis elimination programme for the sub-continent. The treatment is complicated by potential teratogenicity, which means that women of child-bearing age require contraception that must be maintained for 3 months post-treatment due to the long half-life of the drug. Also, the four weeks of oral daily treatment leads to poor compliance, which gives a potential for relapses and the potential selection of resistant forms.[24,8]

Sitamaquine (WR6026), an 8-aminoquinoline, has undergone extensive clinical trials as an oral treatment for visceral leishmaniasis. GSK first in-licensed sitamaquine from the United States Army Medical Research and Material Command in the late 1990s, with the view that sitamaquine would provide an accessible treatment for patients as an affordable, oral monotherapy option, representing a significant advance on available therapies. However, the subsequent clinical development programme demonstrated that sitamaquine provided only 85% efficacy in monotherapy in India and East Africa (Phase 2b)[25,26] as part of a 21-day dosing regimen and would require weight-based dosing and Glucose-6-Phosphate-Dehydrogenase (G6PD) testing prior to therapy. Therefore, as therapies for visceral leishmaniasis have evolved over the past decade it now appears sitamaquine will not offer the significant advances originally hoped for.[27] The mechanism of action of the 8-aminoquinolines is not known. However, other members of this family are active in experimental models of visceral leishmaniasis, including tafenoquine, which is currently in phase II development for *P. vivax* malaria with Medicines for Malaria Venture and GSK,[26] and NPC-1161B from the University of Mississippi, which is in pre-clinical development.[25] Other molecules targeting visceral leishmaniasis are in the discovery and early clinical development phases are discussed by Jacobs in this volume (Chapter 7).

Drug combinations have proved to be a successful strategy to shorten the course of therapy, reduce toxicities through lower dosage and reduce the selection of resistant mutations for several infectious diseases, most notably malaria and tuberculosis. Although co-formulation is not possible for visceral leishmaniasis drugs, a strategy of co-administration (either concomitant or sequential) of available anti-leishmanial drugs has been pursued. DNDi developed a programme of potential combinations in 2005, following on from experimental studies,[28] to provide efficacy and safety data. These studies have shown that, for Indian visceral leishmaniasis, sequential co-administrations of AmBisome (single dose, intravenous) with either miltefosine (7 days oral) or paromomycin (10 days intramuscular) or a concomitant administration of 10 days oral miltefosine and intramuscular paromomycin can achieve a 98% cure rate.[27] This important reduction in treatment time (30 days compared to potential 8 days) is important for both patient treatment and control programmes with benefits to compliance[29] and cost.[16] In Sudan, further to experimental studies[30] and clinical studies by Médecins Sans Frontières,[31] a trial on a combination of sodium stibogluconate and paromomycin (17 days) has recently been concluded with DNDi and the Leishmaniasis East Africa Platform. The Indian studies have not included antimonials in the co-administrations, due to obvious reasons around resistance.

The post-treatment manifestation, post kala-azar leishmaniasis (PKDL), remains a poorly understood manifestation of visceral leishmaniasis, with parasite persistence in the face of an immunological response.[32] Again, there are differences in patient profile between Indian and Sudanese PKDL in relation to frequency, age profile and rate of self cure.[33] Whether this phenomenon is related to specific types of drug treatment is not clear.[34] Treatment based upon long courses with antimonial medication and small studies with miltefosine and AmBisome have been described.[11] The potential of an immunotherapeutic approach, using antimonial drug plus vaccine plus BCG (attenuated Bacillus Calmette-Guérin as an immune stimulus), showed higher cure rate than drug alone[35] and there are further opportunities for research in this area.[36]

6.2.2 HIV/Leishmaniasis Co-Infections

Since the first reported case of HIV and visceral leishmaniasis in 1985, 35 countries have reported co-infections. There are increasing numbers of cases in East Africa, especially in northeastern Ethiopia, where 23% of visceral leishmaniasis cases are co-infections. Various treatment regimes with all standard drugs have been described to treat these co-infection cases, with relapse rates as high as 85%. This has been recently documented by Alvar *et al.*,[37] who state: "HIV infection increases the risk of developing VL by 100–2320 times in areas of endemicity, reduces the likelihood of a therapeutic response and greatly increases the probability of relapse." There is no successful current therapy for patients with co-infections. Different policies are followed by countries; some have adopted a regime of treatment with anti-retrovirals followed by treatment with anti-leishmanial drugs. Other countries have adopted a policy of maintenance therapy, for example in Southern Europe, where lipid amphotericin B formulations are often used.

6.2.3 Cutaneous Leishmaniasis (CL)

Cutaneous leishmaniasis exists in many different forms.[38] In most cases, patients self-cure after 6–18 months of infection, leaving scarred tissue. As most lesions are on exposed skin on the face, arms and legs, it is a stigmatizing disease.[39] In comparison to visceral leishmaniasis there are limited proven options for the cutaneous disease (see Table 6.2). Pentavalent antimonials have proved inconsistent in their effectiveness across the different *Leishmania* species, and pentamidine and amphotericin B are limited to specific types of cutaneous leishmaniasis.[11] Paromomycin in various topical formulations has variable efficacy,[40,41] and there is a continuing search for more effective and less irritant topical creams and gels. A recent formulation of 12% paromomycin, containing also gentamicin (another aminoglycoside antibiotic) and surfactants, showed efficacy in *L. major* cutaneous leishmaniais in Tunisia.[42] Oral miltefosine also has some variable, species-dependent effectiveness[43–45] and is registered for this indication in Colombia.

However, there are questions over the reliability of data on treatments for cutaneous leishmaniasis, and an absence of randomized placebo controlled trials.[38,46] Standardized clinical protocols have been proposed, so that clinical trial results can be compared.[47] The key questions include whether to use patients with new or old lesions, and whether the endpoint should be a resolution of the lesion or complete re-epithelialisation. Two recent Cochrane analyses concluded that most clinical studies could not be included in the analysis as they did not meet the correct standards. For *L. major* and *L. tropica* infections in the Old World, some evidence for the activity of antifungal azoles, fluconazole and itraconazole was shown,[48] whilst in the New World, in addition to antimonials, miltefosine and ketoconazole, and oral allopurinol showed activity in a limited number of trials.[49] A more recently developed triazole, posoconazole, which has potential for treatment of Chagas disease, has activity against cutaneous leishmaniasis in experimental models[50] and also in one patient.[51] The immune system plays a key role in the resolution of cutaneous leishmaniasis. One of the high-quality trials from the Cochrane analyses showed a positive effect of the anti-inflammatory drug pentoxyphylline, as adjunct therapy. As one of the aims of treatment of cutaneous leishmaniasis is

Table 6.2 Available treatments and compounds in trials for cutaneous leishmaniasis.

Structure	Name	Properties	Comment
Available treatments			
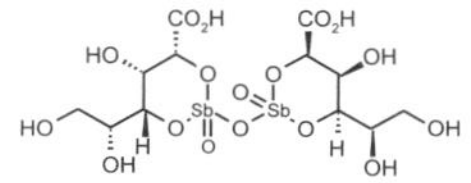 (representative structure) 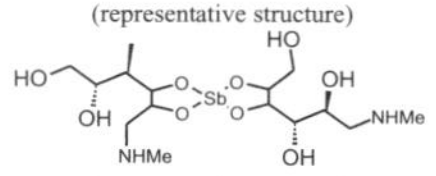 (representative structure)	Sodium stibogluconate (Pentostam), first line treatment. Meglumine antimoniate (Glucantime)	See above	Variable susceptibilty of different *Leishmania species* that cause CL.
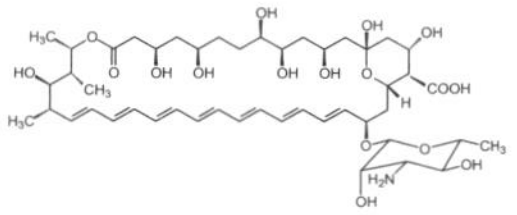	Amphotericin B (Fungizone)	See above	Important for mucocutaneous and other specific forms in South America
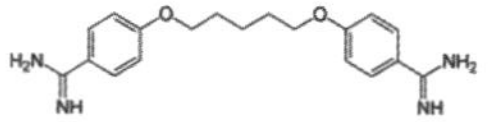	Pentamidine	Diamidine, as isethionate salt. Intramuscular	For specific forms in South America only
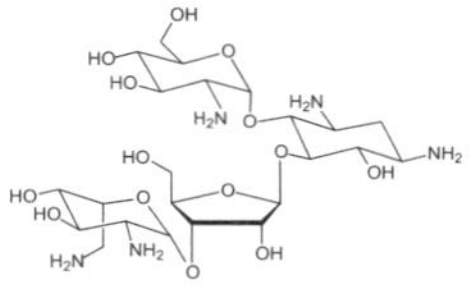	Paromomycin	See above for paromomcyin. Topical formulation containing methyl benzythonium chloride.	

Table 6.2 (*Continued*)

Structure	*Name*	*Properties*	*Comment*
Compounds in clinical trials			
(see above)	Paromomycin	A topical formulation comtaining gentamycin and surfactants	Being tested clinically in a trial against *L. major* infection
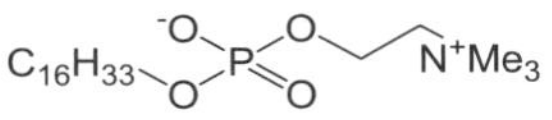	Miltefosine	Hexadecylphophocholine, oral	Registered in Colombia, but undergoing clinical studies against various forms of CL. There is variation in species susceptibility.
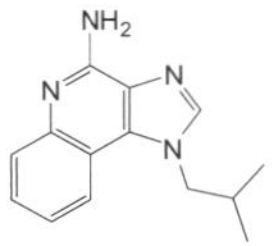	Imiquimod (topical immunomodulator, Phase II)	Imidazoquinoline, used as a topical treatment	Randomized Double Blind Clinical Trial of Imiquimod or placebo used in combination Glucantime (an antimonial) was completed in an 80 patient phase III in Peru in 2006. It is ineffective though in Iran.
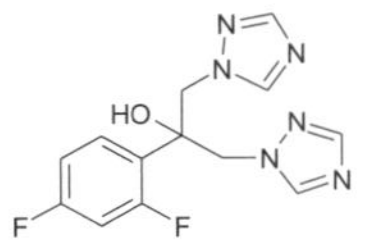 (Fluconazole shown as representative structure)	Fluconazole, intraconazole, ketaconazole		Studies against *Leishmania major*

to accelerate self-cure,[40] approaches including immunomodulators have been tried for many years, from the attenuated Bacillus Calmette-Guérin (BCG), the adjuvant TDM (trehalose-6,6'-dimycolate) (Cord Factor) and the TLR7 agonist imiquimod. Studies on cutaneous leishmanisasis patients in Peru showed 75% cure for imiquimod plus antimonials compared to 58% for antimonials alone.[52] There have also been extensive studies using BCG with pentavalent antimonials in Venezuela.[53] The need for a new drug discovery and development agenda for cutaneous leishmaniasis has been raised, and target product profiles produced for key forms of CL caused by *L. tropica* and *L. braziliensis*.[46]

6.3 Human African Trypanosomiasis

Human African trypanosomiasis (HAT), also known as sleeping sickness, is restricted to the regions of sub-Saharan Africa where the vector, the tsetse fly, is present (Figure 6.2). HAT is caused by two parasites. *Trypanosoma brucei gambiense*, which causes about 95% of reported infections and is essentially anthropnotic and is distributed in West and Central Africa. *Trypanosoma brucei rhodesiense*, is found in East and South Africa and is a zoonotic infection, infecting wild ungulates and cattle. The disease caused by *T. b. gambiense* is chronic whereas that caused by *T. b. rhodesiense* is acute. Recent estimates from WHO are that, following improved surveillance, numbers of cases were

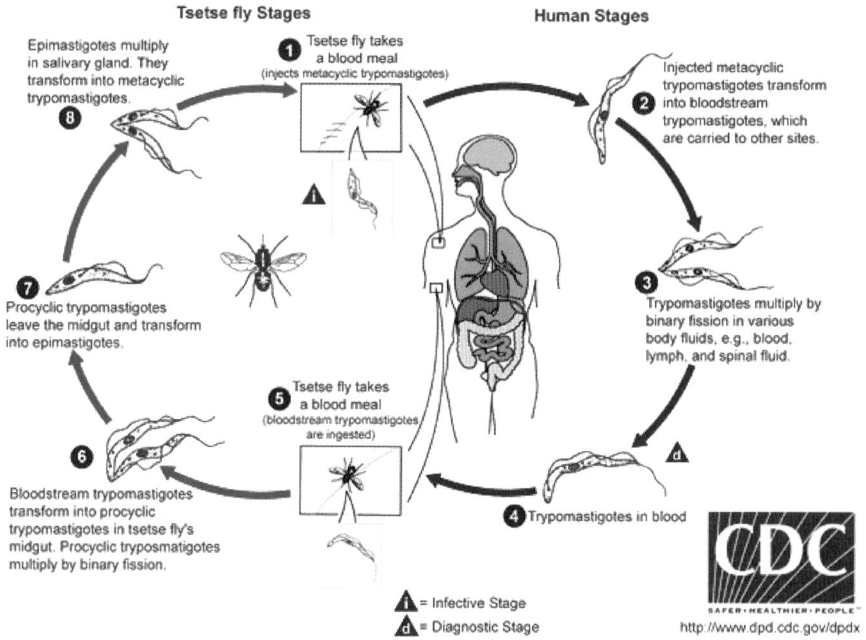

Figure 6.2 Life cycle for human African trypanosomiasis (reproduced from the CDC website).

down to less than 10,000 per annum in 2009 for the first time in 50 years (www.who.int). The disease progresses through a haemolymphatic stage (also called early or stage 1 disease) to a CNS infection (late or stage 2 disease). Most patients come for treatment with second-stage disease, often following a relatively asymptomatic first stage. The second-stage CNS infection is accompanied by an immunopathological response in the brain and a breakdown of neurological function. Although the bloodstream trypomastigote form is well characterized in relation to biochemistry, molecular biology and immunogenicity, little is known about the form and biochemical status of the parasite in the CNS.

There are three key factors about the biology of this parasite that impinge on drug discovery and development programmes. First, that the two subspecies that infect humans may have different drug sensitivities, as exemplified by eflornithine.[54] Second, most of the activity data are obtained using the bloodstream trypomastigote, and this may not be relevant to the very poorly characterized CNS form. Third, for stage 2 disease, drugs must be able to cross the blood–brain barrier, although little is known about the integrity of this barrier during the disease. Improved diagnostic methods could have a profound effect on treatment.[55] There have been significant advances in our understanding of the molecular biology, genome, the use of molecular techniques to identify targets and mechanisms[5,6] and the basis of antigenic variation of these pathogens[56] over the past two decades. However, significant advances in our understanding of the biology of *Trypanosoma brucei* have not yet led to new drugs.

The current drugs for treatment of both haemolymphatic and CNS infections are listed in Table 6.3. An excellent summary of the history and mechanisms of action of these drugs is given in Barrett *et al.*[57] In brief, all current drugs have issues of toxicity and require parenteral routes of administration, not suited to resource-poor settings. Suramin and pentamidine are only useful for stage 1 disease, whilst melarsoprol and eflornithine are recommended for treatment of stage 2 disease. There are reports of the increased incidence of treatment failure with melarsoprol in some areas[58,59] although the cause of this is not known. Eflornithine is a mechanism-based inhibitor of the enzyme ornithine decarboxylase. Since the registration of the ornithine decarboxylase inhibitor, eflornithine, in 1990 for late stage *gambiense* disease there have been few clinical advances. The lower incidence and severity of adverse effects of eflornithine when compared to melarsoprol, led some to advocate that this drug should become the first-line treatment for late stage HAT.[60,61] The requirement for high doses and prolonged intravenous infusion, however, make the drug expensive and difficult to distribute and administer in rural Africa. Its availability as a trypanocide is dependent upon commitments to supply made by Médecins Sans Frontières, the World Health Organisation and Sanofi-Aventis.

There have been two approaches that have led to improvement of the treatment for *T. b. gambiense* stage 2 infections using currently registered drugs. Firstly, pharmacokinetic studies of melarsoprol have enabled the dose to be shortened from 3-day dosing with 1 week intervals spread over 21–35 days,

Table 6.3 Available treatments and in trials for human African trypanosomiasis and Chagas disease.

Structure	*Name*	*Properties*	*Comment*
Human African trypanosomiasis			
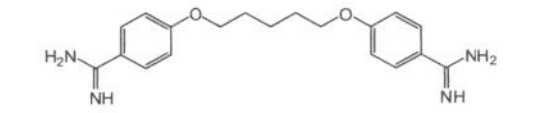	Pentamidine	Haemolymphatic stage, first line drug. Diamidine, as suphate salt. Intramuscular.	Mainly for *T. b. gambiense* infections
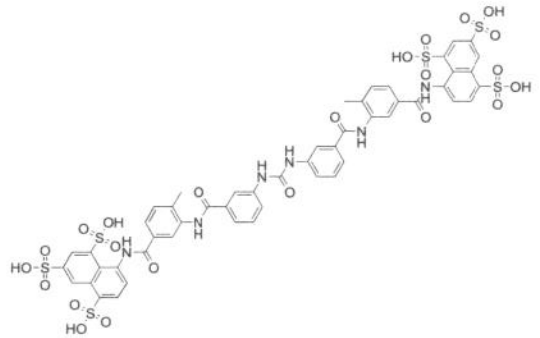	Suramin	As above. Sulfated naphthylamine. Injection.	Mainly for *T. b. rhodesiense* infections
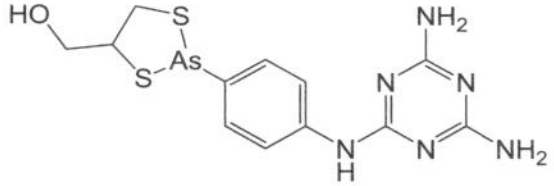	Melarsoprol	CNS stage, first-line drug.Organometal, formulated in 3.6% propylene glycol, intravenous	Rapidly metabolized to melarsen oxide which is active in CNS infection
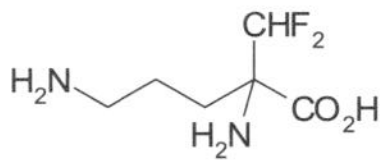	Eflornithine	As above, α-difluoromethylornithine, intravenous	*T. b. gambiense* is sensitive, *T. b. rhodesiense* is innately insensitive. Parenteral administration and high doses make it less than ideal.
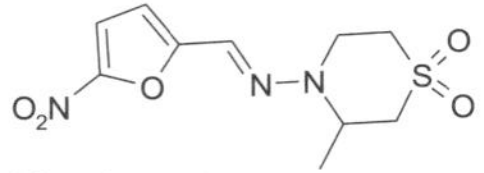 (Nifurtimox)	Nifurtimox and Eflornithine	As above, co-administration, details as above and below	Co-administration, included in WHO recommended treatment list
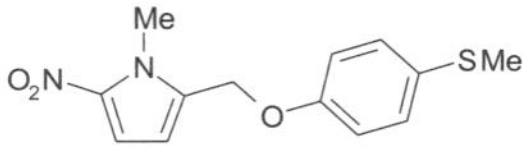	fexinidazole	5-nitroimidiazole, oral	Phase I clinical trials. Rapidly metabolized to active sulfoxide and sulphone derivatives

Table 6.3 (*Continued*)

Structure	*Name*	*Properties*	*Comment*
Chagas disease			
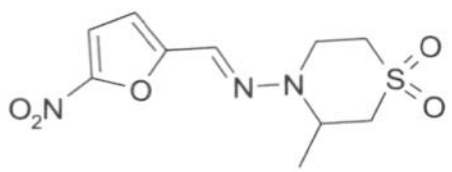	nifurtimox	First-line drug, acute stage. Nitrofuran, oral	
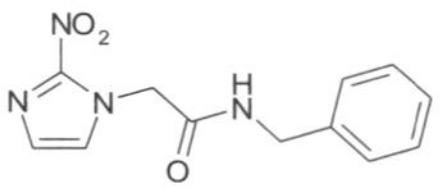	benznidazole	First line drug, acute stage. Nitroimidazole, oral	
(see above)	benznidazole	As above	In clinical studies for treatment of indeterminate/chronic disease (BENEFIT trial)
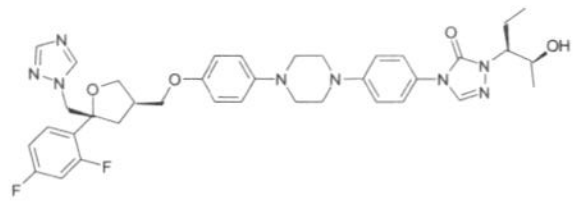	posoconazole	Triazole, oral	Long plasma T 1/2 and large volume of distribution suggest good properties for chronic disease. Trials planned; one successful treatment of chronic case reported in 2010

down to a shortened 10-day course, with no evidence of accumulation.[62] The new short regime, with similar efficacy and toxicity profile to the earlier regime, improves patient compliance and reduces hospital costs.[63] Secondly, studies aimed to modify dosing with eflornithine have also been undertaken, with clinical studies on co-administrations of eflornithine plus melarsoprol[64] or nifurtimox.[65] Nifurtimox, a nitroherocyclic oral drug more commonly associated with the treatment of Chagas disease, was first used for treatment of HAT with limited efficacy in the 1980s.[66] In clinical studies conducted by a partnership of Médecins Sans Frontières and the Drugs for Neglected Disease Initiative over the last few years, the co-administration of an intravenous infusion of 200 mg/kg eflornithine every 12 hours for 7 days plus nifurtimox, oral, 3 times/day for 10 days, was shown to be as effective and safe as eflornithine monotherapy, 200 mg/kg infusion every 6 hours for 14 days.[65] Although this is still a complex regime for administration in resource-poor settings, it does significantly reduce both the time and cost of treatment and is considered an improved therapy.[67] The potential benefit of this combination to delay the selection of eflornithine resistant parasites is more difficult to prove.

The priority for new medicines is for something that will treat the stage 2 disease. A programme to develop diamidine derivatives for treatment of both stages did identify a lead compound, the oral drug parfuramidine, which had stage 1 activity only, with a view to replacing pentamidine (a drug requiring parenteral administration and with known toxicities). Parfuramidine did undergo extensive clinical trials in central Africa for treatment of stage 1 *T. b. gambiense* infections with some success; but the clinical studies were halted when unpredicted toxicities were observed.[68] There have been ongoing efforts to re-evaluate pentamidine to reduce the length of treatment of stage 1 disease from 7 to 3 days; after 60 years there is still sufficient interest in this drug to ensure updated guidelines.[69]

As treatment of patients with melarsoprol is followed by a reactive encephalopathy in 10% of patients, with a 50% death rate in these cases, there have been a limited number of studies to evaluate co-administration of anti-inflammatory drugs, in particular prednisolone,[70,71] though this failed to ameliorate the condition. Other different approaches could come from improved understanding of the immunopathology and immunopharmacology of the CNS infection. During stage 2 there is heavy infiltration of immune cells (plasma cells, lymphocytes), causing perivascular cuffing. The role of astrocytes in antigen presentation, astrocyte activation and the balance of cytokines and prostaglandins in determining the course of CNS disease could be considered as a basis for other interventions.[72]

There are currently two compounds in pre-clinical development, all with activity in CNS rodent models of infection with *T. brucei* species An oxaborozole, SCYX-7158 is being developed in a partnership between Drugs for Neglected Diseases, SCYNEXIS and Anacor, and a diamidine derivative, CPD-802, with the Consortium for Parasitic Drug Development, led from the University of North Carolina. In addition, the nitroimidazole, fexinidazole, originally developed by Hoechst, is now in phase I trials as part of a collaboration between the Drugs for Neglected Diseases initiative and Sanofi-Aventis.[73] Further details of the development of these drugs are given in Chapter 7.[68] The very presence of three drugs in development for human African trypanosomiasis is a significant change compared with 10 years ago, coming with the advent of Product Development Partnerships and increased funding from the public sector and philanthropy.[4] However, there are still many critical hurdles to pass, before new safe oral short course treatments are available for this disease.

6.4 South American Trypanosomiasis (Chagas Disease)

Chagas disease is caused by the parasite *Trypanosoma cruzi*, and is considered to be restricted to Central and South America, and is transmitted by triatomid insects (also called kissing bugs). The parasite lifecycle is mainly zoonotic (Figure 6.3). Transmission to humans follows when the bug defecates following feeding; infective trypanosomes in the faeces enter the host through the mucous membranes, or through the bite wound. There are an estimated 8–10 million people infected with the causative pathogen, with about 50,000 new cases each year. Recently, populations outside the Americas have been shown to be at risk, both in Europe and the US, where migrants with indeterminate disease may be a source of infection via transfusion.[74] *Trypanosoma cruzi* is divided into six discrete lineages, Tc 1, Tc IIa, Tc IIb, Tc IIc, Tc IId and Tc IIe defined by multiple genetic markers.[75,76] These discrete typing units have different distributions, different

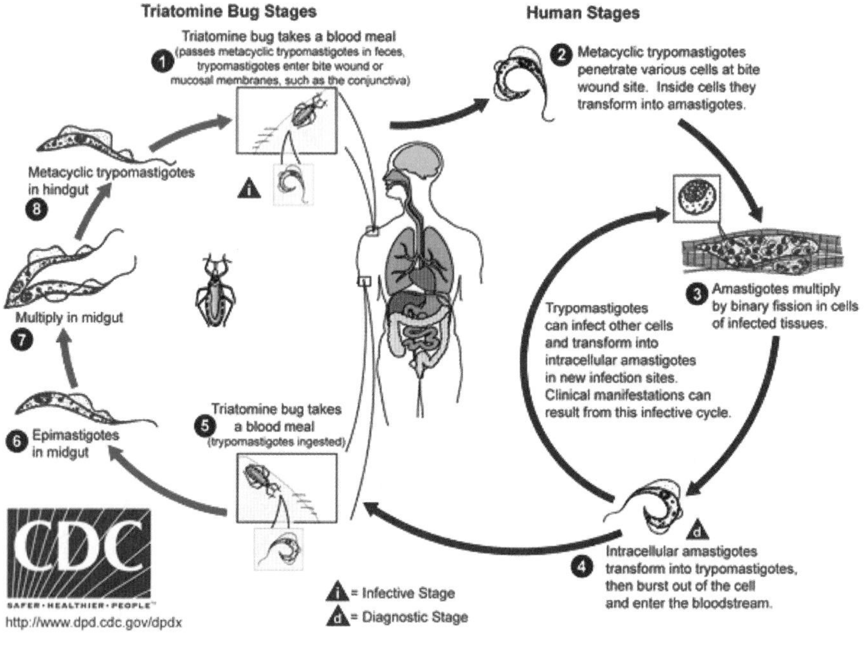

Figure 6.3 Life cycle for Chagas disease (reproduced from the CDC website).

ecological niches and different vector species. How they relate to different disease pathologies and different drug sensitivities is a key area for research.[77]

The disease is characterized by acute, indeterminate and chronic phases. In the acute phase, following infection, the parasite spreads throughout the mammalian host. This stage may be associated with inflammatory lesions at the site of infection, followed by a self-limited febrile illness, although it may be asymptomatic in some cases. The indeterminate stage is generally asymptomatic and may last 5–25 years, during which there is a very low parasite burden, making definitive confirmation of infection extremely difficult. The last, or chronic, phase is associated with various pathologies due to damage to muscle or nerve tissues; 30–40% develop cardiac pathology and 10% of patients develop digestive tract pathology. An important component of this pathology is sustained and diffuse inflammation in the infected tissues. It was postulated for many years that this resulted from an autoimmune response.[78,79] Research over the past decade by Tarleton and others has indicated that persistence of low numbers of parasites in tissues maintains this inflammatory process, with specific CD8+ and CD4+ T cell infiltration, resulting in a Th-1 biased inflammatory response.[80]

Despite the impressive advances in our knowledge about the biology of *T. cruzi*,[1] the only drugs currently available for the treatment of Chagas disease are the nitrofuran nifurtimox and the nitroimidazole benznidazole, both of which were developed in the 1960s and 1970s (Table 6.3). These drugs are approved for use in the treatment of the acute stage of the disease, where they demonstrate up to 80% efficacy, However, they are of limited efficacy against the indeterminate and early chronic-stage disease. They require long courses of treatment (typically 60 days of once per day treatment) and are accompanied by defined, sometimes severe, side effects in up to 40% of patients. With the reduction of transmission of Chagas disease in several foci in South America, resulting in fewer new acute cases of disease, there has been greater focus on the needs for the treatment of indeterminate and early chronic phases, which ultimately are responsible for the high morbidity of this disease.

There are four key features of the biology of *T. cruzi* to note that are relevant to drug discovery and development studies. First, the target is the intracellular amastigote which divides in the mammalian host (in cell cytoplasm not in intracellular vacuoles); the bloodstream trypomastigote in the acute phase is a non-dividing invasive form. Second, the amastigote is found in clusters, pseudocysts, mainly in muscle (cardiac and skeletal) cells as well as nerve and other types of cells. Third, in chronic and indeterminate infections, amastigotes found in many tissues have a low division but are responsible for the immunopathology. Fourthly, the broad taxonomic classification, "*T. cruzi*", hides the fact of defined lineages (as above),[76] which might have different drug sensitivities. Certainly it is well established that different strains of *T. cruzi* demonstrate great variability in drug sensitivity to the standard drugs nifurtimox and benznidazole.[81,82] So an ideal drug, in addition to meeting the criteria of safety, cost, *etc.*, has to be active against acute and chronic infections caused by a number of different strains.

As no new drugs have been registered for the treatment of Chagas disease and as a significant reduction of *T. cruzi* in infected patients appears to be essential to prevent disease progression and to avert its irreversible long-term consequences, a re-examination of the activity of the established drugs has been undertaken. Benznidazole has some efficacy in mouse models of chronic infections and, importantly, early chronic infections in humans.[83] A long-term clinical trial (BENEFIT; Clinical trials.gov identifier NCT00123916) is now underway to determine the efficacy of benzinidazole in reducing parasite burden in patients with cardiomyopathy, as well as safety and tolerability.[84] The trial started in 2004 and data should become available in 2011.

In comparison to leishmaniasis and human African trypanosomiasis, several rational approaches to the treatment of Chagas disease, based upon a clear determination of biochemical pathways in the various stages of the *T. cruzi* parasite, have lead to clinical studies based upon novel compounds or therapeutic switching. In the 1980s, studies on the purine salvage pathway led to the identification of pyrazolopyrimidine compounds that disrupted nucleic acid biosynthesis. A well-established member of this class, allopurinol was highly active in experimental studies, but was not efficacious in clinical trials,[85] most probably due to pharmacokinetic limitations. The potential of specific ergosterol biosynthesis inhibitors that act at the level of C14α sterol demethylase, was also established in the 1980s in studies by Riley and Urbina. The similarities in the sterol biosynthesis pathway in *T. cruzi* and fungi led to clinical studies with well-known antimycotic drugs, ketoconazole and itraconazole, which were active but had limited efficacy.[86,87] However, studies by Urbina and colleagues have shown that new anti-fungal triazole derivatives, for example posoconazole and ravuconazole, have very high potency against *T. cruzi*, with nM level activity *in vitro*, and the capability of curing chronic infections in mice.[88–90] These compounds also have a favourable pharmacokinetic profile of high tissue distribution/long half-life, in line with requirements for treatment of chronic disease.[91] Posoconazole has already proved efficacious in clinical trials [92] and ravuconazole is in pre-clinical development with DNDi.

Inhibitors of cruzipain, an essential protease specific to the parasite are known and one particular vinyl sulphone, K777 (also known as K11777), is in pre-clinical development having proved effective in chronic rodent models of infection.[93] The recent centenary of the discovery of *T. cruzi* as the causative pathogen by Carlos Chagas has provided some excellent detailed reviews of the background to chemotherapy and to these specific developments[91,93] including suggestions for drug combinations,[94] whilst a more recent update is provided by Buckner and Navadi.[95]

6.5 Conclusions

The publication of the genome sequences of the pathogens that cause leishmaniasis and trypanosomiasis, and a wealth of quality research on the molecular biology and biochemistry, have helped us to characterize potential drug targets in the related pathogens that cause these diseases. The subtle differences between the parasites in their metabolic adaptations, the large differences in the required pharmacokinetic properties of drugs to distribute to the different sites of infection, the challenges for approaches required for both acute and chronic infections – all these factors indicate that we need to discover and develop several new drugs for the kinetoplastid diseases and each will require several drugs or formulations of drugs for the treatment of all their manifestations. The discovery and development of these new drugs will require (i) increased input from the disciplines of chemistry, pharmacology, toxicology and pharmaceutics, (ii) further development of suitable predictive disease models and methods for progressing leads and candidate drugs through pre-clinical studies, and (iii) improved diagnostic methods for test of cure to measure clinical efficacy. The limited progress in drug development over the past decades is part of history. Increase in interest and funding for neglected diseases, in incentives in the new not-for-profit model(s) of drug development,[3,96] and in the engagement of the pharmaceutical industry bode well for the future. But this should not be taken for granted and the responsibility for a sustained effort in this field requires effective teams, prioritization where necessary and clear coordinated decision-making.

References

1. N. M. El-Sayed, P. J. Myler, G. Blandin, M. Berriman, J. Crabtree, G. Aggarwal, E. Caler, H. Renauld, E. A. Worthey, C. Hertz-Fowler, E. Ghedin, C. Peacock, D. C. Bartholomeu, B. J. Haas, A. N. Tran, J. R. Wortman, U. C. Alsmark, S. Angiuoli, A. Anupama, J. Badger, F. Bringaud, E. Cadag, J. M. Carlton, G. C. Cerqueira, T. Creasy, A. L. Delcher, A. Djikeng, T. M. Embley, C. Hauser, A. C. Ivens, S. K. Kummerfeld, J. B. Pereira-Leal, D. Nilsson, J. Peterson, S. L. Salzberg, J. Shallom, J. C. Silva, J. Sundaram, S. Westenberger, O. White, S. E. Melville, J. E. Donelson, B. Andersson, K. D. Stuart and N. Hall, *Science*, 2005, **309**, 404.
2. D. F. Smith, C. S. Peacock and A. K. Cruz, *Int. J. Parasitol.*, 2007, **37**, 1173.
3. B. Liese, M. Rosenberg and A. Schratz, *Lancet*, 2010, **375**, 67.
4. M. Moran, J. Guzman, A. L. Ropars, A. McDonald, N. Jameson, B. Omune, S. Ryan and L. Wu, *PLoS Med.*, 2009, **6**, e30.
5. G. Schumann Burkard, P. Jutzi and I. Roditi, *Mol. Biochem. Parasitol.*, 2011, **175**, 91.
6. N. Baker, S. Alsford and D. Horn, *Mol. Biochem. Parasitol.*, 2011, **176**, 55.
7. J. L. Siqueira-Neto, O. R. Song, H. Oh, J. H. Sohn, G. Yang, J. Nam, J. Jang, J. Cechetto, C. B. Lee, S. Moon, A. Genovesio, E. Chatelain, T. Christophe and L. H. Freitas-Junior, *PLoS Negl. Trop. Dis.*, 2010, **4**, e675.
8. S. L. Croft, S. Sundar and A. H. Fairlamb, *Clin. Microbiol. Rev.*, 2006, **19**, 111.
9. J. Lukes, I. L. Mauricio, G. Schonian, J. C. Dujardin, K. Soteriadou, J. P. Dedet, K. Kuhls, K. W. Tintaya, M. Jirku, E. Chocholova, C.

Haralambous, F. Pratlong, M. Obornik, A. Horak, F. J. Ayala and M. A. Miles, *Proc. Natl. Acad. Sci. U. S. A.*, 2007, **104**, 9375.
10. E. E. Zijlstra, A. M. Musa, E. A. Khalil, I. M. el-Hassan and A. M. el-Hassan, *Lancet Infect. Dis.*, 2003, **3**, 87.
11. J. Alvar, S. Croft and P. Olliaro, *Adv. Parasitol.*, 2006, **61**, 223.
12. S. Sundar, *Trop. Med. Int. Health*, 2001, **6**, 849.
13. H. Veeken, K. Ritmeijer, J. Seaman and R. Davidson, *Trop. Med. Int. Health*, 2000, **5**, 312.
14. C. Bern, J. Adler-Moore, J. Berenguer, M. Boelaert, M. den Boer, R. N. Davidson, C. Figueras, L. Gradoni, D. A. Kafetzis, K. Ritmeijer, E. Rosenthal, C. Royce, R. Russo, S. Sundar and J. Alvar, *Clin. Infect. Dis.*, 2006, **43**, 917.
15. S. Sundar, J. Chakravarty, D. Agarwal, M. Rai and H. W. Murray, *N. Engl. J. Med.*, 2010, **362**, 504.
16. F. Meheus, M. Balasegaram, P. Olliaro, S. Sundar, S. Rijal, M. A. Faiz and M. Boelaert, *PLoS Negl. Trop. Dis.*, 2010, **4**.
17. J. D. Berman, R. Badaro, C. P. Thakur, K. M. Wasunna, K. Behbehani, R. Davidson, F. Kuzoe, L. Pang, K. Weerasuriya and A. D. Bryceson, *Bull. World Health Organ.*, 1998, **76**, 25.
18. S. Sundar, T. K. Jha, C. P. Thakur, P. K. Sinha and S. K. Bhattacharya, *N. Engl. J. Med.*, 2007, **356**, 2571.
19. A. Hailu, A. Musa, M. Wasunna, M. Balasegaram, S. Yifru, G. Mengistu, Z. Hurissa, W. Hailu, T. Weldegebreal, S. Tesfaye, E. Makonnen, E. Khalil, O. Ahmed, A. Fadlalla, A. El-Hassan, M. Raheem, M. Mueller, Y. Koummuki, J. Rashid, J. Mbui, G. Mucee, S. Njoroge, V. Manduku, A. Musibi, G. Mutuma, F. Kirui, H. Lodenyo, D. Mutea, G. Kirigi, T. Edwards, P. Smith, L. Muthami, C. Royce, S. Ellis, M. Alobo, R. Omollo, J. Kesusu, R. Owiti and J. Kinuthia, *PLoS Negl. Trop. Dis.*, 2010, **4**, e709.
20. S. L. Croft, R. A. Neal, W. Pendergast and J. H. Chan, *Biochem. Pharmacol.*, 1987, **36**, 2633.
21. S. L. Croft and J. Engel, *Trans. R. Soc. Trop. Med. Hyg.*, 2006, **100** (Suppl 1), S4.
22. S. Sundar, T. K. Jha, C. P. Thakur, J. Engel, H. Sindermann, C. Fischer, K. Junge, A. Bryceson and J. Berman, *N. Engl. J. Med.*, 2002, **347**, 1739.
23. S. K. Bhattacharya, P. K. Sinha, S. Sundar, C. P. Thakur, T. K. Jha, K. Pandey, V. R. Das, N. Kumar, C. Lal, N. Verma, V. P. Singh, A. Ranjan, R. B. Verma, G. Anders, H. Sindermann and N. K. Ganguly, *J. Infect. Dis.*, 2007, **196**, 591.
24. F. J. Perez-Victoria, M. P. Sanchez-Canete, K. Seifert, S. L. Croft, S. Sundar, S. Castanys and F. Gamarro, *Drug Resist. Updat.*, 2006, **9**, 26.
25. N. P. D. Nanayakkara, A. L. A. Jr, M. S. Bartlett, V. Yardley, S. L. Croft, I. A. Khan, J. D. McChesney and L. A. Walker, *Antimicrob. Agents Chemother.*, 2008, **52**, 2130.
26. V. Yardley, F. Gamarro and S. L. Croft, *Antimicrob. Agents Chemother.*, 2010, **54**, 5356.

27. S. Sundar, P. K. Sinha, M. Rai, D. K. Verma, K. Nawin, S. Alam, J. Chakravarty, M. Vaillant, N. Verma, K. Pandey, P. Kumari, C. S. Lal, R. Arora, B. Sharma, S. Ellis, N. Strub-Wourgaft, M. Balasegaram, P. Olliaro, P. Das and F. Modabber, *Lancet*, 2011, **377**, 477.
28. K. Seifert and S. L. Croft, *Antimicrob. Agents Chemother.*, 2006, **50**, 73.
29. P. L. Olliaro, *Curr. Opin. Infect. Dis.*, 2010, **23**, 595.
30. R. A. Neal, S. Allen, N. McCoy, P. Olliaro and S. L. Croft, *J. Antimicrob. Chemother.*, 1995, **35**, 577.
31. Y. Melaku, S. M. Collin, K. Keus, F. Gatluak, K. Ritmeijer and R. N. Davidson, *Am. J. Trop. Med. Hyg.*, 2007, **77**, 89.
32. A. M. Musa, E. A. G. Khalil, M. A. Raheem, E. E. Zijlstra, M. E. Ibrahim, I. M. El Hassan, M. M. Mukhtar and A. M. E. Hassan, *Ann. Trop. Med. Parasitol.*, 2003, **96**, 765.
33. E. E. Zijlstra, A. M. Musa, E. A. G. Khalil, I. M. E. Hassan and A. M. El-Hassan, *The Lancet Infect. Dis.*, 2003, **3**, 87.
34. S. L. Croft, *Indian J. Med. Res.*, 2008, **128**, 10.
35. A. M. Musa, E. A. Khalil, F. A. Mahgoub, S. H. Elgawi, F. Modabber, A. E. Elkadaru, M. H. Aboud, S. Noazin, H. W. Ghalib and A. M. El-Hassan, *Trans. R. Soc. Trop. Med. Hyg.*, 2008, **102**, 58.
36. A. M. Musa, S. Noazin, E. A. Khalil and F. Modabber, *Trans. R. Soc. Trop. Med. Hyg.*, 2010, **104**, 1.
37. J. Alvar, P. Aparicio, A. Aseffa, M. Den Boer, C. Canavate, J. P. Dedet, L. Gradoni, R. Ter Horst, R. Lopez-Velez and J. Moreno, *Clin. Microbiol. Rev.*, 2008, **21**, 334.
38. R. Reithinger, J. C. Dujardin, H. Louzir, C. Pirmez, B. Alexander and S. Brooker, *Lancet Infect. Dis.*, 2007, **7**, 581.
39. M. Kassi, A. K. Afghan, R. Rehman and P. M. Kasi, *PLoS Negl. Trop. Dis.*, 2008, **2**, e259.
40. T. Garnier and S. L. Croft, *Curr. Opin. Investig. Drugs*, 2002, **3**, 538.
41. J. El-On, S. Halevy, M. H. Grunwald and L. Weinrauch, *J. Am. Acad. Dermatol.*, 1992, **27**, 227.
42. A. Ben Salah, P. A. Buffet, G. Morizot, N. Ben Massoud, A. Zaatour, N. Ben Alaya, N. B. Haj Hamida, Z. El Ahmadi, M. T. Downs, P. L. Smith, K. Dellagi and M. Grogl, *PLoS Negl. Trop. Dis.*, 2009, **3**, e432.
43. V. Yardley, S. L. Croft, S. D. Doncker, J.-C. Dujardin, S. Koirala, S. Rijal, C. Miranda, A. Llanos-Cuentas and F. Chappuis, *Am. J. Trop. Med. Hyg.*, 2005, **73**, 272.
44. J. Soto, B. A. Arana, J. Toledo, N. Rizzo, J. C. Vega, A. Diaz, M. Luz, P. Gutierrez, M. Arboleda, J. D. Berman, K. Junge, J. Engel and H. Sindermann, *Clin. Infect. Dis.*, 2004, **38**, 1266.
45. P. R. Machado, J. Ampuero, L. H. Guimaraes, L. Villasboas, A. T. Rocha, A. Schriefer, R. S. Sousa, A. Talhari, G. Penna and E. M. Carvalho, *PLoS Negl. Trop. Dis.*, 2010, **4**, e912.
46. F. Modabber, P. A. Buffet, E. Torreele, G. Milon and S. L. Croft, *Kinetoplastid Biol. Dis.*, 2007, **6**, 3.

47. U. Gonzalez, M. Pinart, L. Reveiz, M. Rengifo-Pardo, J. Tweed, A. Macaya and J. Alvar, *Clin. Infect. Dis.*, 2010, **51**, 409.
48. U. Gonzalez, M. Pinart, L. Reveiz and J. Alvar, *Cochrane Database Syst. Rev.*, 2008, CD005067.
49. U. Gonzalez, M. Pinart, M. Rengifo-Pardo, A. Macaya, J. Alvar and J. A. Tweed, *Cochrane Database Syst. Rev.*, 2009, CD004834.
50. H. M. Al-Abdely, J. R. Graybill, D. Loebenberg and P. C. Melby, *Antimicrob. Agents Chemother.*, 1999, **43**, 2910.
51. A. E. Paniz Mondolfi, C. Stavropoulos, T. Gelanew, E. Loucas, A. M. Perez Alvarez, G. Benaim, B. Polsky, G. Schoenian and E. M. Sordillo, *Antimicrob. Agents Chemother.*, 2011.
52. C. Miranda-Verastegui, G. Tulliano, T. W. Gyorkos, W. Calderon, E. Rahme, B. Ward, M. Cruz, A. Llanos-Cuentas and G. Matlashewski, *PLoS Negl. Trop. Dis.*, 2009, **3**, e491.
53. J. Convit, M. Ulrich, O. Zerpa, R. Borges, N. Aranzazu, M. Valera, H. Villarroel, Z. Zapata and I. Tomedes, *Trans. R. Soc. Trop. Med. Hyg.*, 2003, **97**, 469.
54. M. Iten, E. Matovu, R. Brun and R. Kaminsky, *Trop. Med. Parasitol.*, 1995, **46**, 190.
55. F. Chappuis, L. Loutan, P. Simarro, V. Lejon and P. Buscher, *Clin. Microbiol. Rev.*, 2005, **18**, 133.
56. R. McCulloch and D. Horn, *Trends Parasitol.*, 2009, **25**, 359.
57. M. P. Barrett, D. W. Boykin, R. Brun and R. R. Tidwell, *Br. J. Pharmacol.*, 2007, **152**, 1155.
58. R. Brun, R. Schumacher, C. Schmid, C. Kunz and C. Burri, *Trop. Med. Int. Health*, 2001, **6**, 906.
59. E. Matovu, J. C. Enyaru, D. Legros, C. Schmid, T. Seebeck and R. Kaminsky, *Trop. Med. Int. Health*, 2001, **6**, 407.
60. C. Burri and R. Brun, *Parasitol. Res.*, 2003, **90** (Suppl 1), S49.
61. F. o. Chappuis, N. Udayraj, K. Stietenroth, A. Meussen and P. A. Bovier, *Clin. Infect. Dis.*, 2005, **41**, 748.
62. C. Burri and J. Keiser, *Trop. Med. Int. Health*, 2001, **6**, 412.
63. C. Schmid, S. Nkunku, A. Merolle, P. Vounatsou and C. Burri, *Lancet*, 2004, **364**, 789.
64. G. Priotto, C. Fogg, M. Balasegaram, O. Erphas, A. Louga, F. Checchi, S. Ghabri and P. Piola, *PLoS Clin. Trials*, 2006, **1**, e39.
65. G. Priotto, S. Kasparian, W. Mutombo, D. Ngouama, S. Ghorashian, U. Arnold, S. Ghabri, E. Baudin, V. Buard, S. Kazadi-Kyanza, M. Ilunga, W. Mutangala, G. Pohlig, C. Schmid, U. Karunakara, E. Torreele and V. Kande, *Lancet*, 2009, **374**, 56.
66. J. Pepin, F. Milord, B. Mpia, F. Meurice, L. Ethier, D. DeGroof and H. Bruneel, *Trans. R. Soc. Trop. Med. Hyg.*, 1989, **83**, 514.
67. V. Lutje, J. Seixas and A. Kennedy, *Cochrane Database Syst. Rev.*, 2010, CD006201.
68. M. P. Barrett, *Curr. Opin. Infect. Dis.*, 2010, **23**, 603.
69. T. P. Dorlo and P. A. Kager, *PLoS Negl. Trop. Dis.*, 2008, **2**, e225.

70. J. Pepin, F. Milord, C. Guern, B. Mpia, L. Ethier and D. Mansinsa, *Lancet*, 1989, **1**, 1246.
71. J. Pepin, F. Milord, A. N. Khonde, T. Niyonsenga, L. Loko, B. Mpia and P. De Wals, *Trans. R. Soc. Trop. Med. Hyg.*, 1995, **89**, 92.
72. P. G. Kennedy, *Int. J. Parasitol.*, 2006, **36**, 505.
73. E. Torreele, B. Bourdin Trunz, D. Tweats, M. Kaiser, R. Brun, G. Mazue, M. A. Bray and B. Pecoul, *PLoS Negl. Trop. Dis.*, 2010, **4**, e923.
74. A. Perez-Ayala, J. A. Perez-Molina, F. Norman, M. Navarro, B. Monge-Maillo, M. Diaz-Menendez, J. Peris-Garcia, M. Flores, C. Canavate and R. Lopez-Velez, *Clin. Microbiol. Infect.*, 2010.
75. M. A. Miles, M. D. Feliciangeli and A. R. de Arias, *BMJ*, 2003, **326**, 1444.
76. M. A. Miles, M. S. Llewellyn, M. D. Lewis, M. Yeo, R. Baleela, S. Fitzpatrick, M. W. Gaunt and I. L. Mauricio, *Parasitology*, 2009, **136**, 1509.
77. R. del Puerto, J. E. Nishizawa, M. Kikuchi, N. Iihoshi, Y. Roca, C. Avilas, A. Gianella, J. Lora, F. U. Velarde, L. A. Renjel, S. Miura, H. Higo, N. Komiya, K. Maemura and K. Hirayama, *PLoS Negl. Trop. Dis.*, 2010, **4**, e687.
78. J. Kalil and E. Cunha-Neto, *Parasitol. Today*, 1996, **12**, 396.
79. D. M. Engman and J. S. Leon, *Acta Trop.*, 2002, **81**, 123.
80. R. L. Tarleton, *Int. J. Parasitol.*, 2001, **31**, 550.
81. L. S. Filardi and Z. Brener, *Trans. R. Soc. Trop. Med. Hyg.*, 1987, **81**, 755.
82. S. G. Andrade, J. B. Magalhaes and A. L. Pontes, *Bulletin of the World Health Organisation*, 1985, **63**, 7219726.
83. S. Sosa Estani, E. L. Segura, A. M. Ruiz, E. Velazquez, B. M. Porcel and C. Yampotis, *Am. J. Trop. Med. Hyg.*, 1998, **59**, 526.
84. J. A. Marin-Neto, A. Rassi, Jr., A. Avezum, Jr., A. C. Mattos, A. Rassi, C. A. Morillo, S. Sosa-Estani and S. Yusuf, *Mem. Inst. Oswaldo Cruz*, 2009, **104** (Suppl 1), 319.
85. R. H. Gallerano, J. J. Marr and R. R. Sosa, *Am. J. Trop. Med. Hyg.*, 1990, **43**, 159.
86. Z. Brener, J. R. Cancado, L. M. Galvao, Z. M. da Luz, S. Filardi Lde, M. E. Pereira, L. M. Santos and C. B. Cancado, *Mem. Inst. Oswaldo Cruz*, 1993, **88**, 149.
87. W. Apt, A. Arribada, I. Zulantay, A. Solari, G. Sanchez, K. Mundaca, X. Coronado, J. Rodriguez, L. C. Gil and A. Osuna, *Ann. Trop. Med. Parasitol.*, 2005, **99**, 733.
88. J. Molina, O. Martins-Filho, Z. Brener, A. J. Romanha, D. Loebenberg and J. A. Urbina, *Antimicrob. Agents Chemother.*, 2000, **44**, 150.
89. J. A. Urbina, G. Payares, L. M. Contreras, A. Liendo, C. Sanoja, J. Molina, M. Piras, R. Piras, N. Perez, P. Wincker and D. Loebenberg, *Antimicrob. Agents Chemother.*, 1998, **42**, 1771.
90. J. A. Urbina, *Mem. Inst. Oswaldo Cruz*, 2009, **104** (Suppl 1), 311.
91. J. A. Urbina, *Acta Trop.*, 2009, **115**, 55.

92. M. J. Pinazo, G. Espinosa, M. Gallego, P. L. Lopez-Chejade, J. A. Urbina and J. Gascon, *Am. J. Trop. Med. Hyg.*, 2010, **82**, 583.
93. J. H. McKerrow, P. S. Doyle, J. C. Engel, L. M. Podust, S. A. Robertson, R. Ferreira, T. Saxton, M. Arkin, I. D. Kerr, L. S. Brinen and C. S. Craik, *Mem. Inst. Oswaldo Cruz*, 2009, **104** (Suppl 1), 263.
94. J. R. Coura, *Mem. Inst. Oswaldo Cruz*, 2009, **104**, 549.
95. F. S. Buckner and N. Navabi, *Curr. Opin. Infect. Dis.*, 2010, **23**, 609.
96. M. Moran, *PLoS Med.*, 2005, **2**, e302.

CHAPTER 7

Drug Discovery for Kinetoplastid Diseases

ROBERT T. JACOBS

SCYNEXIS, Inc.; P. O. Box 12878, Research Triangle Park, NC 27709-2878, US

7.1 Introduction

As described in the preceding chapter, diseases caused by trypanosomatid parasites present a significant disease burden, predominantly across the developing world. As such, efforts to discover new treatments for these diseases have been somewhat limited, as it is unlikely that the costs of research and development can be recovered through the sale of new drugs in disease endemic areas. Consequently, much of the research efforts have been executed through academic groups and partnerships, with funding obtained either from governments or philanthropic organizations. While substantial advances in understanding of the fundamental biology of the trypanosomatids have been realized through these efforts, the progression of viable drug candidates to clinical trials has lagged behind. Over the past decade, a number of public–private partnerships have emerged, which have sought to address this gap through integration of fundamental biological research in the academic sector with translational drug discovery efforts from the private sector. The growth of these partnerships, coupled with a renewed engagement from medium to large pharmaceutical companies, in part motivated by public pressure for such organizations to address neglected diseases in a more substantial way, has resulted in the emergence of a significant number of new drug candidates for the treatment of these diseases. This chapter will review recent advances in understanding of biochemical targets for new drugs, as well as new drug classes which have emerged from screening for parasiticidal activity in whole cell assays.

RSC Drug Discovery Series No. 14
Neglected Diseases and Drug Discovery
Edited by Michael J. Palmer and Timothy N. C. Wells

Published by the Royal Society of Chemistry, www.rsc.org

7.2 Background Biology and Genetics

The three kinetoplastid parasite families which will be the focus of this chapter are *Trypanosoma brucei* spp. (causative of human African trypanosomiasis), *Leishmania* spp. (causative of visceral, cutaneous and mucocutaneous leishmaniasis) and *Trypanosoma cruzi* spp. (causative of Chagas disease). The genomes of these parasites have been completed, and as a result, considerable information is available for selection and validation of relevant biochemical targets.[1–5] In addition, a variety of genetic tools, including RNAi, are available in *T. brucei* to evaluate the essentiality of individual target proteins. A regulated, tetracycline-inducible knockout system is operable in *T. brucei*, hence most genetic validation of targets has been performed in this parasite.[6] Representative strains of the three major parasite families can be grown in large numbers using conventional cell culture techniques, and as such, whole cell assays have been developed to evaluate the ability of test compounds to impact parasite growth and viability. The most commonly employed assays use fluorescent oxidation-reduction reagents such as Alamar Blue as indicators of parasite viabilility.[7,8] Alternative methods of detection of parasite viability have emerged in recent years, including an ATP-bioluminescence approach which correlates ATP release from *T. brucei* with parasite viability,[9] a luciferase-linked assay using genetically modified *L. donovani* parasites,[10] and a β-galactosidase-linked assay in genetically modified *T. cruzi*.[11] With advances in liquid handling robotics, most of these assays have been adapted to high-throughput format in 96 384 and 1536-well plates.[12]

7.3 Identification of Parasiticidal Compounds through Whole-cell Assays

A significant number of chemotypes have been discovered as having anti-parasitic activity through utilization of the whole-cell assays described above, and in many cases, little or no information as to mechanism of action for these compounds has been described. In other cases, while initial hits and leads were discovered in this manner, structural similarities of the hits to compounds with know mechanisms of action, either as antiparasitics or against homologous mammalian proteins have facilitated pursuit of SAR through the biochemical assay. In this section, only those classes of compounds where little or no information regarding mechanism of action will be covered, subsequent sections will address compound classes on a mechanistic basis.

7.3.1 Benzoxaboroles

One of the more interesting classes of potential antiparasitic drugs to emerge in the past several years has been a series of benzoxaboroles (Figure 7.1). These compounds were originally explored by Anacor Pharmaceuticals for a number of anti-fungal and anti-bacterial applications, with several analogs progressed

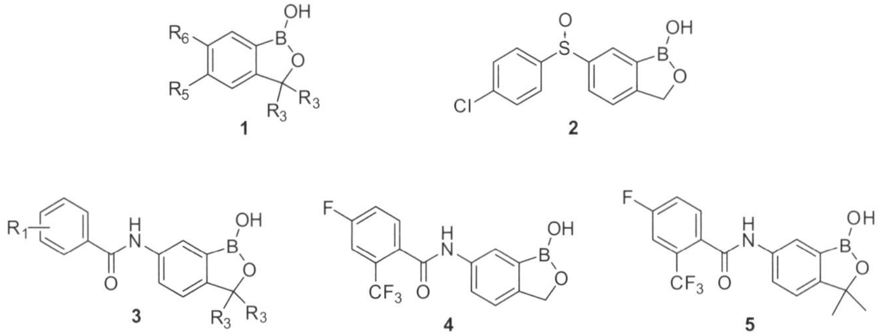

Figure 7.1 Anti-trypanosomal benzoxaboroles.

to clinical trials in these areas.[13–15] Screening of a focused library of benzoxaboroles (**1**) in a whole cell *T. brucei* assay identified a number of 6-substituted analogs as potent trypanocides, with a 6-sulfoxide linker (**2**) exhibiting *in vivo* activity.[16] Optimization of benzoxaborole-6-carboxamides (**3**) to improve potency and pharmacokinetic properties has been reported.[17] These efforts have identified SCYX-6759 (**4**) as an orally active benzoxaborole with good efficacy in a model of stage 1 HAT and modest activity in a stage 2 HAT model. Further improvement of permeability of the benzoxaboroles across the blood–brain barrier has resulted in the identification of SCYX-7158 (**5**), which has been shown to cure mice infected with a strain of *T. brucei* that is resident in the brain.[18] Recently several mechanisms of action have been described for individual compounds in this class, included inhibition of leucyl tRNA-synthase[19] and phosphodiesterase,[20] although the relevance of these mechanisms to the observed anti-trypanosomal activity is unclear.

7.3.2 Lipophilic Amines

Originally based on the observation that the antiviral agent rimantidine (**6**) exhibited parasiticidal activity against the kinetoplastids,[21] a number of lipophilic polycyclic amine scaffolds including adamantanes (**7**–**10**) have been explored, and SAR developed based on whole cell activity (Figure 7.2).[22–25] Some improvements in the *in vitro* potency of these compounds were obtained through cyclization of the amine moiety and addition of lipophilic substituents and some reduction of parasitemia has been reported *in vivo*, but cures of *in vivo* *T. brucei* infections have not been achieved.

Closely related to the adamantine scaffold, several series of aminobicyclo[2.2.2]octane (**11**) and azabicyclo[3.3.2]nonane (**12**) analogs have been synthesized and evaluated against the kinetoplastids.[26–34] Like the adamantanes, a general trend correlating *in vitro* activity with lipophilicity of the molecule has been observed for these series, and *in vivo* activity has been modest, with prolongation of survival, but not cure of *T. brucei* infection in mice.[35]

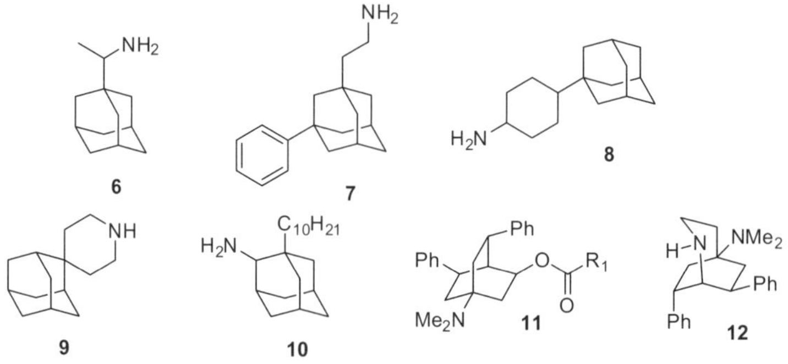

Figure 7.2 Lipophilic amines.

7.3.3 Nitroheterocycles

It can be argued that the mechanism of action of the broad class of nitroheterocycles known to be active against the kinetoplastids is known, as it is generally accepted that they are activated by a nitroreductase in the parasite to generate reactive oxygen species which damage DNA and related biomolecules, though detailed characterization of this activation process has only appeared recently.[36] The nitroheterocycles nifurtimox (**13**) and benznidazole (**14**) have been used clinically in the treatment of Chagas disease, and interest has remained high in the past decade for exploration of this class of antiparasitic compounds.[37] In particular, a large number of nitrofurans (**15**),[38] nitrothiophenes (**16**),[39] nitroindazoles (**17**),[40,41] and nitroimidazoles (**18**)[42] have been synthesized and evaluated in whole cell trypanosomal and leishmanial assays, with potent activity observed across these series (Figure 7.3). Additionally, re-evaluation of older nitroheterocycles, such as fexinidazole (**19**) as a potential treatment for stage 2 HAT has been reported. Fexinidazole has recently been progressed to clinical trials for HAT.[43,44]

Despite these efforts and interest, the nitroheterocycles are potentially compromised by genotoxicity, which has been reported for both nifurtimox and benznidazole.[45,46] There have been studies where SAR for this toxicity has been developed within series of compounds,[47] but to date no definitive strategy for elimination of this liability has emerged. Alternatively, it has been argued that the consequence of differences in the level of expression and activity of nitroreductases between bacteria and mammals is that the classical bacteria-based assays (e.g. Ames assay) may overestimate the potential for genotoxicity of this class of compounds. Supportive of this viewpoint is the observation that, when evaluated in mammalian cell-based genotoxicity assays (e.g. mouse lymphoma or chromosomal aberration assays), some nitroheterocycles such as metronidazole (**20**) have exhibited little evidence of genotoxicity.[48]

Another area of concern for the nitroheterocycle class that may limit utility moving forward is development of resistance. A direct consequence of the reliance on the parasitic nitroreductase enzyme to generate toxic metabolites

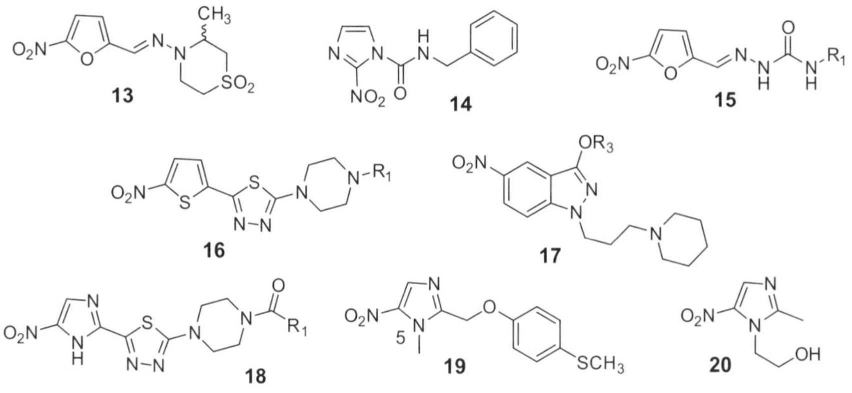

Figure 7.3 Nitroheterocycles.

for activity, it has been shown that genetic knockout of these enzymes in either *T. cruzi*[49] or *T. brucei*[50] affords parasites which are resistant to several nitroheterocycles. Critically, drug resistant parasites could also be generated by *in vitro* selection techniques by continuous exposure to sub-lethal concentrations of nifurtimox (**13**); these parasites were also found to be less sensitive to fexinidazole (**19**).[50]

7.3.4 Metal-based Parasiticides

Classical treatments for kinetoplastid infections include administration of organic complexes of the metalloids arsenic (e.g. melarsoprol, **21**) and antimony (e.g. sodium stibogluconate, **22**) as described in the preceding chapter. The mechanism of action of these drugs has been elucidated to varying degrees. While it has been known for some time that the primary targets of these drugs are enzymes in the polyamine pathway,[51,52] only recently have mechanistic and structural details of the interaction of these metals with the enzymatic targets become available.[53]

These experiences, coupled with the utility of metal-based cytotoxic drugs in the treatment of cancer have prompted examination of this class of drugs as potential antiparasitics (Figure 7.4). For example, it has been shown that cisplatin (**23**) and related platinum compounds are active against *Leishmania* spp.[54–56] In addition, complexation of metals (e.g. Pt, Ru, Pd, V and Au) to a variety of known antiparasitics and antifungals (**24**–**27**) has been explored with some success.[57–63] Some of the most interesting progress in this area has been in the exploration of ruthenium-based nitric oxide (NO) donor complexes (**28**–**31**) focused on treatment of Chagas disease.[64–67] Initially directed at modulation of NO levels in the macrophage of host to facilitate parasite lysis,[68] more recent data suggest that biochemical targets such as glyceraldehyde-3-phosphate dehydrogenase (GAPDH) in the parasite may also be implicated in the activity of these compounds.[67] While additional studies are required to more fully

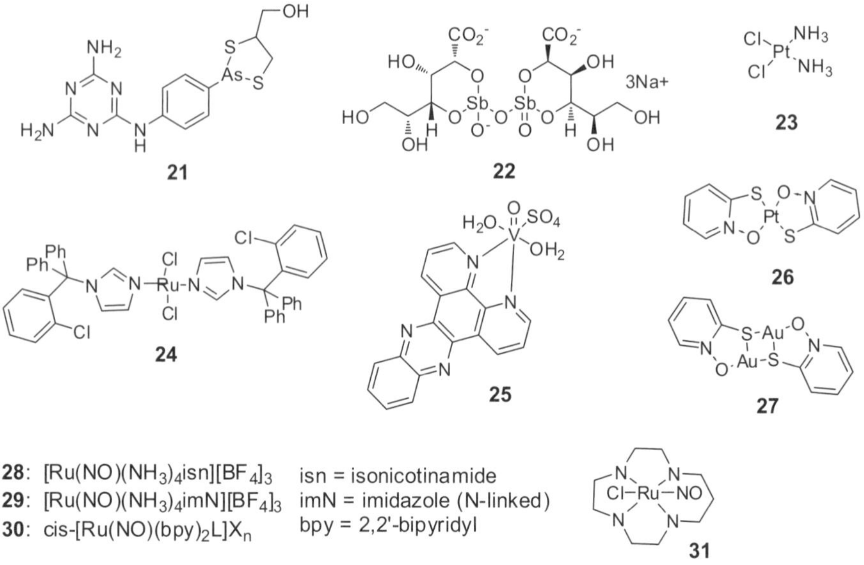

Figure 7.4 Metal-based antiparasitc compounds.

understand the mechanisms of actions of these compounds, the potency observed in both *in vitro* and *in vivo* models of Chagas disease are quite impressive.

7.4 Polyamine Pathway

One of the most well-studied biochemical pathways in kinetoplastids involves the uptake and biosynthesis of polyamines, a class of low molecular weight metabolites that are essential to cell growth and differentiation in both mammals and protozoa.[69–72] In the kinetoplastids, polyamines play an important role in regulation of parasite proliferation, and also protect the parasite from reactive oxygen species through formation of a parasite-specific metabolite, trypanothione.[73] Most of the enzymes in this pathway are conserved across *T. brucei, Leishmania* spp., and *T. cruzi*, although the first enzyme in the pathway, ornithine decarboxylase (ODC), is absent in *T. cruzi*. The full biosynthetic pathway is summarized in Figure 7.5. The enzymes in this pathway are tightly regulated, and differ from mammalian homologs in a number of ways. As a consequence, inhibition of one or more of these enzymes presents an attractive strategy for discovery of parasiticidal compounds. Eflornithine (DFMO, **32**), a clinically relevant anti-trypanosomal compound used for treatment of stage 2 African sleeping sickness, is a suicide inhibitor of ODC, further validating the importance of this pathway.

In addition to efforts to find new inhibitors of ODC, drug discovery programs focused on other enzymes in the pathway, specifically AdoMet DC, trypanothione synthetase (TrpSyn), trypanothione reductase (TrpRed) and

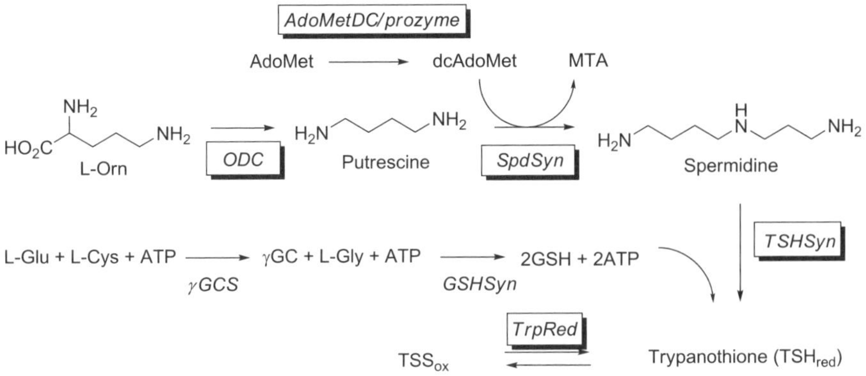

Figure 7.5 Polyamine pathway in *T. brucei* and *Leishmania* spp. Reprinted with permission from ref. 326.

spermidine synthetase (SpdSyn) have been reported. Due to the greater availability of genetic tools for enzymes in the *T. brucei* pathway, most of these efforts have focused on this parasite.

The essentiality of the polyamine pathway biosynthetic enzymes has been demonstrated by a variety of gene knockout and RNAi methods. All of the enzymes in the pathway, with the exception of glutathione synthase, have been characterized by recombinant or biochemical methods. For example, in *T. brucei*, it has been shown that depletion of ODC leads to loss of putrescine and trypanothione, resulting in reduced cell proliferation.[74,75] Also, knockdown of AdoMetDC, its catalytically inactive homolog prozyme or spermidine synthase in blood form parasites leads to partial reduction of spermidine levels, complete depletion of trypanothione and cell death.[75–77] The essentiality of γ-glutamylcysteine synthetase, TrpRed and TrpSyn has been demonstrated by genetic methods.[78–80]

7.4.1 Ornithine Decarboxylase (ODC)

The first committed step in the polyamine pathway is decarboxylation of ornithine by ODC. Most early efforts for discovery of ODC inhibitors focused on preparation of analogs of DFMO (**32**), and did not result in discovery of compounds that exhibited significant improvements over DFMO. No high throughput screening programs for novel inhibitors of ODC were reported until very recently, presumably due to the difficulty in development of a suitable assay. This challenge has been overcome through use of a linked enzyme assay approach, where the carbon dioxide produced from ornithine by ODC is utilized by phosphoenolpyruvate carboxylase to generate oxaloacetate, which is in turn reduced to malate by malate dehydrogenase in an NADH-dependent step.[81] Using this assay system, a library of over 300 000 commercially available compounds was evaluated, leading to the discovery and selection of four novel chemotypes for further exploration (Figure 7.6).[82]

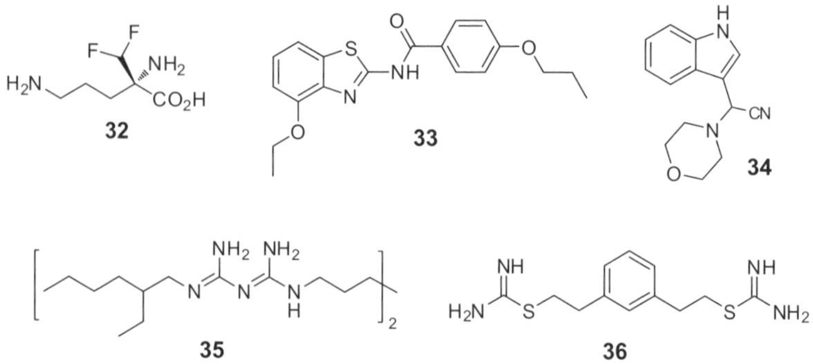

Figure 7.6 Inhibitors of ornithine decarboxylase (ODC).

The benzothiazole (**33**) and indole (**34**) were found to be non-competitive inhibitors of ODC, and also inhibited the human ODC homolog. In contrast, alexidine (**35**) and the bis-thioamidine (**36**) were found to be competitive inhbitors, with the latter class exhibiting good selectivity for the trypanosomal ODC enzyme. Whether these hits can be optimized to provide clinically effective drug candidates is unknown.

7.4.2 S-Adenosylmethionine Decarboxylase (SAM-DC, AdoMet-DC)

One of the most intriguing and well-characterized enzymes in the polyamine pathway is AdoMet-DC, which effects the decarboxylation of S-adenosylmethionine, the product of which is used by spermidine synthetase. In *T. brucei*, AdoMet-DC is regulated by a novel mechanism, with the enzymatically active complex present as a heterodimer consisting of a functional subunit and a catalytically inactive paralog.[83] The catalytically inactive prozyme is unique to the kinetoplastids, with similar homologs identified in both *Leishmania* and *T. cruzi*.[84] It has been shown that prozyme in *T. brucei* acts as an allosteric modulator of the AdoMet-DC complex, and that expression of prozyme is regulated in response to depletion of AdoMet-DC.[76] As this enzyme is present in both kinetoplastids and mammals, inhibitors of AdoMet-DC such as MDL-73,811 (**37**) were initially explored as potential anti-cancer drugs, but were also found to be trypanocidal.[85] Though a potent inhibitor of SAM-DC, and active against *T. brucei* in culture, poor pharmacokinetic properties of MDL-73,811 precluded further development.[86] Similarly, exploration of a series of diamidine analogs (**38**–**40**) also identified compounds with good *in vitro* activities, but only modest activity in mouse models of HAT was observed for **40**.[87]

More recently, efforts to identify improved analogs of MDL-73,811 (**37**) have been reported, with a 8-methyl adenosine analog, Genz-644131 (**41**) found to exhibit improved affinity for inhibition of the AdoMet-DC enzyme and better pharmacokinetic properties (Figure 7.7).[88,89] Subsequent evaluation of **41** in

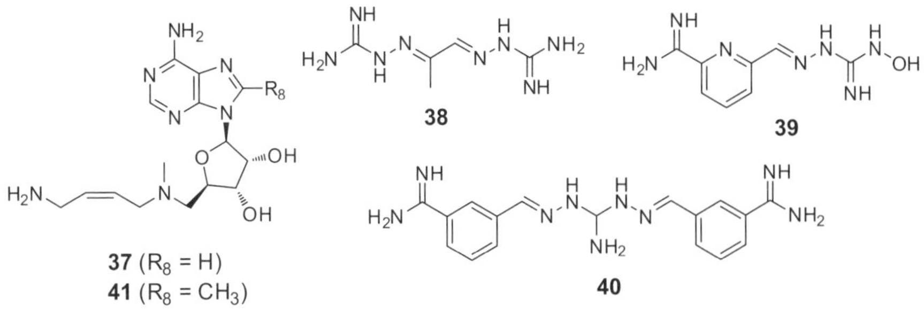

Figure 7.7 Inhibitors of S-adenosylmethionine decarboxylase (AdoMet-DC).

mouse models of HAT demonstrated that this compound is able to cure bloodborne *T. brucei* infections but, despite improved brain penetration of **41**, the compound is not effective in a mouse model of CNS HAT.[90] Compounds in this class show good selectivity for trypanocidal activity relative to cytotoxicity in cell-based assays, but not at the enzyme inhibiton level. A possible explanation for this cellular selectivity is preferential uptake into the parasite by a specific AdoMet transporter.

7.4.3 Spermidine Synthase (SpdSyn)

The third enzyme in the kinetoplastid polyamine pathway, spermidine synthase, has also been validated as a potential target through RNAi in *T. brucei*[77] and gene replacement in *L. donovani*.[91] In contrast to ODC and AdoMet-DC knockout experiments, however, the rate of cell death in the SpdSyn knockout experiment was less rapid, suggesting that this enzyme might be a less attractive target than either ODC or AdoMetDC.[75] To date, no efforts to identify small molecule inhibitors of kinetoplastid spermidine synthases have been reported, although crystal structures of both the human[92] and *P. falciparum*[93] enzymes may provide opportunities for structure-based drug design programs.

7.4.4 Trypanothione Synthetase (TrpSyn)

Unique to the kinetoplastids, trypanothione synthetase catalyzes the condensation of spermidine and glutathione to generate trypanothione, which is essential to parasite survival.[94–96] The essentiality of this enzyme for parasite survival has been established through genetic means in *T. brucei*.[80,97,98] Mechanism-based inhibitors (**42, 43**) of TrpSyn have been described, though translation of enzyme inhibition to trypanocidal activity has not been reported for these compounds (Figure 7.8).[99] Very recently, chemical validation of *T. brucei* TrpSyn has been achieved based on a high throughput screening approach which identified several series of TrpSyn inhibitors (**44–46**) which are also trypanocidal.[100] Further target validation for these inhibitors was confirmed through use of *T. brucei* TrpSyn knock-out and overexpressing cell lines.

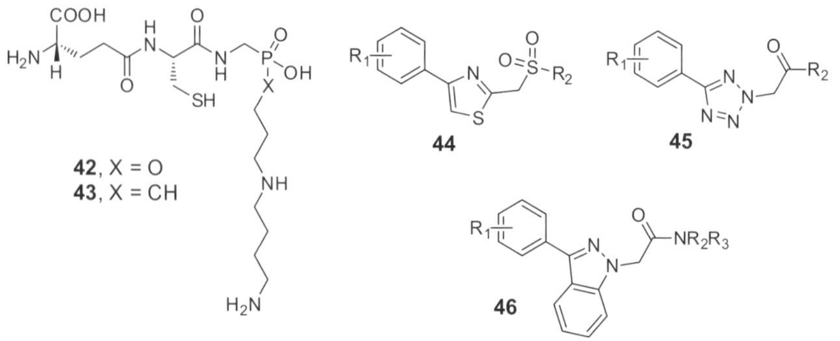

Figure 7.8 Inhibitors of trypanothione synthetase.

7.4.5 Trypanothione Reductase (TrpRed)

Perhaps the most thoroughly explored enzyme in the polyamine pathway is trypanothione reductase, which maintains trypanothione in the reduced state to protect the parasite from damage by oxidative insults.[73] Genetic validation of TrpRed was established in *T. brucei* through generation of parasites containing only one TrpRed gene under the control of a tetracycline-inducible promoter.[79] Essentiality of TrpRed in *L. donovani* has also been established by genetic methods.[101] Representative TrpRed enzymes from *T. brucei, Leishmania* and *T. cruzi* have been cloned and expressed,[102–104] and crystal structures of the *L. infantum* and *T. cruzi* enzymes have been determined.[105,106] Consequently, numerous high throughput screening campaigns, virtual screening and structure-based drug design programs have been implemented against this target.[107–113] To date, however, none of these efforts have progressed through lead optimization to afford a clinical candidate. It has been suggested that possible reasons for these failures are the need for very potent inhibition of the enzyme or a poorly defined large, solvent-exposed binding site.[108]

7.5 Energy Metabolism

The kinetoplastids are solely dependent on glycolysis to provide ATP and energy, and share a highly conserved glycolytic pathway.[114] Subtle differences do exist, not only between parasites, but between individual species of parasite dependent upon their environment in either the insect vector or the mammalian host.[115,116] Despite these differences, the enzymes in the glycolytic pathway are attractive targets for trypanocidal drugs.[117–119] As is the case with other biochemical pathways, the majority of genetic validation of individual enzyme targets in the glycolytic pathway has been peformed in *T. brucei*, and through this work, all of the individual enzymes have been described, isolated and purified, either through classical or recombinant means, allowing a comprehensive understanding of the kinetics and flux of the pathway to be developed (Figure 7.9).[120] The enzymes in the glycolytic pathway are localized in an

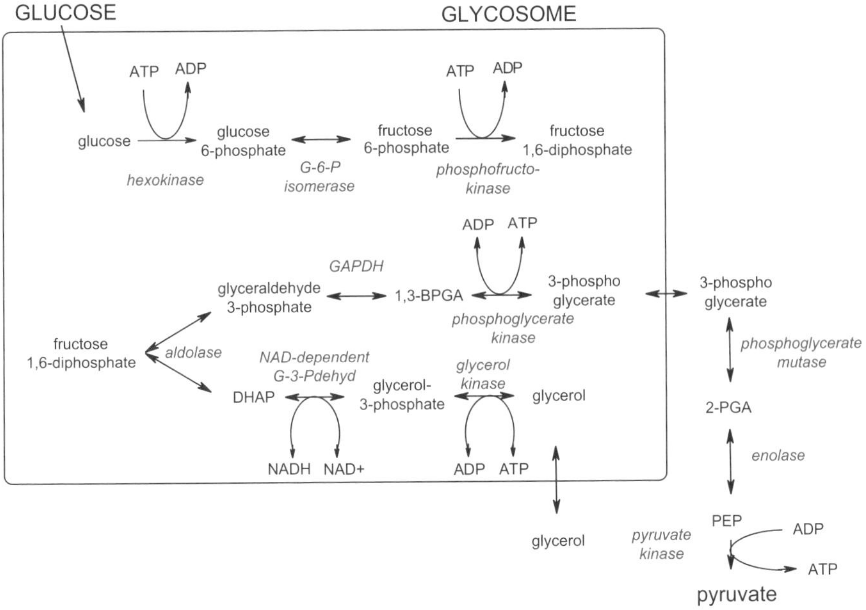

Figure 7.9 Glycolytic pathway in kinetoplastids. Reprinted with permission from ref. 326.

organelle called the glycosome, which renders some enzymes in the pathway differentially sensitive to inhibitors of mammalian homologs.[118]

7.5.1 Hexokinase (HK)

The first enzyme in the glycolytic pathway, hexokinase, transfers phosphate from ATP to glucose. In *T. brucei*, two hexokinases (TbHK1 and TbHK2) that differ primarily in the C-terminus are present, with heterooligomer formation between TbHK1 and TbHK2 believed to be an important regulatory mechanism of activity.[121,122] The essentiality of hexokinase in *T. brucei* has been validated genetically through RNAi methods, and the anti-cancer drug lonidamine (**47**), an inhibitor of mammalian hexokinases, has been suggested to exert its trypanocidal activity via inhibition of the *T. brucei* enzyme.[123] The hexokinase enzymes present in *L. mexicana* and *T. cruzi* have also been cloned, expressed and characterized.[124,125] A series of bisphosphonates (**48, 49**) have been described which inhibit *T. cruzi* hexokinase (Figure 7.10).[126,127]

7.5.2 Phosphoglucose Isomerase (PGI) and Phosphofructokinase (PFK)

The next two enzymes in the glycolysis pathway, phosphoglucose isomerase and phosphofructokinase have been cloned and expressed from *T. brucei*, and their crystal structures have been determined.[128,129] In contrast, from *Leishmania* and *T. cruzi*, only the phosphofructokinase enzymes have been characterized.[130,131] The crystal structures of TbPFK reveal features which distinguish the trypanosomal enzyme from other protozoal, bacterial and mammalian orthologs, and provide opportunities for design of species-specific inhibitors. The synthesis and evaluation of a series of 2,5-anhydro-D-mannitol derivatives (**50**–**52**) as inhibitors of TbPFK have been described (Figure 7.10).[132] Several of

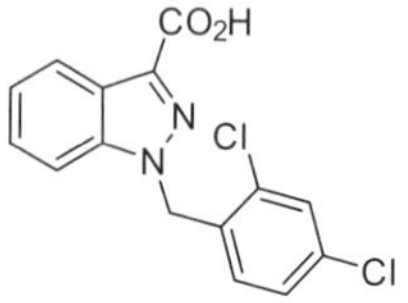

47, rTBHK1 IC_{50} = 850 uM

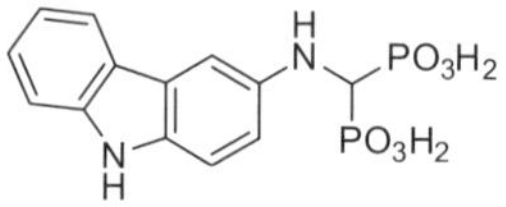

48, TcHK IC_{50} = 0.81 uM

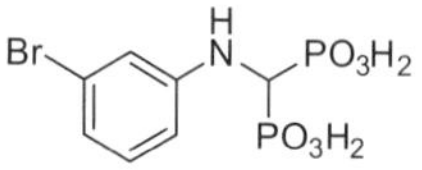

49, TcHK IC_{50} = 0.95 uM

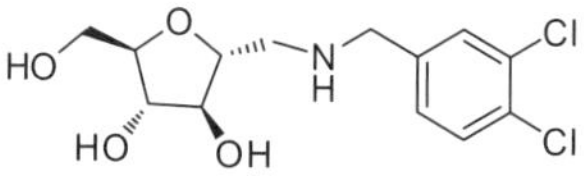

50, TbPFK IC_{50} = 0.41 mM
T. brucei IC_{50} = 0.13 mM

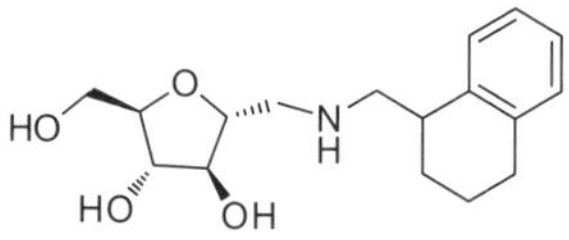

51, Lm PyK IC_{50} = 1.2 mM

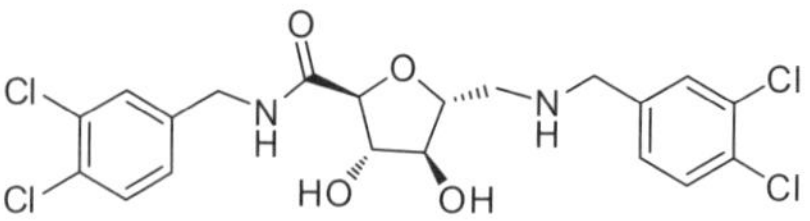

52, TbPFK IC_{50} = 0.023 mM
T. brucei IC_{50} = 0.030 mM

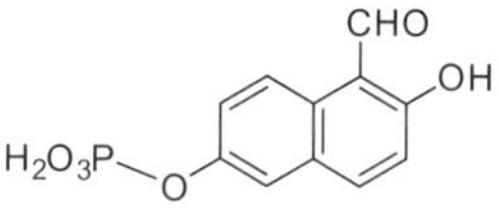

53, TbFBPA K_i = 23 uM

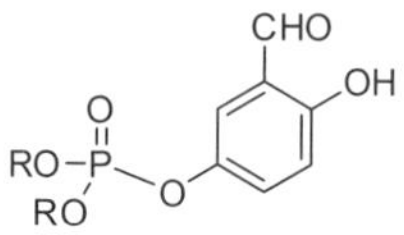

54, R = H
55, R = CH_2OCO^tBu
56, R = $CH_2CH_2SCO^tBu$
57, R = $CH_2CH_2S_2CH_2CH_2OAc$

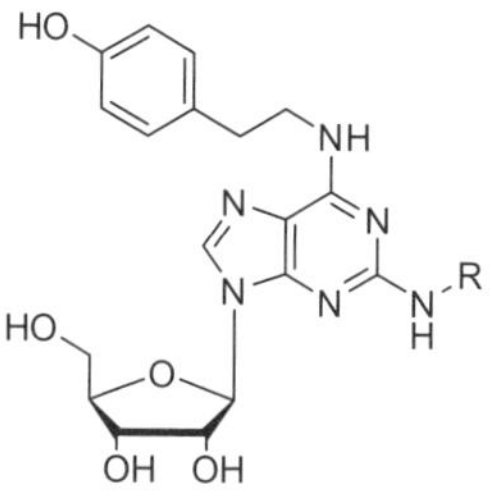

58, R = H, TbPGK IC_{50} = 0.13 mM
59, R = CH_2CH_2Ph, TbPGK IC_{50} = 0.03 mM

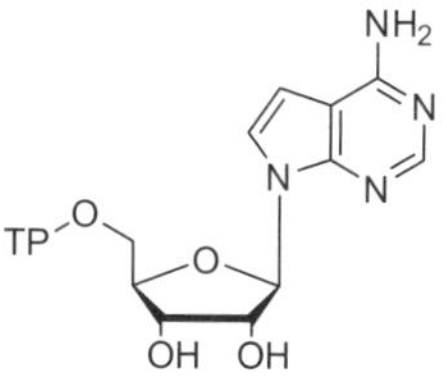

60, TbPGK IC_{50} = 7.5 uM

Figure 7.10 Inhibitors of kinetoplastid glycolytic pathway enzymes.

the more potent analogs in this series also were shown to be trypanocidal in a whole-cell *T. brucei* assay. These compounds were also found to be inhibitors of *Leishmania* pyruvate kinase,[133–135] though the translation of this activity to anti-leishmanial activity was not reported.

7.5.3 Fructose-1,6-Bisphosphate Aldolase

The aldolase enzymes which effect reversible aldol cleavage of fructose 1,6-bisphosphate to dihydroxyacetone phosphate and D-glyceraldehyde-3-phosphate have been cloned, expressed and crystallized from both *T. brucei* and *L. mexicana*.[136,137] The crystal structure of the *T. brucei* enzyme has been used to design inhibitors based on the 1,6-dihydroxy-2-naphthaldehyde (**53**) and 2,5-dihydroxybenzaldehyde scaffolds (**54**), which were selective for this enzyme relative to the rabbit homolog (Figure 7.10).[138] These compounds were not active in a whole-cell *T. brucei* assay, presumably due to poor membrane permeability. This shortcoming was addressed by phosphate ester prodrugs (**55**–**57**), which were found to exhibit modest whole-cell efficacy (Figure 7.10).[139]

7.5.4 Phosphoglycerate Kinase (PGKB)

Several isoforms of the last glycosomal enzyme in the glycolysis pathway, phosphoglycerate kinase, have been identified and characterized from *T. brucei*, *L. mexicana* and *T. cruzi*.[140–142] One of the *T. brucei* isoforms has been crystallized,[143] and this information has enabled a structure-based drug design program to discover a number of purine-based inhibitors (**58, 59**) of this enzyme.[144] In addition, the adenosine analog tubercidin triphosphate (**60**), which is trypanocidal, has been shown to inhibit TbPGKB (Figure 7.10).[145]

7.5.5 Phosphoglycerate Mutase (PGAM), Enolase and Pyruvate Kinase (PyK)

The final three enzymes in the glycolytic pathway, PGAM, enolase and PyK, are not localized in the glycosome, rather are distributed in the cytosol of the parasite. Phosphoglycerate mutase from *L. mexicana* has been cloned, expressed and crystallized.[146] Recently, mechanistic details of this metal-dependent enzyme have emerged, suggesting opportunities for design of parasite-selective inhibitors.[147] The *T. brucei* enzyme is highly homologous to the *L. mexicana* enzyme, and has recently been shown to be essential by RNAi methods.[148] Enolase has been isolated from both *L. mexicana*[149] and *T. brucei*, and a crystal structure of the *T. brucei* enzyme in complex with the inhibitor 2-fluoro-2-phosphonoacetohydroxamate has been reported.[150,151] Finally, mechanistic details of PyK from *Leishmania* have begun to emerge,[152] and inhibitors of this enzyme have been reported.[132]

7.6 Lipid Biosynthesis and Utilization

7.6.1 Fatty Acids

Biosynthesis of fatty acids is critical for kinetoplastids, with *T. brucei, Leishmania* and *T. cruzi* sharing similar enzymatic pathways for this purpose, which are distinct in many ways from mammalian counterparts.[153–156] A particularly important metabolite from this pathway in these parasites is myristate, which is a key component in the variable surface glycoprotein (VSG), which assists the parasites in evasion of immunity response in the host. A unique system of enzymes named elongases are used by the parasite to synthesize myristate and other fatty acids. The essentiality of these enzymes in *T. brucei* has been demonstrated by RNAi methods.[157] In addition to this elongase pathway, a more classical type II fatty acid synthase pathway localized to the mitochondrion has been described in *T. brucei.*[158–160]

Natural products (**61–65**) known to inhibit fatty acid biosynthesis in other parasites (e.g. *Plasmodium falciparum*) have been shown to be active in whole cell *T. brucei* and *L. donovani* assays, though proof of their biochemical targets has not been demonstrated (Figure 7.11).[161–164]

Once synthesized, myristate is attached to a glycophosphatidylinositol (GPI) anchor protein to facilitate localization to the outer membrane of the parasite. The enzyme which effects this transfer is N-myristoyltransferase (NMT), which has been shown to be essential to parasite viability by gene-targetting in *Leishmania* and by RNAi in *T. brucei.*[165] Inhibitors of fungal NMTs (**66–71**) have been described which inhibit both the trypanosomal enzyme and parasite growth, but have been perceived to be of limited potential due to poor drug-like properties (Figure 7.12).[166–168] A structure-based drug design program based on these small molecules and homology models of the *T. brucei* NMT has been reported.[169] Quite recently, the *L. donovani* NMT has been expressed and a crystal structure of the enzyme complexed to a non-hydrolyzable substrate analog has been determined.[170]

Chemical validation of *T. brucei* NMT has recently been demonstrated through a combination of high-throughput screening, crystallography and medicinal chemistry with the discovery of a series of pyrazole sulfonamides.[171]

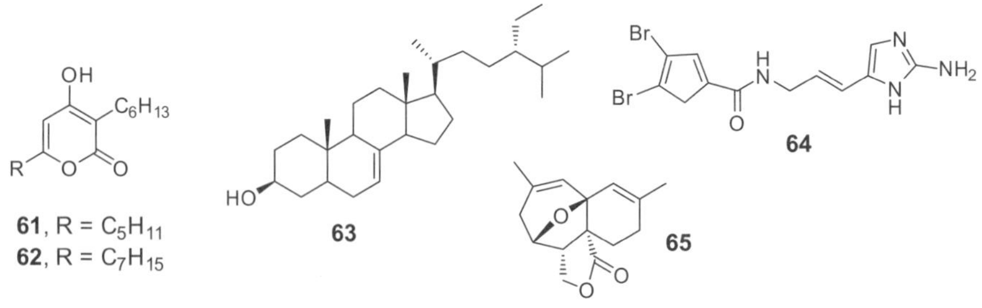

Figure 7.11 Natural product inhibitors of fatty acid biosynthesis.

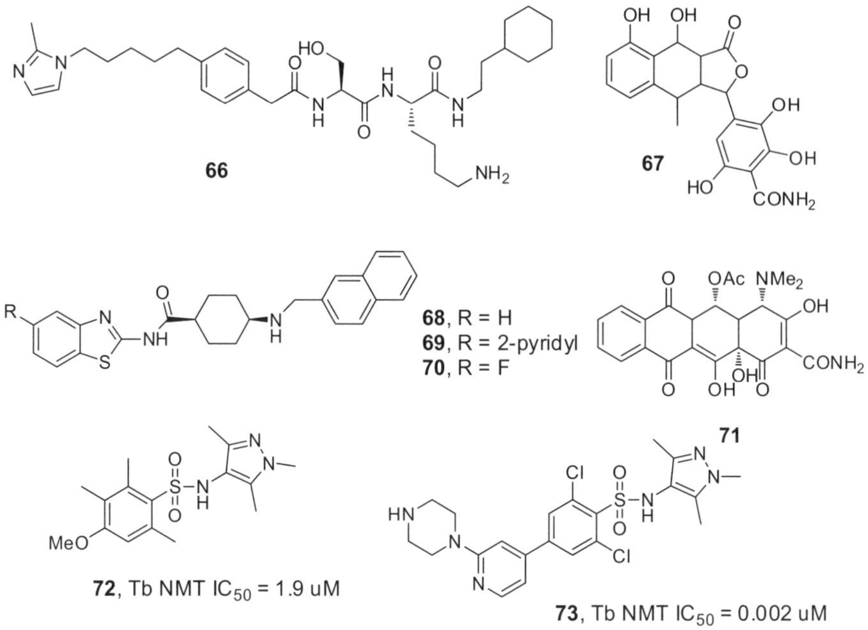

Figure 7.12 Inhibitors of kinetoplastid N-myristoyl transferase (NMT).

In this study, a library of 62 000 compounds was screened against the TbNMT, a number of lead series were identified, including the pyrazole sulfonamide **72**. Optimization of this screening hit through several iterative cycles led to DDD85646 (**73**), which was shown to be trypanocidal both in an *in vitro* whole cell assay and an *in vivo* model of acute trypanosomiasis (Figure 7.12). A crystal structure of LmNMT complexed with DDD85646 was determined, which, along with a number of mechanistic biochemical experiments, demonstrated the link between enzyme inhibition and trypanocidal activity.

7.6.2 Sphingolipids

A second class of lipids critically important to viability of the kinetoplastid parasites is the sphingolipids, which are estimated to account for approximately 10% of the phospholipids in these parasites. The importance of these biomolecules with regard to the structure and function of the GPI membrane anchor system of kinetoplastids, and the potential of enzyme targets in this pathway have been extensively reviewed.[172–175] In *T. brucei*, the key enzyme for biosynthesis of the molecules, sphingolipid synthase, has been shown to be essential based on RNAi and chemical inhibition by aureobasidin A (**74**) (Figure 7.13).[176,177] Aureobasidin A has also been shown to inhibit the related inositol phosphorylceramide synthase from *L. major*.[178] Sphingolipid salvage pathways appear to be more important than *de novo* biosynthesis in *Leishmania*.[179]

7.6.3 Isoprenoids

Like mammalian cells, post-translational modification of a variety of proteins by attachment of isoprenoid groups (e.g. farnesyl and geranyl-geranyl) is an important regulatory mechanism in the kinetoplastids.[180–182] The majority of the enzymes employed in biosynthesis of isoprenoids in kinetoplastids are homologous to those found in mammalian cells, particularly those in the mevalonate pathway, although subtle differences in structure, function and localization have been described.[183–186] Very little has been reported on approaches to inhibit these enzymes as a basis for development of anti-parasitics, with the exception of HMG-CoA reductase.[187] Known inhibitors of the mammalian enzymes have been shown to have some activity against *T. cruzi* growth when administered in combination with sterol biosynthesis inhibitors.[188] Additionally, a single report of antileishmanial activity of the terpene nerolidol (**75**) has appeared, which suggests inhibition of one or more enzymes in the biosynthesis of farnesyl pyrophosphate as a target (Figure 7.14).[189]

In contrast, farnesyl pyrophosphate synthase has been extensively characterized,[190–192] has been shown to be essential in *T. brucei*,[193] and a number of bisphosphonate inhibitors (e.g. **76**–**78**) known to exhibit antiparasitic activity have been evaluated opposite both *T. brucei* and *T. cruzi* FPPS.[194–197]

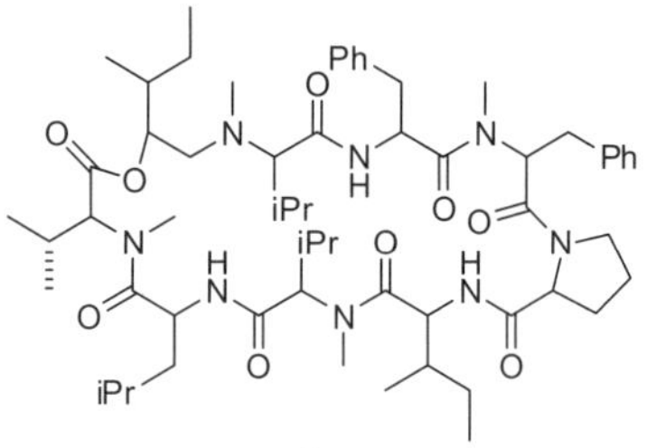

74

Figure 7.13 Aureobasidin A, inhibitor of sphingolipid synthase.

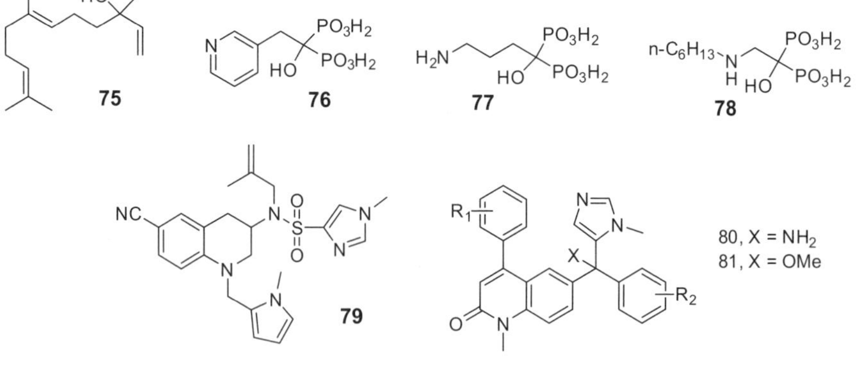

Figure 7.14 Inhibitors of isoprenoid biosynthesis in kinetoplastids.

Once synthesized attachment of farnesyl groups to proteins is accomplished by the enzyme protein farnesyltransferase (PFT), which has been a popular target for screening efforts and subsequent medicinal chemistry programs.[198,199] One class of inhibitors which has exhibited activity against *T. brucei* PFT is tetrahydroquinolines (e.g. **79**).[200] More recently, analogs of a known inhibitor of human PFT, tipifarnib (e.g. **80, 81**), have been found to have potent antiparasitic activity against *T. cruzi* (Figure 7.14).[201,202]

7.6.4 Sterol Biosynthesis

The kinetoplastids also express a reasonably complete set of enzymes which convert farnesylpyrophosphate to sterols, with the ergosterol skeleton as the predominant pathway in both *Leishmania* and *T. cruzi*. By contrast, sterol biosynthesis in the bloodstream form of *T. brucei* appears to be less important, as cholesterol obtained from the host appears to predominate.[203] Of the enzymes in this pathway studied, most effort has focused either on squalene synthase[204] or sterol-14-demethylase.[205]

Early efforts on these targets were based on the availability of inhibitors (**82, 83**) of the human homologs of these enzymes which exhibited activity in whole

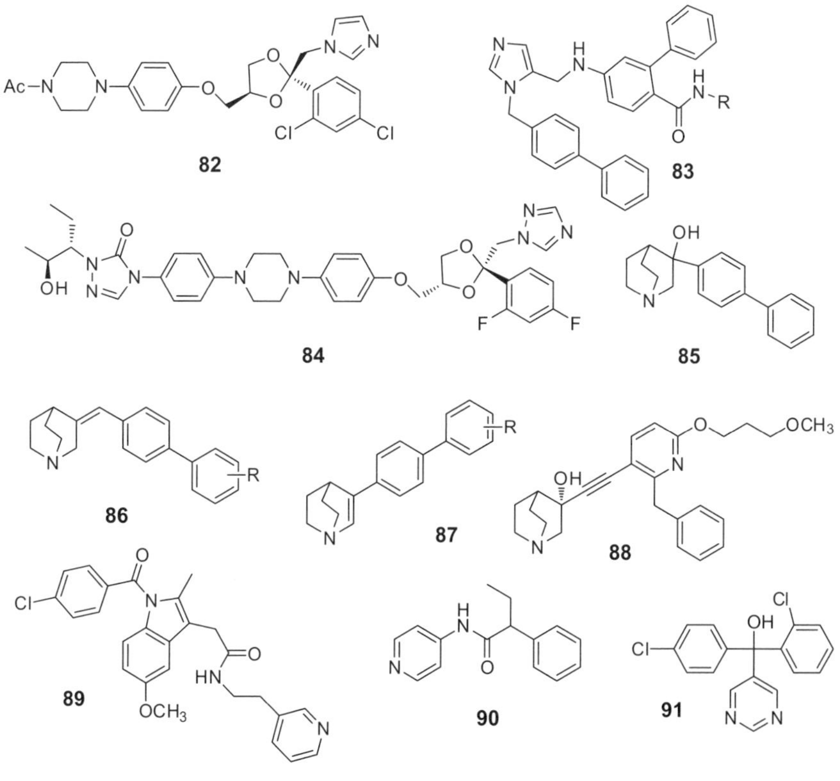

Figure 7.15 Inhibitors of sterol biosynthesis in kinetoplastids.

cell antiparasitic assays (Figure 7.15).[206–208] In addition, it has been reported that posaconazole (**84**), an inhibitor of sterol biosynthesis in fungi is active *in vitro* against *T. cruzi* and mouse models of Chagas disease.[209,210] More recently, further optimization of quinuclidine-based inhibitors (**85–88**) of squalene synthase focused on discovery of selective inhibitors of the *T. cruzi* and *Leishmania* enzymes has been reported (Figure 7.15).[211–214] A somewhat broader array of drug candidates (Figure 7.15). have been identified and optimized for sterol-14-demethylase, including imidazoles (**83**),[207] indoles (**89**),[215] and pyridyl amides (**90**).[216,217] These efforts have been supported by information from crystal structures obtained for the *T. brucei* and *T. cruzi* enzymes.[218,219] It has also been reported that fenarimol (**91**), a compound with anti-leishmanial activity, disrupts sterol biosynthesis in a manner consistent with inhibition of sterol-14-demethylase, and a model for binding of this compound to the catalytic site of the *L. major* enzyme has been developed.[220]

7.7 Signal Transduction Pathways

7.7.1 Phosphodiesterases

Like their mammalian hosts, the kinetoplastids express a number of phosphodiesterases (PDEs), which regulate cAMP levels and impact cell differentiation.[221] Four families of PDEs are conserved across the kinetoplastids, with multiple members of each family known in *T. cruzi* and *T. brucei*.[222] In *T. brucei*, at least five different PDEs are coded by the genome, and two, designated TbPDEB1 and TbPDEB2, have been demonstrated to be essential to parasite growth.[223,224] Similarly, in *Leishmania*, five PDEs are known, though the essentiality of these enzymes has not been established.[225,226] In *T. cruzi*, several PDEs have been identified and characterized,[227–229] and recently one of these enzymes, TcPDEC, has been validated as a therapeutic target through a combination of virtual screening and evaluation of compounds (e.g. **92–94**) in both recombinant enzyme and whole-cell parasiticidal assays (Figure 7.16).[230] Key to further exploitation of the PDEs will be development of parasite-specific compounds, which may be facilitated through structure-based drug design approaches using recently obtained crystallographic information.[231]

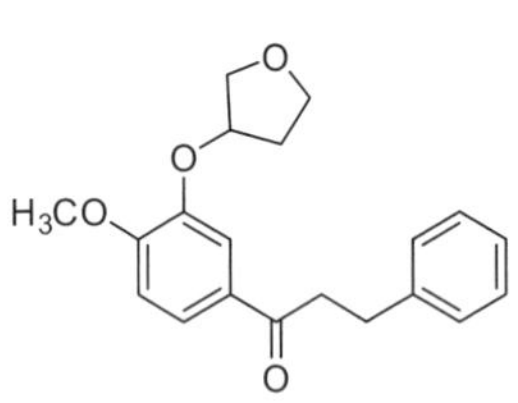

92
TcPDEC IC_{50} = 4.9 uM
T. cruzi IC_{50} = 1.7 uM

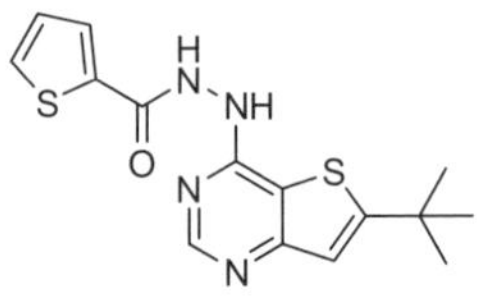

93
TcPDEC IC_{50} = 6.3 uM
T. cruzi IC_{50} = 4.8 uM

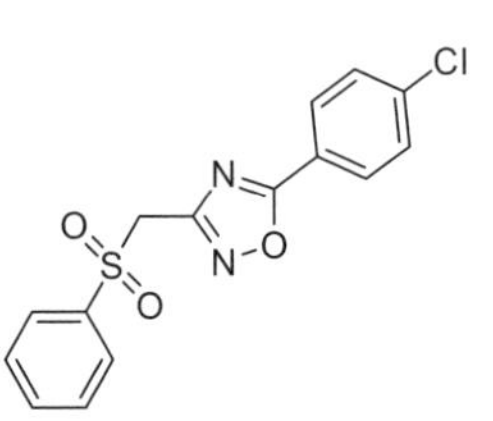

94
TcPDEC IC_{50} = 2.4 uM
T. cruzi IC_{50} < 1 uM (cytotoxic)

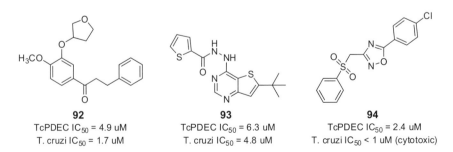

Figure 7.16 Inhibitors of kinetoplastid phosphodiesterases.

7.7.2 Kinases

Several factors contribute to the potential attractiveness of cell cycle kinases as targets for antiparasitic drugs – availability of comparative genetics of the kinetoplastid kinomes,[232] availability of tool compounds from homologous mammalian kinase targets,[233] and the divergence of the cell cycle pathways between these parasites and the mammalian host.[234,235] As a consequence, many kinase targets have been validated by genetic means; fewer have been chemically validated. A number of HTS campaigns have been initiated or planned for particular targets, most notably by the Drug Discovery Unit at Dundee University.[236]

An example where knowledge from a mammalian homolog has been used to initiate interest in the homologs protozoan target is glycogen synthase kinase 3 (GSK3). This kinase has been shown to play a key role in multiple cellular processes including cell survival and death signaling, and can be inhibited by a multitude of ATP-site directed small molecules.[237] Both the *T. brucei* homolog (TbGSK3) and the *Leishmania* homolog (LdGSK3) can be inhibited by known kinase inhibitors.[238,239] In the case of TbGSK3, a diverse set of inhibitors (**95**–**97**) were also evaluated in a whole-cell viability assay, and good correlation was observed between enzyme inhibition and anitparasite activity (Figure 7.17).

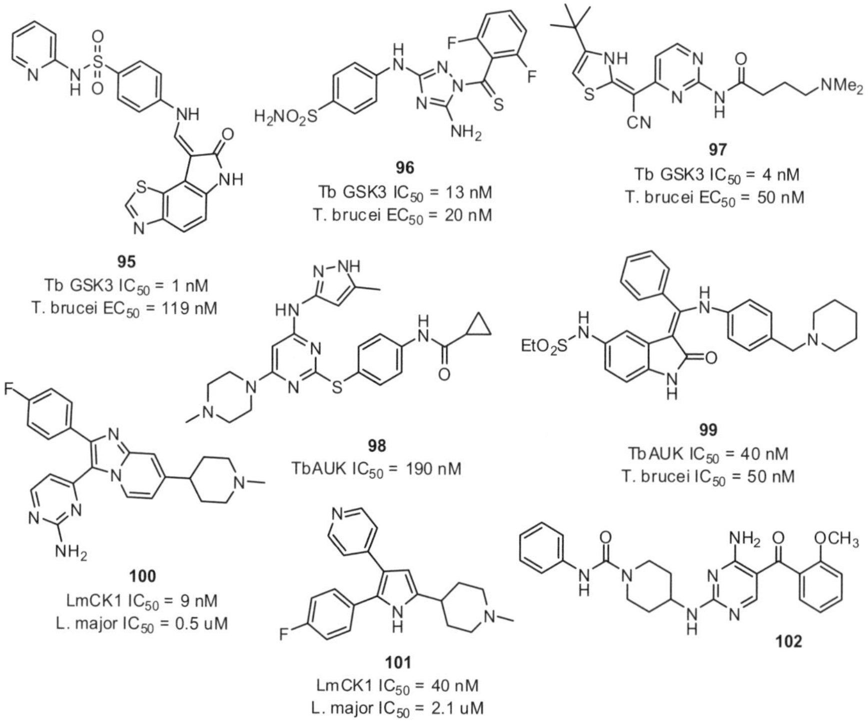

Figure 7.17 Inhibitors of kinetoplastid kinases.

A HTS campaign for this enzyme is planned, and may provide starting points for medicinal chemistry optimization of novel compounds. Similarly, an Aurora kinase homolog from *T. brucei* (TbAUK1) has recently been described as essential for regulation of chromosome segregation and initiation of cytokinesis in the parasite.[240] Inhibition of TbAUK1 by VX-680 (**98**) was shown to inhibit cell cycle progression in *T. brucei.* This enzyme has been further chemically validated by demonstration that hesperadin (**99**), a functional inhibitor of TbAUK1, also inhibits parasite growth (Figure 7.17).[241] A kinase target which has been shown to be essential in *T. brucei* by RNAi is casein kinase 1 (TbCK1).[242] Inhibition of the homologous kinase from *L. major* (LmCK1) with imidazopyridine (**100**) and pyrrole (**101**) analogs has been shown to inhibit parasite growth, further validating this kinase as an attractive target.[243]

Finally, the mitogen-activated protein kinases (MAPKs) have emerged in the last several years as potentially attractive targets, particularly in *Leishmania.*[244] In particular, LmxMPK4 has been shown to be essential in both promastigotes and amastigotes,[245] and is inhibited by staurosporine when activated by co-expressed LmxMKK5, providing the basis for a coupled screening assay for small molecule inhibitors.[246] Very recently, work from our laboratories has demonstrated that several MAPKs (e.g MPK2, MPK9) from both *T. brucei* and *L. major* are likely targets of a series of diaminopyrimidines (**102**) (Figure 7.17).[247]

7.7.3 Proteases

The importance of proteases, and in particular, cysteine proteases, for disease progression and pathogenesis of protozoal infections has been established for quite some time, and approaches based on protease inhibition as an approach to treating these diseases, particularly *T. cruzi* infections, have been reviewed.[248–251] In *T. cruzi,* the cysteine protease cruzipain has been implicated as a key to host cell invasion, and a number of peptidic inhibitors have been identified and evaluated in animal models of Chagas disease. The most advanced peptidic inhibitor, the vinyl sulfone K11777 (**103**), has been shown to be curative in mouse models of *T. cruzi* infection (Figure 7.18).[252,253] Cysteine proteases are also found in *T. brucei* (brucipain, rhodesain, TbcatB) and *Leishmania* (CPB2.8), and peptidic inhibitors (**104**–**106**) for these enzymes have also been described.[254–256] As has been frequently observed with peptide-based enzyme inhibitors, high metabolic liability and poor pharmacokinetic properties, exacerbated by the presence of reactive functional groups such as the vinyl sulfone have limited the utility of these compounds. Consequently, efforts to discover non-peptidic inhibitors of these enzymes have been reported, including several non-peptidic inhibitors containing a vinyl sulfone (**107**), which inhibit cruzain, rhodesain and TbcatB.[257] Additionally, a group of alpha-substituted ketones based on a triazole scaffold (**108**) have been described which inhibit cruzain and are efficacious in a mouse model of Chagas disease (Figure 7.18).[258,259]

Using a combination of information from design of non-peptidic inhibitors for mammalian cysteine proteases[260] and the output of a HTS campaign[261] targeting cruzain, purine (**109**) and triazine (**110**) nitrile inhibitors of TbcatB

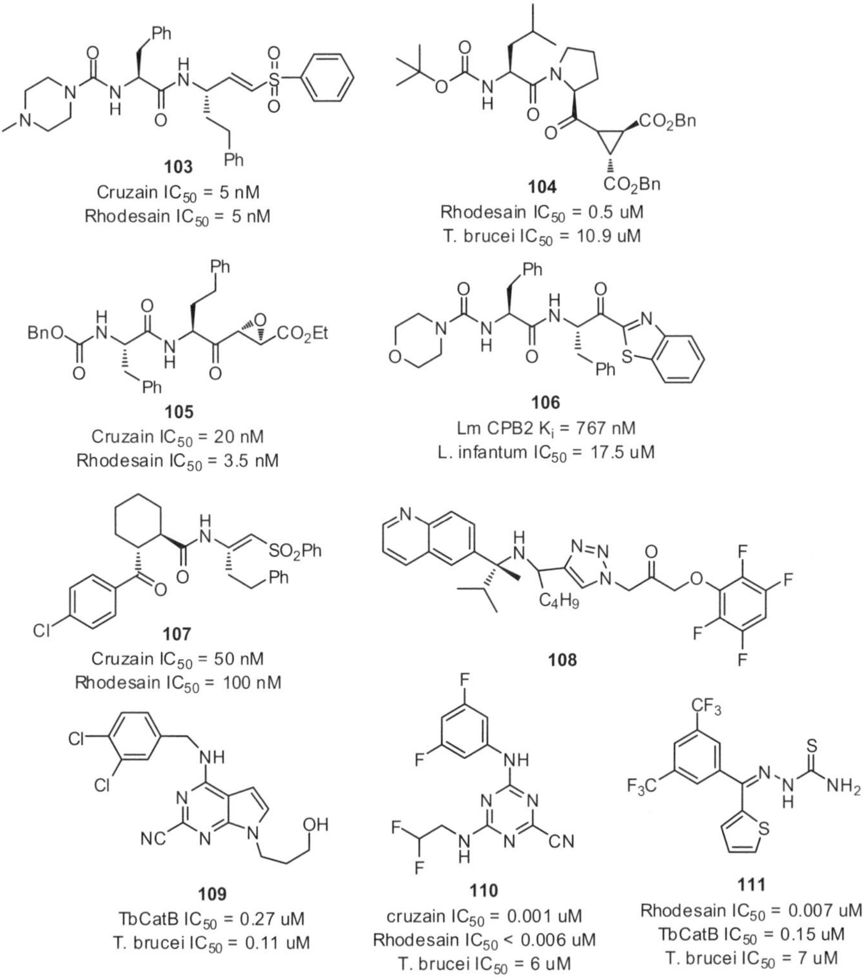

Figure 7.18 Inhibitors of kinetoplastid proteases.

and cruzain have been described (Figure 7.18).[262–264] Non-peptide thiosemicarbazone inhibitors (**111**) of both rhodesain and TbcatB have also been reported, though no correlation between enzyme inhibition and cell-based activity was observed.[265]

7.8 Nucleic Acids

7.8.1 Purine Uptake and Metabolism

The kinetoplastids, like many obligate intracellular parasites, have a limited ability to synthesize purines and pyrimidines *de novo*, and hence are dependent upon salvaging these important heterocycles from the host.[266–268] Consequently, unique sets of transporters and enzymes to salvage a variety of purine-containing precursors from the host and convert them to essential nucleotides are present.[269] A number of approaches based on inhibition of these transporters or enzymes have been pursued, which aim to capitalize on differences between the parasite and host.[270] One such approach has emerged from the examination of cordycepin (**112**), which is actively transported into *T. brucei* through the purine transporter TbAT1 and terminates DNA and RNA synthesis (Figure 7.19).[271] While cordycepin is potent *in vitro*, it is not active *in vivo* unless an adenosine deaminase inhibitor, such as coformycin is administered.[272] The activity of cordycepin and the related adenosine arabinoside (**113**) is dependent on the adenosine kinase (TbAK1), which has been shown to play a role in transport of purines.[273] Inhibitors of this enzyme such as ABT-702 (**114**) have exhibited weak activity against *T. brucei*, but are synergistic with cordycepin.[274]

A second enzyme important for utilization of scavenged purines is the purine-specific nucleoside hydrolase IAG-NH.[275,276] Extensive exploration of the structural features of the *T. vivax* and *T. brucei* enzymes (TvNH and TbNH, respectively), in complex with the iminoribitol-based immucillins (**115**) have allowed for design of simplified N-aryl analogs (**116–119**) which are potent inhibitors of these enzymes (Figure 7.19).[277–279] Translation of enzyme inhibition to efficacy against *T. brucei* has been difficult in this series, although the analog UAMC-00363 (**120**) has recently been shown to be active in both *in vitro* and *in vivo T. b. brucei* assays.[280]

An alternative approach to take advantage of the dependence of the kinetoplastids on active transport of purines for survival is to use the same transporters to facilitate accumulation of toxic compounds in the parasite. Some

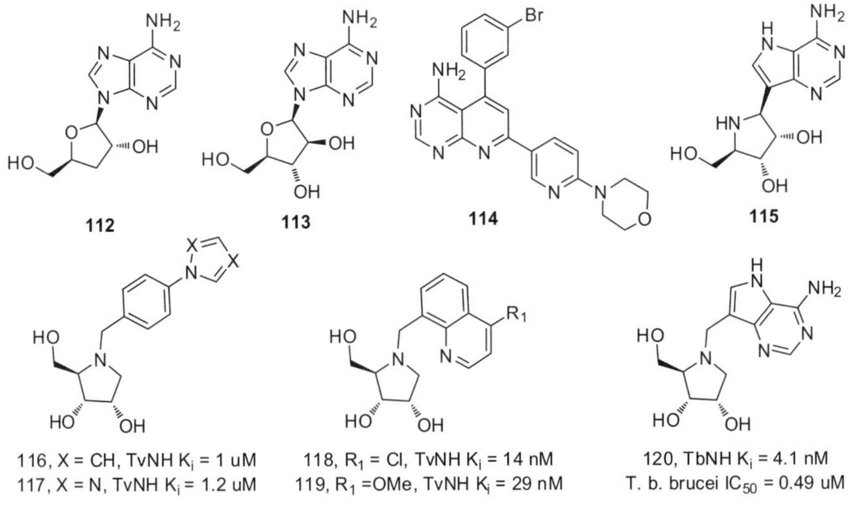

Figure 7.19 Inhibitors of kinetoplastid purine transport and metabolism.

known antiparasitics contain structural features recognized by these transporters, and hence their activity is enhanced by active transport (Figure 7.20).[281–283] More recently, knowledge of such structural features have prompted linkage of groups such as melamine to increase activity of otherwise modestly active nitroheterocycles (**121, 122**).[284–286] Increased activity of an ODC inhibitor (**123**) has also been observed by use of the melamine unit to assist transport.[287] In constrast, an attempt to use this strategy to improve delivery of eflornithine or a fluoroquinoline into *T. brucei* was less successful than anticipated.[288] Similarly, attachment of melamine to a series of Mannich bases (**124**) has not proven to increase activity of this class of compounds as parasiticides.[289]

7.8.2 DNA Topoisomerases

In the kinetoplastids, the DNA topoisomerases are distinct in their structure, as they are present as heteromultimers, as opposed to the single polypeptide motif found in mammals.[290] Despite this difference, inhibitors of mammalian DNA topoisomerases such as doxorubicin (**125**) have been shown to have

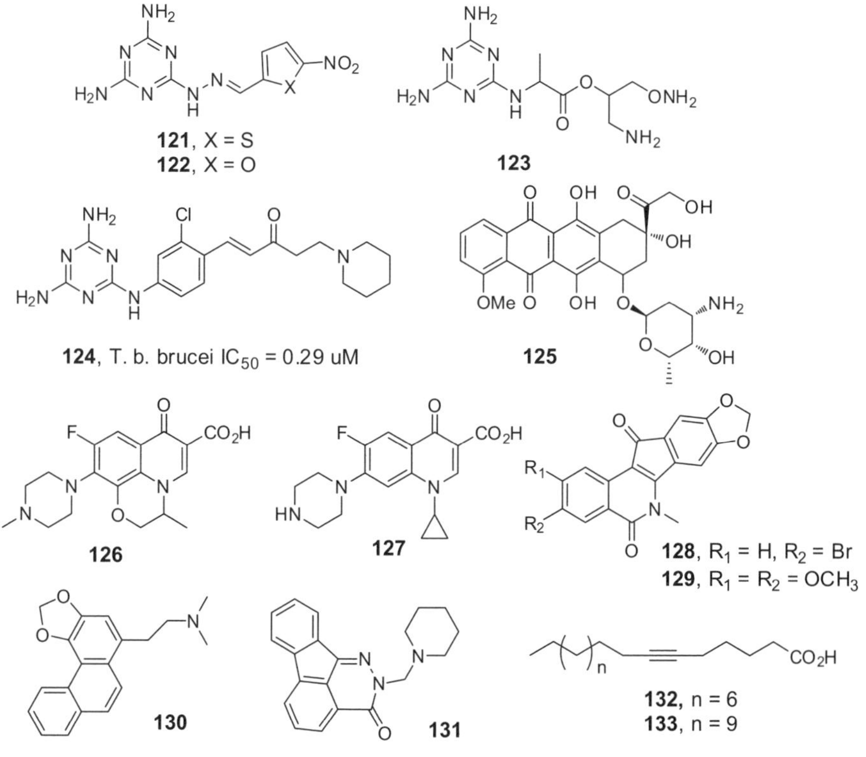

Figure 7.20 Melamine directed trypanocidal agents and DNA topoisomerase inhibitors.

trypanocidal activity.[291] In addition, antibacterial fluoroquinolones such as ofloxacin (**126**) and ciprofloxacin (**127**) have exhibited trypanocidal activity (Figure 7.20).[292,293] Recently, a series of indenoisoquinolines (e.g. **128, 129)** were shown to have trypanocidal activity both *in vitro* and *in vivo*.[294] While current evidence suggests that these compounds act predominantly via topoisomerase-1B in *T. brucei*, other compounds identified from a bacterial topoisomerase-1A screen (**130, 131**) have been found to exert trypanocidal activity through topoisomerase-2 (Figure 7.20).[295] Several unsaturated fatty acid analogs (**132, 133**) have been shown to inhibit the *L. donovani* topoisomerase and have *in vitro* activity in a whole-cell assay.[296,297]

7.8.3 DNA Binding Agents – Diamidines

The SAR of the diamidine class of anti-parasitic drugs, as exemplified by pentamidine, a clinically relevant drug for stage 1 HAT, has been extensively explored for potential new compounds that might exhibit superior potency and lowered toxicity, has been extensively reviewed.[298,299] As highlighted in the preceeding chapter, the candidate drug DB289 (**134**), which was progressed to clinical trials for stage 1 HAT based on efficacy in both murine and non-human models of HAT emerged from these efforts (Figure 7.21).[300,301] Unfortunately, clinical development of DB289 was terminated in 2008 due to safety concerns.

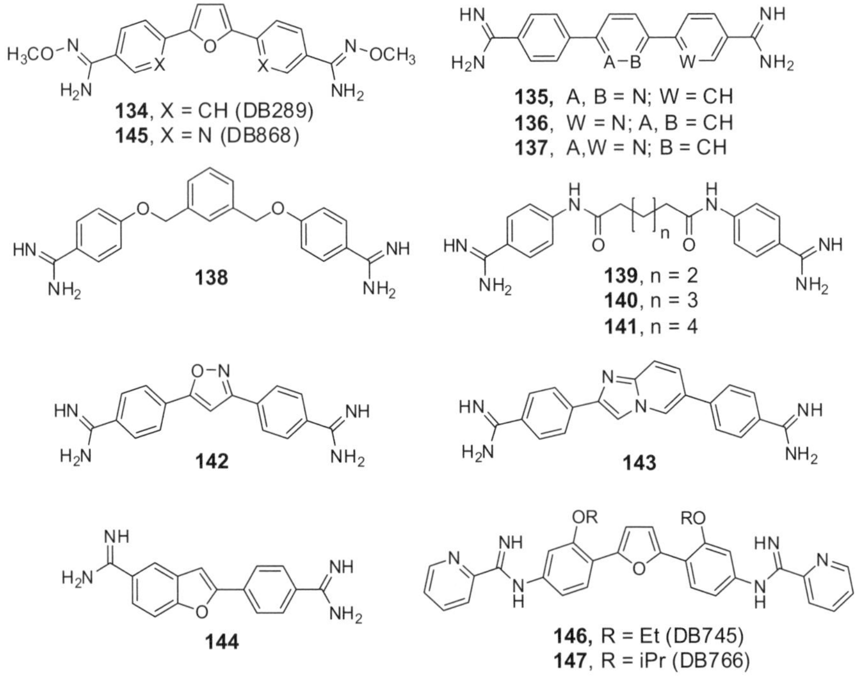

Figure 7.21 Anti-kinetoplastid diamidines and related structures.

Over the past decade, variants of the core scaffold in the diamidine series have been reported. These include linear triaryl (**135–137**),[302] alkoxyaromatic (**138**),[303] diamide (**139–14**),[304] and numerous heterocyclic (**142–144**) frameworks (Figure 7.21).[305–311] As expected, based on the mechanism of action of these molecules, activity has been observed against all three kinetoplastids, and many of the compounds described have exhibited *in vivo* activity. One of the limitation of many of the diamidines described to date has been poor permeability across the blood–brain barrier, which limits their usefulness as treatments for stage 2 HAT. This limitation has been a focus of effort in this class, and has resulted in the identification of a close analog of DB289, the dipyridyl DB868 (**145**), which is active in animal models of CNS disease.[312]

Structurally related to the diamidines, but with divergent physicochemical properties and SAR are a series of arylimidamides (**146, 147**), initially reported as efficacious against *T. cruzi.*[313,314] More recently, this class of molecules has been progressed as antileishmanial agents.[315]

7.9 Tubulin

Polymerization of tubulin to microtubules in kinetoplastids is essential for chromosomal segregation and motility, suggesting that agents capable of disrupting this polymerization should exhibit antiparasitic activity. Classical tubulin inhibitors have been shown to inhibit parasite growth.[316] As tubulin is expressed in both the parasites and the mammalian host, development of compounds that are selective for disruption of parasite tubulin has been a key focus of efforts in this area since the discovery that the herbicide trifluralin (**148**) exhibited this selectivity (Figure 7.22).[317] A related class of dinitrosulfanilides exemplied by oryzalin (**149**), have been extensively explored as both anti-leishmanial and anti-trypanosomal agents.[318–320]

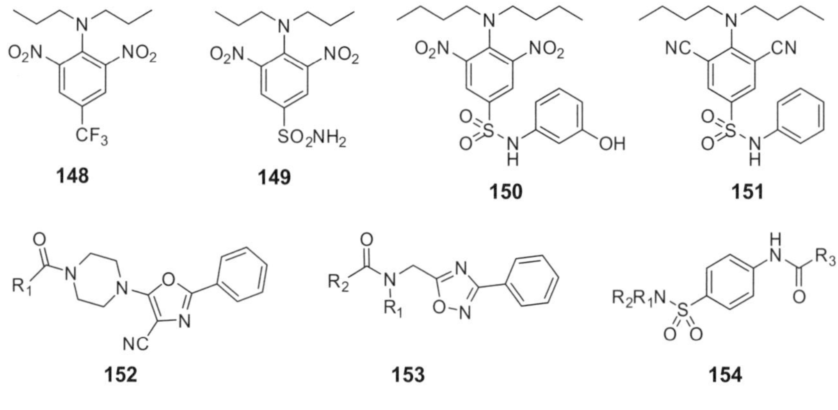

Figure 7.22 Inhibitors of kinetoplastid tubulin assembly.

Metabolic stability of early compounds in this series was suggested as a potential contributor to the poor *in vivo* activity observed in this class.[321] Efforts to improve metabolic stability through variation of substituents on the dinitroaniline and sulfonamide moieties (e.g. **150**) have met with limited success.[322] Concerns over potential genotoxicity of these dinitroaromatics have been addressed by demonstration that dicyano analogs (e.g. **151**) are essentially equipotent in whole cell *in vitro* assays (Figure 7.22).[323] Continued challenges with obtaining *in vivo* activity in this class have prompted efforts to identify alternative starting points for lead optimization. A recently reported HTS for compounds which inhibit assembly of tubulin isolated from *L. tarentolae* has provided several classes (**152**–**154**) of compounds for this purpose.[324,325]

7.10 Conclusions

In the past several years, exploration of new approaches to discovery of drugs to treat kinetoplastid diseases has seen a resurgence of effort, with several new classes of drug molecules discovered through a combination of whole cell antiparasitic assays, high-throughput screening and target-directed screening approaches. While no single class of compounds has yet emerged that could be considered a "silver bullet" for treatment of the diseases caused by these parasites, several promising classes have been described, including the benzoxaboroles, diamidines and nitroheterocycles. From target-based approaches, discovery of inhibitors of key enzymes in the polyamine pathway (e.g. ODC, SAM-DC), lipid biosynthesis and utilization (e.g. NMT), and cell cycle kinases (e.g. MAPKs, CKs) appear to offer attractive starting points for lead optimization programs. Like all drug discovery programs, considerable challenges remain for progression of compounds from these approaches to clinical utility, although it is encouraging that an increasing emphasis on integration of pharmacokinetics, medicinal chemistry and pharmacology in these programs has emerged.

References

1. N. M. El-Sayed, P. J. Myler, G. Blandin, M. Berriman, J. Crabtree, G. Aggarwal, E. Caler, H. Renauld, E. A. Worthey, C. Hertz-Fowler, E. Ghedin, C. Peacock, D. C. Bartholomeu, B. J. Haas, A. N. Tran, J. R. Wortman, U. C. Alsmark, S. Angiuoli, A. Anupama, J. Badger, F. Bringaud, E. Cadag, J. M. Carlton, G. C. Cerqueira, T. Creasy, A. L. Delcher, A. Djikeng, T. M. Embley, C. Hauser, A. C. Ivens, S. K. Kummerfeld, J. B. Pereira-Leal, D. Nilsson, J. Peterson, S. L. Salzberg, J. Shallom, J. C. Silva, J. Sundaram, S. Westenberger, O. White, S. E. Melville, J. E. Donelson, B. Andersson, K. D. Stuart and N. Hall, *Science*, 2005, **309**, 404.

2. M. Berriman, E. Ghedin, C. Hertz-Fowler, G. Blandin, H. Renauld, D. C. Bartholomeu, N. J. Lennard, E. Caler, N. E. Hamlin, B. Haas, U. Bohme, L. Hannick, M. A. Aslett, J. Shallom, L. Marcello, L. Hou, B. Wickstead, U. C. Alsmark, C. Arrowsmith, R. J. Atkin, A. J. Barron, F. Bringaud, K. Brooks, M. Carrington, I. Cherevach, T. J. Chillingworth, C. Churcher, L. N. Clark, C. H. Corton, A. Cronin, R. M. Davies, J. Doggett, A. Djikeng, T. Feldblyum, M. C. Field, A. Fraser, I. Goodhead, Z. Hance, D. Harper, B. R. Harris, H. Hauser, J. Hostetler, A. Ivens, K. Jagels, D. Johnson, J. Johnson, K. Jones, A. X. Kerhornou, H. Koo, N. Larke, S. Landfear, C. Larkin, V. Leech, A. Line, A. Lord, A. Macleod, P. J. Mooney, S. Moule, D. M. Martin, G. W. Morgan, K. Mungall, H. Norbertczak, D. Ormond, G. Pai, C. S. Peacock, J. Peterson, M. A. Quail, E. Rabbinowitsch, M. A. Rajandream, C. Reitter, S. L. Salzberg, M. Sanders, S. Schobel, S. Sharp, M. Simmonds, A. J. Simpson, L. Tallon, C. M. Turner, A. Tait, A. R. Tivey, S. Van Aken, D. Walker, D. Wanless, S. Wang, B. White, O. White, S. Whitehead, J. Woodward, J. Wortman, M. D. Adams, T. M. Embley, K. Gull, E. Ullu, J. D. Barry, A. H. Fairlamb, F. Opperdoes, B. G. Barrell, J. E. Donelson, N. Hall, C. M. Fraser, S. E. Melville and N. M. El-Sayed, *Science*, 2005, **309**, 416.
3. A. C. Ivens, C. S. Peacock, E. A. Worthey, L. Murphy, G. Aggarwal, M. Berriman, E. Sisk, M. A. Rajandream, E. Adlem, R. Aert, A. Anupama, Z. Apostolou, P. Attipoe, N. Bason, C. Bauser, A. Beck, S. M. Beverley, G. Bianchettin, K. Borzym, G. Bothe, C. V. Bruschi, M. Collins, E. Cadag, L. Ciarloni, C. Clayton, R. M. Coulson, A. Cronin, A. K. Cruz, R. M. Davies, J. D. Gaudenzi, D. E. Dobson, A. Duesterhoeft, G. Fazelina, N. Fosker, A. C. Frasch, A. Fraser, M. Fuchs, C. Gabel, A. Goble, A. Goffeau, D. Harris, C. Hertz-Fowler, H. Hilbert, D. Horn, Y. Huang, S. Klages, A. Knights, M. Kube, N. Larke, L. Litvin, A. Lord, T. Louie, M. Marra, D. Masuy, K. Matthews, S. Michaeli, J. C. Mottram, S. Muller-Auer, H. Munden, S. Nelson, H. Norbertczak, K. Oliver, S. O'Neil, M. Pentony, T. M. Pohl, C. Price, B. Purnelle, M. A. Quail, E. Rabbinowitsch, R. Reinhardt, M. Rieger, J. Rinta, J. Robben, L. Robertson, J. C. Ruiz, S. Rutter, D. Saunders, M. Schafer, J. Schein, D. C. Schwartz, K. Seeger, A. Seyler, S. Sharp, H. Shin, D. Sivam, R. Squares, S. Squares, V. Tosato, C. Vogt, G. Volckaert, R. Wambutt, T. Warren, H. Wedler, J. Woodward, S. Zhou, W. Zimmermann, D. F. Smith, J. M. Blackwell, K. D. Stuart, B. Barrell and P. J. Myler, *Science*, 2005, **309**, 436.
4. N. M. El-Sayed, P. J. Myler, D. C. Bartholomeu, D. Nilsson, G. Aggarwal, A. N. Tran, E. Ghedin, E. A. Worthey, A. L. Delcher, G. Blandin, S. J. Westenberger, E. Caler, G. C. Cerqueira, C. Branche, B. Haas, A. Anupama, E. Arner, L. Aslund, P. Attipoe, E. Bontempi, F. Bringaud, P. Burton, E. Cadag, D. A. Campbell, M. Carrington, J. Crabtree, H. Darban, J. F. da Silveira, P. de Jong, K. Edwards, P. T. Englund, G. Fazelina, T. Feldblyum, M. Ferella, A. C. Frasch,

K. Gull, D. Horn, L. Hou, Y. Huang, E. Kindlund, M. Klingbeil, S. Kluge, H. Koo, D. Lacerda, M. J. Levin, H. Lorenzi, T. Louie, C. R. Machado, R. McCulloch, A. McKenna, Y. Mizuno, J. C. Mottram, S. Nelson, S. Ochaya, K. Osoegawa, G. Pai, M. Parsons, M. Pentony, U. Pettersson, M. Pop, J. L. Ramirez, J. Rinta, L. Robertson, S. L. Salzberg, D. O. Sanchez, A. Seyler, R. Sharma, J. Shetty, A. J. Simpson, E. Sisk, M. T. Tammi, R. Tarleton, S. Teixeira, S. Van Aken, C. Vogt, P. N. Ward, B. Wickstead, J. Wortman, O. White, C. M. Fraser, K. D. Stuart and B. Andersson, *Science*, 2005, **309**, 409.
5. A. P. Jackson, M. Sanders, A. Berry, J. McQuillan, M. A. Aslett, M. A. Quail, B. Chukualim, P. Capewell, A. MacLeod, S. E. Melville, W. Gibson, J. D. Barry, M. Berriman and C. Hertz-Fowler, *PLoS Negl. Trop. Dis.*, 2010, **4**, e658.
6. E. Ullu, C. Tschudi and T. Chakraborty, *Cellular Microbiology*, 2004, **6**, 509.
7. B. Räz, M. Iten, Y. Grether-Bühler, R. Kaminsky and R. Brun, *Acta Tropica*, 1997, **68**, 139.
8. J. Mikus and D. Steverding, *Parasitol. Int.*, 2000, **48**, 265.
9. Z. B. Mackey, A. M. Baca, J. P. Mallari, B. Apsel, A. Shelat, E. J. Hansell, P. K. Chiang, B. Wolff, K. R. Guy, J. Williams and J. H. McKerrow, *Chem. Biol. Drug. Des.*, 2006, **67**, 355.
10. C. J. Thalhofer, J. W. Graff, L. Love-Homan, S. M. Hickerson, N. Craft, S. M. Beverley and M. E. Wilson, *J. Vis. Exp.*, 2010.
11. F. Buckner, C. Verlinde, A. L. Flamme and W. Van Voorhis, *Antimicrob. Agents Chemother.*, 1996, **40**, 2592.
12. J. L. Siqueira-Neto, O. R. Song, H. Oh, J. H. Sohn, G. Yang, J. Nam, J. Jang, J. Cechetto, C. B. Lee, S. Moon, A. Genovesio, E. Chatelain, T. Christophe and L. H. Freitas-Junior, *PLoS Negl. Trop. Dis.*, 2010, **4**, e675.
13. S. J. Baker, Y. K. Zhang, T. Akama, A. Lau, H. Zhou, V. Hernandez, W. Mao, M. R. Alley, V. Sanders and J. J. Plattner, *J. Med. Chem.*, 2006, **49**, 4447.
14. S. J. Baker, T. Akama, Y. K. Zhang, V. Sauro, C. Pandit, R. Singh, M. Kully, J. Khan, J. J. Plattner, S. J. Benkovic, V. Lee and K. R. Maples, *Bioorg. Med. Chem. Lett.*, 2006, **16**, 5963.
15. T. Akama, S. J. Baker, Y. K. Zhang, V. Hernandez, H. Zhou, V. Sanders, Y. Freund, R. Kimura, K. R. Maples and J. J. Plattner, *Bioorg. Med. Chem. Lett.*, 2009, **19**, 2129.
16. D. Ding, Y. Zhao, Q. Meng, D. Xie, B. Nare, D. Chen, C. J. Bacchi, N. Yarlett, Y.-K. Zhang, V. Hernandez, Y. Xia, Y. Freund, M. Abdulla, K.-H. Ang, J. Ratnam, J. H. McKerrow, R. T. Jacobs, H. Zhou and J. J. Plattner, *ACS Medicinal Chemistry Letters*, 2010, **1**, 165.
17. B. Nare, S. Wring, C. Bacchi, B. Beaudet, T. Bowling, R. Brun, D. Chen, C. Ding, Y. Freund, E. Gaukel, A. Hussain, K. Jarnagin, M. Jenks, M. Kaiser, L. Mercer, E. Mejia, A. Noe, M. Orr, R. Parham, J. Plattner, R. Randolph, D. Rattendi, C. Rewerts, J. Sligar, N. Yarlett, R. Don and R. Jacobs, *Antimicrob. Agents Chemother.*, 2010, **54**, 4379.

18. R. T. Jacobs, B. Nare, S. A. Wring, M. D. Orr, D. Chen, J. M. Sligar, M. G. Jenks, R. A. Noe, T. S. Bowling, L. T. Mercer, C. Rewerts, E. Gaukel, J. Owens, R. Parham, R. Randolph, B. Beaudet, C. Bacchi, N. Yarlett, J. J. Plattner, Y. R. Freund, C. Ding, K. Jarnagin, T. Akama, Y.-K. Zhang, R. Brun, M. Kaiser, I. Scandale and R. Don, *PLoS Negl. Trop. Dis.*, 2011, **5**, e1151.
19. F. L. Rock, W. Mao, A. Yaremchuk, M. Tukalo, T. Crepin, H. Zhou, Y. K. Zhang, V. Hernandez, T. Akama, S. J. Baker, J. J. Plattner, L. Shapiro, S. A. Martinis, S. J. Benkovic, S. Cusack and M. R. Alley, *Science*, 2007, **316**, 1759.
20. Y.-K. Zhang, J. J. Plattner, T. Akama, S. J. Baker, V. S. Hernandez, V. Sanders, Y. Freund, R. Kimura, W. Bu, K. M. Hold and X.-S. Lu, *Bioorg. Med. Chem. Lett.*, 2010, **20**, 2270.
21. J. M. Kelly, M. A. Miles and A. C. Skinner, *Antimicrob. Agents Chemother.*, 1999, **43**, 985.
22. J. M. Kelly, G. Quack and M. M. Miles, *Antimicrob. Agents Chemother.*, 2001, **45**, 1360.
23. N. Kolocouris, G. Zoidis, G. B. Foscolos, G. Fytas, S. R. Prathalingham, J. M. Kelly, L. Naesens and E. D. Clercq, *Bioorg. Med. Chem. Lett.*, 2007, **17**, 4358.
24. I. Papanastasiou, A. Tsotinis, N. Kolocouris, S. R. Prathalingam and J. M. Kelly, *J. Med. Chem.*, 2008, **51**, 1496.
25. G. Zoidis, N. Kolocouris, J. M. Kelly, S. R. Prathalingam, L. Naesens and E. D. Clercq, *Eur. J. Med. Chem.*, 2010, **45**, 5022.
26. J. Faist, W. Seebacher, M. Kaiser, R. Brun, R. Saf and R. Weis, *Eur. J. Med. Chem.*, 2010, **45**, 179.
27. C. Schlapper, W. Seebacher, J. Faist, M. Kaiser, R. Brun, R. Saf and R. Weis, *Eur. J. Med. Chem.*, 2009, **44**, 736.
28. C. Schlapper, W. Seebacher, M. Kaiser, R. Brun, R. Saf and R. Weis, *Eur. J. Med. Chem.*, 2008, **43**, 800.
29. H. Berger, R. Weis, M. Kaiser, R. Brun, R. Saf and W. Seebacher, *Bioorg. Med. Chem.*, 2008, **16**, 6371.
30. C. Schlapper, W. Seebacher, M. Kaiser, R. Brun, R. Saf and R. Weis, *Bioorg. Med. Chem.*, 2007, **15**, 5543.
31. W. Seebacher, R. Weis, M. Kaiser, R. Brun and R. Saf, *J. Pharm. Pharm. Sci.*, 2005, **8**, 578.
32. W. Seebacher, C. Schlapper, R. Brun, M. Kaiser, R. Saf and R. Weis, *Eur. J. Pharm. Sci.*, 2005, **24**, 281.
33. W. Seebacher, R. Brun and R. Weis, *Eur. J. Pharm. Sci.*, 2004, **21**, 225.
34. R. Weis, R. Brun, R. Saf and W. Seebacher, *Monatshefte für Chemie/Chemical Monthly*, 2003, **134**, 1019.
35. R. Weis and W. Seebacher, *Curr. Med. Chem.*, 2009, **16**, 1426.
36. B. S. Hall, X. Wu, L. Hu and S. R. Wilkinson, *Antimicrob. Agents Chemother.*, 2010, **54**, 1193.
37. H. Cerecetto and M. Gonzalez, *Mini. Rev. Med. Chem.*, 2008, **8**, 1355.

38. E. Cabrera, M. G. Murguiondo, M. G. Arias, C. Arredondo, C. Pintos, G. Aguirre, M. Fernandez, Y. Basmadjian, R. Rosa, J. P. Pacheco, S. Raymondo, R. D. Maio, M. Gonzalez and H. Cerecetto, *Eur. J. Med. Chem.*, 2009, **44**, 3909.
39. A. Foroumadi, S. Pournourmohammadi, F. Soltani, M. Asgharian-Rezaee, S. Dabiri, A. Kharazmi and A. Shafiee, *Bioorg. Med. Chem. Lett.*, 2005, **15**, 1983.
40. J. Rodriguez, V. J. Aran, L. Boiani, C. Olea-Azar, M. L. Lavaggi, M. Gonzalez, H. Cerecetto, J. D. Maya, C. Carrasco-Pozo and H. S. Cosoy, *Bioorg. Med. Chem.*, 2009, **17**, 8186.
41. J. Rodriguez, A. Gerpe, G. Aguirre, U. Kemmerling, O. E. Piro, V. J. Aran, J. D. Maya, C. Olea-Azar, M. Gonzalez and H. Cerecetto, *Eur. J. Med. Chem.*, 2009, **44**, 1545.
42. F. Poorrajab, S. K. Ardestani, S. Emami, M. Behrouzi-Fardmoghadam, A. Shafiee and A. Foroumadi, *Eur. J. Med. Chem.*, 2009, **44**, 1758.
43. M. P. Barrett, *Curr. Opin. Infect. Dis.*, 2010, **23**, 603.
44. C. Burri, *Parasitology*, 2010, **137**, 1987.
45. R. C. Ferreira, U. Schwarz and L. C. Ferreira, *Mutat. Res.*, 1988, **204**, 577.
46. A. Buschini, L. Ferrarini, S. Franzoni, S. Galati, M. Lazzaretti, F. Mussi, C. Northfleet de Albuquerque, T. Maria Araujo Domingues Zucchi and P. Poli, *J. Parasitol. Res.*, 2009, **2009**, 463–575.
47. A. Buschini, F. Giordani, C. N. de Albuquerque, C. Pellacani, G. Pelosi, C. Rossi, T. M. Zucchi and P. Poli, *Biochem. Pharmacol.*, 2007, **73**, 1537.
48. A. Bendesky, D. Menéndez and P. Ostrosky-Wegman, *Mutation Research/Reviews in Mutation Research*, 2002, **511**, 133.
49. S. R. Wilkinson, M. C. Taylor, D. Horn, J. M. Kelly and I. Cheeseman, *Proc. Natl. Acad. Sci. U. S. A.*, 2008, **105**, 5022.
50. A. Y. Sokolova, S. Wyllie, S. Patterson, S. L. Oza, K. D. Read and A. H. Fairlamb, *Antimicrob. Agents Chemother.*, 2010, **54**, 2893.
51. A. H. Fairlamb, N. S. Carter, M. Cunningham and K. Smith, *Mol. Biochem. Parasitol.*, 1992, **53**, 213.
52. M. L. Cunningham and A. H. Fairlamb, *Eur. J. Biochem.*, 1995, **230**, 460.
53. P. Baiocco, G. Colotti, S. Franceschini and A. Ilari, *J. Med. Chem.*, 2009, **52**, 2603.
54. P. A. Nguewa, M. A. Fuertes, S. Iborra, Y. Najajreh, D. Gibson, E. Martinez, C. Alonso and J. M. Perez, *J. Inorg. Biochem.*, 2005, **99**, 727.
55. J. Tavares, M. Ouaissi, A. Ouaissi and A. Cordeiro-da-Silva, *Acta Trop.*, 2007, **103**, 133.
56. S. Kaur, H. Sachdeva, S. Dhuria, M. Sharma and T. Kaur, *Parasitol. Int.*, 2010, **59**, 62.
57. R. A. Sanchez-Delgado, K. Lazardi, L. Rincon and J. A. Urbina, *J. Med. Chem.*, 1993, **36**, 2041.
58. R. A. Sanchez-Delgado and A. Anzellotti, *Mini. Rev. Med. Chem.*, 2004, **4**, 23.

59. M. Vieites, P. Smircich, B. Parajon-Costa, J. Rodriguez, V. Galaz, C. Olea-Azar, L. Otero, G. Aguirre, H. Cerecetto, M. Gonzalez, A. Gomez-Barrio, B. Garat and D. Gambino, *J. Biol. Inorg. Chem.*, 2008, **13**, 723.
60. J. J. Nogueira Silva, W. R. Pavanelli, F. R. Gutierrez, F. C. Alves Lima, A. B. Ferreira da Silva, J. Santana Silva and D. Wagner Franco, *J. Med. Chem.*, 2008, **51**, 4104.
61. M. Vieites, P. Smircich, L. Guggeri, E. Marchan, A. Gomez-Barrio, M. Navarro, B. Garat and D. Gambino, *J. Inorg. Biochem.*, 2009, **103**, 1300.
62. M. Pagano, B. Demoro, J. Toloza, L. Boiani, M. Gonzalez, H. Cerecetto, C. Olea-Azar, E. Norambuena, D. Gambino and L. Otero, *Eur. J. Med. Chem.*, 2009, **44**, 4937.
63. J. Benitez, L. Guggeri, I. Tomaz, J. C. Pessoa, V. Moreno, J. Lorenzo, F. X. Aviles, B. Garat and D. Gambino, *J. Inorg. Biochem.*, 2009, **103**, 1386.
64. J. J. Silva, A. L. Osakabe, W. R. Pavanelli, J. S. Silva and D. W. Franco, *Br. J. Pharmacol.*, 2007, **152**, 112.
65. J. J. Silva, W. R. Pavanelli, J. C. Pereira, J. S. Silva and D. W. Franco, *Antimicrob. Agents Chemother.*, 2009, **53**, 4414.
66. P. M. Guedes, F. S. Oliveira, F. R. Gutierrez, G. K. da Silva, G. J. Rodrigues, L. M. Bendhack, D. W. Franco, M. A. Do Valle Matta, D. S. Zamboni, R. S. da Silva and J. S. Silva, *Br. J. Pharmacol.*, 2010, **160**, 270.
67. J. J. Silva, P. M. Guedes, A. Zottis, T. L. Balliano, F. O. Nascimento Silva, L. G. Franca Lopes, J. Ellena, G. Oliva, A. D. Andricopulo, D. W. Franco and J. S. Silva, *Br. J. Pharmacol.*, 2010, **160**, 260.
68. J. S. Silva, F. S. Machado and G. A. Martins, *Front. Biosci.*, 2003, **8**, s314.
69. A. E. Pegg and D. J. Feith, *Biochem. Soc. Trans.*, 2007, **35**, 295.
70. R. A. Casero, Jr and L. J. Marton, *Nat. Rev. Drug Discov.*, 2007, **6**, 373.
71. O. Heby, L. Persson and M. Rentala, *Amino Acids*, 2007, **33**, 359.
72. G. Colotti and A. Ilari, *Amino Acids*, 2010, 1.
73. A. H. Fairlamb, P. Blackburn, P. Ulrich, B. T. Chait and A. Cerami, *Science*, 1985, **227**, 1485.
74. F. Li, S. B. Hua, C. C. Wang and K. M. Gottesdiener, *Exp. Parasitol.*, 1998, **88**, 255.
75. Y. Xiao, D. E. McCloskey and M. A. Phillips, *Eukaryot. Cell*, 2009, **8**, 747.
76. E. K. Willert and M. A. Phillips, *PLoS Pathog.*, 2008, **4**, e1000183.
77. M. C. Taylor, H. Kaur, B. Blessington, J. M. Kelly and S. R. Wilkinson, *Biochem. J.*, 2008, **409**, 563.
78. T. T. Huynh, V. T. Huynh, M. A. Harmon and M. A. Phillips, *J. Biol. Chem.*, 2003, **278**, 39794.
79. S. Krieger, W. Schwarz, M. R. Ariyanayagam, A. H. Fairlamb, R. L. Krauth-Siegel and C. Clayton, *Mol. Microbiol.*, 2000, **35**, 542.

80. S. Wyllie, S. L. Oza, S. Patterson, D. Spinks, S. Thompson and A. H. Fairlamb, *Mol. Microbiol.*, 2009, **74**, 529.
81. D. C. Smithson, A. A. Shelat, J. Baldwin, M. A. Phillips and R. K. Guy, *Assay Drug. Dev. Technol.*, 2010.
82. D. C. Smithson, J. Lee, A. A. Shelat, M. A. Phillips and R. K. Guy, *J. Biol. Chem.*, 2010, **285**, 16771.
83. E. K. Willert, R. Fitzpatrick and M. A. Phillips, *Proc. Natl. Acad. Sci. U. S. A.*, 2007, **104**, 8275.
84. E. K. Willert and M. A. Phillips, *Mol. Biochem. Parasitol.*, 2009, **168**, 1.
85. A. J. Bitonti, T. L. Byers, T. L. Bush, P. J. Casara, C. J. Bacchi, A. B. Clarkson, Jr., P. P. McCann and A. Sjoerdsma, *Antimicrob. Agents Chemother.*, 1990, **34**, 1485.
86. C. J. Bacchi, H. C. Nathan, N. Yarlett, B. Goldberg, P. P. McCann, A. J. Bitonti and A. Sjoerdsma, *Antimicrob. Agents Chemother.*, 1992, **36**, 2736.
87. C. J. Bacchi, R. Brun, S. L. Croft, K. Alicea and Y. Buhler, *Antimicrob. Agents Chemother.*, 1996, **40**, 1448.
88. R. H. Barker, Jr., H. Liu, B. Hirth, C. A. Celatka, R. Fitzpatrick, Y. Xiang, E. K. Willert, M. A. Phillips, M. Kaiser, C. J. Bacchi, A. Rodriguez, N. Yarlett, J. D. Klinger and E. Sybertz, *Antimicrob. Agents Chemother.*, 2009, **53**, 2052.
89. B. Hirth, R. H. Barker, Jr., C. A. Celatka, J. D. Klinger, H. Liu, B. Nare, A. Nijjar, M. A. Phillips, E. Sybertz, E. K. Willert and Y. Xiang, *Bioorg. Med. Chem. Lett.*, 2009, **19**, 2916.
90. C. J. Bacchi, R. H. Barker, Jr., A. Rodriguez, B. Hirth, D. Rattendi, N. Yarlett, C. L. Hendrick and E. Sybertz, *Antimicrob. Agents Chemother.*, 2009, **53**, 3269.
91. S. C. Roberts, Y. Jiang, A. Jardim, N. S. Carter, O. Heby and B. Ullman, *Mol. Biochem. Parasitol.*, 2001, **115**, 217.
92. H. Wu, J. Min, H. Zeng, D. E. McCloskey, Y. Ikeguchi, P. Loppnau, A. J. Michael, A. E. Pegg and A. N. Plotnikov, *J. of Biol. Chem.*, 2008, **283**, 16135.
93. V. T. Dufe, W. Qiu, I. B. Müller, R. Hui, R. D. Walter and S. Al-Karadaghi, *J. Mol. Biol.*, 2007, **373**, 167.
94. D. V. Lueder and M. A. Phillips, *J. Biol. Chem.*, 1996, **271**, 17485.
95. M. R. Ariyanayagam, S. L. Oza, A. Mehlert and A. H. Fairlamb, *J. Biol. Chem.*, 2003, **278**, 27612.
96. S. L. Oza, M. P. Shaw, S. Wyllie and A. H. Fairlamb, *Mol. Biochem. Parasitol.*, 2005, **139**, 107.
97. M. R. Ariyanayagam, S. L. Oza, M. L. Guther and A. H. Fairlamb, *Biochem. J.*, 2005, **391**, 425.
98. M. A. Comini, S. A. Guerrero, S. Haile, U. Menge, H. Lunsdorf and L. Flohe, *Free Radic. Biol. Med.*, 2004, **36**, 1289.
99. S. L. Oza, S. Chen, S. Wyllie, J. K. Coward and A. H. Fairlamb, *FEBS J.*, 2008, **275**, 5408.

100. L. S. Torrie, S. Wyllie, D. Spinks, S. L. Oza, S. Thompson, J. R. Harrison, I. H. Gilbert, P. G. Wyatt, A. H. Fairlamb and J. A. Frearson, *J. Biol. Chem.*, 2009, **284**, 36137.
101. J. Tovar, S. Wilkinson, J. C. Mottram and A. H. Fairlamb, *Mol. Microbiol.*, 1998, **29**, 653.
102. F. X. Sullivan, S. L. Shames and C. T. Walsh, *Biochemistry*, 1989, **28**, 4986.
103. M. K. Mittal, S. Misra, M. Owais and N. Goyal, *Protein Expr. Purif.*, 2005, **40**, 279.
104. F. X. Sullivan and C. T. Walsh, *Mol. Biochem. Parasitol.*, 1991, **44**, 145.
105. P. Baiocco, S. Franceschini, A. Ilari and G. Colotti, *Protein Pept. Lett.*, 2009, **16**, 196.
106. Y. Zhang, C. S. Bond, S. Bailey, M. L. Cunningham, A. H. Fairlamb and W. N. Hunter, *Protein Sci.*, 1996, **5**, 52.
107. S. K. Venkatesan, A. K. Shukla and V. K. Dubey, *J. Comput. Chem.*, 2010, **31**, 2463.
108. D. Spinks, E. J. Shanks, L. A. Cleghorn, S. McElroy, D. Jones, D. James, A. H. Fairlamb, J. A. Frearson, P. G. Wyatt and I. H. Gilbert, *ChemMedChem*, 2009, **4**, 2060.
109. J. L. Richardson, I. R. Nett, D. C. Jones, M. H. Abdille, I. H. Gilbert and A. H. Fairlamb, *ChemMedChem*, 2009, **4**, 1333.
110. R. Perez-Pineiro, A. Burgos, D. C. Jones, L. C. Andrew, H. Rodriguez, M. Suarez, A. H. Fairlamb and D. S. Wishart, *J. Med. Chem.*, 2009, **52**, 1670.
111. G. A. Holloway, W. N. Charman, A. H. Fairlamb, R. Brun, M. Kaiser, E. Kostewicz, P. M. Novello, J. P. Parisot, J. Richardson, I. P. Street, K. G. Watson and J. B. Baell, *Antimicrob. Agents Chemother.*, 2009, **53**, 2824.
112. A. Cavalli, F. Lizzi, S. Bongarzone, R. Brun, R. Luise Krauth-Siegel and M. L. Bolognesi, *Bioorg. Med. Chem. Lett.*, 2009, **19**, 3031.
113. D. C. Martyn, D. C. Jones, A. H. Fairlamb and J. Clardy, *Bioorg. Med. Chem. Lett.*, 2007, **17**, 1280.
114. M. P. Barrett, *Parasitology Today*, 1997, **13**, 11.
115. A. G. M. Tielens and J. J. Van Hellemond, *Parasitology Today*, 1998, **14**, 265.
116. D. Rosenzweig, D. Smith, F. Opperdoes, S. Stern, R. W. Olafson and D. Zilberstein, *FASEB J.*, 2008, **22**, 590.
117. C. L. M. J. Verlinde, V. Hannaert, C. Blonski, M. Willson, J. J. Périé, L. A. Fothergill-Gilmore, F. R. Opperdoes, M. H. Gelb, W. G. J. Hol and P. A. M. Michels, *Drug Res. Upd.*, 2001, **4**, 50.
118. F. Lakhdar-Ghazal, C. Blonski, M. Willson, P. Michels and J. Perie, *Curr. Top. Med. Chem.*, 2002, **2**, 439.
119. M. Parsons, *Mol. Microbiol.*, 2004, **53**, 717.
120. M. A. Albert, J. R. Haanstra, V. Hannaert, J. Van Roy, F. R. Opperdoes, B. M. Bakker and P. A. Michels, *J. Biol. Chem.*, 2005, **280**, 28306.

121. M. T. Morris, C. DeBruin, Z. Yang, J. W. Chambers, K. S. Smith and J. C. Morris, *Eukaryot. Cell*, 2006, **5**, 2014.
122. J. W. Chambers, M. T. Kearns, M. T. Morris and J. C. Morris, *J. Biol. Chem.*, 2008, **283**, 14963.
123. J. W. Chambers, M. L. Fowler, M. T. Morris and J. C. Morris, *Mol. Biochem. Parasitol.*, 2008, **158**, 202.
124. A. J. Caceres, R. Portillo, H. Acosta, D. Rosales, W. Quinones, L. Avilan, L. Salazar, M. Dubourdieu, P. A. Michels and J. L. Concepcion, *Mol. Biochem. Parasitol.*, 2003, **126**, 251.
125. M. A. Pabon, A. J. Caceres, M. Gualdron, W. Quinones, L. Avilan and J. L. Concepcion, *Parasitol. Res.*, 2007, **100**, 803.
126. M. P. Hudock, C. E. Sanz-Rodriguez, Y. Song, J. M. Chan, Y. Zhang, S. Odeh, T. Kosztowski, A. Leon-Rossell, J. L. Concepcion, V. Yardley, S. L. Croft, J. A. Urbina and E. Oldfield, *J. Med. Chem.*, 2006, **49**, 215.
127. C. E. Sanz-Rodríguez, J. L. Concepción, S. Pekerar, E. Oldfield and J. A. Urbina, *J. Biol. Chem.*, 2007, **282**, 12377.
128. D. Arsenieva, B. L. Appavu, G. H. Mazock and C. J. Jeffery, *Proteins*, 2009, **74**, 72.
129. I. W. McNae, J. Martinez-Oyanedel, J. W. Keillor, P. A. Michels, L. A. Fothergill-Gilmore and M. D. Walkinshaw, *J. Mol. Biol.*, 2009, **385**, 1519.
130. C. Lopez, N. Chevalier, V. Hannaert, D. J. Rigden, P. A. Michels and J. L. Ramirez, *Eur. J. Biochem.*, 2002, **269**, 3978.
131. E. Rodriguez, N. Lander and J. L. Ramirez, *Mem. Inst. Oswaldo Cruz*, 2009, **104**, 745.
132. M. W. Nowicki, L. B. Tulloch, L. Worralll, I. W. McNae, V. Hannaert, P. A. Michels, L. A. Fothergill-Gilmore, M. D. Walkinshaw and N. J. Turner, *Bioorg. Med. Chem.*, 2008, **16**, 5050.
133. L. A. Fothergill-Gillmore, D. J. Rigden, P. A. Michels and S. E. Phillips, *Biochem. Soc. Trans.*, 2000, **28**, 186.
134. D. J. Rigden, S. E. V. Phillips, P. A. M. Michels and L. A. Fothergill-Gilmore, *J. Mol. Biol.*, 1999, **291**, 615.
135. I. Ernest, M. Callens, F. R. Opperdoes and P. A. Michels, *Mol. Biochem. Parasitol.*, 1994, **64**, 43.
136. D. M. Chudzik, P. A. Michels, S. de Walque and W. G. Hol, *J. Mol. Biol.*, 2000, **300**, 697.
137. S. de Walque, F. R. Opperdoes and P. A. Michels, *Mol. Biochem. Parasitol.*, 1999, **103**, 279.
138. C. Dax, F. Duffieux, N. Chabot, M. Coincon, J. Sygusch, P. A. M. Michels and C. Blonski, *J. Med. Chem.*, 2006, **49**, 1499.
139. L. Azema, C. Lherbet, C. Baudoin and C. Blonski, *Bioorg. Med. Chem. Lett.*, 2006, **16**, 3440.
140. O. Misset and F. R. Opperdoes, *Eur. J. Biochem.*, 1987, **162**, 493.
141. G. McKoy, M. Badal, Q. Prescott, H. Lux and D. T. Hart, *Mol. Biochem. Parasitol.*, 1997, **90**, 169.

142. J. L. Concepcion, C. A. Adje, W. Quinones, N. Chevalier, M. Dubourdieu and P. A. Michels, *Mol. Biochem. Parasitol.*, 2001, **118**, 111.
143. B. E. Bernstein and W. G. Hol, *Biochemistry*, 1998, **37**, 4429.
144. J. C. Bressi, J. Choe, M. T. Hough, F. S. Buckner, W. C. Van Voorhis, C. L. Verlinde, W. G. Hol and M. H. Gelb, *J. Med. Chem.*, 2000, **43**, 4135.
145. M. E. Drew, J. C. Morris, Z. Wang, L. Wells, M. Sanchez, S. M. Landfear and P. T. Englund, *J. Biol. Chem.*, 2003, **278**, 46596.
146. B. Poonperm, D. G. Guerra, I. W. McNae, L. A. Fothergill-Gilmore and M. D. Walkinshaw, *Acta Crystallogr. D. Biol. Crystallogr.*, 2003, **59**, 1313.
147. M. W. Nowicki, B. Kuaprasert, I. W. McNae, H. P. Morgan, M. M. Harding, P. A. Michels, L. A. Fothergill-Gilmore and M. D. Walkinshaw, *J. Mol. Biol.*, 2009, **394**, 535.
148. A. Djikeng, S. Raverdy, J. Foster, D. Bartholomeu, Y. Zhang, N. M. El-Sayed and C. Carlow, *Parasitol. Res.*, 2007, **100**, 887.
149. W. Quinones, P. Pena, M. Domingo-Sananes, A. Caceres, P. A. Michels, L. Avilan and J. L. Concepcion, *Exp. Parasitol.*, 2007, **116**, 241.
150. A. S. N. M. V. de, S. M. Gomes Dias, L. V. Mello, M. T. da Silva Giotto, S. Gavalda, C. Blonski, R. C. Garratt and D. J. Rigden, *FEBS J.*, 2007, **274**, 5077.
151. M. T. da Silva Giotto, V. Hannaert, D. Vertommen, M. V. d. A. S. Navarro, M. H. Rider, P. A. M. Michels, R. C. Garratt and D. J. Rigden, *J. Mol. Biol.*, 2003, **331**, 653.
152. H. P. Morgan, I. W. McNae, M. W. Nowicki, V. Hannaert, P. A. Michels, L. A. Fothergill-Gilmore and M. D. Walkinshaw, *J. Biol. Chem.*, 2010, **285**, 12892.
153. V. I. Livore, K. E. Tripodi and A. D. Uttaro, *FEBS J.*, 2007, **274**, 264.
154. S. H. Lee, J. L. Stephens and P. T. Englund, *Nat. Rev. Microbiol.*, 2007, **5**, 287.
155. J. J. van Hellemond and A. G. Tielens, *FEBS Lett.*, 2006, **580**, 5552.
156. K. E. Tripodi, L. V. Buttigliero, S. G. Altabe and A. D. Uttaro, *FEBS J.*, 2006, **273**, 271.
157. S. H. Lee, J. L. Stephens, K. S. Paul and P. T. Englund, *Cell*, 2006, **126**, 691.
158. J. L. Guler, E. Kriegova, T. K. Smith, J. Lukes and P. T. Englund, *Mol. Microbiol.*, 2008, **67**, 1125.
159. K. J. Autio, J. L. Guler, A. J. Kastaniotis, P. T. Englund and J. K. Hiltunen, *FEBS Lett.*, 2008, **582**, 729.
160. J. L. Stephens, S. H. Lee, K. S. Paul and P. T. Englund, *J. Biol. Chem.*, 2007, **282**, 4427.
161. A. Karioti, H. Skaltsa, A. Linden, R. Perozzo, R. Brun and D. Tasdemir, *J. Org. Chem.*, 2007, **72**, 8103.
162. A. C. Giddens, L. Nielsen, H. I. Boshoff, D. Tasdemir, R. Perozzo, M. Kaiser, F. Wang, J. C. Sacchettini and B. R. Copp, *Tetrahedron*, 2008, **64**, 1242.
163. D. Tasdemir, B. Topaloglu, R. Perozzo, R. Brun, R. O'Neill, N. M. Carballeira, X. Zhang, P. J. Tonge, A. Linden and P. Ruedi, *Bioorg. Med. Chem.*, 2007, **15**, 6834.

164. D. Tasdemir, N. D. Guner, R. Perozzo, R. Brun, A. A. Donmez, I. Calis and P. Ruedi, *Phytochemistry*, 2005, **66**, 355.
165. H. P. Price, M. R. Menon, C. Panethymitaki, D. Goulding, P. G. McKean and D. F. Smith, *J. Biol. Chem.*, 2003, **278**, 7206.
166. P. W. Bowyer, E. W. Tate, R. J. Leatherbarrow, A. A. Holder, D. F. Smith and K. A. Brown, *ChemMedChem*, 2008, **3**, 402.
167. P. W. Bowyer, R. S. Gunaratne, M. Grainger, C. Withers-Martinez, S. R. Wickramsinghe, E. W. Tate, R. J. Leatherbarrow, K. A. Brown, A. A. Holder and D. F. Smith, *Biochem. J.*, 2007, **408**, 173.
168. C. Panethymitaki, P. W. Bowyer, H. P. Price, R. J. Leatherbarrow, K. A. Brown and D. F. Smith, *Biochem. J.*, 2006, **396**, 277.
169. C. Sheng, H. Ji, Z. Miao, X. Che, J. Yao, W. Wang, G. Dong, W. Guo, J. Lu and W. Zhang, *J. Comput. Aided. Mol. Des.*, 2009, **23**, 375.
170. J. A. Brannigan, B. A. Smith, Z. Yu, A. M. Brzozowski, M. R. Hodgkinson, A. Maroof, H. P. Price, F. Meier, R. J. Leatherbarrow, E. W. Tate, D. F. Smith and A. J. Wilkinson, *J. Mol. Biol.*, 2010, **396**, 985.
171. J. A. Frearson, S. Brand, S. P. McElroy, L. A. Cleghorn, O. Smid, L. Stojanovski, H. P. Price, M. L. Guther, L. S. Torrie, D. A. Robinson, I. Hallyburton, C. P. Mpamhanga, J. A. Brannigan, A. J. Wilkinson, M. Hodgkinson, R. Hui, W. Qiu, O. G. Raimi, D. M. van Aalten, R. Brenk, I. H. Gilbert, K. D. Read, A. H. Fairlamb, M. A. Ferguson, D. F. Smith and P. G. Wyatt, *Nature*, 2010, **464**, 728.
172. M. Ferguson, *J. Cell. Sci.*, 1999, **112**, 2799.
173. K. Zhang, F. F. Hsu, D. A. Scott, R. Docampo, J. Turk and S. M. Beverley, *Mol. Microbiol.*, 2005, **55**, 1566.
174. T. K. Smith and P. Bütikofer, *Mol. Biochem. Parasitol.*, 2010, **172**, 66.
175. S. Chandra, D. Ruhela, A. Deb and R. A. Vishwakarma, *Expert Opin. Ther. Targets*, 2010, **14**, 739.
176. S. S. Sutterwala, F. F. Hsu, E. S. Sevova, K. J. Schwartz, K. Zhang, P. Key, J. Turk, S. M. Beverley and J. D. Bangs, *Mol. Microbiol.*, 2008, **70**, 281.
177. J. G. Mina, S.-Y. Pan, N. K. Wansadhipathi, C. R. Bruce, H. Shams-Eldin, R. T. Schwarz, P. G. Steel and P. W. Denny, *Mol. Biochem. Parasitol.*, 2009, **168**, 16.
178. P. W. Denny, H. Shams-Eldin, H. P. Price, D. F. Smith and R. T. Schwarz, *J. Biol. Chem.*, 2006, **281**, 28200.
179. K. Zhang and S. M. Beverley, *Mol. Biochem. Parasitol.*, 2010, **170**, 55.
180. K. Yokoyama, Y. Lin, K. D. Stuart and M. H. Gelb, *Mol. Biochem. Parasitol.*, 1997, **87**, 61.
181. M. H. Gelb, W. C. Van Voorhis, F. S. Buckner, K. Yokoyama, R. Eastman, E. P. Carpenter, C. Panethymitaki, K. A. Brown and D. F. Smith, *Mol. Biochem. Parasitol.*, 2003, **126**, 155.
182. E. Oldfield, *Acc. Chem. Res.*, 2010, **43**, 1216.
183. I. Coppens and P.-J. Courtoy, *Exp. Parasitol.*, 1996, **82**, 76.
184. T. Sgraja, T. Smith and W. Hunter, *BMC Structural Biology*, 2007, **7**, 20.

185. E. Byres, M. S. Alphey, T. K. Smith and W. N. Hunter, *J. Mol. Biol.*, 2007, **371**, 540.
186. J. Carrero-Lérida, G. Pérez-Moreno, V. M. Castillo-Acosta, L. M. Ruiz-Pérez and D. González-Pacanowska, *Int. J. Parasitol.*, 2009, **39**, 307.
187. R. Hurtado-Guerrrero, J. Pena-Diaz, A. Montalvetti, L. M. Ruiz-Perez and D. Gonzalez-Pacanowska, *FEBS Lett.*, 2002, **510**, 141.
188. J. A. Urbina, K. Lazardi, E. Marchan, G. Visbal, T. Aguirre, M. M. Piras, R. Piras, R. A. Maldonado, G. Payares and W. de Souza, *Antimicrob. Agents Chemother.*, 1993, **37**, 580.
189. D. C. Arruda, F. L. D'Alexandri, A. M. Katzin and S. R. B. Uliana, *Antimicrob. Agents Chemother.*, 2005, **49**, 1679.
190. P. Mukherjee, P. V. Desai, A. Srivastava, B. L. Tekwani and M. A. Avery, *J. Chem. Inf. Model.*, 2008, **48**, 1026.
191. L. Sigman, V. M. Sanchez and A. G. Turjanski, *J. Mol. Graph. Model.*, 2006, **25**, 345.
192. J. M. Sanders, A. O. Gomez, J. Mao, G. A. Meints, E. M. Van Brussel, A. Burzynska, P. Kafarski, D. Gonzalez-Pacanowska and E. Oldfield, *J. Med. Chem.*, 2003, **46**, 5171.
193. A. Montalvetti, A. Fernandez, J. M. Sanders, S. Ghosh, E. Van Brussel, E. Oldfield and R. Docampo, *J. Biol. Chem.*, 2003, **278**, 17075.
194. S. B. Gabelli, J. S. McLellan, A. Montalvetti, E. Oldfield, R. Docampo and L. M. Amzel, *Proteins*, 2006, **62**, 80.
195. L. R. Garzoni, A. Caldera, N. Meirelles Mde, S. L. de Castro, R. Docampo, G. A. Meints, E. Oldfield and J. A. Urbina, *Int. J. Antimicrob. Agents*, 2004, **23**, 273.
196. S. H. Szajnman, A. Montalvetti, Y. Wang, R. Docampo and J. B. Rodriguez, *Bioorg. Med. Chem. Lett.*, 2003, **13**, 3231.
197. V. Yardley, A. A. Khan, M. B. Martin, T. R. Slifer, F. G. Araujo, S. N. Moreno, R. Docampo, S. L. Croft and E. Oldfield, *Antimicrob. Agents Chemother.*, 2002, **46**, 929.
198. K. Yokoyama, P. Trobridge, F. S. Buckner, W. C. Van Voorhis, K. D. Stuart and M. H. Gelb, *J. Biol. Chem.*, 1998, **273**, 26497.
199. K. Yokoyama, P. Trobridge, F. S. Buckner, J. Scholten, K. D. Stuart, W. C. Van Voorhis and M. H. Gelb, *Mol. Biochem. Parasitol.*, 1998, **94**, 87.
200. R. T. Eastman, F. S. Buckner, K. Yokoyama, M. H. Gelb and W. C. Van Voorhis, *J. Lipid. Res.*, 2006, **47**, 233.
201. J. M. Kraus, C. L. Verlinde, M. Karimi, G. I. Lepesheva, M. H. Gelb and F. S. Buckner, *J. Med. Chem.*, 2009, **52**, 1639.
202. J. M. Kraus, H. B. Tatipaka, S. A. McGuffin, N. K. Chennamaneni, M. Karimi, J. Arif, C. L. M. J. Verlinde, F. S. Buckner and M. H. Gelb, *J. Med. Chem.*, 2010, **53**, 3887.
203. C. W. Roberts, R. McLeod, D. W. Rice, M. Ginger, M. L. Chance and L. J. Goad, *Mol. Biochem. Parasitol.*, 2003, **126**, 129.
204. J. A. Urbina, J. L. Concepcion, S. Rangel, G. Visbal and R. Lira, *Mol. Biochem. Parasitol.*, 2002, **125**, 35.

205. F. S. Buckner, L. N. Nguyen, B. M. Joubert and S. P. Matsuda, *Mol. Biochem. Parasitol.*, 2000, **110**, 399.
206. L. J. Goad, R. L. Berens, J. J. Marr, D. H. Beach and G. G. Holz Jr, *Mol. Biochem. Parasitol.*, 1989, **32**, 179.
207. F. Buckner, K. Yokoyama, J. Lockman, K. Aikenhead, J. Ohkanda, M. Sadilek, S. Sebti, W. Van Voorhis, A. Hamilton and M. H. Gelb, *Proc. Natl. Acad. Sci. U. S. A.*, 2003, **100**, 15149.
208. M. V. Braga, J. A. Urbina and W. de Souza, *Int. J. Antimicrob. Agents*, 2004, **24**, 72.
209. J. Molina, O. Martins-Filho, Z. Brener, A. J. Romanha, D. Loebenberg and J. A. Urbina, *Antimicrob. Agents Chemother.*, 2000, **44**, 150.
210. M. L. Ferraz, R. T. Gazzinelli, R. O. Alves, J. A. Urbina and A. J. Romanha, *Antimicrob. Agents Chemother.*, 2007, **51**, 1359.
211. S. Orenes Lorente, R. Gomez, C. Jimenez, S. Cammerer, V. Yardley, K. de Luca-Fradley, S. L. Croft, L. M. Ruiz Perez, J. Urbina, D. Gonzalez Pacanowska and I. H. Gilbert, *Bioorg. Med. Chem.*, 2005, **13**, 3519.
212. S. B. Cammerer, C. Jimenez, S. Jones, L. Gros, S. O. Lorente, C. Rodrigues, J. C. Rodrigues, A. Caldera, L. M. Ruiz Perez, W. da Souza, M. Kaiser, R. Brun, J. A. Urbina, D. Gonzalez Pacanowska and I. H. Gilbert, *Antimicrob. Agents Chemother.*, 2007, **51**, 4049.
213. M. Sealey-Cardona, S. Cammerer, S. Jones, L. M. Ruiz-Perez, R. Brun, I. H. Gilbert, J. A. Urbina and D. Gonzalez-Pacanowska, *Antimicrob. Agents Chemother.*, 2007, **51**, 2123.
214. J. C. Fernandes Rodrigues, J. L. Concepcion, C. Rodrigues, A. Caldera, J. A. Urbina and W. de Souza, *Antimicrob. Agents Chemother.*, 2008, **52**, 4098.
215. M. E. Konkle, T. Y. Hargrove, Y. Y. Kleshchenko, J. P. von Kries, W. Ridenour, M. J. Uddin, R. M. Caprioli, L. J. Marnett, W. D. Nes, F. Villalta, M. R. Waterman and G. I. Lepesheva, *J. Med. Chem.*, 2009, **52**, 2846.
216. L. M. Podust, J. P. von Kries, A. N. Eddine, Y. Kim, L. V. Yermalitskaya, R. Kuehne, H. Ouellet, T. Warrier, M. Altekoster, J. S. Lee, J. Rademann, H. Oschkinat, S. H. Kaufmann and M. R. Waterman, *Antimicrob. Agents Chemother.*, 2007, **51**, 3915.
217. C. K. Chen, P. S. Doyle, L. V. Yermalitskaya, Z. B. Mackey, K. K. Ang, J. H. McKerrow and L. M. Podust, *PLoS Negl. Trop. Dis.*, 2009, **3**, e372.
218. C. K. Chen, S. S. Leung, C. Guilbert, M. P. Jacobson, J. H. McKerrow and L. M. Podust, *PLoS Negl. Trop. Dis.*, 2010, **4**, e651.
219. G. I. Lepesheva, H. W. Park, T. Y. Hargrove, B. Vanhollebeke, Z. Wawrzak, J. M. Harp, M. Sundaramoorthy, W. D. Nes, E. Pays, M. Chaudhuri, F. Villalta and M. R. Waterman, *J. Biol. Chem.*, 2010, **285**, 1773.
220. E. Zeiman, C. L. Greenblatt, S. Elgavish, I. Khozin-Goldberg and J. Golenser, *J. Parasitol.*, 2008, **94**, 280.
221. T. Seebeck, R. Schaub and A. Johner, *Curr. Mol. Med.*, 2004, **4**, 585.

222. S. Laxman and J. A. Beavo, *Mol. Interv.*, 2007, **7**, 203.
223. R. Zoraghi and T. Seebeck, *Proc. Natl. Acad. Sci. U. S. A.*, 2002, **99**, 4343.
224. M. Oberholzer, G. Marti, M. Baresic, S. Kunz, A. Hemphill and T. Seebeck, *FASEB J.*, 2007, **21**, 720.
225. S. Kunz, J. A. Beavo, M. A. D'Angelo, M. M. Flawia, S. H. Francis, A. Johner, S. Laxman, M. Oberholzer, A. Rascon, Y. Shakur, L. Wentzinger, R. Zoraghi and T. Seebeck, *Mol. Biochem. Parasitol.*, 2006, **145**, 133.
226. A. Johner, S. Kunz, M. Linder, Y. Shakur and T. Seebeck, *BMC Microbiol.*, 2006, **6**, 25.
227. M. A. D'Angelo, S. Sanguineti, J. M. Reece, L. Birnbaumer, H. N. Torres and M. M. Flawia, *Biochem. J.*, 2004, **378**, 63.
228. G. D. Alonso, A. C. Schoijet, H. N. Torres and M. M. Flawia, *Mol. Biochem. Parasitol.*, 2006, **145**, 40.
229. G. D. Alonso, A. C. Schoijet, H. N. Torres and M. M. Flawia, *Mol. Biochem. Parasitol.*, 2007, **152**, 72.
230. S. King-Keller, M. Li, A. Smith, S. Zheng, G. Kaur, X. Yang, B. Wang and R. Docampo, *Antimicrob. Agents Chemother.*, 2010, **54**, 3738.
231. H. Wang, Z. Yan, J. Geng, S. Kunz, T. Seebeck and H. Ke, *Mol. Microbiol.*, 2007, **66**, 1029.
232. M. Parsons, E. A. Worthey, P. N. Ward and J. C. Mottram, *BMC Genomics*, 2005, **6**, 127.
233. T. C. Hammarton, S. Kramer, L. Tetley, M. Boshart and J. C. Mottram, *Mol. Microbiol.*, 2007, **65**, 1229.
234. C. Doerig, *Biochim. Biophys. Acta*, 2004, **1697**, 155.
235. C. Naula, M. Parsons and J. C. Mottram, *Biochim. Biophys. Acta*, 2005, **1754**, 151.
236. R. Brenk, A. Schipani, D. James, A. Krasowski, I. H. Gilbert, J. Frearson and P. G. Wyatt, *ChemMedChem*, 2008, **3**, 435.
237. L. Meijer, M. Flajolet and P. Greengard, *Trends in Pharmacological Sciences*, 2004, **25**, 471.
238. K. K. Ojo, J. R. Gillespie, A. J. Riechers, A. J. Napuli, C. L. Verlinde, F. S. Buckner, M. H. Gelb, M. M. Domostoj, S. J. Wells, A. Scheer, T. N. Wells and W. C. Van Voorhis, *Antimicrob. Agents Chemother.*, 2008, **52**, 3710.
239. E. Xingi, D. Smirlis, V. Myrianthopoulos, P. Magiatis, K. M. Grant, L. Meijer, E. Mikros, A. L. Skaltsounis and K. Soteriadou, *Int. J. Parasitol.*, 2009, **39**, 1289.
240. Z. Li, T. Umeyama and C. C. Wang, *PLoS Pathog.*, 2009, **5**, e1000575.
241. N. Jetton, K. G. Rothberg, J. G. Hubbard, J. Wise, Y. Li, H. L. Ball and L. Ruben, *Mol. Microbiol.*, 2009, **72**, 442.
242. M. D. Urbaniak, *Mol. Biochem. Parasitol.*, 2009, **166**, 183.
243. J. J. Allocco, R. Donald, T. Zhong, A. Lee, Y. S. Tang, R. C. Hendrickson, P. Liberator and B. Nare, *Int. J. Parasitol.*, 2006, **36**, 1249.

244. M. Wiese, *Int. J. Parasitol.*, 2007, **37**, 1053.
245. Q. Wang, I. M. Melzer, M. Kruse, C. Sander-Juelch and M. Wiese, *Kinetoplastid Biol. Dis.*, 2005, **4**, 6.
246. S. J. von Freyend, H. Rosenqvist, A. Fink, I. M. Melzer, J. Clos, O. N. Jensen and M. Wiese, *Int. J. Parasitol.*, 2010, **40**, 969.
247. L. Mercer, T. Bowling, J. Perales, J. Freeman, T. Nguyen, R. Don, R. Jacobs and B. Nare, *PLoS Negl. Trop. Dis.*, 2010, **5**, e956.
248. J. H. McKerrow, C. Caffrey, B. Kelly, P. Loke and M. Sajid, *Annu. Rev. Pathol.*, 2006, **1**, 497.
249. J. H. McKerrow, P. J. Rosenthal, R. Swenerton and P. Doyle, *Curr. Opin. Infect. Dis.*, 2008, **21**, 668.
250. J. H. McKerrow, P. S. Doyle, J. C. Engel, L. M. Podust, S. A. Robertson, R. Ferreira, T. Saxton, M. Arkin, I. D. Kerr, L. S. Brinen and C. S. Craik, *Mem. Inst. Oswaldo Cruz*, 2009, **104** (Suppl 1), 263.
251. C. R. Caffrey, S. Scory and D. Steverding, *Curr. Drug Targets*, 2000, **1**, 155.
252. J. C. Engel, P. S. Doyle, I. Hsieh and J. H. McKerrow, *J. Exp. Med.*, 1998, **188**, 725.
253. P. S. Doyle, Y. M. Zhou, J. C. Engel and J. H. McKerrow, *Antimicrob. Agents Chemother.*, 2007, **51**, 3932.
254. K. Steert, M. Berg, J. C. Mottram, G. D. Westrop, G. H. Coombs, P. Cos, L. Maes, J. Joossens, P. Van der Veken, A. Haemers and K. Augustyns, *ChemMedChem*, 2010, **5**, 1734.
255. F. V. Gonzalez, J. Izquierdo, S. Rodriguez, J. H. McKerrow and E. Hansell, *Bioorg. Med. Chem. Lett.*, 2007, **17**, 6697.
256. R. Vicik, V. Hoerr, M. Glaser, M. Schultheis, E. Hansell, J. H. McKerrow, U. Holzgrabe, C. R. Caffrey, A. Ponte-Sucre, H. Moll, A. Stich and T. Schirmeister, *Bioorg. Med. Chem. Lett.*, 2006, **16**, 2753.
257. C. Bryant, I. D. Kerr, M. Debnath, K. K. Ang, J. Ratnam, R. S. Ferreira, P. Jaishankar, D. Zhao, M. R. Arkin, J. H. McKerrow, L. S. Brinen and A. R. Renslo, *Bioorg. Med. Chem. Lett.*, 2009, **19**, 6218.
258. K. Brak, P. S. Doyle, J. H. McKerrow and J. A. Ellman, *J. Am. Chem. Soc.*, 2008, **130**, 6404.
259. K. Brak, I. D. Kerr, K. T. Barrett, N. Fuchi, M. Debnath, K. Ang, J. C. Engel, J. H. McKerrow, P. S. Doyle, L. S. Brinen and J. A. Ellman, *J. Med. Chem.*, 2010, **53**, 1763.
260. E. Altmann, S. W. Cowan-Jacob and M. Missbach, *J. Med. Chem.*, 2004, **47**, 5833.
261. J. Inglese, D. S. Auld, A. Jadhav, R. L. Johnson, A. Simeonov, A. Yasgar, W. Zheng and C. P. Austin, *Proc. Natl. Acad. Sci. U. S. A.*, 2006, **103**, 11473.
262. J. P. Mallari, A. A. Shelat, T. Obrien, C. R. Caffrey, A. Kosinski, M. Connelly, M. Harbut, D. Greenbaum, J. H. McKerrow and R. K. Guy, *J. Med. Chem.*, 2008, **51**, 545.
263. J. P. Mallari, A. A. Shelat, A. Kosinski, C. R. Caffrey, M. Connelly, F. Zhu, J. H. McKerrow and R. K. Guy, *J. Med. Chem.*, 2009, **52**, 6489.

264. B. T. Mott, R. S. Ferreira, A. Simeonov, A. Jadhav, K. K. Ang, W. Leister, M. Shen, J. T. Silveira, P. S. Doyle, M. R. Arkin, J. H. McKerrow, J. Inglese, C. P. Austin, C. J. Thomas, B. K. Shoichet and D. J. Maloney, *J. Med. Chem.*, 2010, **53**, 52.
265. J. P. Mallari, A. Shelat, A. Kosinski, C. R. Caffrey, M. Connelly, F. Zhu, J. H. McKerrow and R. K. Guy, *Bioorg. Med. Chem. Lett.*, 2008, **18**, 2883.
266. M. Berg, P. Van der Veken, A. Goeminne, A. Haemers and K. Augustyns, *Curr. Med. Chem.*, 2010, **17**, 2456.
267. N. S. Carter, P. Yates, C. S. Arendt, J. M. Boitz and B. Ullman, *Adv. Exp. Med. Biol.*, 2008, **625**, 141.
268. S. Gudin, N. B. Quashie, D. Candlish, M. I. Al-Salabi, S. M. Jarvis, L. C. Ranford-Cartwright and H. P. de Koning, *Exp. Parasitol.*, 2006, **114**, 118.
269. M. H. el Kouni, *Pharmacol. Ther.*, 2003, **99**, 283.
270. A. Luscher, H. P. de Koning and P. Maser, *Curr. Pharm. Des.*, 2007, **13**, 555.
271. P. Maser, D. Vogel, C. Schmid, B. Raz and R. Kaminsky, *J. Mol. Med.*, 2001, **79**, 121.
272. M. E. Rottenberg, W. Masocha, M. Ferella, F. Petitto-Assis, H. Goto, K. Kristensson, R. McCaffrey and H. Wigzell, *J. Infect. Dis.*, 2005, **192**, 1658.
273. M. Vodnala, A. Fijolek, R. Rofougaran, M. Mosimann, P. Maser and A. Hofer, *J. Biol. Chem.*, 2008, **283**, 5380.
274. A. Luscher, P. Onal, A. M. Schweingruber and P. Maser, *Antimicrob. Agents Chemother.*, 2007, **51**, 3895.
275. R. W. Miles, P. C. Tyler, G. B. Evans, R. H. Furneaux, D. W. Parkin and V. L. Schramm, *Biochemistry*, 1999, **38**, 13147.
276. W. Versees, K. Decanniere, R. Pelle, J. Depoorter, E. Brosens, D. W. Parkin and J. Steyaert, *J. Mol. Biol.*, 2001, **307**, 1363.
277. W. Versees, J. Barlow and J. Steyaert, *J. Mol. Biol.*, 2006, **359**, 331.
278. A. Goeminne, M. Berg, M. McNaughton, G. Bal, G. Surpateanu, P. Van der Veken, S. D. Prol, W. Versees, J. Steyaert, A. Haemers and K. Augustyns, *Bioorg. Med. Chem.*, 2008, **16**, 6752.
279. M. Berg, G. Bal, A. Goeminne, P. Van der Veken, W. Versees, J. Steyaert, A. Haemers and K. Augustyns, *ChemMedChem*, 2009, **4**, 249.
280. M. Berg, L. Kohl, P. Van der Veken, J. Joossens, M. I. Al-Salabi, V. Castagna, F. Giannese, P. Cos, W. Versees, J. Steyaert, P. Grellier, A. Haemers, M. Degano, L. Maes, H. P. de Koning and K. Augustyns, *Antimicrob. Agents. Chemother.*, 2010, **54**, 1900.
281. P. G. Bray, M. P. Barrett, S. A. Ward and H. P. de Koning, *Trends Parasitol.*, 2003, **19**, 232.
282. E. Matovu, M. L. Stewart, F. Geiser, R. Brun, P. Maser, L. J. Wallace, R. J. Burchmore, J. C. Enyaru, M. P. Barrett, R. Kaminsky, T. Seebeck and H. P. de Koning, *Eukaryot. Cell*, 2003, **2**, 1003.
283. H. P. de Koning, L. F. Anderson, M. Stewart, R. J. Burchmore, L. J. Wallace and M. P. Barrett, *Antimicrob. Agents Chemother.*, 2004, **48**, 1515.

284. M. L. Stewart, G. J. Bueno, A. Baliani, B. Klenke, R. Brun, J. M. Brock, I. H. Gilbert and M. P. Barrett, *Antimicrob. Agents Chemother.*, 2004, **48**, 1733.
285. A. Baliani, G. J. Bueno, M. L. Stewart, V. Yardley, R. Brun, M. P. Barrett and I. H. Gilbert, *J. Med. Chem.*, 2005, **48**, 5570.
286. A. Baliani, V. Peal, L. Gros, R. Brun, M. Kaiser, M. P. Barrett and I. H. Gilbert, *Org. Biomol. Chem.*, 2009, **7**, 1154.
287. N. Klee, P. E. Wong, B. Baragana, F. E. Mazouni, M. A. Phillips, M. P. Barrett and I. H. Gilbert, *Bioorg. Med. Chem. Lett.*, 2010, **20**, 4364.
288. C. Chollet, A. Baliani, P. E. Wong, M. P. Barrett and I. H. Gilbert, *Bioorg. Med. Chem.*, 2009, **17**, 2512.
289. I. N. Wenzel, P. E. Wong, L. Maes, T. J. Muller, R. L. Krauth-Siegel, M. P. Barrett and E. Davioud-Charvet, *ChemMedChem*, 2009, **4**, 339.
290. R. Balana-Fouce, C. M. Redondo, Y. Perez-Pertejo, R. Diaz-Gonzalez and R. M. Reguera, *Drug Discov. Today*, 2006, **11**, 733.
291. A. Deterding, F. A. Dungey, K. A. Thompson and D. Steverding, *Acta Trop.*, 2005, **93**, 311.
292. E. Nenortas, T. Kulikowicz, C. Burri and T. A. Shapiro, *Antimicrob. Agents Chemother.*, 2003, **47**, 3015.
293. E. Nenortas, C. Burri and T. A. Shapiro, *Antimicrob. Agents Chemother.*, 1999, **43**, 2066.
294. R. P. Bakshi, D. Sang, A. Morrell, M. Cushman and T. A. Shapiro, *Antimicrob. Agents Chemother.*, 2009, **53**, 123.
295. S. C. Tang and T. A. Shapiro, *Antimicrob. Agents Chemother.*, 2010, **54**, 620.
296. N. Carballeira, M. Cartagena, C. Prada, C. Rubio and R. Balaña-Fouce, *Lipids*, 2009, **44**, 953.
297. N. M. Carballeira, N. Montano, R. Balana-Fouce and C. F. Prada, *Chem. Phys. Lipids*, 2009, **161**, 38.
298. K. Werbovetz, *Curr. Opin. Investig. Drugs*, 2006, **7**, 147.
299. M. F. Paine, M. Z. Wang, C. N. Generaux, D. W. Boykin, W. D. Wilson, H. P. De Koning, C. A. Olson, G. Pohlig, C. Burri, R. Brun, G. A. Murilla, J. K. Thuita, M. P. Barrett and R. R. Tidwell, *Curr. Opin. Investig. Drugs*, 2010, **11**, 876.
300. J. K. Thuita, S. M. Karanja, T. Wenzler, R. E. Mdachi, J. M. Ngotho, J. M. Kagira, R. Tidwell and R. Brun, *Acta Trop.*, 2008, **108**, 6.
301. R. E. Mdachi, J. K. Thuita, J. M. Kagira, J. M. Ngotho, G. A. Murilla, J. M. Ndung'u, R. R. Tidwell, J. E. Hall and R. Brun, *Antimicrob. Agents Chemother.*, 2009, **53**, 953.
302. R. K. Arafa, M. A. Ismail, M. Munde, W. D. Wilson, T. Wenzler, R. Brun and D. W. Boykin, *Eur. J. Med. Chem.*, 2008, **43**, 2901.
303. D. A. Patrick, S. A. Bakunov, S. M. Bakunova, E. V. K. Suresh Kumar, H. Chen, S. K. Jones, T. Wenzler, T. Barzcz, K. A. Werbovetz, R. Brun and R. R. Tidwell, *Eur. J. Med. Chem*, 2009, **44**, 3543.
304. T. L. Huang, J. J. Vanden Eynde, A. Mayence, M. S. Collins, M. T. Cushion, D. Rattendi, I. Londono, L. Mazumder, C. J. Bacchi and N. Yarlett, *Bioorg. Med. Chem. Lett.*, 2009, **19**, 5884.

305. J. L. Gonzalez, C. E. Stephens, T. Wenzler, R. Brun, F. A. Tanious, W. D. Wilson, T. Barszcz, K. A. Werbovetz and D. W. Boykin, *Eur. J. Med. Chem.*, 2007, **42**, 552.
306. S. M. Bakunova, S. A. Bakunov, T. Wenzler, T. Barszcz, K. A. Werbovetz, R. Brun, J. E. Hall and R. R. Tidwell, *J. Med. Chem.*, 2007, **50**, 5807.
307. D. A. Patrick, S. A. Bakunov, S. M. Bakunova, E. V. Kumar, R. J. Lombardy, S. K. Jones, A. S. Bridges, O. Zhirnov, J. E. Hall, T. Wenzler, R. Brun and R. R. Tidwell, *J. Med. Chem.*, 2007, **50**, 2468.
308. S. A. Bakunov, S. M. Bakunova, T. Wenzler, T. Barszcz, K. A. Werbovetz, R. Brun and R. R. Tidwell, *J. Med. Chem.*, 2008, **51**, 6927.
309. M. A. Ismail, R. K. Arafa, T. Wenzler, R. Brun, F. A. Tanious, W. D. Wilson and D. W. Boykin, *Bioorg. Med. Chem.*, 2008, **16**, 683.
310. S. A. Bakunov, S. M. Bakunova, T. Wenzler, M. Ghebru, K. A. Werbovetz, R. Brun and R. R. Tidwell, *J. Med. Chem.*, 2010, **53**, 254.
311. D. Branowska, A. A. Farahat, A. Kumar, T. Wenzler, R. Brun, Y. Liu, W. D. Wilson and D. W. Boykin, *Bioorg. Med. Chem.*, 2010, **18**, 3551.
312. T. Wenzler, D. W. Boykin, M. A. Ismail, J. E. Hall, R. R. Tidwell and R. Brun, *Antimicrob. Agents Chemother.*, 2009, **53**, 4185.
313. C. E. Stephens, F. Tanious, S. Kim, W. D. Wilson, W. A. Schell, J. R. Perfect, S. G. Franzblau and D. W. Boykin, *J. Med. Chem.*, 2001, **44**, 1741.
314. C. E. Stephens, R. Brun, M. M. Salem, K. A. Werbovetz, F. Tanious, W. D. Wilson and D. W. Boykin, *Bioorg. Med. Chem. Lett.*, 2003, **13**, 2065.
315. M. Z. Wang, X. Zhu, A. Srivastava, Q. Liu, J. M. Sweat, T. Pandharkar, C. E. Stephens, E. Riccio, T. Parman, M. Munde, S. Mandal, R. Madhubala, R. R. Tidwell, W. D. Wilson, D. W. Boykin, J. E. Hall, D. E. Kyle and K. A. Werbovetz, *Antimicrob. Agents Chemother.*, 2010, **54**, 2507.
316. D. O. Ochola, R. K. Prichard and G. W. Lubega, *J. Parasitol.*, 2002, **88**, 600.
317. M. M. Chan and D. Fong, *Science*, 1990, **249**, 924.
318. G. Bhattacharya, M. M. Salem and K. A. Werbovetz, *Bioorg. Med. Chem. Lett.*, 2002, **12**, 2395.
319. K. A. Werbovetz, D. L. Sackett, D. Delfin, G. Bhattacharya, M. Salem, T. Obrzut, D. Rattendi and C. Bacchi, *Mol. Pharmacol.*, 2003, **64**, 1325.
320. G. Bhattacharya, J. Herman, D. Delfin, M. M. Salem, T. Barszcz, M. Mollet, G. Riccio, R. Brun and K. A. Werbovetz, *J. Med. Chem.*, 2004, **47**, 1823.
321. D. Wu, T. G. George, E. Hurh, K. A. Werbovetz and J. T. Dalton, *Life Sci.*, 2006, **79**, 1081.
322. T. G. George, J. Johnsamuel, D. A. Delfin, A. Yakovich, M. Mukherjee, M. A. Phelps, J. T. Dalton, D. L. Sackett, M. Kaiser, R. Brun and K. A. Werbovetz, *Bioorg. Med. Chem.*, 2006, **14**, 5699.

323. T. G. George, M. M. Endeshaw, R. E. Morgan, K. V. Mahasenan, D. A. Delfin, M. S. Mukherjee, A. J. Yakovich, J. Fotie, C. Li and K. A. Werbovetz, *Bioorg. Med. Chem.*, 2007, **15**, 6071.
324. R. E. Morgan, S. Ahn, S. Nzimiro, J. Fotie, M. A. Phelps, J. Cotrill, A. J. Yakovich, D. L. Sackett, J. T. Dalton and K. A. Werbovetz, *Chem. Biol. Drug Des.*, 2008, **72**, 513.
325. R. E. Morgan and K. A. Werbovetz, *Adv. Exp. Med. Biol.*, 2008, **625**, 33.
326. R. T. Jacobs, B. Nare and M. A. Phillips, *Curr. Top. Med. Chem.*, 2011, **11**, 1255.

CHAPTER 8

The Challenges of Flavivirus Drug Discovery

PEI-YONG SHI,[a] QING-YIN WANG[a] AND THOMAS H. KELLER*[b]

[a] Novartis Institute for Tropical Diseases, 10 Biopolis Road, #05-01 Chromos, Singapore 138670; [b] Experimental Therapeutics Center, 31 Biopolis Way, #03-01 Nanos, Singapore 138669

8.1 Introduction

Many flaviviruses are human pathogens of global importance, but no antiviral medicines are currently available to manage these diseases. Over the past decade, significant progress has been made towards the understanding of flavivirus biology. Structural information is available for all of the important proteins of the dengue virus, which has provided important insight into the mechanistic aspects of replication. Mechanistic and structural studies on the envelope protein have provided information on the cell entry mechanism of the flavivirus genus. All of these advances have established the foundation for the development of pharmacological treatments of flavivirus infections using modern drug discovery. While both industry and academia have invested considerable effort in the development of flavivirus antivirals and a number of compounds with efficacies in cell culture and animal models have been obtained, very few drug candidates have reached the stage of clinical development. In this review we summarize the current status of flavivirus drug discovery, focusing on the challenges and discussing potential paths forward.

RSC Drug Discovery Series No. 14
Neglected Diseases and Drug Discovery
Edited by Michael J. Palmer and Timothy N. C. Wells

Published by the Royal Society of Chemistry, www.rsc.org

8.2 Flaviviral Diseases

The *Flaviviridae* family consists of three genera: *Flavivirus, Pestivirus*, and *Hepacivirus*. There are more than 70 viruses in the flavivirus genus, many of which are arthropod-borne human pathogens. Yellow fever virus (YFV), the prototype flavivirus, is the first human virus that was shown to be experimentally transferred *via* the filtered serum of an infected individual, while naturally the infectious agent was transmitted to humans by mosquitoes.[1] Besides YFV, several flaviviruses are important human pathogens, including the four serotypes of dengue virus (DENV), West Nile virus (WNV), and Japanese encephalitis virus (JEV). Infection with these flaviviruses causes significant human disease, including fevers, encephalitis, and hemorrhage; some of the infected individuals develop life-threatening symptoms, which can result in death. Due to reduced mosquito control, an increase in global transportation, and dense urbanization, flavivirus infections have increased significantly in tropical and sub-tropical regions of the world in the past 40 years. DENV alone was estimated to cause 100 million human infections annually, with 500 000 cases of dengue hemorrhagic fever and 22 000 deaths globally.[2] There are an estimated 200 000 cases of yellow fever (causing 30 000 deaths) globally each year.[3] Since an outbreak in New York City in 1999, WNV has caused thousands of human infections in North America.[4] In Asia, approximately 50 000 people (mostly young, healthy children) develop Japanese encephalitis, causing about 10 000 fatalities annually.[5] There is no established drug therapy for any flavivirus infection. Treatment is symptomatic, aimed at preventing complications and reducing symptoms for the comfort of the patients.

Human vaccines for flaviviruses are only available for YFV, JEV, and tick-borne encephalitis virus (TBEV). Two live-attenuated YFV vaccines, 17D and FNV (French neurotropic vaccine), were developed almost concurrently in the 1930s. Due to its association with a high incidence of encephalitis in children, FNV was discontinued in the 1980s. The 17D vaccine strain, on the other hand, was well tolerated and is regarded as one of the safest and most effective live-attenuated viral vaccines ever developed. Two sub-strains, 17D-204 and 17DD, are currently used as vaccines.[3] A single dose of vaccine gives protective immunity for at least 10 years, after which the WHO recommends revaccination.

The first-generation inactivated JEV vaccines made from infected mouse brain or primary hamster kidney cells were routinely used for decades until 2005. Recently, the development and licensure of second-generation, non-mouse-brain-derived JEV vaccines has changed the landscape. SA14-14-2, a live attenuated vaccine developed in China has an excellent safety profile and efficacy. It is now becoming available to many Asian countries. IC51, a formalin-inactived JEV vaccine developed by Intercell AG was licensed for use in the United States, Australia, and Europe in 2009.[6]

Development of a dengue vaccine is challenging because a successful vaccine requires a balanced immune response to all four serotypes of DENV; an unbalanced vaccine could lead to enhanced virus replication mediated by the serotype-cross-reactive antibodies.[7] As a consequence, antiviral therapy is urgently needed for the treatment of DENV and WNV infections, while the availability of a drug for other flavivirus pathogens would open important treatment options for infected patients.

8.3 Anti-flavivirus Strategies

Flaviviruses contain a single-strand genomic RNA of plus-sense polarity. The genomic RNA is approximately 11 kb in length, encoding a 5' untranslated region (UTR), a single open-reading-frame, and a 3' UTR (Figure 8.1). The single open-reading-frame encodes three structural proteins (capsid [C], pre-membrane [PrM], envelope [E] protein) and seven nonstructural proteins (NS1, NS2A, NS2B, NS3, NS4A, NS4B, and NS5). The structural proteins form the viral particle and the non-structural proteins participate in RNA replication, virion assembly,[8,9] and invasion of the innate immune response.[10–13] Among the ten viral proteins, only NS3 and NS5 have enzymatic activities. The N-terminal domain of NS3 contains a serine protease activity; the C-terminal domain functions as an RNA helicase, an RNA triphosphatase (RTPase), and an NTP phosphatase (NTPase).[14–16] The N-terminal domain of NS5 has a methyltransferase activity; the C-terminal domain serves as an RNA-dependent RNA polymerase (RdRp).[17–20] Other proteins are also required for RNA replication, among which NS2A, NS2B, NS4A, and NS4B are transmembrane proteins that form the scaffold of the viral replication complex.[21–23] The exact topology of the replication complex remains to be defined.

In principle, three strategies are available for flavivirus drug discovery.

(i) Viral proteins are essential for the flavivirus infection cycle, and represent the most attractive antiviral targets. Since crystal structures of many flavivirus structural proteins and enzymes have been solved, a target-based approach could be pursued, including enzyme activity-based high-throughput screening (HTS), fragment-based screening, structure-based rational design, and virtual screening. This approach has been highly successful for the human immunodeficiency virus (HIV) where a number of antiviral drugs are available.[24]

(ii) Host proteins required for viral replication are potential antiviral targets. A number of cell-based assays have been developed to identify

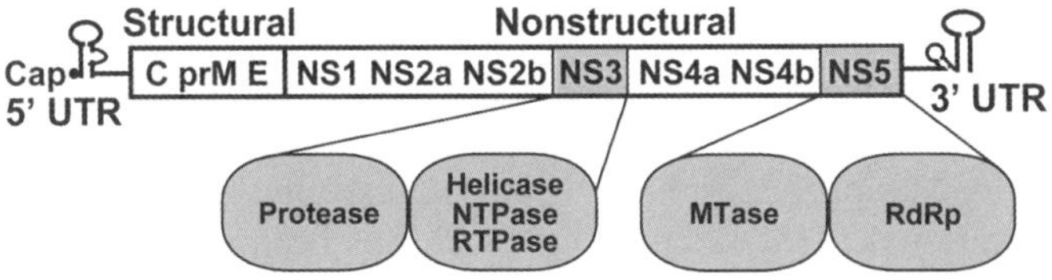

Figure 8.1 Flavivirus genome structure and viral proteins. Enzymatic activities of NS3 and NS5 are indicated. NTPase: NTP phosphatase; RTPase: RNA triphosphatase; MTase: methyltransferase; RdRp: RNA-dependent RNA polymerase. The sizes of genome and proteins are not drawn to scale.

inhibitors of both host and viral targets: replicon-based assay,[25] virus-like particle infection-based assay,[26] and reporting virus infection-based assay.[27] While the minimization of side effects will be a challenge for such a strategy, there is precedent from HIV drug discovery.[28]

(iii) Host pathways that lead to disease exacerbation could be targeted to block/mitigate the severity of flaviviral diseases. Such approaches would be especially attractive for diseases with a hemorrhagic component, but they require a detailed understanding of the molecular mechanism of disease development.

The following sections review key antiviral targets and the challenges encountered with the current strategies. The technical details of the methodology and assays have recently been reviewed.[29]

8.4 Inhibition of Viral Proteins

The inhibition of the viral enzymes required for replication is conceptually the most attractive approach. Since these proteins often are unique to the virus, or differ substantially from related proteins in the human genome, it should be quite straightforward to craft selective inhibitors. Furthermore, the history of antiviral drug discovery has clearly shown that at least the viral protease and polymerase are enzyme classes that can be inhibited by drug-like compounds. Not surprisingly, the drug discovery effort in the flavivirus area has been concentrated on these two enzymes, while very little work has been done on the helicase and methyltransferase domains. It is important to stress that so far all of the structure-based design has been performed on protein domains. Since any drug that targets the replication of the virus needs to interact with the targeted enzyme in the replication complex, the absence of knowledge about the structure of this multi-protein assembly poses a challenge for all antiviral projects.

8.4.1 NS3 Protease

The flavivirus genome encodes a trypsin-like serine protease in the N-terminal region of NS3, which requires association with NS2B for activity.[30] Similar to other viruses, this protease is responsible (together with host proteases) for processing of the polyprotein, an essential step in the replication of the virus as well as the assembly of the virion.

Virally encoded proteases have been successfully targeted for HIV and HCV.[31] Traditionally, the favored approach has been to model inhibitors on the sequences of substrate peptides and then optimize pharmacokinetics by eliminating metabolic weak spots and changing the physical properties of the inhibitor so that it can be administered by the oral route. While this is a very demanding strategy, nine out of ten registered HIV protease inhibitors have been designed using this approach (e.g. Atazanavir Scheme 8.1).[31] The only

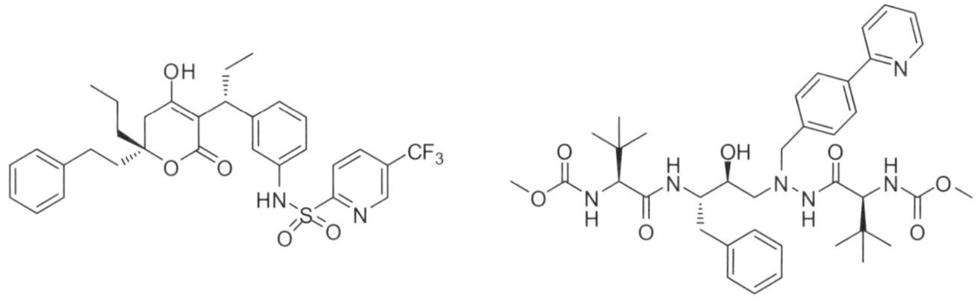

Tipranavir
optimized from HTS hit

Atazanavir
peptidomimetic

Scheme 8.1

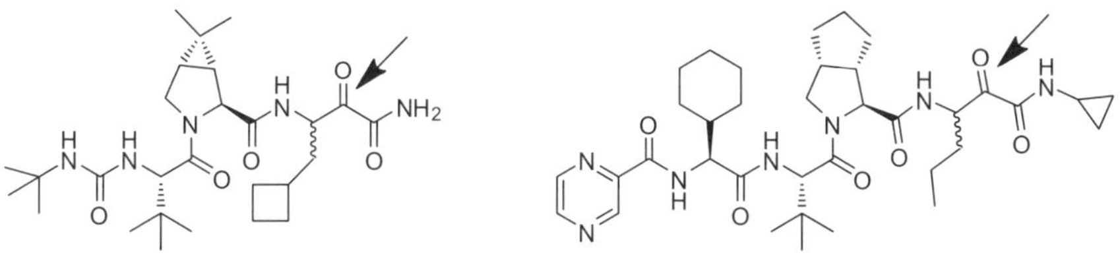

Boceprevir

Telaprevir

Scheme 8.2

exception is Tipranavir,[32] which was optimized from an HTS hit and therefore has no resemblance to the substrate peptide.

The success of structure-based design in the HIV protease area has inspired the search for HCV and flavivirus protease inhibitors. Unfortunately, these serine protease enzymes (HIV protease is an aspartic protease) have proven to be extremely taxing drug discovery targets, mainly due to the flat and solvent exposed active site (see Figure 8.3).[31] In order to overcome these challenges, HCV protease inhibitors have been designed with either macroyclic rings and/or electrophilic warheads to increase the affinity of the inhibitors (e.g. clinical HCV protease inhibitors telaprevir and boceprevir).[31] Electrophilic warheads, which can be aldehydes, activated ketones or boronic acids, are designed to form a reversible, covalent bond with the active site serine (the electrophilic ketone in telaprevir and boceprevir is designated with an arrow in Scheme 8.2).[33,34]

Yin and coworkers demonstrated the importance of electrophilic warheads for flavivirus NS3 protease inhibitors.[34] By modifying an optimized substrate of DENV NS2B-NS3 with different electrophilic functional groups, the activity of the inhibitor could be increased more than 1000-fold (Table 8.1). This suggests that similar to the HCV protease development compounds, such a warhead may be an important component for a flavivirus protease inhibitor.

Table 8.1 Electrophilic warheads for DENV NS2B-NS3.[a]

	Bz-Nle-Lys-Arg-Arg-R					
	$R = OH$	$R = NH_2$	$R = H$	$R = CF_3$	$R = CONH_2$	$R = B(OH)_2$
Ki (μM)	>500	128	5.8	0.85	Ns	0.04

[a]Data from reference 34; NS = not stable.

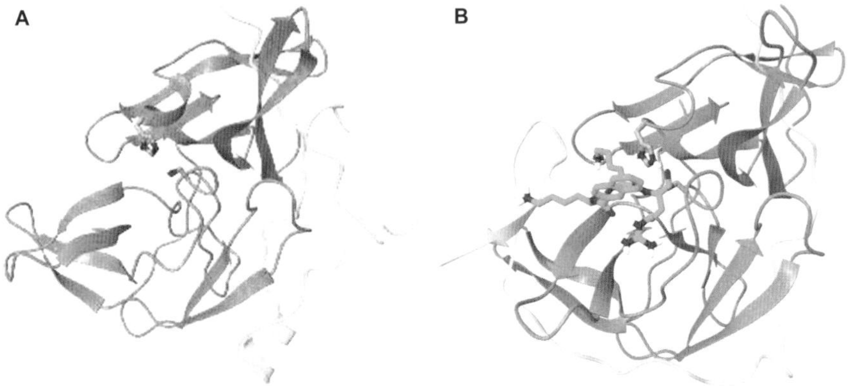

Figure 8.2 Inactive (A) and active (B) form of a flavivirus serine protease. The bound peptide (B) and, the active site serine and histidine (A and B) are shown in green with red oxygen and blue nitrogen atoms; NS2B is shown in yellow. Crystal structures 2FOM[35] (A) and 3E90[37] (B) were used for the illustration.

The best-characterized flavivirus protease is WNV NS2B-NS3 protease. Three independent groups have reported crystal structures of the enzyme in complex with peptide ligands[35–37] and the interactions of inhibitors with the protein have been analyzed.[38] Recently, a detailed investigation of the conformational dynamics by NMR spectroscopy has been reported.[39] All of the biophysical data is in agreement with the observation by Falgout et al.[30] that the protease activity is dependent on association with NS2B. In fact, NS2B serves as a co-factor that participates in the formation of the active site, forming a part of the S2 and S3 binding pockets which are required for the processing of the enzyme substrates (Figure 8.2B).[35] Both X-ray crystallography and NMR spectroscopy suggest that in the absence of inhibitors, NS2B can exist in conformations that do not contribute to the active site (e.g. Figure 8.2A), leading to protein that presumably cannot function as an enzyme. For WNV the NMR spectroscopy results show that the open form (Figure 8.2A) plays only a minor role in the conformational equilibrium,[39] which simplifies all aspects of drug discovery with this enzyme.

Table 8.2 Potent peptidic inhibitors of WNV NS2B-NS3 protease.[a]

Peptide	IC_{50} (nM)	K_i (nM)
Phenacetyl-KKR-aldehyde	51	9
4-Phenyl-4-phenacetyl-KKR-aldehyde	32	6

[a]Data from reference 40.

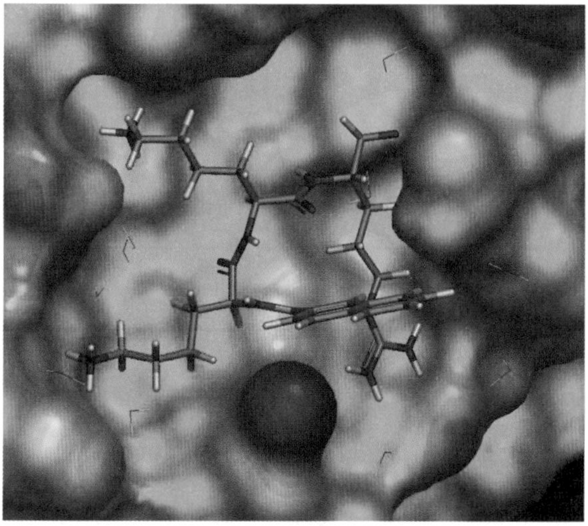

Figure 8.3 Peptide inhibitor 2-naphthoyl-KKR-aldehyde[37] bound to WNV NS2B-NS3 protease. Protein is shown with a solvent accessible surface colored according to polarity. Red-yellow = hydrophobic, blue-green = hydrophilic.

Fairlie and coworkers have reported highly potent peptide aldehyde inhibitors for WNV NS2B-NS3 protease (Table 8.2) in conjunction with structural information.[37,40] In the X-ray crystal structure[37] the peptide is bound to the flat and polar active site through the expected serine-aldehyde bond and the basic side chains form strong interactions with the protein (Figure 8.3).

While a lot of work is needed to turn such peptides into drug candidates, and the highly charged nature of these high-affinity ligands poses a challenge to the medicinal chemists, it should be feasible to design peptidomimetics with better physical properties based on the available structural information.

A number of research groups have attempted to find chemical starting points by virtual screening or HTS of WNV NS2B-NS3 protease.[41–43] These activities have identified compounds that bind in the active site, but all are rather weak (μM) and some have charged substituents (Scheme 8.3).

While it cannot be excluded that such leads can be optimized to potent inhibitors with good physical properties and acceptable PK, the experience

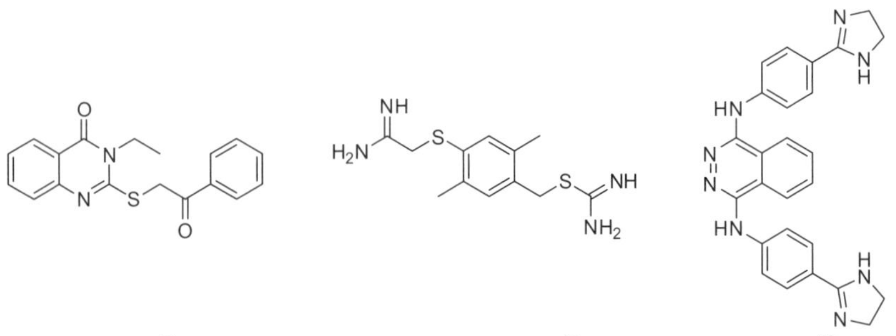

Scheme 8.3

Table 8.3 The best currently existing inhibitor of DENV NS2B-NS3 protease.[a]

Peptide	*IC_{50} (nM)*
Bz- Lys-Arg-Arg-H	1500

[a]Data from reference 34.

from HCV, where all clinical candidates are peptidomimetics, suggests that the hurdles are very high. To illustrate this point we have examined the active site of the WNV NS2B-NS3 protease (crystal structure 3E90) with Sitemap,[44] a software that assesses binding site druggability. The calculated Dscore of 0.88 is in the undruggable region, which indicates that an oral small molecular weight inhibitor may be difficult to obtain for this target. It needs to be stressed that this druggability score does not apply to peptidomimetics with warheads since such inhibitors form a covalent bond with the serine protease.[44]

DENV NS2B-NS3 protease has so far resisted all attempts to produce a crystal structure with a peptide ligand. While several structures have been reported in the literature,[35,45,46] all of them show NS2B in the dissociated form (e.g. Figure 8.2A). Since structural information on the catalytically active dengue enzyme is very important for progress in drug discovery, it is interesting to speculate why DENV NS2B-NS3 protease seems to behave differently from WNV NS2B-NS3 protease. One of the key differences seems to be that the dengue enzyme is much less ordered. Again NMR spectroscopy provides the most conclusive evidence, suggesting that DENV NS2B, in the absence of inhibitors, is 90% of the time dissociated from NS3 (Figure 8.2A).[39] Even in the presence of inhibitors, NS2B seems to remain largely dissociated, which may explain the persistent difficulties to obtain structural information with an enzyme bound inhibitor. This hypothesis may also explain why the most potent peptide-aldehyde inhibitor for DENV NS2B-NS3 protease is 180-fold weaker (Table 8.3) than comparable inhibitors on WN NS2B-NS3 protease (Table 8.2).

It is clear that we urgently need more information on the structure of the DENV NS2B-NS3 protease so that we can develop a better understanding. (i) under what conditions the closed, active form of the enzyme exists; (ii) how important the apparent differences in the dynamics of the different flavivirus proteases are. Before tangible progress can be made on this important drug target, we need to know whether potent (low nanomolar) peptidomimetics can be designed for dengue, or whether a more promising avenue would be to search for allosteric inhibitors that stabilize one of the inactive conformations of DENV NS2B-NS3 protease.

8.4.2 NS3 Helicase

The C-terminal domain of the NS3 protein contains a RNA helicase/NTPase domain that is essential for the replication of flaviviruses. The available crystal structures of the DENV helicase domain (Figure 8.4),[47] and the full-length NS3[48] proteins of DENV and Murray Valley encephalitis virus[49] provide interesting insights into the role of the NS3 protein in flavivirus replication. Since the protease and helicase domain are connected by a flexible linker which allows the protein to adopt different conformations, it has been proposed that the RNA processing by the helicase/NTPase domain is facilitated by the movement of the two catalytic domains.[48] The implication of these findings for drug discovery is currently unclear. So far, flavivirus helicases have not been

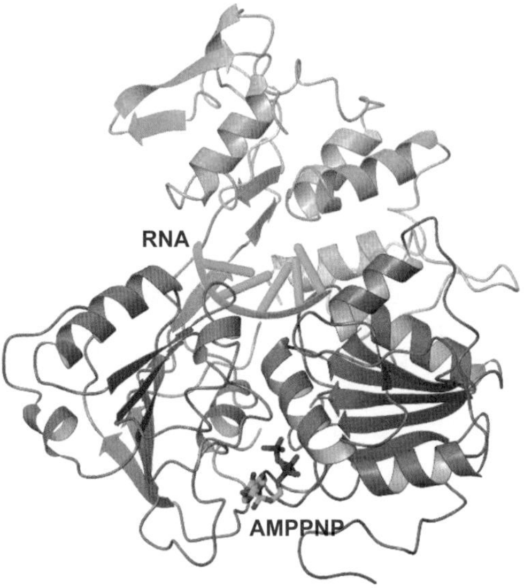

Figure 8.4 Ternary complex of flaviviral NS3 helicase with bound RNA and non-hydrolysable ATP analogue 5-adenylyl-imidodiphosphate (AMPPNP);[46] RNA in green and AMPPNP colored according to atom type.

extensively explored for drug discovery and even for HCV there has been very little activity in the helicase area.[50] One can think of several avenues to inhibit the activity of this important enzyme: (i) inhibit the binding of ATP to its binding site; (ii) block the RNA tunnel; (iii) inhibit the domain movement of the helicase.

Unfortunately, all the crystal structures suggest that the ATP binding site is solvent exposed and the hydrophobic part of the ATP-molecule interacts only weakly with the protein. This suggests that finding potent ATP binding site inhibitors could be quite challenging.

While allosteric inhibitors may be an attractive drug discovery option for the future, further data about the dynamics of the protein are needed before allosteric inhibitors can be designed. At present, the best avenue forward would be to perform an HTS with one of the flavivirus helicase enzymes. Unfortunately, helicase assays have proven to be very challenging, and it has been very difficult to generate a robust assay system that is clearly ATP-dependent and can accurately simulate the RNA-processing that occurs inside the cell.

8.4.3 NS5 Polymerase

The RNA-dependent RNA polymerase (RdRp) domain in the C-terminal part of the NS5 protein synthesizes the RNA that is required for the replication of flaviviruses. It is thought that the RNA-template binds to the enzyme and then the synthesis of the complementary RNA-strand is initiated by binding of a nucleoside-triphosphate followed by elongation, termination and dissociation of the finished nucleic acid from the protein. Since the synthesis of new RNA is a key step in the replication of such viruses, polymerases should be ideal targets for drug discovery.[51]

Polymerase inhibitors have been widely used as antivirals against herpes and hepatitis B virus, and they have become the cornerstone in the treatment of HIV infection.[52] HIV drug discovery has shown us that there are two different ways to inhibit the polymerase enzyme. Zidovudine (Scheme 8.4), the first drug licensed to treat HIV patients, is a nucleoside that is transformed inside the cell to its triphosphate analog, which then competes with endogenous nucleoside triphosphates for incorporation into the nucleic acid. Once incorporated, it serves as a chain terminator of viral transcripts.[52] Nevirapine (Scheme 8.4)

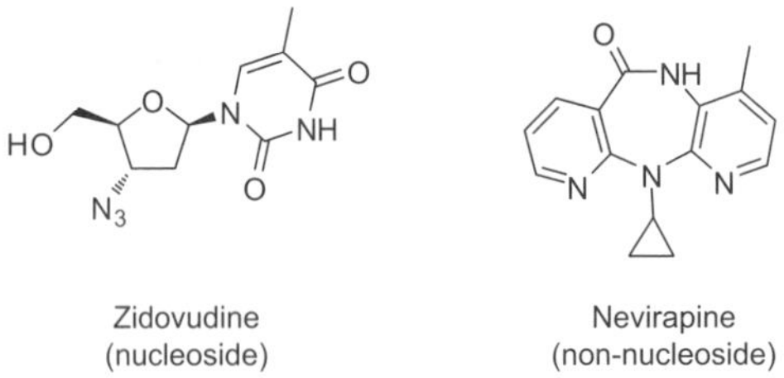

Scheme 8.4

which was approved in 1996 interacts with the same enzyme as Zidovudine, however, by a mechanism that relies on allosteric disruption of the enzymatic activity.[53] While the above-mentioned drugs target the reverse transciptase in HIV, the two approaches should also be applicable to the inhibition of *Flaviviridae* RdRp. Indeed, several representatives from both classes are currently in development for HCV.[54]

Purine nucleoside **1** (Scheme 8.5) was the first compound that was reported to have activity on a flavivirus RdRp. While working towards an inhibitor of HCV polymerase, Migliaccio *et al.*[55] identified a chain terminating ribonucleoside that showed activity on several viruses of the *Flaviviridae* family (Table 8.4).

The compound, however, was not suitable for *in vivo* animal studies, since it was rapidly metabolized by cellular adenosine deaminase. Further structure activity studies identified more stable derivatives, with **2** showing the best combination of HCV activity and pharmacokinetic properties.[56]

Based on these nucleosides (mainly designed for their HCV activity), Yin and co-workers[57,58] synthesized compound **3**. Similar to the two compounds discovered by the Merck scientists (**1** and **2**), ribonucleoside derivative **3** is a pan-*Flaviviridae* inhibitor (Table 8.5).

In addition to its broad spectrum of activity, nucleoside **3** exhibits good pharmacokinetics and is active in DENV animal models.[57,58] In order to prove that the compound indeed works as a chain-terminating RdRp inhibitor, Yin *et al.* demonstrated that the triphosphate of **3** is a potent inhibitor of DENV RdRp, and studies with radioactively labeled **3** demonstrated that the triphosphate is indeed formed *in vivo*.[57] While the preclinical profile of this compound

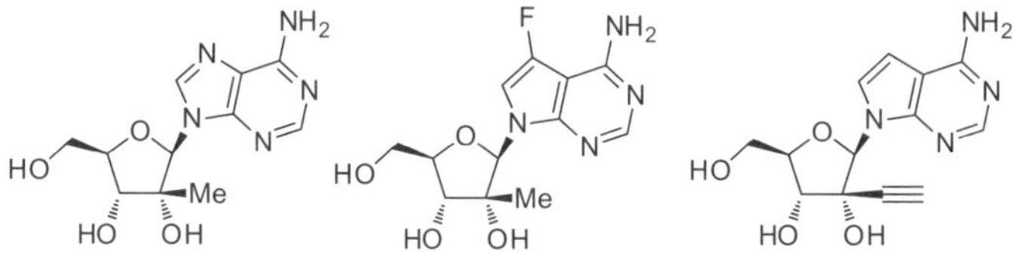

1 **2** **3**

Scheme 8.5

Table 8.4 Activity of **1** against *Flaviviridae*.[a]

Virus	*EC50 (μM)*	*CC50 (μM)*
Hepatitis C	0.25	n.a.
Bovine Viral Diarrhea	1.6	n.a.
West Nile	5.1	25
Dengue 2	4	18
Yellow Fever	3.2	13

[a]Data from reference 55; n.a. = data not available.

Table 8.5 Activity of ribonucleoside derivative **3.**[a]

Virus	*Cell type*	*Assay*	*EC50 (μM)*
HCV	Huh7	Replicon	0.11
DENV1	BHK	CFI	0.16
DENV2	BHK	CFI	0.65
DENV2	BHK	Replicon	0.23
DENV2	Huh7	CPE	1.25
DENV3	BHK	CFI	0.46
DENV4	BHK	CFI	0.22
JEV	BHK	CPE	1.25
WNV	BHK	CPE	3.75
YFV	A549	CPE	0.85

[a]Data from reference 57; CFI = cell-based flavivirus immunodetection assay; CPE = see reference 29.

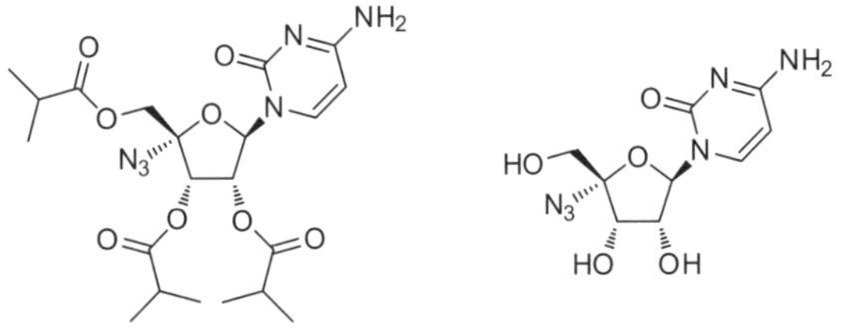

Scheme 8.6

looked encouraging, its development was halted when a two-week toxicological study could not establish an appropriate safety window.[57]Attempts to improve the toxicological profile with prodrugs[59] did not lead to the desired results and, therefore, development of **3** was abandoned.

Another nucleoside that has caught the attention of the flavivirus community is Balapiravir (**4**), a prodrug of 4-azidocytidine (**5**) (Scheme 8.6) that was first developed as an HCV treatment.[60] Balapiravir was successfully tested in a phase IIa study in treatment-naïve patients with HCV infections,[61] however, the development was later abandoned due to unspecified safety findings.[54] It appears now that Roche has decided to test the compound in a clinical trial in patients with DENV infections,[62] however, no scientific data on the effectiveness of **4** as a DENV antiviral have been published.

Allosteric or non-nucleoside inhibitors of RdRp inhibit the conformational flexibility of the polymerase and, therefore, block the domain movement that is necessary for RNA synthesis.[51] In HCV, four different allosteric sites have been identified, largely by high-throughput screening of small molecular weight compound libraries.[54] Rational design of such inhibitors is still challenging since the dynamics of RdRp enzymes are poorly understood and, therefore, the *de novo* design of functional inhibitors is difficult. Nevertheless the experience in HCV drug discovery suggests that the chances of obtaining a viable non-nucleoside inhibitor is greater than for chain terminators,[54] mainly because of the multitude of binding sites (nucleoside structure-activity space is very limited) and the more predictable pharmacokinetics of allosteric inhibitors.[63]

So far, only one allosteric inhibitor for a flavivirus RdRp has been described. HTS of a DENV polymerase assay lead to the identification of **6** (Scheme 8.7).[64] The lead compound was optimized to **7**, a compound that shows excellent selectivity against HCV, WNV and human DNA polymerases.[64] The mechanism of action of **7** was investigated by a combination of photo-affinity labeling[64] and biochemical experiments.[65] The results of both of these experiments suggest that **7** binds to the entrance of the RNA tunnel and competes with RNA for its binding site in the DENV RdRp enzyme (Figure 8.5). While **7** is an interesting tool compound for the study of flavivirus RdRp, it will need

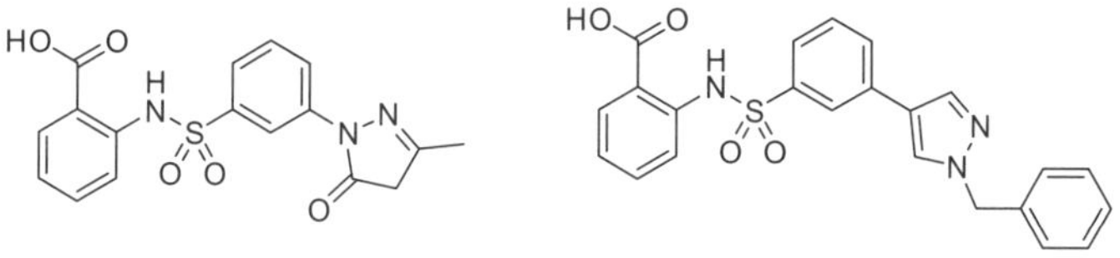

6 **7**

Scheme 8.7

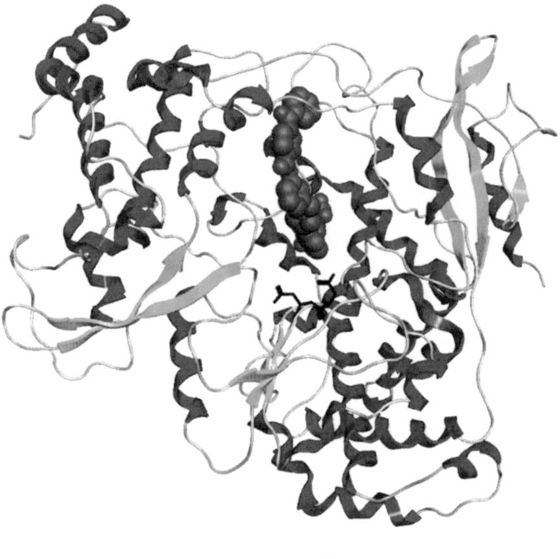

Figure 8.5 Crystal structure of DENV RdRp[66] with docked allosteric inhibitor (cpk). The active site residues Gly-Asp-Asp are depicted in grey stick presentation.

further optimization to become a viable drug candidate. Such a lead optimization would be greatly facilitated by a co-crystal structure of the allosteric inhibitor with the polymerase,[66] however, so far this has not been achieved.

8.4.4 NS5 Methyltransferase

The 5'-end of all flavivirus genomes contains a type 1 cap structure (m7GpppAmG).[67,68] The flavivirus methyltransferase (MTase) domain, located at the N-terminus of NS5, catalyzes N-7 and 2'-O methylations of the viral RNA cap in a sequential manner, GpppA-RNA→m7GpppA-RNA→m7GpppAm-RNA (Figure 8.6).[18,19]

The biological importance of flavivirus MTase has been validated in the context of the full-length virus. In both WNV and DENV, the N-7 methylation was shown to be essential for viral replication, whereas the 2'-O methylation is not critical, but facilitates replication.[69,19] Although the exact function of 2'-O methylation remains to be determined, the available data clearly indicate that antiviral drug discovery on this target should be focused on the inhibition of

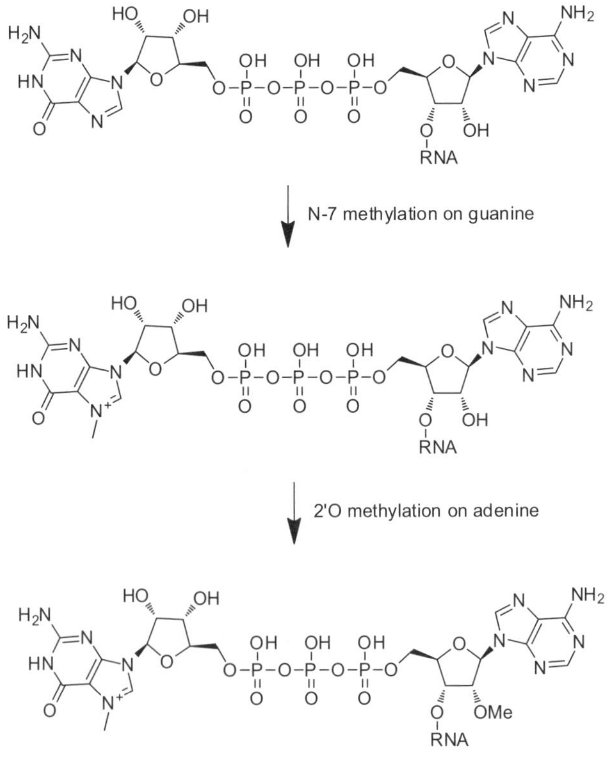

Figure 8.6 Sequential N-7 and 2'-O methylations of the viral RNA cap.

N-7 methylation activity. Scintillation proximity assays (SPA) have been developed for HTS of DENV 2'-O and N-7 MTase.[70] It should be noted that N-7 methylation requires the 5'-terminal stem-loop RNA structure of the viral genome as an active substrate,[69] making an N-7 MTase HTS very expensive and labor-intensive (preparing RNA substrate).

One obvious challenge in targeting the flaviviral MTase is the design of inhibitors that selectively suppress viral MTase without affecting host MTases. Because both viral and host MTases share a common core structure, a non-specific inhibitor would suppress host MTases, leading to toxicity. A recent report describes a hydrophobic pocket that is unique[73] to the flavivirus MTase, pointing to a potential solution to this selectivity problem (Figure 8.7). Since the pocket is located next to the SAH (S-adenosyl-L-homocysteine, the by-product after transfer of methyl group)-binding site, it was proposed that SAH analogs with substituents binding to the newly identified pocket should selectively inhibit flavivirus MTase. Indeed, such SAH derivatives specifically suppressed flaviviral MTases, but not human MTases (Lim *et al.*, submitted for publication). Although crystal structures have been reported for many flavivirus MTases, none of the current structures captures the RNA cap in a catalytic conformation. Therefore, caution should be taken when using the structural information for rational design and data interpretation.

Another potential challenge for development of MTase antivirals is resistance. In the case of WNV, it was shown that the replication of a mutant virus defective in both N-7 and 2'-O methylations could be rescued by a single amino acid mutation (W751R) in the RdRp domain.[74] The W751R substitution, located at the entrance of the RNA template tunnel of the polymerase, improved the polymerase activity of the recombinant full-length NS5 by greater than 5-fold. The result implies that resistance to MTase inhibitors may arise from mutations in the RdRp domain.

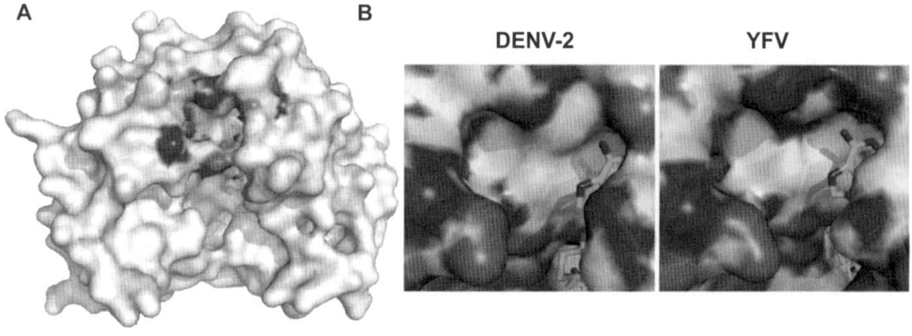

Figure 8.7 Flavivirus MTase structure. (A) WNV MTase structure (PDB code: 2OY0).[71] The flavivirus-conserved hydrophobic pocket, located next to the SAH-binding site, is labeled in color. (B) The hydrophobic pockets from the DENV-2 (PDB code: 1L9K)[18] and YFV MTases (PDB code: 3EVA).[72] The SAH molecule is shown in stick presentation.

8.5 Host Targets

8.5.1 Host Targets Required for Viral Replication

All viruses rely on host resources, including nucleic acids, proteins, and membrane components, to complete their life cycle. This dependency offers an attractive strategy for therapeutic intervention. This approach is exemplified by the HCV drug ribavirin (RBV) and the recently identified cyclophilin inhibitors. Ribavirin, in combination with pegylated interferon, is the current standard therapy for HCV infections (Scheme 8.8). Various mechanisms of action have been suggested for the broad spectrum antiviral activity of RBV.[75] These include (i) depletion of intracellular GTP pools (the 5'-monophosphate metabolite of RBV competitively inhibits the host enzyme inosine-monophosphate-dehydrogenase (IMPDH) essential for the *de novo* synthesis of GTP);[76] (ii) inhibition of viral polymerase activity by the 5'-triphosphate metabolite of RBV;[77] (iii) immunomodulation via a Th2 to Th1 shift in the immune response;[78] and (iv) induction of an error catastrophe by increasing the error rate during viral replication.[79] RBV has been reported to have weak *in vitro* activity against WNV and DENV, and mechanistic studies have shown that inhibition of IMPDH activity is the predominant antiviral mechanism of RBV for flaviviruses.[80] *In vivo* studies of RBV in rodent models were, however, less promising, and questionable clinical benefit was observed in a small patient population during a WNV outbreak in 2000.[81] Similar to RBV, inhibitors of the host enzyme dihydroorotate dehydrogenase (DHODH) (e.g. brequinar, BQR) exhibit activities against flaviviruses.[82] BQR exerts its antiviral activity through depletion of intracellular pyrimidine pools; it has an anti-DENV EC_{50} value of 78 nM when tested in cell culture, but this *in vitro* potency failed to translate into *in vivo* efficacy in the dengue mouse model.[82] It is likely that BQR could not completely suppress the pyrimidine level below the threshold required for viral replication under physiological condition *in vivo*.

Non-immunosuppressive cyclophilin (Cyp) inhibitors represent the other successful anti-HCV therapy that functions by targeting host proteins. Three inhibitors, Alisporivir (DEBIO-025), NIM811 and SCY-635, are currently in exploratory phase I and II trials.[83] Cyps are a family of cellular enzymes possessing peptidyl-prolyl isomerase (PPIase) activity, which have been shown

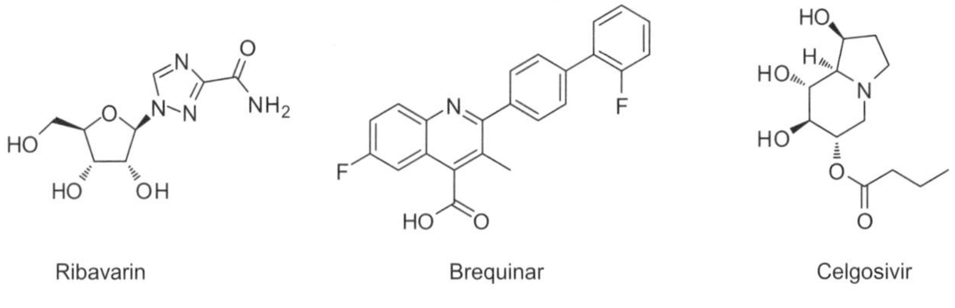

Scheme 8.8

to be critical for HCV replication.[84] Specifically, CypA was reported to bind to several viral proteins, including NS2, NS5A, and NS5B. It is speculated that CypA supports polyprotein processing, and viral RNA synthesis. Recently, the role of CypA in flavivirus replication has been demonstrated.[85] Knockdown of Cyps in Huh-7.5 cells reduced the replication of WNV, YFV, and DENV, and this decrease could be restored by exogenous expression of a wild-type but not a PPIase-defective CypA. Direct interaction between CypA with WNV NS5 protein was blocked by cyclosporin A, an immunosuppressive CypA inhibitor. These results suggest that, similar to HCV, CypA is a component of the flavivirus replication complex and could be a potential host target for anti-flavivirus development.

Glucosidase, a host enzyme responsible for the proper folding of many different glycoproteins, is another potential cellular target for therapeutic intervention. A number of glucosidase inhibitors were studied extensively in various virus systems including DENV[86] and HCV.[87] Celgosivir developed by MIGENIX Inc. is currently in a phase II trial to investigate its anti-HCV effect in combination with pegylated INFα2b plus ribavirin.[87] While celgosivir alone was not effective as a monotherapy in HCV, the compound may have potential as an anti-dengue treatment.

Cellular receptors for virus entry have proven to be interesting targets in HIV drug discovery. Maraviroc, a selective CCR5 antagonist, blocks the interaction between HIV envelop protein (gp120) and its coreceptor (CCR5) thereby inhibiting HIV entry. It was approved in 2007 by the FDA for treating HIV-infected patients with viral replication and HIV strains resistant to multiple anti-retroviral agents.[88] There are two main reasons for the success of HIV entry inhibitors: (i) the discovery of specific cellular receptors for HIV entry; and (ii) the identification of a subset of Caucasian population that is homozygous for CCR5 deletion and resistant to HIV-1 infection. In contrast, no specific cellular receptor(s) has been identified for flavivirus entry, and the viral–host membrane fusion occurs in the endosome, which makes the receptor-targeted approach challenging. Efforts for targeting flavivirus entry have been focused on the envelope protein (E protein). The crystal structure of the DENV E protein revealed a hydrophobic pocket that is occupied by a small detergent molecule (*N*-octyl-β-D-glucoside), suggesting the potential for identifying small-molecule inhibitors of membrane fusion.[89] Computational high-throughput screening of chemical libraries against this pocket has since been applied and potentially interesting compounds have been identified (e.g. **8, 9** and **10**) (Scheme 8.9).[90–93] Biological testing using various cell-based assays suggested that some of these molecules indeed have antiviral and fusion-inhibitory activities (e.g. compound **10**). However, much of the chemical matter disclosed displays only micromolar activity and, therefore, requires optimization.

The biggest challenge for targeting host proteins is toxicity. Nevertheless, assuming that upregulation of specific genes in certain cellular pathways is required to ensure productive viral infection, a partial inhibition of the gene function may already be sufficient to disrupt the virus-host interaction with few/limited side effects. Genome-wide screening technologies are expanding our

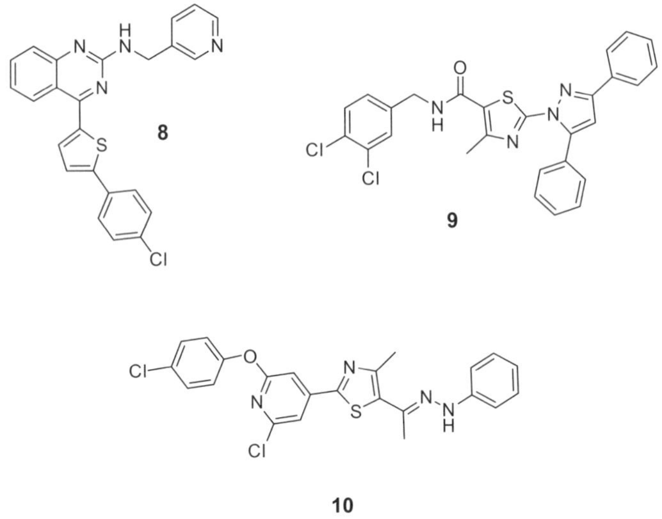

Scheme 8.9

understanding of the complex battle between the virus and the host, and the identification and validation of the genes/cellular pathways exploited exclusively by viruses may eventually lead to new anti-flaviviral therapies.

8.5.2 Host Targets Involved in Disease Exacerbation

Besides the host proteins that facilitate virus replication, there is another class of proteins that may be responsible for the severe response to viral infection. For example, a small percentage of DENV-infected patients (3–5%) progress to life-threatening dengue hemorrhagic fever (DHF) or dengue shock syndrome (DSS). Conceptually, pharmacological intervention in the pathways responsible for the severe form of dengue would be expected to result in clinical benefit. While our understanding of the immunopathogenesis of DHF/DSS is in its infancy, a recent study suggests that this area of research may have a bright future.

Chen *et al.* showed that CLEC5A (C-type lectin domain family 5, member A) interacts directly with DENV, resulting in the release of proinflammatory cytokines.[93] Anti-CLEC5A antibodies preventing DENV binding to the receptor reduced vascular leakage and mortality of STAT1$^{-/-}$ knockout mice infected with DENV. An observational study in patients with or after dengue infection was launched in 2007 with the aim to analyze the influence of single nucleotide polymorphism of the CLEC5A receptor on disease progression.[94] This study will provide valuable information on the role of this receptor in the development of severe disease. Humanized anti-CLEC5A monoclonal antibodies may be useful to mitigate vascular leakage in DENV-infected patients.[93]

8.6 Cell-based Screening and Optimization

Phenotypic cell-based screening approaches have been employed by both pharmaceutical companies and academic groups to identify antiviral agents during the past decade. This approach is attractive when specific protein targets are not available, or when chemical genomics approaches to identify novel targets with small molecular-weight compounds are desirable.

Despite intensive efforts, the return has so far been limited in the *Flaviviridae* area. The only tangible success is an HCV NS5A inhibitor that was identified in a HCV replicon assay. Following the identification of the hit **11** (Scheme 8.10), the target protein was identified through the use of resistance selection.[95] Compound **11** formed the basis of an extensive series of chemical refinements that resulted in the development compound BMS-790052 (**12**), which has shown proof-of-concept in clinical trial for treating HCV infection.[96]

The reasons behind the limited success of cellular screening when compared to target-based approaches are threefold. (i) Viruses as absolute parasites are extremely sensitive to compound-mediated cytotoxicity. The measurement of compound-mediated cytotoxicity is therefore paramount for accurately determining antiviral activity. However, this measurement is often ignored in primary screening due to practical issues (e.g. large-scale cell material preparation and limited compound library supplies). This can greatly impact the quality of the screening effort and lower the chances of success. (ii) Development of inhibitors of viral targets is usually preferred over that of inhibitors of host targets to avoid potential side effects. The genome of flaviviruses encodes 10 genes, whereas hundreds or thousands of host factors are involved in virus replication. The chance of identifying inhibitors of viral targets in a random phenotypic screen is therefore very slim. (iii) Related to the preceding point, target deconvolution of the cell-based screening hits is challenging and costly. Typically, active compounds are passaged with the virus to generate escape

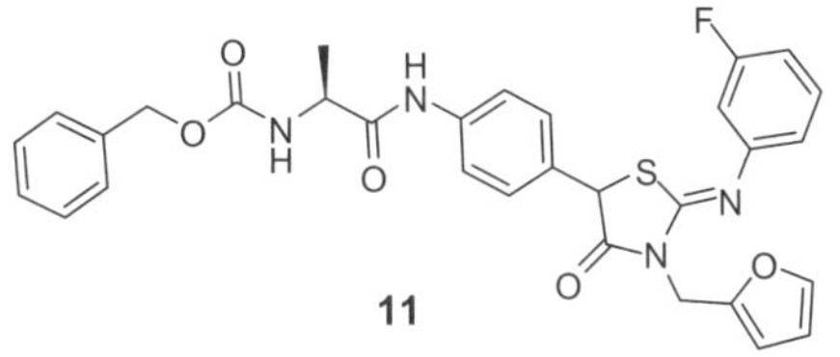

11

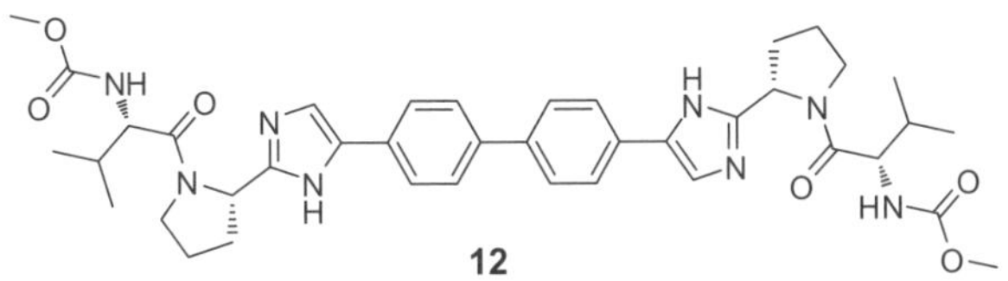

12
BMS-790052

Scheme 8.10

mutants, which might give clues to the mechanism of action by identifying the viral determinant(s) in conferring resistance. However, this does not exclude the possibility of the viral protein being an indirect binder. Chemical proteomics has nowadays become a powerful tool for target profiling of small molecules. This postgenomic version of classical drug affinity chromatography benefits from the tremendous advances in high-resolution mass spectrometry technology, but it is important to note that implementation of this technology requires a huge amount of resources, including dedicated teams and purpose-fitted laboratories.[97]

8.7 Conclusions

The last 10 years have produced a major leap forward in the understanding of flavivirus biology and we have seen the first attempts to find antivirals for the treatment of WNV and DENV. Because most flaviviral diseases belong to the commercially unattractive indications, the resources that have so far been invested pale in comparison with HCV and HIV. So while the comparison between HCV, HIV and flaviviral diseases is useful to gauge the drug discovery challenges, it is unrealistic to expect the neglected diseases to progress at a similar pace. The genome of the flaviviruses is rather small providing a limited number of targets for antiviral drug discovery. Of the few available enzyme targets, only the protease and polymerase are clearly druggable, while the helicase and methyltransferase belong to enzyme classes that so far have not been cracked in any diseases.

Research on host targets is still at a very early stage, so that rational, target-based drug discovery is currently not feasible (maybe with the exception of CLEC5A). This leaves cell-based screening as the only alternative to standard structure-based drug discovery on viral enzymes. While this is a feasible approach, it needs a sizable investment in chemistry (hit to lead) and screening technologies (e.g. orthogonal and counter screens, and high content screening), as well as access to a state-of-the-art target identification platform.

Overall, the field of flavivirus drug discovery stands on a solid scientific basis, but progress in several areas is needed to increase the likelihood of success. The field needs:

1. more information on the structure and dynamics of the viral enzymes;
2. proteins that are amenable to reproducible crystallization with ligands (co-crystallization and soaking) and biophysical measurements (e.g. isothermal calorimetry);
3. information on the replication complex, including assays that can assess the activity of compounds on the enzymatic activity in the replication complex; and
4. increased research on host factors that influence disease progression.

References

1. B. D. Lindenbach, H. J. Thiel, C. M. Rice, *Fields Virology*, ed. D. M. Knipe and P. M. Howley, Lippincott William & Wilkins, Philadelphia, 2007, **Vol 1**, p. 1101.
2. http://www.who.int/csr/disease/dengue/impact/en/index.html (last accessed 20. October 2010).
3. C. L. Gardner and K. D. Ryman, *Clin. Lab. Med.*, 2010, **30**, 237.
4. D. Gubler, G. Kuno, L. Markoff in *Fields Virology*, ed. D. M. Knipe and P. M. Howley, Lippincott William & Wilkins, Philadelphia, 2007, **Vol 1**, p. 1153.
5. S. B. Halstead, J. Jacobson in *Vaccines*, ed. S. A. Plotkin, W. A. Orenstein and P. A. Offit, Saunders Elsevier, Philadelphia, 2008, p. 311.
6. S. B. Halstead and S. Thomas, *Clin. Infect. Dis.*, 2010, **50**, 1155.
7. S. B. Halstead, *Adv. Virus Res.*, 2003, **60**, 421–67.
8. B. M. Kummerer and C. M. Rice, *J. Virol.*, 2002, **76**, 4773.
9. W. J. Liu, H. B. Chen and A. A. Khromykh, *J. Virol.*, 2003, **77**, 7804.
10. J. Guo, J. Hayashi and C. Seeger, *J. Virol.*, 2005, **79**, 1343.
11. W. Liu, X. Wang, V. Mokhonov, P. Y. Shi, R. Randall and A. Khromykh, *J. Virol.*, 2005, **79**, 1934.
12. J. L. Munoz-Jordan, M. Laurent-Rolle, J. Ashour, L. Martinez-Sobrido, M. Ashok, W. I. Lipkin and A. Garcia-Sastre, *J. Virol.*, 2005, **79**, 8004.
13. J. L. Munoz-Jordan, G. G. Sanchez-Burgos, M. Laurent-Rolle and A. Garcia-Sastre, *Proc. Natl. Acad. Sci. USA*, 2003, **100**, 14333.
14. B. Falgout, R. H. Miller and C. J. Lai, *J. Virol.*, 1993, **67**, 2034.
15. G. Wengler and G. Wengler, *Virology*, 1991, **184**, 707.
16. G. Wengler and G. Wengler, *Virology*, 1993, **197**, 265.
17. M. Ackermann and R. Padmanabhan, *J. Biol. Chem.*, 2001, **276**, 39926.
18. M. P. Egloff, D. Benarroch, B. Selisko, J. L. Romette and B. Canard, *EMBO J.*, 2002, **21**, 2757.
19. D. Ray, A. Shah, M. Tilgner, Y. Guo, Y. Zhao, H. Dong, T. Deas, Y. Zhou, H. Li and P. Y. Shi, *J. Virol.*, 2006, **80**, 8362.
20. B. H. Tan, J. Fu, R. J. Sugrue, E. H. Yap, Y. C. Chan and Y. H. Tan, *Virology*, 1996, **216**, 317–25.
21. B. Lindenbach and C. Rice, *J. Virol.*, 1997, **71**, 9608.
22. S. Miller, S. Kastner, J. Krijnse-Locker, S. Buhler and R. Bartenschlager, *J. Biol. Chem.*, 2007, **282**, 8873.
23. S. Miller, S. Sparacio and R. Bartenschlager, *J. Biol. Chem.*, 2006, **281**, 8854.
24. J. J. Tan, X. J. Cong, L. M. Hu, C. X. Wang, L. Jia and X. J. Liang, *Drug Discov. Today*, 2010, **15**, 186.
25. L. Lo, M. Tilgner and P. Y. Shi, *J. Virol.*, 2003, **77**, 12901.
26. M. Qing, W. Liu, Z. Yuan, F. Gu and P. Y. Shi, *Antiviral Res.*, 2010, **86**, 163.
27. F. Puig-Basagoiti, T. S. Deas, P. Ren, M. Tilgner, D. M. Ferguson and P. Y. Shi, *Antimicrob. Agent. Chemother.*, 2005, **49**, 4980.

28. J. A. Esté and A. Telenti, *Lancet*, 2007, **370**, 81.
29. C. G. Noble, Y. L. Chen, H. Dong, F. Gu, S. P. Lim, W. Schul, Q. Y. Wang and P. Y. Shi, *Antiviral Res.*, 2010, **85**, 450.
30. B. Falgout and M. Pethel, *J. Virol.*, 1991, **65**, 2467.
31. Y. S. Tsantrizos, *Accounts Chem. Res.*, 2008, **41**, 1252.
32. S. R. Turner, J. W. Strohbach, R. A. Tommasi, P. A. Aristoff, P. D. Johnson, H. I. Skulnick, L. A. Dolak, E. P. Seest, P. K. Tomich, M. J. Bohanon, M. M. Horng, J. C. Lynn, K. T. Chong, R. R. Hinshaw, K. D. Watenpough, M. N. Janakiraman and S. Thaisrivongs, *J. Med. Chem.*, 1998, **41**, 3467.
33. F. Narjes, K. F. Koehler, U. Koch, B. Gerlach, S. Colarusso, C. Steinkuehler, M. Brunetti, S. Altamura, R. D. Francesco and V. G. Matassa, *Bioorg. Med. Chem. Lett.*, 2002, **12**, 701.
34. Z. Yin, S. J. Patel, W. L. Wang, G. Wang, W. L. Chan, K. R. R. Rao, J. Alam, D. A. Jeyaraj, X. Ngew, V. Patel, D. Beer, S. P. Lim, S. G. Vasudevan and T. H. Keller, *Bioorg. Med. Chem. Lett.*, 2006, **16**, 36.
35. P. Erbel, N. Schiering, F. Villard, S. P. Lim, Z. Yin, T. H. Keller, S. G. Vasudevan and U. Hommel, *Nat. Struct. Mol. Biol.*, 2006, **13**, 372.
36. A. E. Aleshin, S. A. Shiryaev, A. Y. Strongin and R. C. Liddington, *Prot. Sci.*, 2007, **16**, 795.
37. G. Robin, K. Chappell, M. J. Stoermer, S. H. Hu, P. R. Young, D. P. Fairlie and J. L. Martin, *J. Mol. Biol.*, 2009, **385**, 1568.
38. J. E. Knox, N. L. Ma, Z. Yin, S. J. Patel, W. L. Wang, W. L. Chan, K. R. R. Rao, G. Wang, X. Ngew, V. Patel, D. Beer, S. P. Lim, S. G. Vasudevan and T. H. Keller, *J. Med. Chem.*, 2006, **49**, 6585.
39. X. C. Su, K. Ozawa, R. Qi, S. G. Vasudevan, S. P. Lim and G. Otting, *PLOS Neglect. Trop. D.*, 2009, **3**, e561.
40. M. J. Stoermer, K. J. Chappell, S. Liebscher, C. M. Jensen, C. H. Gan, P. K. Gupta, W. J. Xu, P. R. Young and D. P. Fairlie, *J. Med. Chem.*, 2008, **51**, 5714.
41. N. H. Meller, N. Pattabiraman, C. Ansarah-Sobrinho, P. Viswanathan, T. C. Pierson and R. Padmanabhan, *Antimicrob. Agents Ch.*, 2008, **52**, 3385.
42. D. Ekonomiuk, X. C. Su, K. Ozawa, C. Bodenreider, S. P. Lim, Z. Yin, T. H. Keller, D. Beer, V. Patel, G. Otting, A. Caflish and D. Huang, *PLOS Neglect. Trop. D.*, 2009, **3**, e356.
43. C. Bodenreider, D. Beer, T. H. Keller, S. Sonntag, D. Wen, L. Yap, Y. H. Yau, S. G. Shochat, D. Huang, T. Zhou, A. Caflisch, X. C. Su, K. Ozawa, G. Otting, S. G. Vasudevan, J. Lescar and S. P. Lim, *Anal. Biochem.*, 2009, **395**, 195.
44. T. A. Halgren, *J. Chem. Inf. Model.*, 2009, **49**, 377.
45. S. Chandramouli, J. S. Joseph, S. Daudenarde, J. Gatchalian, C. Cornillez-Ty and P. Kuhn, *J. Virol.*, 2010, **84**, 3059.
46. D. Luo, T. Xu, C. Hunke, G. Gruber, S. G. Vasudevan and J. Lescar, *J. Virol.*, 2008, **82**, 173.
47. T. Xu, A. Sampath, A. Chao, D. Wen, M. Nanao, P. Chene, S. G. Vasudevan and J. Lescar, *J. Virol.*, 2005, **79**, 10278.

48. D. Luo, T. Xu, R. P. Watson, D. Scherer-Becker, A. Sampath, W. Jahnke, S. S. Yeong, C. H. Wang, S. P. Lim, A. Strongin, S. G. Vasudevan and J. Lescar, *EMBO J.*, 2008, **27**, 3209.
49. R. Assenberg, E. Mastrangelo, T. S. Walter, A. Verma, M. Milani, R. J. Owens, D. I. Stuart, J. M. Grimes and E. J. Mancini, *J. Virol.*, 2009, **83**, 12895.
50. C. A. Belon, Y. D. High, T. I. Lin, F. Pauwels and D. N. Frick, *Biochemistry*, 2010, **49**, 1822.
51. H. Malet, N. Masse, B. Selisko, J. L. Romette, K. Alvarez, J. C. Guillemot, H. Tolou, T. L. Yap, S. Vasudevan, J. Lescar and B. Canard, *Antivir. Res.*, 2008, **80**, 23.
52. T. Cilar and A. S. Ray, *Antivir. Res.*, 2010, **85**, 39.
53. M. P. de Bethune, *Antivir. Res.*, 2010, **85**, 75.
54. H. Li and S. T. Shi, *Future Med. Chem.*, 2010, **2**, 121.
55. G. Migliaccio, J. E. Tomassini, S. S. Carroll, L. Tomei, S. Altamura, B. Bhat, L. Bartholomew, M. R. Bosserman, A. Ceccacci, L. F. Colwell, R. Cortese, R. D. Francesco, A. B. Eldrup, K. L. Getty, X. S. Hou, R. L. LaFemina, S. W. Ludmerer, M. MacCoss, D. R. McMasters, M. W. Stahlhut, D. B. Olsen, D. J. Hazuda and O. A. Flores, *J. Biol. Chem.*, 2003, **49**, 49164.
56. A. B. Eldrup, M. Prhavc, J. Brooks, B. Bhat, T. P. Prakash, Q. Song, S. Bera, N. Bhat, P. Dande, P. D. Cook, C. F. Bennett, S. S. Carroll, R. G. Ball, M. R. Bosserman, C. Burlein, L. F. Colwell, J. F. Fay, O. A. Flores, K. Getty, R. L. LaFemina, J. Leone, M. MacCoss, D. R. McMasters, J. E. Tomassini, D. Von Langen, B. Wolanski and D. B. Olsen, *J. Med. Chem.*, 2004, **47**, 5284.
57. Z. Yin, Y. L. Chen, W. Schul, Q. Y. Wang, F. Gu, J. Duraiswamy, R. Reddy Kondreddi, P. Niyomrattanakit, S. B. Lakshminarayana, A. Goh, H. Y. Xu, W. Liu, B. Liu, J. Y. H. Lim, C. Y. Ng, M. Qing, C. C. Lim, A. Yip, G. Wang, W. L. Chan, H. P. Tan, K. Lin, B. Zhang, K. A. Bernard, C. Garrett, K. Beltz, M. Dong, M. Weaver, H. He, A. Pichota, V. Dartois, T. H. Keller and P. Y. Shi, *Proc. Natl. Acad. Sci. USA*, 2009, **106**, 20435.
58. Y. L. Chen, Z. Yin, J. Duraiswamy, W. Schul, C. C. Lim, B. Liu, H. Y. Xu, M. Qing, A. Yip, G. Wang, W. L. Chan, H. P. Tan, M. Lo, S. Liung, R. Reddy Kondreddi, R. Rao, H. Gu, H. He, T. H. Keller and P. Y. Shi, *Antimicrob. Agents Ch.*, 2010, **54**, 2932.
59. Y. L. Chen, Z. Yin, S. B. Lakshminarayana, M. Qing, W. Schul, J. Duraiswamy, R. Reddy Kondreddi, A. Goh, H. Y. Xu, A. Yip, B. Liu, M. Weaver, V. Dartois, T. H. Keller and P. Y. Shi, *Antimicrob. Agents Ch.*, 2010, **54**, 3255.
60. D. B. Smith, J. A. Martin, K. Klumpp, S. J. Baker, P. A. Blomgren, R. Devos, C. Granycome, J. Hang, C. J. Hobbs, W. R. Jiang, C. Laxton, S. L. Pogam, V. Leveque, H. Ma, G. Maile, J. H. Merrett, A. Pichota, K. Sarma, M. Smith, S. Swallow, J. Symons, D. Vesey, I. Najera and N. Cammack, *Bioorg. Med. Chem. Lett.*, 2007, **17**, 2570.

61. P. J. Pockros, D. Nelson, E. Godofsky, M. Rodrigez-Torres, G. T. Everson, M. W. Fried, R. Ghalib, S. Harrison, L. Nyberg, M. L. Shiffman, I. Najera, A. Chan and G. Hill, *Hepatology*, 2008, **48**, 385.
62. www.rochetrials.com/trialDetailsGet.action?studyNumber = PP22799 &diseaseCategoryId = 17&divisionName = PHA (last accessed 7. November 2010).
63. M. Pastor-Anglada, P. Cano-Soldadoa, M. Molina-Arcasa, M. P. Lostaob, I. Larrayoz, J. Martinez-Picadoc and F. J. Casado, *Virus Res.*, 2005, **107**, 151.
64. Z. Yin, Y. L. Chen, R. Reddy Kondreddi, W. L. Chan, G. Wang, R. H. Ng, J. Y. H. Lim, W. Y. Lee, D. A. Jeyaraj, P. Niyomrattanakit, D. Wen, A. Chao, J. F. Glickman, H. Voshol, D. Mueller, C. Spanka, S. Dressler, S. Nilar, S. G. Vasudevan, P. Y. Shi and T. H. Keller, *J. Med. Chem.*, 2009, **52**, 7934.
65. P. Niyomrattanakit, Y. L. Chen, H. Dong, Z. Yin, M. Qing, J. F. Glickman, K. Lin, D. Mueller, H. Voshol, J. Y. H. Lim, S. Nilar, T. H. Keller and P. Y. Shi, *J. Virol.*, 2010, **84**, 5678.
66. T. L. Yap, T. Xu, Y. L. Chen, H. Malet, M. P. Egloff, B. Canard, S. G. Vasudevan and J. Lescar, *J Virol.*, 2007, **81**, 4753.
67. G. R. Cleaves and D. T. Dubin, *Virology*, 1979, **96**, 159.
68. G. Wengler, *Virology*, 1981, **113**, 544.
69. H. Dong, D. C. Chang, X. Xie, Y. X. Toh, K. Y. Chung, G. Zou, J. Lescar, S. P. Lim and P. Y. Shi., *Virology*, 2010, **405**, 568.
70. K. Y. Chung, H. Dong, A. T. Chao, P. Y. Shi, J. Lescar and S. P. Lim, *Virology*, 2010, **402**, 52.
71. Y. Zhou, D. Ray, Y. Zhao, H. Dong, S. Ren, Z. Li, Y. Guo, K. Bernard, P.-Y. Shi and H. Li. Structure and function of flavivirus NS5 methyltransferase. *J. Virol.*, 2007, **81**, 3891.
72. B. J. Geiss, A. A. Thompson, A. J. Andrews, R. L. Sons, H. H. Gari, S. M. Keenan and O. B. Peersen. Analysis of flavivirus NS5 methyltransferase cap binding. *J. Mol. Biol.*, 2009, **385**, 1643.
73. H. Dong, L. Liu, G. Zou, Y. Zhao, Z. Li, S. P. Lim, P. Y. Shi and H. Li, *J. Biol. Chem.*, 2010, **285**, 32586.
74. B. Zhang, H. Dong, Y. Zhou and P. Y. Shi, *J. Virol.*, 2008, **82**, 7047.
75. N. M. Dixit and A. S. Perelson, *Cell Mol Life Sci.*, 2006, **63**, 832.
76. J. Y. Lau, R. C. Tam, T. J. Liang and Z. Hong, *Hepatology*, 2002, **35**, 1002.
77. D. Maag, C. Castro, Z. Hong and C. E. Cameron, *J. Biol. Chem.*, 2001, **276**, 46094.
78. R. C. Tam, B. Pai, J. Bard, C. Lim, D. R. Averett, U. T. Phan and T. Milovanovic, *J. Hepatol.*, 1999, **30**, 376.
79. Z. Hong, *Hepatology*, 2003, **38**, 807.
80. P. Leyssen, J. Balzarini, E. D. Clercq and J. Neyts, *J. Virol.*, 2005, **79**, 1943.
81. T. Parkinson and D. Pryde, *Future Med. Chem.*, 2010, **2**, 1181.
82. M. Qing, G. Zou, Q. Y. Wang, H. Y. Xu, H. Dong, Z. Yuan and P. Y. Shi, *Antimicrob Agents Chemother.*, 2010, **54**, 3686.

83. G. Fischer, P. Gallay and S. Hopkins, *Curr. Opin. Investig. Drugs*, 2010, **11**, 911.
84. Z. Liu, F. Yang, J. M. Robotham and H. Tang, *J. Virol.*, 2009, **83**, 6554.
85. M. Qing, F. Yang, B. Zhang, G. Zou, J. M. Robida, Z. Yuan, H. Tang and P. Y. Shi, *Antimicrob. Agents Chemother.*, 2009, **53**, 3226.
86. J. Chang, L. Wang, D. Ma, X. Qu, H. Guo, X. Xu, P. M. Mason, N. Bourne, R. Moriarty, B. Gu, J. T. Guo and T. M. Block, *Antimicrob. Agents Chemother.*, 2009, **53**, 1501.
87. D. Durantel, *Curr. Opin. Investig. D.*, 2009, **10**, 860.
88. J. C. Tilton and R. W. Doms, *Antiviral Res.*, 2010, **85**, 91.
89. Y. Modis, S. Ogata, D. Clements and S. C. Harrison, *Proc. Natl. Acad. Sci. USA*, 2003, **100**, 6986.
90. Q. Y. Wang, S. J. Patel, E. Vangrevelinghe, H. Y. Xu, R. Rao, D. Jaber, W. Schul, F. Gu, O. Heudi, N. L. Ma, M. K. Poh, W. Y. Phong, T. H. Keller, E. Jacoby and S. G. Vasudevan, *Antimicrob. Agents Chemother.*, 2009, **53**, 1823.
91. Z. Li, M. Khaliq, Z. Zhou, C. B. Post, R. J. Kuhn and M. Cushman, *J. Med. Chem.*, 2008, **51**, 4660.
92. T. Kampmann, R. Yennamalli, P. Campbell, M. J. Stoermer, D. P. Fairlie, B. Kobe and P. R. Young, *Antivir. Res.*, 2009, **84**, 234.
93. S. T. Chen, Y. L. Lin, M. T. Huang, M. F. Wu, S. C. Cheng, H. Y. Lei, C. K. Lee, T. W. Chiou, C. H. Wong and S. L. Hsieh, *Nature*, 2008, **453**, 672.
94. http://clinicaltrials.gov/ct2/show/NCT00688389?term=CLEC5A&rank=1 (last accessed 12. November 2010).
95. M. Gao, R. E. Nettles, M. Belema, L. B. Snyder, V. N. Nguyen, R. A. Fridell, M. H. Serrano-Wu, D. R. Langley, J. H. Sun, D. R. O'Boyle, J. A. Lemm, C. Wang, J. O. Knipe, C. Chien, R. J. Colonno, D. M. Grasela, N. A. Meanwell and L. G. Hamann, *J. Virol.*, 2010, **84**, 482.
96. M. Gao, R. E. Nettles, M. Belema, L. B. Snyder, V. N. Nguyen, R. A. Fridell, M. H. Serrano-Wu, D. R. Langley, J. H. Sun, D. R. O'Boyle, J. A. Lemm, C. Wang, J. O. Knipe, C. Chien, R. J. Colonno, D. M. Grasela, N. A. Meanwell and L. G. Hamann, *Nature*, 2010, **465**, 96.
97. S. M. A. Huang, Y. M. Mishina, S. Liu, A. Cheung, F. Stegmeier, G. A. Michaud, O. Charlat, E. Wiellette, Y. Zhang, S. Wiessner, M. Hild, X. Shi, C. J. Wilson, C. Mickanin, V. Myer, A. Fazal, R. Tomlinson, F. Serluca, W. Shao, H. Cheng, M. Shultz, C. Rau, M. Schirle, J. Schlegl, S. Ghidelli, S. Fawell, C. Lu, D. Curtis, M. W. Kirschner, C. Lengauer, P. M. Finan, J. A. Tallarico, T. Bouwmeester, J. A. Porter, A. Bauer, F. Cong and Feng, *Nature*, 2009, **461**, 614.

CHAPTER 9

Current Approaches to Tuberculosis Drug Discovery and Development

MARK J. MITTON-FRY*[a] AND DEBRA HANNA[b]

[a] Pfizer, Antibacterials Medicinal Chemistry, PharmaTherapeutics, Eastern Point Road, Groton, CT 06340, MS 8220-2464, US; [b] Critical Path Institute, 1730 E River Road, Tucson, Arizona 85718, USA

9.1 The Global Problem of Tuberculosis and Current State of Affairs

Tuberculosis (TB) remains a dominant infectious disease causing morbidity and mortality worldwide.[1,2] It is estimated that one-third of the world's population is latently infected with *Mycobacterium tuberculosis* (MTB).[3,4] In addition, approximately 9 million incident cases of active disease occur annually, resulting in approximately 2 million deaths worldwide when considering both HIV-positive and negative persons.[5] TB, HIV and Malaria exemplify the continued threat posed by infectious diseases in the modern age. Despite the ongoing battle with TB, the effective eradication and control of this pathogen remain elusive.

There has been a slow reduction in the incidence rates of TB, less than 1% per year, as a result of intensive global efforts to implement the Stop TB strategy and its predecessor, directly observed therapy short-course (DOTS). The decline in TB burden has been slowed by the overall increase in population,

RSC Drug Discovery Series No. 14
Neglected Diseases and Drug Discovery
Edited by Michael J. Palmer and Timothy N. C. Wells

Published by the Royal Society of Chemistry, www.rsc.org

and the spread of TB is negatively impacted by factors such as the overcrowded conditions observed in prison populations in the developed and developing world.[6] Drug-susceptible TB is still the most common form of the disease, but the Centers for Disease Control (CDC) and Infectious Disease Society of America (IDSA) have published their growing concerns regarding the overall increase in the number of drug-resistant cases despite the overall decline in new TB cases reported.[7]

Two major obstacles to the control of global tuberculosis (TB) have emerged. The first is the high prevalence of HIV among TB patients in some regions,[8] including sub-Saharan Africa. The increasing problem of anti-TB drug resistance constitutes the second. Beginning in the 1990s, TB drug resistance emerged as a significant problem in the United States as well as other parts of the world with the development of multidrug-resistant (MDR) TB. MDR-TB has since spread in both the developing and developed world, largely due to difficulties with treatment compliance. MDR-TB is defined as TB caused by MTB strains which are proven to be resistant to at least two of the most potent first line anti-TB drugs, isoniazid and rifampin.[9,10] It is estimated that more than 90% of all MDR-TB cases are not being treated by international guidelines, further exacerbating this issue. Reports of MDR-TB also resistant to many second-line drugs were reported in 2000.[11,12] At this time, 10% of MDR-TB isolates were found to be resistant to three of six classes of second line drugs. These isolates were termed extensively drug resistant (XDR) TB. XDR-TB strains are resistant to isoniazid or rifampin, any fluoroquinolone, and at least one of three second line anti-TB injectable drugs such as capreomycin, kanamycin, and amikacin.[13,14,15] In 2006, investigators reported on an XDR-TB outbreak in KwaZulu Natal where most patients were co-infected with HIV. 221 patients in this study had MDR-TB, 53 of which had the XDR form. 98% of the XDR patients died, and a majority of the patients had not received prior TB treatment, suggesting that the majority of drug resistance was transmitted versus being generated during treatment itself.[16] In 2007, the WHO reported treatment outcomes from all sites reporting complete data for new and previously treated MDR-TB patients. The overall treatment success for MDR-TB was 60% (8% mortality). For new cases, treatment success averaged to 64%, but fell to 58% (13% mortality) for previously treated cases.[17] Although overall treatment success rates of MDR and XDR-TB ranging from 40–80% have been observed in many global settings, this is significantly lower than the 85–99% cure rate for susceptible TB.[10,11,18–20] Treatment outcomes are found to be especially poor with cases of XDR-TB, especially in settings of high HIV prevalence.[16,21,22] To address this health crisis, first-line and second-line drugs are being combined to treat both MDR- and XDR-TB, and the selection of an appropriate combination may be guided by strain susceptibility testing. A major problem for rapid treatment is that conventional culture-based methods take 3–4 weeks to identify drug resistance from patient sputum samples. The use of second-line therapy agents only upon patient failure to first-line therapy is consequently a common outcome. Second-line options include aminoglycosides, cycloserine, terizidone, ethionamide, prothionamide, capreomycin, para-aminosalicylic acid and fluoroquinolones. Although drug-sensitive TB remains the most common form of the disease, the treatment of TB becomes more complicated as the resistance profile for this pathogen broadens. Additionally, treatment of MDR-TB is up to 100 times more costly than drug-sensitive disease, requires extensive case management of 18–24 months, and is more toxic.[14,18,20,23,24] This has underscored the urgency for the discovery, development, and delivery of new therapies for this disease, ideally including a shortened treatment duration of two months or less.

The discovery and development of novel chemical entities for the treatment of TB is no small task given the complexity of the disease, the need for multidrug regimens, and the infrastructure required to evaluate new compounds. The challenge of bringing forward new agents and regimens is reflected by the fact that, with the exception of the fluoroquinolones, no new anti-TB drug has been introduced in the past 45 years.[25,26] The first major breakthrough in TB treatment was the discovery and introduction of streptomycin in 1944, enabling the initial therapy for this disease. Combination of streptomycin with isoniazid was introduced in the 1950s and represented the primary line of therapy of this disease for two decades. Pyrazinamide was widely used between 1950 and 1970 to treat patients with TB resistant to isoniazid or streptomycin. During this period, the combination of ethionoamide, cycloserine, and pyrazinamide showed excellent cure rates and bacteriological conversion, largely due to the ability of pyrazinamide to remain active during the entire treatment duration.[27,28,29] The addition of rifampin to the multidrug regimen in 1965[30] contributed to the delivery of short-course therapy, which, when effectively administered, is bactericidal and offers sterilizing activity for 95% of drug-susceptible disease.[31,32] Drug-susceptible pulmonary TB is currently treated, per recommendation of the World Health Organization for newly diagnosed smear positive pulmonary TB,[33] with standard short-course chemotherapy. This regimen consists of rifampin[34] and isoniazid[35] given in combination for 6 months, supplemented with pyrazinamide[36] and ethambutol[37] in the first 2 months of therapy (Table 9.1). Although this regimen is effective against susceptible disease, it is often compromised by compliance issues driven by the long duration of treatment, lack of health care infrastructure, and difficult side effects. The implementation of DOTS,[22] and the subsequent STOP TB strategy, has helped to improve the overall rate of compliance driven by treatment duration and infrastructure issues. The potential for drug-drug interactions with antiretroviral agents, including non-nucleoside reverse transcriptase and protease inhibitors, further complicates the effective use of this regimen.[38] These challenges, taken in totality, exemplify the need for shorter duration therapy if control and elimination of TB is to be achieved.

There has been a significant re-engagement in the field of TB drug discovery and development in the past decade, and unprecedented collaborative efforts focused on streamlining the process to bring new compounds forward are beginning to emerge. The Critical Path to TB Drug Regimens (CPTR) initiative exemplifies this renewed focus on delivering an innovative solution for this public health crisis. The CPTR represents a broad collaboration of

Table 9.1 Standard short-course chemotherapy agents.

Drug	*Structure*	*Mechanism of Action*
Rifampin	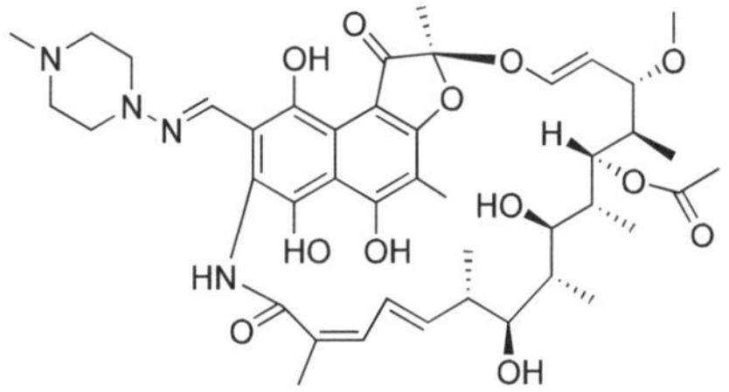	rpoB gene product (RNA polymerase)[34]
Isoniazid	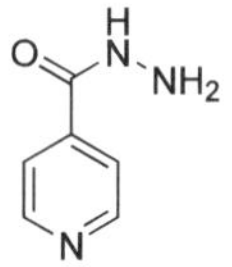	InhA (fatty acid biosynthesis)[35]
Pyrazinamide	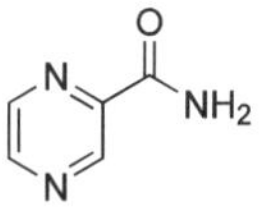	Not well understood[36]
Ethambutol	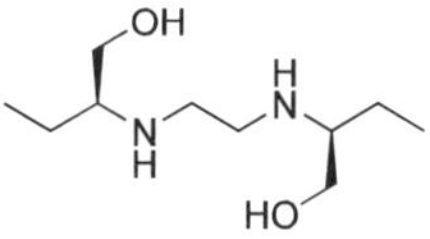	Arabinosyl transferases (cell wall biosynthesis)[37]

pharmaceutical companies, government interests, regulatory bodies, academic investigators, patient advocates, and non-profit organizations. The mission of this initiative involves innovating drug development by qualifying new regulatory science tools and ensuring the optimal infrastructure necessary for the efficient evaluation of novel drug regimens. In this chapter we seek to provide an assessment of the currently available tools for preclinical drug discovery and early clinical development, specifically highlighting the advances and innovations in this area where possible. These efforts, partnered with the ongoing work of the Stop TB strategy and the Global Plan to Stop TB, will be critical for attaining the WHO goal to reduce the global burden of TB as part of the Millennium health initiative.

9.2 The Preclinical Path to Developing New Agents

Having presented an overview of TB incidence and current therapy, one can identify several important opportunities for the development of new molecular entities (NMEs), specifically: 1) a reduction in the duration of treatment; 2) the ability to treat MDR- and XDR-TB with equivalent efficacy to drug-susceptible disease; 3) an improvement in the safety/toleration profile of new

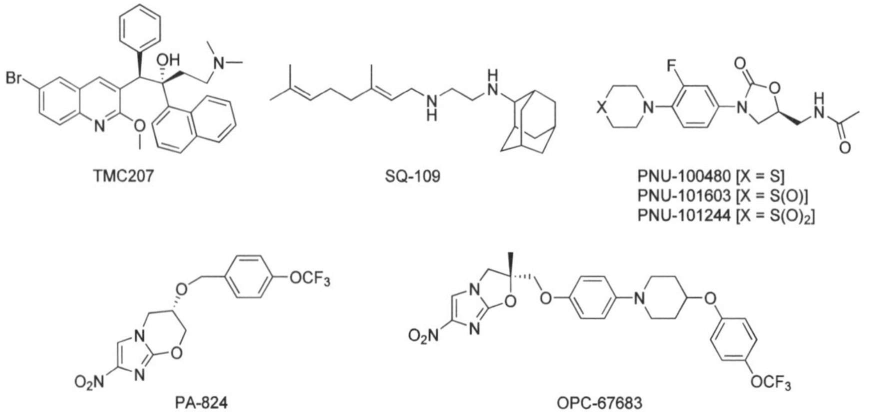

Figure 9.1 Selected novel agents currently in development.

medicines; 4) the possibility of intermittent rather than daily-administered therapy; and 5) enhanced treatment for the eradication of latent TB from asymptomatic individuals. This chapter will focus largely on the first two issues, both in the context of novel entities and in the loftier objective of delivering a novel therapeutic regimen.

This chapter will describe many of the assays and tools currently being employed in TB drug discovery and development, with the intention to make the reader familiar with their scientific rationale and utility in decision-making. As the authors seek to aid those involved in the discovery and development of new molecular entities (NMEs), examples will be drawn from 5 novel agents currently in development, specifically nitroimidazo derivatives (PA-824[39] and OPC-67683[40]), the diarylquinoline TMC207[41–43](R207910), the diamine SQ109[44] and a novel oxazolidinone, PNU-100480[45] (Figure 9.1). A large number of general reviews of agents in clinical development have been published recently,[i,47–56] and the excitement surrounding these compounds has naturally stimulated additional medicinal chemistry efforts.[57–67] Other novel agents with anti-TB activity also continue to be discovered and described on a regular basis (Table 9.2),[68–77] but this chapter will focus on the 5 agents mentioned above. While preclinical safety is a key component of any drug development program, this chapter will generally be limited to reviewing the pharmacology literature.

Before examining non-clinical and clinical assays, it is useful to reflect on the origin of these novel compounds and the diversity of methods used in their initial discovery (Table 9.3). The diarylquinoline class was first identified by means of a high-throughput screen using a surrogate organism for MTB,

[i] An excellent review of state-of-the-art methodologies for evaluating novel anti-TB agents with additional focus given to currently used therapies has been published recently. Please see the review by Mitchison and Davies, 2008.[46]

Table 9.2 Additional new anti-TB agents in development.

Drug (Phase of development)	*Structure*	*Mechanism of Action*
Moxifloxacin (Phase III)	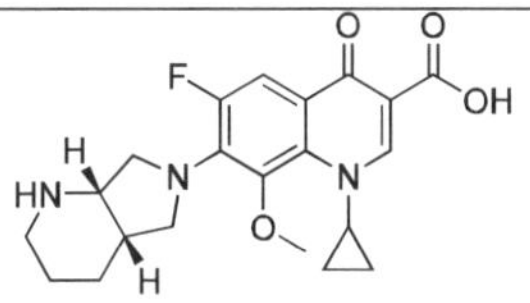	DNA gyrase[74]
Gatifloxaxin (Phase III)	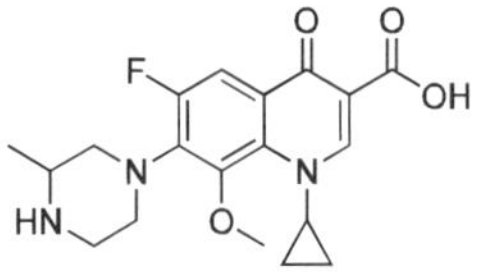	DNA gyrase[72]
LL-3858 (Phase II)	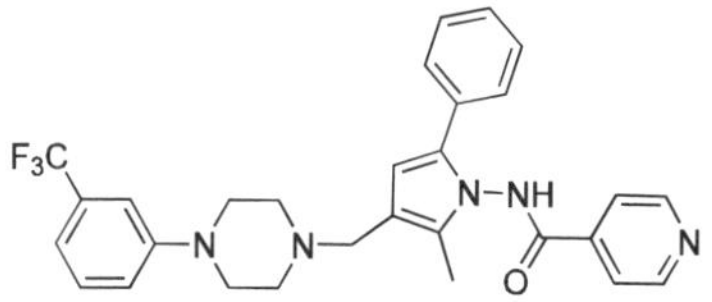	Unknown[71]
Linezolid (Phase II)	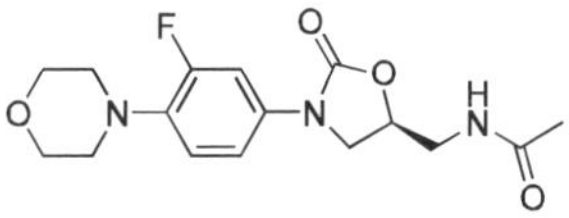	Protein synthesis initiation complex[73,75]
Meropenem and clavulanic acid (under clinical investigation)	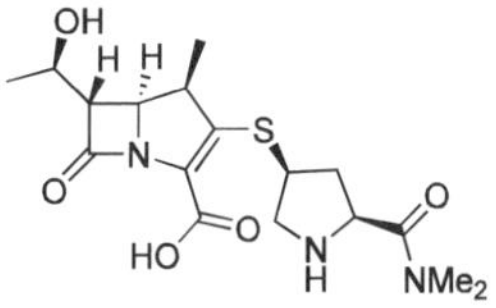	Penicillin binding proteins (cell wall synthesis)[77]
ACH-702 (preclinical)	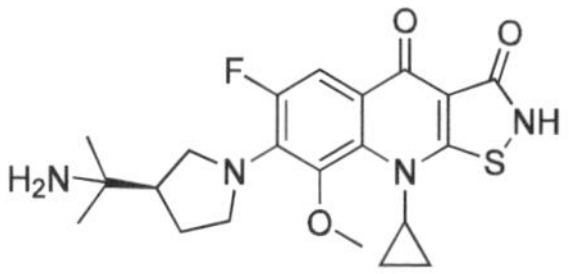	DNA gyrase[69]
BTZ043 (preclinical)	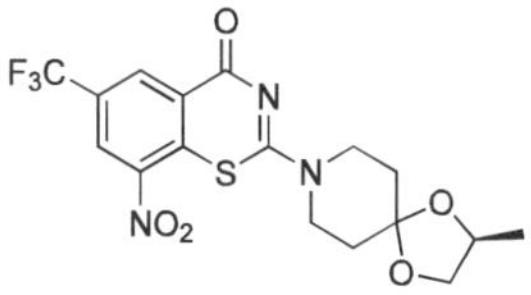	Arabinan synthesis[68]
DC-159a (preclinical)	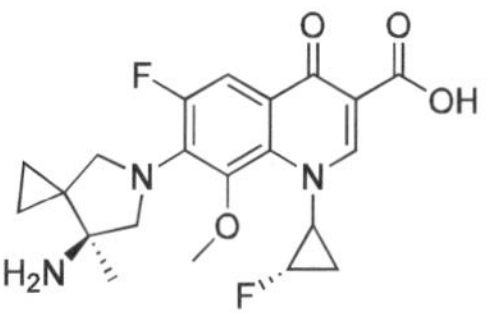	DNA gyrase[70]
AZD5847 (Phase 1)	Undisclosed	Protein synthesis initiation complex[56,76]

Table 9.3 Origin of selected novel agents currently in development.

Drug	*Mechanism of Action*	*Origin of chemical matter*
TMC207	ATP synthase[78]	HTS screen with *M. smegmatisn*
SQ109	Unknown[81]	Combinatorial chemistry[80]
OPC-67683	Likely similar to PA-824[55]	Lead from literature (CGI-17341)[83,85,86]
PA-824	Intracellular NO release (non-replicating MTb);[93] complex mechanism, including inhibition of respiration and cell wall synthesis[88]	Lead from literature (CGI-17341)[84–86]
PNU-100480	Protein synthesis initiation complex[75]	Analog from linezolid discovery effort[45,94]

Mycobacterium smegmatis.[78] Additional medicinal chemistry efforts led to the identification of TMC207. The novel mechanism of action, disruption of ATP synthase activity, was identified through the generation of drug-resistant mycobacteria and subsequent microbiological studies.[79] In contrast, the diamine SQ109 was designed as part of a combinatorial chemistry library of >60,000 analogs of ethambutol.[80] Intriguingly, the mechanism of action appears to be different than that of ethambutol, as evidenced by activity against ethambutol-resistant MTB,[81] and the molecular target is as yet unknown.[82] The nitroimidazo derivatives OPC-67683[83] and PA-824[84] were synthesized following reports of the promising anti-TB activity of a nitroimidazole, CGI-17341.[85,86] Numerous papers directed to understanding the mechanism of action of this class of molecules have subsequently appeared.[87-93] Finally, PNU-100480[45] was identified during medicinal chemistry efforts that led to the discovery and eventual marketing of linezolid, a first-in-class oxazolidinone for the treatment of serious Gram-positive infections.[94] Another oxazolidinone, AZD5847, is also in clinical development,[95] although details of its *in vitro* and *in vivo* activity against TB have not been disclosed. Linezolid itself has been reported to have potential utility in the treatment of TB, although the case studies are limited in size, and its benefits appear to be limited as a result of poor toleration associated with long-term therapy.[96]

The ideal novel TB therapy would be equally efficacious against drug-susceptible, MDR-TB and XDR-TB. In principle, this property could be achieved through the use of a novel mechanism of action. A new medicine will also ideally enable the shortening of TB therapy to 2 months or less.[97] In order to accomplish this task, it will need rapid "sterilizing" activity, i.e. the ability to eradicate slowly replicating (persister) mycobacteria. Such persistent organisms are believed to drive disease relapse after initial treatment, and it is the prevention of this relapse that defines a durable clinical cure for a TB patient.[98] It would also be orally bioavailable, possess no unfavorable drug-drug interactions, provide a favorable safety profile, and have a pharmacokinetic-pharmacodynamic (PK-PD) profile consistent with once-daily or less frequent dosing. Achieving these diverse objectives with a single new entity is highly unlikely, particularly against the backdrop of historical TB studies, where the emergence of resistance to single agents played a key role in preventing a permanent cure for TB.[46] As such, an entirely new therapeutic multidrug regimen may be needed, further escalating the challenge faced both by TB drug researchers and regulatory agencies. This chapter will highlight progress toward this eventual goal, mainly through the identification and individual study of several promising new agents for the treatment of TB.

9.3 *In Vitro* Assays

9.3.1 Minimum Inhibitory Concentration Susceptibility Testing

Operational simplicity, minimal compound requirements, and the lack of confounding variables typically drive pharmaceutical researchers to assess drug activity first using *in vitro* systems. Conventional *in vitro* assays of anti-TB activity, such as the minimum inhibitory concentration (MIC), measure the activity of a new agent against extracellular and actively dividing bacilli. This measure also correlates reasonably well with early bactericidal activity (*vide infra*) demonstrated *in vivo*. The MIC may be viewed as the simplest pharmacodynamic measure. It represents the minimum drug concentration which arrests growth of bacteria for 90–99% of colony forming units (CFU) *in vitro* under conditions of unrestricted growth, at standardized mycobacterial density, and with static drug exposure. This method of evaluating whole cell activity is frequently used as the primary drug discovery screen and functions well in measuring relative activity of multiple agents using a single endpoint measure. The MIC can be loosely related to pharmacokinetic (PK) properties by establishing susceptibility breakpoints associated with human plasma concentrations, and it can aid in the prediction of *in vivo* pharmacodynamics within a series of related chemical agents. One may frequently see a desire to "cover" the MIC or some multiple thereof pharmacokinetically, i.e. to provide free drug concentrations in excess of this variable throughout a given dosing interval. However, extensive PK variability may be observed in TB patients, based on the global nature of the disease, varying levels of severity of the illness, underlying comorbidities, the coadministration of other medicines, or a lack of adherence to a specified treatment regimen. Therefore, the rate of achieving concentrations of drug above the MIC is variable. It has been suggested that PK-PD-derived exposure targets should be integrated with population PK information via Monte Carlo simulation for NMEs as described by Gumbo *et al.* 2009.[99] Moreover, the existence of post antibiotic effects (PAEs), i.e. continued bacterial eradication after elimination of the drug from the system, also helps mitigate against problems from PK insufficiency.[46] The MIC also provides a facile method for demonstrating the lack of preexisting resistance to a NCE as well as its potential for the treatment of MDR- and XDR-TB.

Limitations of the MIC method include the inability to assess sterilizing activity, incapacity to measure efficacy against specific subpopulations, limits imposed by the lack of dynamic drug concentration conditions, and the lack of information correlating drug effect with time course. Furthermore, the MIC does not necessarily represent the concentration at which growth ceases, and it does not establish bacteriostatic *vs.* bactericidal activity. Predictive *in vitro* tools for assessing bactericidal activity of actively growing TB, best representing bacilli in the sputum of smear positive patients originating from liquefied caseum or open cavities, should be considered.

Considering the fundamental role of the MIC in evaluating novel TB drugs, it comes as no surprise that early reports on novel entities often assess this property. Activity versus drug-resistant strains of MTB also plays an important role in the evaluation of novel compounds.[45,78,81,83,84,100–102] While the MIC alone is an insufficient measure of the true value of a novel therapy, it should be noted that all of the analogs under discussion in this chapter possess highly potent MICs against both drug-sensitive and drug-resistant MTB. MIC testing may also reveal the specificity of a particular compound toward MTB versus other bacteria or mycobacteria. In this context, it should be noted that the strategy used to identify the diarylquinoline class (surrogate screening of *M. smegmatis*)[78] cannot be expected to be universally effective, given the limited activity of certain agents against non-tubercular mycobacteria.[84,103,104] Given the important role of drug combinations in TB therapy, assessment of synergistic interactions *via* MIC testing is also relevant. Such studies have been of particular interest with SQ109, both in combination with frontline agents[103] and with TMC207.[82]

9.3.2 Models for Assessing Activity Against Non-replicating Bacteria

The effective control and treatment of TB is significantly hindered by the ability of MTB to reside in human tissues for decades in the absence of replication. These persistent bacilli are not sensitive to anti-TB drugs that kill replicating organisms,[105] are therefore capable of surviving several months of combination therapy, and are responsible for the prolonged regimens needed to treat this disease. The organisms retain the ability to re-emerge and resume growth to produce active disease. Such extreme latency is essentially unparalleled relative to other infectious diseases. The shift to this non-replicating state (dormancy), in which the organism slows or halts replication in an effort to survive adverse conditions, is triggered by the host immune response.[106,107] Several factors can be involved in the shift of the bacilli to a non-replicating state, including nutrient depletion, shifts in pH, production of growth limiting by-products, and oxygen depletion. Reactivation is thought to be triggered by numerous factors, including a shift to aerobic respiration.

To evaluate the efficacy of new compounds against the non-replicating pathogen, culture models that generate persistent organisms by either oxygen depletion with nutrient-rich media (Wayne[108]) or nutrient deprivation with oxygen-rich media (Loebel[109]) have frequently been employed.

9.3.3 Wayne Model of Oxygen Depletion

TB lesions are often avascular, and the oxygen-deprived inflammatory and necrotic regions of the granulomas are described as microaerobic or anaerobic in nature. During natural infection, the bacilli first encounter pulmonary alveolar macrophages. However, the macrophages encountered in the later stages of infection are facultative anaerobes and respire in hypoxic sites. TB requires oxygen for growth, and oxygen deprivation is lethal unless the shift is gradual and the bacilli are able to adapt incrementally. Non-replicating bacilli are also observed to resist the action of many antibacterial agents which are otherwise effective against actively replicating bacteria. Oxygen depletion models have been devised to mimic the hypoxic environment of the granuloma.[108,110] The Wayne model of oxygen depletion was developed to more effectively simulate the environment in which non-replicating TB might reside and to provide a method for the evaluation of new agents. It is based on the gradual adaptation of TB to microaerophilic and ultimately anaerobic growth. In the Wayne model system, TB is slowly added to a sealed liquid culture system with controlled stirring. The bacteria are exposed to a limited headspace volume of air whereby the bacteria gradually deplete the flask of oxygen. This method relies on a temporal, rather than spatial oxygen gradient (as described in early adaptations of the Wayne model) while permitting sampling and characterization of the populations of organism at multiple stages of the process. These organisms are tolerant of the otherwise lethal effect imposed by the anaerobic environment. Non-replicating bacteria cultured by oxygen depletion shift their metabolism to the glyoxylate cycle, with the glyoxlate-to-glycine shunt providing for nicotinamide adenine dinucleotide (NAD) regeneration.[111] Replication of the bacilli can be reactivated upon culture aeration. The utility of oxygen depletion for the study of non-replicating bacteria *in vitro* has also been extended through *in vivo* adaptations of the model.[112] In such models, TB grown using the Wayne model of oxygen depletion have been shown to be virulent in mouse models upon aerosol and intranasal infection.[113]

The *in vitro* Wayne model has been used to assess the activity of both PA-824 and TMC207 versus persistent bacteria. PA-824 showed strong and dose-dependent activity, killing *ca.* 90% of non-replicating TB at a concentration of 10 μg/mL.[100] A potentially more clinically relevant concentration[114] of 2 μg/mL led to *ca.* 85% inhibition of growth. TMC207 has also been studied using this model by two different groups, showing 90% killing in each case at *ca.* 1 μg/mL.[115,116] Additional studies of activity *versus* dormant TB have also been carried out. TMC207 showed similar levels of activity in a drastic hypoxia model as in the Wayne model, with somewhat diminished activity in a model of dormancy induced by NO.[115]

9.3.4 Loebel Model of Nutrient Depletion

The Loebel model of progressive nutrient depletion represents a second method of evaluating efficacy against non-replicating or stationary phase organisms.[109,116] In this model, nutrient-starved non-replicating bacilli are generated by taking an exponential phase culture of TB, washing and re-suspending in phosphate-buffered saline (PBS), and incubating in rolling bottle cultures.[108] Bacilli remain fully viable in the nutrient-starved but aerated environment and maintain stationary phase growth. These organisms have been shown to be tolerant to a spectrum of antimycobacterial agents. Less is known about the genetic state of bacilli generated using the Loebel method as compared to the Wayne model of oxygen depletion. TMC207 lacked cidality even at concentrations up to 100 μM.[116]

9.3.5 Additional *In Vitro* Models

Several variants of the Hu–Coates model against 100-day static cultures have been employed to study PA-824 further.[117] Studies showed that PA-824 had superior sterilizing activity *versus* moxifloxacin in this model at a concentration of 10 μM. Activity at lower concentrations was strongly dose-dependent. OPC-67683 has been examined in the BACTEC model of tolerance and showed dose-dependent killing superior to isoniazid.[118] Most recently, Cole and coworkers have tested several marketed and investigational agents in non-replicating MTB 18b.[119] PA-824 and rifampin showed particularly good activity in this new model of dormant TB. In short, PA-824, OPC-67683, and TMC207 have shown some promise in the treatment of dormant mycobacteria using various *in vitro* models. Nonetheless, the lack of a perfect model for predicting clinical performance[120] makes drawing absolute conclusions impossible.

9.4 Mammalian Cell-based *In Vitro* and *Ex Vivo* Assays

9.4.1 Intracellular Infection Models

The ability to build an informed understanding for the translation of *in vitro* drug efficacy to *in vivo* outcome requires the assessment of drug efficacy in relevant intracellular infection models. As with the *in vitro* models of TB, a variety of cell-based culture models have been developed. Most TB-causing human pulmonary disease resides outside of host macrophages within the infected cavity. However, the periphery of the cavity wall contains a layer of tissue flush with macrophages, lymphocytes, granulocytes, and other host immune cells. During pulmonary infection, alveolar macrophages and alveolar epithelial cells are also the first cells to encounter TB. It is important to consider the ability of new agents to penetrate into tissues and impact intracellular forms of TB to fully assess the efficacy of novel compounds.[43,81,83,121–123]

9.4.2 Macrophage Assays

Given their involvement in phagocytosis of TB, murine or human macrophages are often used for the evaluation of intracellular antimicrobial activity.[124,125] The studies are often performed in 96-well format, and efficacy of new agents is evaluated for up to 7 days. Surviving organisms can be cultured from lysed cells and plated to determine efficacy. The uptake of the compound into the macrophage is determined by spiking cultured cells with fixed concentrations of test drug. The cells are then washed and lysed, and the supernatant can be tested to determine the maximum dilution that is needed to kill the intracellular organisms.

Several of the investigational compounds have been investigated using TB-infected macrophages[122] of either human or murine origin, either as a precursor or a supplement to more laborious *in vivo* studies. Early work with SQ109 demonstrated its ability to kill MTB in RAW 264.7 mouse macrophages.[81,123] Additionally, TMC207 has been reported to possess substantially higher activity than SQ109 in multiple macrophage cell lines.[43] OPC-67683 was deemed superior to PA-824, rifampin, and isoniazid using MTB-infected THP-1 cells with 4 h ("pulsed") exposure,[83] and its potent killing in this short exposure setting was seen as a positive indicator of the potential for intermittent therapy in humans.

9.4.3 Whole Blood Bactericidal Assay

A more recent innovation for the evaluation of compound efficacy is the whole blood bactericidal activity assay (WBA).[115] MTB is added to whole blood culture derived from healthy volunteers. The mycobacteria undergo rapid phagocytosis and remain intracellular for the duration of a 72-hour culture window. Components of the immune system interact with the infected monocytes, mimicking *in vivo* infection. In this way, the WBA model helps build the understanding of the impact of host immune response to the killing of mycobacteria.

This assay has typically explored varying drugs concentrations in the clinical setting.[ii,126–128] In practice, blood samples are derived at various timepoints from healthy human volunteers or patients who have received the test drug. In this way, the levels of drug in whole blood culture accurately reflect drug concentrations dictated by human PK and protein binding. The killing of MTB in these cultures is also reflective of the killing effect rendered by the combination of drug effect, immune response, and virulence of the infecting strain. As a preclinical tool, this assay could potentially be applied to assess the efficacy of a new agent before regulatory approval for human testing. For example, murine whole blood samples could be utilized following oral therapy with investigational agents. Alternatively, human whole blood could be spiked with

[ii] Additional information regarding the use of WBA assay will be discussed in clinical section 9.8.1 of this chapter, where its use in the early evaluation of PNU-100480 will be exemplified.

varying amounts of new drugs, designed in such a way as to reflect projected concentrations in humans. Given the potential correlation of WBA activity with the ability to eradicate MTB from sputum during pulmonary TB treatment, the observation of activity in this model may help identify candidate drugs with higher odds of success in later human clinical trials.

9.5 Resistance Profiling

Resistance development by individual bacilli results from spontaneous genetic mutations which confer reduced drug susceptibility.[129] The spontaneous mutation rate is low for MTB, but patients with advanced TB have high burdens of resistant organisms in the lung and other organs. Drug resistance may also result from prolonged exposures to a single drug or an ineffective combination, leading to the selection and expansion of resistant subpopulations of the organism. This is the driving principle for combination therapy for TB. Resistance development may occur in the clinical setting when patients fail to comply with therapy or when an inadequate regimen is prescribed. In developing countries, drug shortages, interruption in drug supply, or low-quality drugs may further contribute to drug resistance.[33]

It is important to consider the potential of new agents to prevent the emergence of resistance when applied as a single agent or as combination therapy. Early studies with TMC207 described the selection of spontaneously resistant mutants with frequencies similar to rifampin and were critical in understanding the biological target of this drug.[78,79] The selection of resistance to PA-824 was reported to occur with frequencies "slightly less" than those observed with isoniazid.[84] The mutant prevention concentration (MPC) is another parameter that can aid in determining the risk of resistance emergence.[130] Specifically, the MPC describes the concentration of drug that prevents the recovery of mutants from a susceptible population of MTB of 10^{10} or greater.[131] This agar-based method has been successfully applied with agents such as the fluoroquinolones with clinical strains of MTB.[132] TMC207 has also been reported to have an MPC of 3 mg/L.[133] This method can help identify resistance potential, but for maximal clinical utility, the MPC should be below the minimal concentration of drug in serum or tissue in humans at a well-tolerated dose. Notably, whereas the total human C_{max} of TMC207 exceeds the MPC at the 400 mg dose, the C_{min} does not.[163] The high level of plasma protein binding to TMC207[134] also renders such an analysis more complicated. The potential for emergence of resistance should also be monitored during *in vivo* studies (*vide infra*). It may also be investigated with simulated dosing regimens using *in vitro* PK-PD systems as described below.

9.6 *In Vitro* PK-PD Hollow Fiber Systems

There are several limitations to the evaluation of anti-TB agent efficacy under static drug conditions. Static drug exposure in either the MIC or static time kill setting cannot accurately represent dynamic *in vivo* PK-PD relationships. Furthermore, growth conditions in static systems do not adequately represent the growth conditions of the multiple bacterial subpopulations which may exist in patients or *in vivo* infection models. The implication of subpopulations is especially important for TB, where the organism can reside in multiple metabolic states as previously discussed in this chapter. The utility of *in vitro* dynamic systems is appealing for the evaluation of NMEs given that evaluation of the sterilizing effects using *in vivo* models, described later in this chapter, is both compound- and labor- intensive and can be inefficient early in the discovery process. As a result, *in vitro* PK-PD models have been successfully developed to evaluate the efficacy of new agents against TB under dynamic PK conditions relevant to human treatment.[135] The advantages of these systems include the ability to simulate human dosing schedules, evaluate both monotherapy and combination therapies, and assess the potential for resistance emergence in the absence of an immune system.

In vitro hollow fiber systems have been the most successfully employed *in vitro* dynamic model to evaluate the efficacy of anti-TB agents. The hollow fiber model simulates a two-compartment infection where bacteria are seeded within the peripheral compartment of the unit. Fresh medium flows through the unit allowing sustained growth of the organism, which is especially critical for this slow-growing pathogen. The agent or combination of agents being studied can be dosed to simulate human regimens, where they diffuse across the hollow fibers to the site of infection. Clearance is driven by a peristaltic pump. These *in vitro* hollow fiber systems allow for the repeated evaluation of killing under conditions of multiple relevant growth conditions for MTB strains. When using the hollow fiber system, it is important to consider the differing state of disease, and compounds should be evaluated under all relevant conditions (aerobic, anaerobic) to accurately assess activity of an agent or regimen. Furthermore, the emergence of resistance in these long-term studies should be evaluated with caution and balanced with the potential clinical outcomes of TB. Failure of a treatment regimen refers to patients whose sputum cultures fail to convert to negative during treatment (typically by 4 months). In MDR-TB, failure most often occurs due to the emergence of new resistance. In drug-sensitive TB, failure is most likely related to patient non-compliance or to phenotypic drug tolerance. In these cases, drug-sensitive patients do not typically acquire new resistance.

The hollow fiber model approach has been used to study the relationship of drug exposure and efficacy against TB for fluoroquinolones, isoniazid, and rifampin.[135–138] Transferring the environmental conditions of persistence models to the hollow fiber systems is thought to be an ideal approach to study the pharmacodynamics (PD) of sterilizing activity *in vitro*. An advantage of this system is the ability to perform more advanced pharmacodynamic investigation of potential resistance emergence, most important when evaluating monotherapy. Such studies may require exposures of antimycobacterial agents at levels that might be toxic to animals. In these studies performed by Gumbo *et al.*, TB was exposed to the drug by simulating half-life and dose schedules of human patients. The data obtained from their work was mathematically modeled to define drug exposure breakpoints associated with maximum kill and the ability to prevent resistance development. Monte Carlo simulation was further employed to determine the dose of the drug needed to achieve the predicted exposure target in clinical patients.[99] It should be noted that the prolonged experiments in this system required for the evaluation of TB, which has doubling time of 18–22 hours for wild type strains, can provide challenges such as contamination. The advantage of this model system is that it provides sizeable populations of non-multiplying and drug-tolerant organisms for evaluation of both efficacy and resistance potential for new agents and or combinations.[137,138] However, the direct translation to *in vivo* model systems is unclear, and data generated in these systems should be partnered with *in vivo* time course models.

9.7 *In Vivo* Infection Models

There is no single animal model which fully captures the complexities of TB in patients. A variety of murine models have been developed to help answer fundamental questions regarding the bactericidal activity against rapidly multiplying organisms, the propensity to generate resistant mutants, the ability to eradicate persistent bacteria (sterilization), and the delineation of the PD parameters involved in driving efficacy.[139] The discovery and development path for new TB agents has varied, but in the current paradigm, the characterization of *in vivo* antitubercular activity for a new agent is generally assessed using a murine model of TB infection once a compound has shown appropriate *in vitro* efficacy and has demonstrated a lack of cross-resistance to other TB drugs. Although tuberculosis is primarily a pulmonary disease, the bacteria can infect and cause disease in almost all tissues and organs.[140] Nevertheless, most murine infection models involve establishing a pulmonary infection with potential dissemination to other organs, such as the spleen.

Animal models have contributed significantly to the understanding of the pathology, pathogenesis, and immunologic response to this disease. However, the wide spectrum of disease states occurring in humans has made it challenging to identify universal *in vivo* tools to assess efficacy in a way that is robustly translatable to humans. The following section will describe various types of animal infection models and their potential utility and drawbacks. The *in vivo* assessment of new drugs is key in bridging the gap between *in vitro* studies and patients and builds in the dynamic interaction of the host, drug, and pathogen. Interpretation of *in vivo* data should be incorporated into a larger holistic assessment of performance using the full range of preclinical tool sets.

9.7.1 Murine Models

Mice are relatively resistant to TB, but they can be infected *via* respiratory or intravenous routes using a sub-lethal inoculum, resulting in a stable tissue infection with a high organ burden. Disease progression in the mouse model includes eventual immunosuppression of the animal and the occurrence of full respiratory disease.[121] Immunocompetent mice infected by intravenous administration with a burden of 5 x 10^3 organisms develop a non-lethal, but chronic infection in which CFU counts plateau at approximately 1 x 10^6 CFU per lung and spleen. Low-dose aerosol infection produces a similar chronic disease state with equivalent burden in the lung.[141,142] Simple mono-therapy protection experiments using a lethal inoculum produced *via* aerosol or intravenous routes are most common, can provide proof of efficacy, and are useful for preliminary dose selection. Short-term studies in the murine model can estimate the bactericidal activity of single drugs or drug combinations using colony counting of mycobacteria from organ homogenates.[143] The acute models are most useful when screening multiple drug candidates, and the most active agents can be progressed into more chronic TB disease models.

Although there are significant advantages for using the short-course therapy murine model in the drug discovery paradigm,[139] there are disadvantages that should be considered. Specifically, true latent infection is especially difficult to model in mice.[144,145] In humans, the initial infection is usually controlled and reduced to manageable levels by the host (evidence of disease is absent). In the mouse, the bacterial burden is relatively high and is not reduced until chemotherapy is introduced. The pathology of TB in the mouse lung is also significantly different than that of human disease. Mice do not develop caseation necrosis, whereas the caseous granuloma is a hallmark of human TB infection.[145] Hypoxia, nutrient depletion and acidic pH are characteristics of a caseous granuloma and drive bacterial persistence.[144,145] Understanding these features may also be of critical importance in delineating the role of new therapies in the treatment of human TB.

Murine models of tuberculosis have played a pivotal role in the advancement of the novel therapeutics currently in clinical studies as well as the assessment of different drug combinations. Such studies also typically include PK assessments, enabling a deeper understanding of the compounds. Considering the potential limitations of these models, it is useful to explore some specific examples and their impact on the development pathway. Unfortunately, there have been no published studies that provide a comprehensive understanding of the performance of investigational agents. Subtle differences in models employed with respect to route of infection, inoculum size, timing, etc. make cross-study comparisons somewhat challenging. Nevertheless, the inclusion of active controls across the various studies helps to provide some benchmark data. Since there exist a vast variety of murine models with different purposes, a general description of several of the principal formats will be discussed, followed by detailed examples from the 5 novel agents that form the focus of this chapter.

9.7.1.1 Bactericidal Acute

Once the bioavailability of a compound is understood and *in vitro* and *ex vivo* efficacy has been established, the ability of compounds to prevent lung lesions and dissemination of infection to the spleen is often assessed in an acute model of TB infection. The short-term or acute model can help to establish the bactericidal activity of new agents or drug combinations.[139,146] In the acute lethal model, mice are typically infected *via* the intravenous route with an acutely lethal burden. Infection is established over a 1–3-day period, and mice are then treated once-daily with a range of doses for 28 days. At 28 days, surviving animals are sacrificed and organ homogenates from the lungs, spleen, and liver are plated on agar and incubated for 3–4 weeks for CFU determination. The dose that prevents the development of visible lung lesions is identified and is thereby defined as the minimum effective dose. The minimal bactericidal dose can also be determined if the CFU counts are obtained and counted at the beginning and end of the experiment. It is defined as the dose that results in a 99% reduction of CFU over 28-days. The acute model evaluates the ability of a drug to eliminate actively growing bacteria.

Foundational antibacterial efficacy driver assessment (AUC/MIC, %Time > MIC, and C_{max}/MIC) can be performed in the murine model as well. Knowledge of the maximum tolerated dose and minimum bactericidal dose, partnered with an understanding of bioavailability and PK properties, can guide subsequent and more sophisticated *in vivo* studies. There are disadvantages to the acute model. The initial bacterial burden is high, and the untreated controls often expire before the end of study. The loss of control animals early in the experiment may make it difficult to calculate the reduction in CFU.

9.7.1.2 Chronic or Latent TB Models to Assess Sterilizing TB Activity

Murine studies of bactericidal activity are not sufficient to determine the potential sterilizing activity of a new drug and may, in fact, provide different rank orders of activity.[147] Given the critical connection between sterilizing activity, treatment duration, and the prevention of relapse, studies aimed at understanding this feature are of the utmost importance. There are no perfect *in vivo* models, which accurately mimic human latent or chronic TB infection. Immunocompetent mice are able to contain infection via host immune response, and consequently, organism burden plateaus at approximately 10^6 CFU per lung and spleen.[141,148] The Cornell mouse model was the first developed to simulate latent MTB infection and offers an intensive measure of sterilization. The Cornell murine model, developed by McCune *et al.* at Cornell University, employs the use of high dose chemotherapy to control the infection rather than relying on the host immune system.[149] In this model, high-dose chemotherapy (typically isoniazid and pyrazinamide) is given for up to 3 months duration to obtain a cure negative state as defined by negative organ cultures after 3 month follow-up.[149–151]

More recently, this model has been modified to evaluate sterilizing activity of anti-TB drugs. This murine model of TB uses a high bacterial burden challenge with 10^5–10^7 CFU administered intravenously, high-dose burden 10^6–10^7 CFU introduced intranasally,[142,152,153] or a low dose (< 100 CFU) delivered via the aerosol route.[100,122,154] CFU counts in the lungs and spleen are monitored over time and drug efficacy is related to the minimum effective dose required to prevent the development of gross lesions in the lung and the minimal bactericidal dose able to reduce CFU count by 99%.[146] Chemotherapy is often administered for up to 3 months beginning 3 or more weeks post infection, thereby allowing the bacteria to reach a stable infection state. One advantage of this model is that the initial inoculum is low enough that control mice frequently survive the experiment (as compared to the acute model).[146,155] Such studies can assess the ability of a drug to reduce an established TB infection. However, chronic models require more time and labor before obtaining experimental results.

9.7.1.3 Examples of Murine Models in the Development of Novel Agents

One of the most striking examples of the impact of murine models comes from PA-824. After identifying a large number of potent leads using *in vitro* assays (> 50), a "short term" murine model of infection was used to select PA-824 as the lead candidate.[84] This model utilized a strain of MTB engineered to express firefly luciferase (rMTB-*lux*) administered by tail vein injection. Although not the most potent analog *in vitro*, PA-824 demonstrated superior performance in this study. Such screening has the advantage of developing a preliminary understanding of *in vivo* PD effects and removing compounds with seemingly unsuitable PK properties while maintaining reasonably high compound throughput. Additional early studies reported by Stover *et al.*[84] showed that 10 d of therapy with PA-824 had similar efficacy to isoniazid in this model at equivalent doses of 25 mg/kg.

Early *in vivo* efficacy studies using PNU-100480 used an acute infection model in which mice were infected intravenously via the caudal vein and treated commencing 1 d or 7 d later for 1 m on a 5 d/week schedule.[152] Rather than the luciferase assays described above, these experiments relied on quantitation of mycobacteria (CFU count) in both the lung and spleen. These studies demonstrated the potent *in vivo* activity of PNU-100480 and highlighted its superior activity to linezolid. Rifampin (20 mg/kg) and isoniazid (25 mg/kg) were also examined as controls, as were regimens combining PNU-100480 with either rifampin or isoniazid. While all therapeutic regimens demonstrated substantial efficacy *versus* untreated control animals, no statistically significant differences between the various therapeutic regimens were observed. It must also be noted that PNU-100480 is metabolized *in vivo* largely to a sulfoxide metabolite (PNU-101603, Figure 9.1) and, to a lesser extent, to a sulfone (PNU-101244, Figure 9.1).

The continued development of PNU-100480 was substantially enabled by results reported some years later using an "established" model of infection. In this model, therapy began *ca.* 2w following an aerosol infection, and efficacy was ascertained by measurement of lung CFU count and spleen weight.[75] This study, conducted for both 1 m and 2 m duration, revealed several intriguing findings, namely: 1) a minimal bactericidal dose of 50 mg/kg with activity equivalent to 25 mg/kg isoniazid; 2) confirmation of superiority versus linezolid, even under conditions wherein exposure of PNU-100480 and its metabolites did not exceed linezolid exposure; and 3) robust demonstration of synergistic activity in several combinations. The last point was surprising given earlier results in the acute model of infection.

The *in vivo* activity of SQ109 was assayed using a lateral tail vein infection followed 20 d later with treatment 5 d/w. Dose-dependent efficacy, assessed by CFU count in both lung and spleen, was observed for SQ109. While 25 mg/kg of SQ109 was not as efficacious as 25 mg/kg isoniazid, as little as 10 mg/kg SQ109 was as efficacious as 100 mg/kg ethambutol.[123] This study also explored the PK properties of SQ109, including the elucidation of a high volume of distribution, high drug levels in the lung and spleen relative to the plasma, poor oral absorption, and substantial first-pass clearance. A subsequent report explored the activity of SQ109 in combination with front-line drugs in the chronic mouse model described above (1–2-month duration) to evaluate its potential use in the intensive phase of clinical TB treatment.[156] These studies revealed a positive contribution of SQ109 to mycobacterial killing, as well as its superiority in combination as compared to combinations involving ethambutol.

OPC-67683 has been reported to have strong bactericidal activity in a murine TB model utilizing 28 d therapy begun 1 d post intravenous infection.[83,101] Doses as low as 0.5 mg/kg demonstrated efficacy similar to standard agents such as rifampin at 5 mg/kg.[101] In addition, apparent superiority was demonstrated relative to PA-824.[83] Rapid reductions in bacterial counts over 3 months were also observed during a longer study of OPC-67683 in combination with pyrazinamide and rifampin. These results were believed to offer promise in developing a novel regimen for the shortening of TB therapy.[83] This latter point in the context of other drugs in development has been more frequently assessed using models of relapse infection, as described later in this chapter.

The further evaluation of PA-824 has also included assessment at 100 mg/kg in a 3 m model following aerosol infection. Bacterial CFU counts were reduced to very low levels with PA-824 as well as with active comparators moxifloxacin, gatifloxacin, and isoniazid.[100] Results also correlated well with the more exploratory murine studies described earlier.[84] Several subsequent studies using PA-824 alone and in combination have also been published. A 1 m study using aerosol-infected mice provided a minimal bactericidal dose of 100 mg/kg.[146] In a 2 m study, PA-824 alone at 100 mg/kg demonstrated efficacy similar to isoniazid at 25 mg/kg but inferior to a combination of rifampin, isoniazid, and pyrazinamide.[157] Intriguingly, both isoniazid and PA-824 alone selected for drug-resistant mutants in lungs as well. Assessment of the activity of PA-824 in combination with rifampin, pyrazinamide, and rifampin plus pyrazinamide has also been accomplished using the aerosol infection model.[157] In short, PA-824 at 100 mg/kg added to the efficacy observed with rifampin alone, enhanced bacterial killing with a rifampin plus pyrazinamide combination, and synergized very effectively with pyrazinamide. Notably, longer treatment with this combination led to the selection of mutants resistant to PA-824. The authors also raised the question of the relevance of the exposures obtained in mice relative to human exposures, a topic which will receive additional attention later in the chapter. The three drug combination of PA-824, pyrazinamide, and moxifloxacin has also been assessed and shows potential for the development of novel regimens lacking rifampin.[158]

Early studies with TMC207 utilized a non-established model of infection, wherein therapy (5 d/w for 28 d) commenced 1 d post-infection. Bactericidal activity was observed at doses as low as 12.5 mg/kg, and this dose also demonstrated superiority to isoniazid at 25 mg/kg. Moreover, early evidence suggested that dosing as infrequently as once per week may have benefit owing to the relatively long half-life of the drug.[78] A later report expanded upon this finding by studying combinations of TMC207, including one with pyrazinamide and rifapentine in which 9/10 mice dosed once weekly for 2 m showed negative cultures for MTB in the lung, a truly remarkable finding.[159]

TMC207 has also been studied in an established model of infection wherein treatment began 12–14 d post-infection. In the first study,[78] 25 mg/kg of TMC207 demonstrated equivalent efficacy to the standard 3-drug regimen of rifampin (10 mg/kg), isoniazid (25 mg/kg), and pyrazinamide (150 mg/kg)[160] and showed greater reductions in lung CFU counts as part of 3- and 4-drug combinations with these agents. A later study also included moxifloxacin in combination with TMC207 and explored several regimens designed to understand synergy between TMC207 and pyrazinamide.[155] The pyrazinamide plus TMC207 combination proved, in fact, to be the most powerful one studied, permitting a 5.6 log CFU reduction at 1 m and 100% culture negativity in mice treated for 2 m. The addition of rifampin, moxifloxacin, or isoniazid to this combination did not further improve its activity at either 1m or 2m. Nevertheless, the authors did not rule out the potential role of a third drug for clinical therapy, either for guaranteeing sterilization or acting to suppress the selection of drug-resistant mutants during therapy. Additional combination studies for the potential treatment of MDR- TB have also been carried out.[161]

In summary, both non-established ("acute") and established models of infection have provided substantial information on the bactericidal activity of new agents and combinations which employ them. Pyrazinamide offered synergistic activity in combination with either PA-824[157] or TMC207.[155] Given that most of these studies were conducted prior to the understanding of human doses and exposures, the precise selection of dose in each study may impart some limitations to their translational value. Nevertheless, such murine models play a critical role in enabling TB drug development.

It is believed that the overall duration of TB therapy is governed by the ability to eradicate so-called "persister" bacteria.[162] In this context, additional murine models have been developed which explore the sterilizing activity of investigational agents and combination regimens. Ibrahim and coworkers used the Cornell model of sterilizing activity to explore the potential of TMC207 at 25 mg/kg.[155] Stated simply, mice were infected by tail vein, and therapy began after 19 d. Mice were treated for 2 m with various 3–4 drug regimens during the intense phase of therapy followed by 2–4m using a 2–3 drug regimen. Relapse was assessed by the presence of MTB in lung or spleen 3 m after the cessation of therapy and compared with the relapse rate seen for the WHO standard regimen (2 m isonazid, rifampin, and pyrazinamide followed by 4 m of isoniazid and rifampin). TMC207 was able to substitute effectively for isoniazid in this study and to achieve equivalent relapse rates after 4 m of therapy versus 6 m for the standard regimen.

PA-824 has been studied using a related experimental design. An early study demonstrated that the substitution of isoniazid at 25 mg/kg with PA-824 at 100 mg/kg in the standard regimen resulted in reduced time to culture conversion as well as lowering lung CFU counts after 2 m of therapy.[143] Unfortunately, no statistically significant differences in the primary endpoint of relapse[163] were apparent, as both the standard and modified regimens had extremely low relapse rates. Another study showed that a combination of PA-824, moxifloxacin, and pyrazinamide provided sterilization more rapidly (4 m) than the standard regimen (5 m in this case).[158]

Nuermberger and coworkers have also examined the ability of PNU-100480 to shorten TB therapy by reducing the time to stable cure.[162] At a dose of 160 mg/kg, the addition of PNU-100480 to the standard regimen showed an equivalent relapse rate after 4 m therapy (5%) to 5 m of standard therapy. Moreover, the study also provided evidence for the beneficial replacement of isoniazid by PNU-100480 during the continuation phase of treatment. Such a regimen, utilizing only rifampin and PNU-100480 during the continuation phase, had a superior relapse rate to the corresponding rifampin plus isoniazid regimen. In addition, sterilizing activity during the initial phase of treatment for PNU-100480 was also observed.

In summary, murine models of sterilizing activity (prevention of relapse, stable cure) have been extensively used to explore the ability of new agents to shorten TB therapy.[143,146,153] Such studies play a critical role in expanding our understanding, especially in the context of evaluating novel combination regimens. Moreover, rank ordering regimens based on their bactericidal properties alone will not necessarily provide the same results as doing so based on sterilizing ability,[147] further emphasizing the different value provided by each type of model.

9.7.2 Other *In Vivo* Species

9.7.2.1 Guinea Pig

The guinea pig model of infection has been extensively used to study human disease and evaluate the efficacy of compounds against TB.[164,165] The guinea pig is the most easily infected with MTB of all animal models, and the pathology of disease is progressive and more consistent with human disease as compared to the murine model. Specifically, guinea pigs display similar pathogenic features of TB, including the formation of caseating granulomas, necrosis, and lung tissue hypoxia. Low-burden aerosol infection eventually progresses to fatal disease in the guinea pig model within months to a year. Despite the similarity to human disease state and progression, murine infection models are more commonly used in the drug discovery process due to the resource advantage of these models. Guinea pigs are cost- and resource-intensive in terms of animal budget, housing space required, and the quantity of compound required for efficacy evaluation. Therefore, this model is most effectively used as a confirmatory model for murine results with advanced compounds.

The aforementioned limitations of murine models have also been noted in the context of the novel agents in development. In studying PA-824, Nuermberger *et al.* raised a concern that the absence of caseation in the murine model might lead to underestimates of drug potency for compounds which are particularly active in hypoxic environments.[143] The absence of hypoxia in the mouse model has also been highlighted as a limitation by Lenaerts *et al.*[166] Owing to these limitations, a limited amount of experimental work has been carried out with novel agents in the guinea pig model of tuberculosis.

The earliest report on PA-824 included its evaluation in an aerosol infection model using guinea pigs.[84] In that 1 m study, oral doses of 40 mg/kg PA-824 provided comparable reductions in MTB bacilli to isoniazid at 25 mg/kg in both lung and spleen. A more recent study has also explored the use of an inhaled version of PA-824 in this model, with partial success.[167] TMC207 has also been examined in an aerosol infection guinea pig model of TB.[166] After 6w of therapy with either 5, 10, or 15 mg/kg TMC207, dramatic reductions in both lung and spleen CFU counts were observed. These values compared quite favorably to those obtained using standard therapy in this model.

In summary, although underexploited relative to murine models of infection, some experimental work has been carried out in the guinea pig model of infection using novel agents. Such studies may be of value given the potential for the guinea pig model to replicate more accurately certain elements of human disease as well as the diversity of TB subpopulations.[166]

9.7.2.2 Macaque

There has been a recent resurgence of the use of the non-human primate model, developed by Joanne Flynn *et al.*, for the evaluation of anti-TB activity. Cynomolgus macaques are highly susceptible to infection with low numbers of MTB when deposited directly into the lungs *via* bronchoscope. As with human disease, there are variable outcomes of disease progression in this model and the pathology of disease is similar to that of human.[168] Caseous necrotic and solid non-necrotic granulomas can develop and have been shown to correlate with the extent of the disease formed in the animal.[169] Additionally, hypoxic lesions have been confirmed in this model.[170] The macaque model is viewed as the most effective model of true latent tuberculosis, with additional advantages including the availability of reagents for immunologic and pathologic analysis. The model also has the ability to more effectively mimic human pathology, and the spectrum of disease forms is similar to that of humans. Newer imaging methodologies for the evaluation of drug efficacy and impact on specific lesions offer an exciting new opportunity for TB drug discovery and development.

9.8 Clinical Testing of Novel Therapies for TB

9.8.1 Phase 1 Trials

Having completed a wide variety of preclinical assays, novel agents undergo rigorous preclinical animal toxicology testing to demonstrate their suitability for human trials. The subsequent clinical development of new medicines for tuberculosis is particularly challenging owing to a number of factors. The high level of efficacy of the standard regimen when correctly administered for drug-susceptible disease ($<5\%$ relapse after completion of therapy) sets an enormous challenge in demonstrating the viability of a new drug or regimen. As such, and in recognition of the needs of patients, two other objectives once again become paramount. Reducing the length of treatment needed to prevent relapse has potential benefits in patient compliance, convenience, and safety. Such improvements could yield superior outcomes for efficacy and help avoid resistance selection in realistic settings outside of highly controlled clinical trials. In addition, the opportunity to treat MDR- or XDR-TB with efficacy approaching that of drug-susceptible TB would constitute a major advance. TB trials are further complicated by the use of multidrug combinations, making it more difficult to establish the contribution of any one drug. In addition, biomarker development outside of colony counts for viable mycobacteria has been a difficult area.[171] Finally, the selection of an optimal dose must once again be ascertained in the human population. While benefitting from the extensive preclinical work described above, human clinical trials are designed to reveal additional information regarding safety and efficacy.

Phase 1 clinical trials are used to study the safety, tolerability, and PK properties of new agents. Typically, these trials involve a two-stage design and explore a fairly wide range of dose levels. Healthy volunteers are first administered ascending single doses of the study drug, and an initial assessment of tolerability and PK is performed. A subsequent study involves healthy volunteers receiving multiple doses of study drug, with a typical duration of 2 weeks. The initial trials with PA-824 demonstrated oral bioavailability, showed a PK half-life of 16–20 h, and provided drug levels well in excess of MIC concentrations determined *in vitro*.[114] Exposures at all dose levels, however, were lower than those provided by the 50 mg/kg dose frequently used in murine models of efficacy. In addition, the first multiple ascending dose trial was stopped early, owing to observations of increased creatinine clearance and resultant concerns about renal toxicity.[114] Fortunately, additional studies of renal function revealed the mechanism to be the inhibition of renal tubular creatinine secretion.[172] Since this mechanism is believed to be clinically benign, PA-824 has continued its successful progress in clinical development.

Results from the single ascending dose trial of PNU-100480 have also been reported. PK parameters of the study drug and the pharmacologically active major circulating metabolite (PNU-101603, a sulfoxide) were described.[127] Single doses of 1000 mg and 1500 mg provided exposures of PNU-101603 in excess of the MIC for a 24 h period, and the study drug were well tolerated. In addition to pharmacokinetic and safety parameters, this Phase 1 study incorporated an *ex vivo* whole blood bactericidal activity assay (WBA, *vide supra*) using samples from volunteers to study the pharmacodynamic properties. Mean and cumulative whole blood bactericidal activity was assessed and compared to values obtained from 300 mg once daily doses of linezolid. Once-daily doses of 1000 mg and higher showed sustained bactericidal activity at all timepoints, and the superior antimycobacterial activity of PNU-100480 *versus* linezolid previously observed in murine studies was seen again in this setting. It is to be hoped that techniques such as the WBA assay will be useful in refining dose selection prior to more extensive Phase 2 studies, and a subsequent, as-yet-unpublished, Phase 1 multiple ascending dose trial also included this measurement in its design objectives (Clinicaltrials.gov locator NCT00990990).

9.8.2 Phase 2a trials: Early Bactericidal Activity

In the absence of WBA or biomarker-based approaches, the first clinical assessment of efficacy typically comes *via* a study of early bactericidal activity (EBA). Such studies also facilitate the continuing evaluation of safety, now in the context of patients rather than healthy volunteers. An excellent review describing the general methodology has been published recently, and the reader is directed to that publication for comprehensive references to the original literature.[173] Briefly summarized, EBA studies quantify mycobacterial CFU in the sputum of newly diagnosed sputum microscopy smear-positive TB patients. Reductions in CFU counts are evaluated as a measure of the bactericidal activity of investigational agents, and such trials may be used to help select therapeutically active doses for subsequent trials involving larger numbers of patients. Owing to concerns about the selection for resistance, monotherapy trials such as these are of limited duration (up to two weeks). Although the placebo effect has been shown to be essentially zero, these studies frequently include a positive comparator to help validate study outcomes. It has been observed that the period of days 0–2 have particularly utility in discriminating between different drugs and regimens, and isoniazid has been shown repeatedly to have the highest EBA, consistent with its ability to kill rapidly replicating mycobacteria. Caveats of EBA trials include a limited ability to predict sterilizing activity[174] and the difficulties in studying combinations, particularly when using isoniazid (as a result of its overpowering activity).[173]

TMC207 was evaluated at multiple dose levels for 7 d, using both isoniazid 300 mg and rifampin 600 mg as comparators.[174] Whereas activity was lacking in the traditional EBA on days 0–2, mean decreases in CFU counts were statistically significant from day 4 onwards in the 400 mg dose group. Pyrazinamide, a drug with potent sterilizing ability, also fails to show potent activity in the first two days of EBA studies.[46] In addition, daily decreases in CFU counts from days 4–7 were comparable between TMC207 at 400 mg and the positive control arms.

Diacon *et al.* reported the first results from an EBA trial of PA-824, studied for 14 d, compared to a standard 4-drug regimen (pyrazinamide, ethambutol, rifampin, and isoniazid).[175] Total bactericidal activity demonstrated statistically significant difference from zero between days 2 and 3. Intriguingly, the four doses of PA-824 under study (200, 600, 1000, and 1200 mg) demonstrated equivalent EBA in this study. These findings were unexpected based on studies in mice. The exact explanation is not well understood, although it may be related to time above MIC-dependent pharmacodynamic effects. This positive, albeit unexpected observation, has led to a second EBA trial with lower doses, the results of which have not yet been disclosed (Clinicaltrials.gov locator NCT00944021).

9.8.3 Phase 2b Trials

With the exception of TMC207, none of the novel agents reviewed in this chapter has proceeded to Phase 2b trials. TMC207 was evaluated as additive therapy versus placebo in conjunction with a standardized 5-drug regimen of second-line TB therapy.[176] The trial used conversion of sputum cultures in liquid culture from positive to negative as the primary endpoint. The use of TMC207 resulted in both quicker conversion to negative sputum culture status and a higher percentage of negative culture status at 8 weeks (48% versus 9% on placebo). Only modest effects were seen for TMC207 at early timepoints in the study (up to day 7), consistent with other observations of delayed antitubercular activity for this compound.[174] Overall, the study was seen as clinical validation for the efficacy of both TMC207 and the general mechanism of ATP synthesis inhibition for the treatment of MDR-TB.

9.9 Conclusions

This chapter has reviewed the current discovery and development paradigms of 5 novel anti-TB agents, relating those efforts to current tools and methodologies to enable effective decision-making. The optimal use of any new TB agent will require a deep understanding of its utility when combined with either existing TB agents or other new chemical entities. There are a number of methods that have emerged to drive an improved understanding of efficacy against TB and were evident in the discovery and development of the 5 new agents discussed in this chapter. It should be noted, however, that the translation of efficacy amongst preclinical models and from preclinical models to patients continues to present a major challenge to the TB drug discovery and development community. The *in vitro* assessment of whole-cell efficacy is standard for all new antibacterial agents; however, understanding efficacy against the multiple relevant organism states (replicating and non-replicating) is uniquely important for the evaluation of new TB agents and regimens. The assessment of bactericidal capacity is most commonly evaluated using the *in vivo* non-established ("acute") and established models of infection. These models have provided substantial information on the efficacy of new agents and their combinations. Second-tier *in vivo* models are often employed to further evaluate the sterilizing capacity of new agents and their respective combinations, useful information in defining the duration of therapy. *In vitro* hollow fiber modeling has been a more recently applied technology to help bridge the gap between static drug concentration studies and dynamic *in vivo* systems. Importantly, the ability to test novel drug regimens in such systems can bolster the understanding of both efficacy and the potential for resistance development. More cutting-edge technologies include the whole blood *ex vivo* assay, which helps incorporate the impact of host immune response to the killing of mycobacteria and is now being applied as part of clinical investigations. With a variety of novel investigational therapies and experimental methodologies available, substantial progress against TB is anticipated.

Acknowledgement

The authors wish to express their gratitude to Richard Steel for his assistance.

References

1. National Institute of Allergy and Infectious Diseases, *Understanding Microbes in Sickness and Health* 2006, National Institutes of Health (National Institute of Allergy and Infectious Diseases), Bethesda, Maryland, https://scholarworks.iupui.edu/bitstream/handle/1805/747/Understanding%20microbes%2c%20in%20sickness%20and%20in%20health.pdf?sequence=1
2. World Health Organization, *WHO-IULTALD Global Project on Tuberculosis Resistance Survelliance. Anti-tuberculosis Drug Resistance in the World*, 2008, World Health Organization, Geneva, http://www.who.int/tb/publications/2008/drs_report4_26feb08.pdf
3. C. Dye, *Lancet*, 2006, **367**, 938.
4. E. L. Corbett, C. J. Watt, N. Walker, D. Maher, B. G. Williams, M. C. Raviglione and C. Dye, *Arch. Intern. Med.*, 2003, **163**, 1009.
5. World Health Organization, *Global Tuberculosis Control: A short update to the 2009 report*, in *Global Tuberculosis Control*, World Health Organization, Geneva, http://whqlibdoc.who.int/publications/2009/9789241598866_eng.pdf
6. K. Lonnroth, E. Jaramillo, B. G. Williams, C. Dye and M. Raviglione, *Soc Sci Med*, 2009, **68**, 2240.

7. R. Pratt, V. Robison, T. Navin, H. Menzies, *Trends in Tuberculosis — United States, 2007*, in *MMWR Morbidity and Mortality Weekly Report*. 2007. p. 281.
8. A. Wright, M. Zignol, A. Van Deun, D. Falzon, S. R. Gerdes, K. Feldman, S. Hoffner, F. Drobniewski, L. Barrera, D. van Soolingen, F. Boulabhal, C. N. Paramasivan, K. M. Kam, S. Mitarai, P. Nunn and M. Raviglione, *Lancet*, 2009, **373**, 1861.
9. L. Chacon, M. Lainez, E. Rosales, M. Mercado and J. A. Caminero, *Int J Tuberc Lung Dis*, 2009, **13**, 62.
10. C. D. Mitnick, S. S. Shin, K. J. Seung, M. L. Rich, S. S. Atwood, J. J. Furin, G. M. Fitzmaurice, F. A. Alcantara Viru, S. C. Appleton, J. N. Bayona, C. A. Bonilla, K. Chalco, S. Choi, M. F. Franke, H. S. Fraser, D. Guerra, R. M. Hurtado, D. Jazayeri, K. Joseph, K. Llaro, L. Mestanza, J. S. Mukherjee, M. Munoz, E. Palacios, E. Sanchez, A. Sloutsky and M. C. Becerra, *N. Engl. J. Med.*, 2008, **359**, 563.
11. N. S. Shah, A. Wright, G. H. Bai, L. Barrera, F. Boulahbal, N. Martin-Casabona, F. Drobniewski, C. Gilpin, M. Havelkova, R. Lepe, R. Lumb, B. Metchock, F. Portaels, M. F. Rodrigues, S. Rusch-Gerdes, A. Van Deun, V. Vincent, K. Laserson, C. Wells and J. P. Cegielski, *Emerg. Infect. Dis.*, 2007, **13**, 380.
12. Centers for Disease Prevention and Control, *Emergence of Mycobacterium tuberculosis with extensive resistance to to second-line drugs – worldwide, 2000-2004*, in *MMWR Morbidity and Mortality Weekly Repor*, 2006, Centers for Disease Prevention and Control, p. 301.
13. J. A. Caminero, *Eur. Respir. J.*, 2005, **25**, 928.
14. WHO, *Report of the meeting of the WHO Global Task Force on XDR-TB Geneva, Switzerland*, 2007, World Health Organization: Geneva.
15. Centers for Disease Prevention and Control, *MMWR Morbidity and Mortality Weekly Report*, 2006, **55**, 1176.
16. N. R. Gandhi, A. Moll, A. W. Sturm, R. Pawinski, T. Govender, U. Lalloo, K. Zeller, J. Andrews and G. Friedland, *Lancet*, 2006, **368**, 1575.
17. World Health Organization, *Mutidrug and extensively drug resistant TB (MDR/XDR-TB); 2010 Report on Global Survellance and Response*, 2007.
18. E. Nathanson, C. Lambregts-van Weezenbeck, M. L. Rich, R. Gupta, J. Bayona, K. Blondal, J. A. Caminero, J. P. Cegielski, M. Danilovits, M. A. Espinal, V. Hollo, E. Jaramillo, V. Leimane, C. D. Mitnick, J. S. Mukherjee, P. Nunn, A. Pasechnikov, T. Tupasi, C. Wells and M. C. Raviglione, *Emerg. Infect. Dis.*, 2006, **12**, 1389.
19. T. E. Tupasi, R. Gupta, M. I. Quelapio, R. B. Orillaza, N. R. Mira, N. V. Mangubat, V. Belen, N. Arnisto, L. Macalintal, M. Arabit, J. Y. Lagahid, M. Espinal and K. Floyd, *PLoS Med.*, 2006, **3**, e352.
20. V. Leimane, *Lancet*, 2005, **365**, 318.
21. H. R. Kim, S. S. Hwang, H. J. Kim, S. M. Lee, C. G. Yoo, Y. W. Kim, S. K. Han, Y. S. Shim and J. J. Yim, *Clin. Infect. Dis.*, 2007, **45**, 1290.

22. Centers for Disease Prevention and Control. *Emergence of Mycobacterium tuberculosis with extensive resistance to second-line drugs worldwide*, in *MMWR Morbidity and Mortality Weekly Report*, 2006. p. 301.
23. S. S. Rajbhandary, S. M. Marks and N. N. Bock, *Int. J. Tuberc. Lung Dis.*, 2004, **8**, 1012.
24. H. A. Ward, D. D. Marciniuk, V. H. Hoeppner and W. Jones, *Int. J. Tuberc. Lung Dis.*, 2005, **9**, 164.
25. J. A. Caminero, *Int. J. Tuberc. Lung Dis.*, 2006, **10**, 829.
26. L. E. Ziganshina and S. B. Squire, *Cochrane. Database Syst. Rev.*, 2008, CD004795.
27. J. A. Caminero, *Eur. Respir. J.*, 2008, **32**, 1413.
28. J. Tousek, E. Jancik, M. Zelenka and M. Jancikova-Makova, *Tubercle*, 1967, **48**, 27.
29. A. R. Somner and A. A. Brace, *Br. Me. J.*, 1966, **1**, 775.
30. N. Maggi, C. R. Pasqualucci, R. Ballotta and P. Sensi, *Chemotherapy*, 1966, **11**, 285.
31. L. Cavacey and G. Monzali, *Clin. Ter.*, 1966, **39**, 547.
32. M. Luchessi, G. Pallotta, P. Rossi and M. Sbampato, *Ann. 1st Calo. Forlanini*, 1967, **27**, 199.
33. World Health Organization, *Treatment for Tuberculosis, Guidelines for National Programmes*, 2003.
34. Anonymous, *Tuberculosis (Edinb)*, 2008, **88**, 151.
35. Anonymous, *Tuberculosis (Edinb)*, 2008, **88**, 112.
36. Anonymous, *Tuberculosis (Edinb)*, 2008, **88**, 141.
37. Anonymous, *Tuberculosis (Edinb)*, 2008, **88**, 102.
38. Centers for Disease Prevention and Control, *MMWR Morbidity and Mortality Weekly Report*, 2002, **51**, 14.
39. Anonymous, *Tuberculosis (Edinb)*, 2008, **88**, 134.
40. Anonymous, *Tuberculosis (Edinb)*, 2008, **88**, 132.
41. Anonymous, *Tuberculosis (Edinb)*, 2008, **88**, 168.
42. A. Matteelli, A. C. Carvalho, K. E. Dooley and A. Kritski, *Future Microbiol.*, 2010, **5**, 849.
43. N. Lounis, J. Guillemont, N. Veziris, A. Koul, V. Jarlier and K. Andries, *Med. Mal. Infect.*, 2010, **40**, 383.
44. Anonymous, *Tuberculosis (Edinb)*, 2008, **88**, 159.
45. M. R. Barbachyn, D. K. Hutchinson, S. J. Brickner, M. H. Cynamon, J. O. Kilburn, S. P. Klemens, S. E. Glickman, K. C. Grega, S. K. Hendges, D. S. Toops, C. W. Ford and G. E. Zurenko, *J. Med. Chem.*, 1996, **39**, 680.
46. D. A. Mitchison and G. R. Davies, *Open Inf. Dis. J.*, 2008, **2**, 59.
47. C.-M. Tam, *Expert Rev. Clin. Pharmacol.*, 2009, **2**, 405.
48. J. Guillemont, *Current Bioactive Compounds*, 2009, **5**, 137.
49. P. J. Barry and T. M. O'Connor, *Curr. Med. Chem.*, 2007, **14**, 2000.
50. E. C. Rivers and R. L. Mancera, *Drug Discov. Today*, 2008, **13**, 1090.
51. H. D. Showalter and W. A. Denny, *Tuberculosis (Edinb)*, 2008, **88 S3**, S3.

52. E. C. Rivers and R. L. Mancera, *Curr. Med. Chem.*, 2008, **15**, 1956.
53. V. Bhowruth, L. G. Dover and G. S. Besra, *Prog. Med. Chem.*, 2007, **45**, 169.
54. E. Leibert and W. N. Rom, *Expert Rev. Anti. Infect. Ther.*, 2010, **8**, 801.
55. C. E. Barry and J. S. Blanchard, *Curr. Opin. Chem. Biol.*, 2010, **14**, 456.
56. A. M. Ginsberg, *Drugs*, 2010, **70**, 2201.
57. A. Lilienkampf, S. Karkola, S. Alho-Richmond, P. Koskimies, N. Johansson, K. Huhtinen, K. Vihko and K. Wahala, *J. Med. Chem.*, 2009, **52**, 6660.
58. R. S. Upadhayaya, J. K. Vandavasi, R. A. Kardile, S. V. Lahore, S. S. Dixit, H. S. Deokar, P. D. Shinde, M. P. Sarmah and J. Chattopadhyaya, *Eur. J. Med. Chem.*, 2010, **45**, 1854.
59. R. S. Upadhayaya, S. V. Lahore, A. Y. Sayyed, S. S. Dixit, P. D. Shinde and J. Chattopadhyaya, *Org. Biomol. Chem.*, 2010, **8**, 2180.
60. V. Faugeroux, Y. Genisson, Y. Salma, P. Constant and M. Baltas, *Bioorg. Med. Chem.*, 2007, **15**, 5866.
61. O. K. Onajole, P. Govender, P. D. van Helden, H. G. Kruger, G. E. Maguire, I. Wiid and T. Govender, *Eur. J. Med. Chem.*, 2010, **45**, 2075.
62. P. Kim, S. Kang, H. I. Boshoff, J. Jiricek, M. Collins, R. Singh, U. H. Manjunatha, P. Niyomrattanakit, L. Zhang, M. Goodwin, T. Dick, T. H. Keller, C. S. Dowd and C. E. Barry, *J. Med. Chem.*, 2009, **52**, 1329.
63. P. Kim, L. Zhang, U. H. Manjunatha, R. Singh, S. Patel, J. Jiricek, T. H. Keller, H. I. Boshoff, C. E. Barry and C. S. Dowd, *J. Med. Chem.*, 2009, **52**, 1317.
64. J. G. Hurdle, R. B. Lee, N. R. Budha, E. I. Carson, J. Qi, M. S. Scherman, S. H. Cho, M. R. McNeil, A. J. Lenaerts, S. G. Franzblau, B. Meibohm and R. E. Lee, *J. Antimicrob Chemother*, 2008, **62**, 1037.
65. X. Li, U. H. Manjunatha, M. B. Goodwin, J. E. Knox, C. A. Lipinski, T. H. Keller, C. E. Barry and C. S. Dowd, *Bioorg. Med. Chem. Lett.*, 2008, **18**, 2256.
66. R. Villar, E. Vicente, B. Solano, S. Perez-Silanes, I. Aldana, J. A. Maddry, A. J. Lenaerts, S. G. Franzblau, S. H. Cho, A. Monge and R. C. Goldman, *J. Antimicrob. Chemother.*, 2008, **62**, 547.
67. E. Vicente, R. Villar, A. Burguete, B. Solano, S. Perez-Silanes, I. Aldana, J. A. Maddry, A. J. Lenaerts, S. G. Franzblau, S. H. Cho, A. Monge and R. C. Goldman, *Antimicrob. Agents Chemother.*, 2008, **52**, 3321.
68. V. Makarov, G. Manina, K. Mikusova, U. Mollmann, O. Ryabova, B. Saint-Joanis, N. Dhar, M. R. Pasca, S. Buroni, A. P. Lucarelli, A. Milano, E. De Rossi, M. Belanova, A. Bobovska, P. Dianiskova, J. Kordulakova, C. Sala, E. Fullam, P. Schneider, J. D. McKinney, P. Brodin, T. Christophe, S. Waddell, P. Butcher, J. Albrethsen, I. Rosenkrands, R. Brosch, V. Nandi, S. Bharath, S. Gaonkar, R. K. Shandil, V. Balasubramanian, T. Balganesh, S. Tyagi, J. Grosset, G. Riccardi and S. T. Cole, *Science*, 2009, **324**, 801.

69. C. A. Molina-Torres, J. Ocampo-Candiani, A. Rendon, M. J. Pucci and L. Vera-Cabrera, *Antimicrob. Agents Chemother.*, 2010, **54**, 2188.
70. A. Disratthakit and N. Doi, *Antimicrob. Agents Chemother.*, 2010, **54**, 2684.
71. Anonymous, *Tuberculosis (Edinb)*, 2008, **88**, 126.
72. Anonymous, *Tuberculosis (Edinb)*, 2008, **88**, 109.
73. Anonymous, *Tuberculosis (Edinb)*, 2008, **88**, 122.
74. Anonymous, *Tuberculosis (Edinb)*, 2008, **88**, 127.
75. K. N. Williams, C. K. Stover, T. Zhu, R. Tasneen, S. Tyagi, J. H. Grosset and E. Nuermberger, *Antimicrob. Agents Chemother.*, 2009, **53**, 1314.
76. E. L. Nuermberger, M. K. Spigelman and W. W. Yew, *Respirology*, 2010, **15**, 764.
77. J. E. Hugonnet, L. W. Tremblay, H. I. Boshoff, C. E. Barry and J. S. Blanchard, *Science*, 2009, **323**, 1215.
78. K. Andries, P. Verhasselt, J. Guillemont, H. W. Gohlmann, J. M. Neefs, H. Winkler, J. Van Gestel, P. Timmerman, M. Zhu, E. Lee, P. Williams, D. de Chaffoy, E. Huitric, S. Hoffner, E. Cambau, C. Truffot-Pernot, N. Lounis and V. Jarlier, *Science*, 2005, **307**, 223.
79. S. Petrella, E. Cambau, A. Chauffour, K. Andries, V. Jarlier and W. Sougakoff, *Antimicrob. Agents Chemother.*, 2006, **50**, 2853.
80. R. E. Lee, M. Protopopova, E. Crooks, R. A. Slayden, M. Terrot and C. E. Barry, *J. Comb. Chem.*, 2003, **5**, 172.
81. M. Protopopova, C. Hanrahan, B. Nikonenko, R. Samala, P. Chen, J. Gearhart, L. Einck and C. A. Nacy, *J. Antimicrob. Chemother.*, 2005, **56**, 968.
82. V. M. Reddy, L. Einck, K. Andries and C. A. Nacy, *Antimicrob. Agents Chemother.*, 2010, **54**, 2840.
83. M. Matsumoto, H. Hashizume, T. Tomishige, M. Kawasaki, H. Tsubouchi, H. Sasaki, Y. Shimokawa and M. Komatsu, *PLoS Med.*, 2006, **3**, e466.
84. C. K. Stover, P. Warrener, D. R. VanDevanter, D. R. Sherman, T. M. Arain, M. H. Langhorne, S. W. Anderson, J. A. Towell, Y. Yuan, D. N. McMurray, B. N. Kreiswirth, C. E. Barry and W. R. Baker, *Nature*, 2000, **405**, 962.
85. K. Nagarajan, R. G. Shankar, S. Rajappa, S. J. Shenoy and R. Costa-Pereira, *Eur. J. Med. Chem.*, 1989, **24**, 631.
86. D. R. Ashtekar, R. Costa-Pereira, K. Nagarajan, N. Vishvanathan, A. D. Bhatt and W. Rittel, *Antimicrob. Agents Chemother.*, 1993, **37**, 183.
87. A. Maroz, S. S. Shinde, S. G. Franzblau, Z. Ma, W. A. Denny, B. D. Palmer and R. F. Anderson, *Org. Biomol. Chem.*, 2010, **8**, 413.
88. U. Manjunatha, H. I. Boshoff and C. E. Barry, *Commun. Integr. Biol.*, 2009, **2**, 215.
89. K. P. Choi, T. B. Bair, Y. M. Bae and L. Daniels, *J. Bacteriol.*, 2001, **183**, 7058.
90. U. H. Manjunatha, H. Boshoff, C. S. Dowd, L. Zhang, T. J. Albert, J. E. Norton, L. Daniels, T. Dick, S. S. Pang and C. E. Barry, *Proc. Natl. Acad. Sci. U S A*, 2006, **103**, 431.

91. R. F. Anderson, S. S. Shinde, A. Maroz, M. Boyd, B. D. Palmer and W. A. Denny, *Org. Biomol. Chem.*, 2008, **6**, 1973.
92. L. M. Fu and S. C. Tai, *Int. J. Microbiol.*, 2009, **2009**, 879621.
93. R. Singh, U. Manjunatha, H. I. Boshoff, Y. H. Ha, P. Niyomrattanakit, R. Ledwidge, C. S. Dowd, I. Y. Lee, P. Kim, L. Zhang, S. Kang, T. H. Keller, J. Jiricek and C. E. Barry, *Science*, 2008, **322**, 1392.
94. S. J. Brickner, M. R. Barbachyn, D. K. Hutchinson and P. R. Manninen, *J. Med. Chem.*, 2008, **51**, 1981.
95. U. G. Lalloo and A. Ambaram, *Curr. HIV/AIDS Rep.*, 2010, **7**, 143.
96. G. F. Schecter, C. Scott, L. True, A. Raftery, J. Flood and S. Mase, *Clin. Infect. Dis.*, 2010, **50**, 49.
97. A. M. Ginsberg, *Semin. Respir. Crit. Care Med.*, 2008, **29**, 552.
98. D. A. Mitchison, *Am. Rev. Respir. Dis.*, 1993, **147**, 1062.
99. T. Gumbo, C. S. Dona, C. Meek and R. Leff, *Antimicrob. Agents Chemother.*, 2009, **53**, 3197.
100. A. J. Lenaerts, V. Gruppo, K. S. Marietta, C. M. Johnson, D. K. Driscoll, N. M. Tompkins, J. D. Rose, R. C. Reynolds and I. M. Orme, *Antimicrob. Agents Chemother.*, 2005, **49**, 2294.
101. H. Sasaki, Y. Haraguchi, M. Itotani, H. Kuroda, H. Hashizume, T. Tomishige, M. Kawasaki, M. Matsumoto, M. Komatsu and H. Tsubouchi, *J. Med. Chem.*, 2006, **49**, 7854.
102. M. R. Pasca, G. Degiacomi, A. L. Ribeiro, F. Zara, P. De Mori, B. Heym, M. Mirrione, R. Brerra, L. Pagani, L. Pucillo, P. Troupioti, V. Makarov, S. T. Cole and G. Riccardi, *Antimicrob. Agents Chemother.*, 2010, **54**, 1616.
103. P. Chen, J. Gearhart, M. Protopopova, L. Einck and C. A. Nacy, *J. Antimicrob. Chemother.*, 2006, **58**, 332.
104. M. S. Kabir, O. A. Namjoshi, R. Verma, R. Polanowski, S. M. Krueger, D. Sherman, M. A. Rott, W. R. Schwan, A. Monte and J. M. Cook, *Bioorg. Med. Chem.*, 2010, **18**, 4178.
105. Y. Zhang, *Front Biosci.*, 2004, **9**, 1136.
106. H. I. Boshoff and C. E. Barry, *Nat. Rev. Microbiol.*, 2005, **3**, 70.
107. J. Chan and J. Flynn, *Clin. Immunol.*, 2004, **110**, 2.
108. L. G. Wayne and L. G. Hayes, *Infect. Immun.*, 1996, **64**, 2062.
109. J. C. Betts, P. T. Lukey, L. C. Robb, R. A. McAdam and K. Duncan, *Mol. Microbiol.*, 2002, **43**, 717.
110. L. G. Wayne and C. D. Sohaskey, *Annu. Rev. Microbiol.*, 2001, **55**, 139.
111. Y. Hu, J. A. Mangan, J. Dhillon, K. M. Sole, D. A. Mitchison, P. D. Butcher and A. R. Coates, *J. Bacteriol.*, 2000, **182**, 6358.
112. L. Woolhiser, M. H. Tamayo, B. Wang, V. Gruppo, J. T. Belisle, A. J. Lenaerts, R. J. Basaraba and I. M. Orme, *Infect. Immun.*, 2007, **75**, 2621.
113. S. Gaonkar, S. Bharath, N. Kumar, V. Balasubramanian and R. K. Shandil, *International Journal of Microbiology*, 2010, 1.
114. A. M. Ginsberg, M. W. Laurenzi, D. J. Rouse, K. D. Whitney and M. K. Spigelman, *Antimicrob. Agents Chemother.*, 2009, **53**, 3720.

115. A. Koul, L. Vranckx, N. Dendouga, W. Balemans, I. Van den Wyngaert, K. Vergauwen, H. W. Gohlmann, R. Willebrords, A. Poncelet, J. Guillemont, D. Bald and K. Andries, *J. Biol. Chem.*, 2008, **283**, 25273.
116. M. Gengenbacher, S. P. Rao, K. Pethe and T. Dick, *Microbiology*, 2010, **156**, 81.
117. Y. Hu, A. R. Coates and D. A. Mitchison, *Int. J. Tuberc. Lung Dis.*, 2008, **12**, 69.
118. O. Y. Saliu, C. Crismale, S. K. Schwander and R. S. Wallis, *J. Antimicrob. Chemother.*, 2007, **60**, 994.
119. C. Sala, N. Dhar, R. C. Hartkoorn, M. Zhang, Y. H. Ha, P. Schneider and S. T. Cole, *Antimicrob. Agents Chemother.*, 2010, **54**, 4150.
120. Y. Zhang, *Clin. Pharmacol. Ther.*, 2007, **82**, 595.
121. I. M. Orme, A. D. Roberts, S. K. Furney and P. S. Skinner, *Eur. J. Clin. Microbiol. Infect. Dis.*, 1994, **13**, 994.
122. I. Orme, *Antimicrob. Agents Chemother.*, 2001, **45**, 1943.
123. L. Jia, J. E. Tomaszewski, C. Hanrahan, L. Coward, P. Noker, G. Gorman, B. Nikonenko and M. Protopopova, *Br. J. Pharmacol.*, 2005, **144**, 80.
124. A. Khan and D. Sarkar, *J. Microbiol. Methods*, 2008, **73**, 62.
125. K. E. Lougheed, D. L. Taylor, S. A. Osborne, J. S. Bryans and R. S. Buxton, *Tuberculosis (Edinb)*, 2009, **89**, 364.
126. R. S. Wallis, S. Patil, S. H. Cheon, K. Edmonds, M. Phillips, M. D. Perkins, M. Joloba, A. Namale, J. L. Johnson, L. Teixeira, R. Dietze, S. Siddiqi, R. D. Mugerwa, K. Eisenach and J. J. Ellner, *Antimicrob. Agents Chemother.*, 1999, **43**, 2600.
127. R. S. Wallis, W. M. Jakubiec, V. Kumar, A. M. Silvia, D. Paige, D. Dimitrova, X. Li, L. Ladutko, S. Campbell, G. Friedland, M. Mitton-Fry and P. F. Miller, *J. Infect. Dis.*, 2010, **202**, 745.
128. R. S. Wallis, S. A. Vinhas, J. L. Johnson, F. C. Ribeiro, M. Palaci, R. L. Peres, R. T. Sa, R. Dietze, A. Chiunda, K. Eisenach and J. J. Ellner, *J. Infect. Dis.*, 2003, **187**, 270.
129. H. L. David, *Clin. Chest. Med.*, 1980, **1**, 227.
130. K. Drlica and X. Zhao, *Clin. Infect. Dis.*, 2007, **44**, 681.
131. Y. Dong, X. Zhao, J. Domagala and K. Drlica, *Antimicrob. Agents Chemother.*, 1999, **43**, 1756.
132. Y. Dong, X. Zhao, B. N. Kreiswirth and K. Drlica, *Antimicrob. Agents Chemother.*, 2000, **44**, 2581.
133. E. Huitric, P. Verhasselt, A. Koul, K. Andries, S. Hoffner and D. I. Andersson, *Antimicrob. Agents Chemother.*, 2010, **54**, 1022.
134. N. Lounis, T. Gevers, J. Van Den Berg, T. Verhaeghe, R. van Heeswijk and K. Andries, *J. Clin. Microbiol.*, 2008, **46**, 2212.
135. T. Gumbo, A. Louie, M. R. Deziel, L. M. Parsons, M. Salfinger and G. L. Drusano, *J. Infect. Dis.*, 2004, **190**, 1642.
136. T. Gumbo, A. Louie, W. Liu, D. Brown, P. G. Ambrose, S. M. Bhavnani and G. L. Drusano, *Antimicrob. Agents Chemother.*, 2007, **51**, 2329.

137. T. Gumbo, A. Louie, M. R. Deziel, W. Liu, L. M. Parsons, M. Salfinger and G. L. Drusano, *Antimicrob. Agents Chemother.*, 2007, **51**, 3781.
138. T. Gumbo, A. Louie, M. R. Deziel and G. L. Drusano, *Antimicrob. Agents Chemother.*, 2005, **49**, 3178.
139. E. Nuermberger, *Semin. Respir Crit. Care Med.*, 2008, **29**, 542.
140. A. L. Krtiski, F. A. F. de Melo, *Tuberculosis, 2007; from basic science to patient care*, ed. J. C. Palomino, S. C. Leao, V. Ritacco, 2007.
141. B. P. Kelly, S. K. Furney, M. T. Jessen and I. M. Orme, *Antimicrob. Agents Chemother.*, 1996, **40**, 2809.
142. J. Grosset, C. Truffot-Pernot, C. Lacroix and B. Ji, *Antimicrob. Agents Chemother.*, 1992, **36**, 548.
143. E. Nuermberger, I. Rosenthal, S. Tyagi, K. N. Williams, D. Almeida, C. A. Peloquin, W. R. Bishai and J. H. Grosset, *Antimicrob. Agents Chemother.*, 2006, **50**, 2621.
144. S. Aly, K. Wagner, C. Keller, S. Malm, A. Malzan, S. Brandau, F. C. Bange and S. Ehlers, *J. Pathol.*, 2006, **210**, 298.
145. D. F. Warner and V. Mizrahi, *Clin. Microbiol. Rev.*, 2006, **19**, 558.
146. S. Tyagi, E. Nuermberger, T. Yoshimatsu, K. Williams, I. Rosenthal, N. Lounis, W. Bishai and J. Grosset, *Antimicrob. Agents Chemother.*, 2005, **49**, 2289.
147. K. Andries, T. Gevers and N. Lounis, *Antimicrob. Agents Chemother.*, 2010, **54**, 1236.
148. J. Grosset, B. Ji, *Experimental chemotherapy of mycobacterial diseases*, Mycobacteria, ed. P. R. J. Gangadharam, Jenkins, P. A., **Vol. 2**, 1998: Chapman and Hall.
149. R. M. McCune, R. Tompsett and W. McDermott, *J. Exp. Med.*, 1956, **104**, 763.
150. R. M. McCune, F. M. Feldmann and W. McDermott, *J. Exp. Med.*, 1966, **123**, 469.
151. R. M. McCune, F. M. Feldmann, H. P. Lambert and W. McDermott, *J. Exp. Med.*, 1966, **123**, 445.
152. M. H. Cynamon, S. P. Klemens, C. A. Sharpe and S. Chase, *Antimicrob. Agents Chemother.*, 1999, **43**, 1189.
153. A. M. Lenaerts, S. E. Chase, A. J. Chmielewski and M. H. Cynamon, *Antimicrob. Agents Chemother.*, 1999, **43**, 2356.
154. E. Nuermberger, S. Tyagi, K. N. Williams, I. Rosenthal, W. R. Bishai and J. H. Grosset, *Am. J. Respir Crit. Care Med.*, 2005, **172**, 1452.
155. M. Ibrahim, K. Andries, N. Lounis, A. Chauffour, C. Truffot-Pernot, V. Jarlier and N. Veziris, *Antimicrob. Agents Chemother.*, 2007, **51**, 1011.
156. B. V. Nikonenko, M. Protopopova, R. Samala, L. Einck and C. A. Nacy, *Antimicrob. Agents Chemother.*, 2007, **51**, 1563.
157. R. Tasneen, S. Tyagi, K. Williams, J. Grosset and E. Nuermberger, *Antimicrob. Agents Chemother.*, 2008, **52**, 3664.
158. E. Nuermberger, S. Tyagi, R. Tasneen, K. N. Williams, D. Almeida, I. Rosenthal and J. H. Grosset, *Antimicrob. Agents Chemother.*, 2008, **52**, 1522.

159. N. Veziris, M. Ibrahim, N. Lounis, A. Chauffour, C. Truffot-Pernot, K. Andries and V. Jarlier, *Am. J. Respir Crit. Care Med.*, 2009, **179**, 75.
160. F. R. Almeida, S. Novak and G. R. Foxcroft, *Theriogenology*, 2000, **53**, 1389.
161. N. Lounis, N. Veziris, A. Chauffour, C. Truffot-Pernot, K. Andries and V. Jarlier, *Antimicrob. Agents Chemother.*, 2006, **50**, 3543.
162. K. N. Williams, B. S. J., S. C. K., T. Zhu, A. Ogden, R. Tasneen, S. Tyagi, G. J. H. and E. L. Nuermberger, *Am. J. Respir Crit. Care Med.*, 2009, **180**, 371.
163. G. R. Davies, A. S. Pym, D. A. Mitchison, E. L. Nuermberger and J. H. Grosset, *Antimicrob. Agents Chemother.*, 2007, **51**, 403.
164. J. Flynn, A. Cooper, W. Bishais, *Animal models of tuberculosis, In: Tuberculosis and the Tubercle bacillus*, 2005, Washington, DC: ASM Press.
165. U. D. Gupta and V. M. Katoch, *Tuberculosis (Edinb)*, 2005, **85**, 277.
166. A. J. Lenaerts, D. Hoff, S. Aly, S. Ehlers, K. Andries, L. Cantarero, I. M. Orme and R. J. Basaraba, *Antimicrob. Agents Chemother.*, 2007, **51**, 3338.
167. L. Garcia-Contreras, J. C. Sung, P. Muttil, D. Padilla, M. Telko, J. L. Verberkmoes, K. J. Elbert, A. J. Hickey and D. A. Edwards, *Antimicrob. Agents Chemother.*, 2010, **54**, 1436.
168. P. L. Lin, S. Pawar, A. Myers, A. Pegu, C. Fuhrman, T. A. Reinhart, S. V. Capuano, E. Klein and J. L. Flynn, *Infect. Immun.*, 2006, **74**, 3790.
169. J. L. Flynn, S. V. Capuano, D. Croix, S. Pawar, A. Myers, A. Zinovik and E. Klein, *Tuberculosis (Edinb)*, 2003, **83**, 116.
170. L. E. Via, P. L. Lin, S. M. Ray, J. Carrillo, S. S. Allen, S. Y. Eum, K. Taylor, E. Klein, U. Manjunatha, J. Gonzales, E. G. Lee, S. K. Park, J. A. Raleigh, S. N. Cho, D. N. McMurray, J. L. Flynn and C. E. Barry, *Infect. Immun.*, 2008, **76**, 2333.
171. G. Walzl, K. Ronacher, J. F. Djoba Siawaya and H. M. Dockrell, *J. Infect.*, 2008, **57**, 103.
172. A. M. Ginsberg, M. W. Laurenzi, D. J. Rouse, K. D. Whitney and M. K. Spigelman, *Antimicrob. Agents Chemother.*, 2009, **53**, 3726.
173. P. R. Donald and A. H. Diacon, *Tuberculosis (Edinb)*, 2008, **88** (Suppl 1), S75.
174. R. Rustomjee, A. H. Diacon, J. Allen, A. Venter, C. Reddy, R. F. Patientia, T. C. Mthiyane, T. De Marez, R. van Heeswijk, R. Kerstens, A. Koul, K. De Beule, P. R. Donald and D. F. McNeeley, *Antimicrob. Agents Chemother.*, 2008, **52**, 2831.
175. A. H. Diacon, R. Dawson, M. Hanekom, K. Narunsky, S. J. Maritz, A. Venter, P. R. Donald, C. van Niekerk, K. Whitney, D. J. Rouse, M. W. Laurenzi, A. M. Ginsberg and M. K. Spigelman, *Antimicrob. Agents Chemother.*, 2010, **54**, 3402.
176. A. H. Diacon, A. Pym, M. Grobusch, R. Patientia, R. Rustomjee, L. Page-Shipp, C. Pistorius, R. Krause, M. Bogoshi, G. Churchyard, A. Venter, J. Allen, J. C. Palomino, T. De Marez, R. P. van Heeswijk, N. Lounis, P. Meyvisch, J. Verbeeck, W. Parys, K. de Beule, K. Andries and D. F. McNeeley, *N. Engl. J. Med.*, 2009, **360**, 2397.

CHAPTER 10
Diarrhoeal Diseases

DAVID BROWN, PhD, CChem, FRSC

Alchemy Biomedical Consulting, Cambridge, UK

10.1 Disease Burden

10.1.1 Morbidity and Mortality Rates

Diarrhoea is the second leading cause of death in children globally. Just under 9 million children aged under 5 years died in 2008 and 1.5 million of these deaths were due to diarrhoea.[1–4] The disease kills more children than AIDS, malaria, and measles combined (see Figure 10.1).

According to the World Health Organization (WHO), the morbidity and mortality associated with diarrhoeal diseases is very high indeed. In 2009, in addition to 1.5 million child deaths, diarrhoea was estimated to have caused an additional 1.1 million deaths in people aged 5 and over,[5] making a total of over 2.5 million deaths per annum caused by diarrhoeal diseases. Measured in Disability Adjusted Life Years (DALYs), which consider factors such as the loss of healthy years because of disability or death, diarrhoeal diseases were responsible for over 72 million years lost in 2004.[2] Less than half this number (34 million) of years is lost due to malaria, a disease which receives much more global attention.[2]

Despite this burden of disease, funding for prevention and treatment of diarrhoea over the past decade has been insufficient to address the global burden.[6]

The potential to reduce death rates was demonstrated in the 1970s and 1980s when funding directed toward childhood diarrhoea resulted in a major

RSC Drug Discovery Series No. 14
Neglected Diseases and Drug Discovery
Edited by Michael J. Palmer and Timothy N. C. Wells

Published by the Royal Society of Chemistry, www.rsc.org

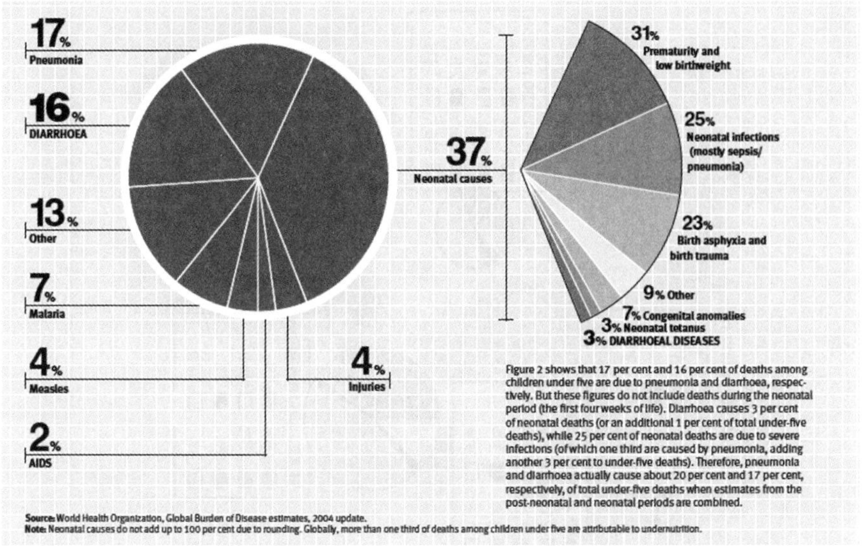

Figure 10.1 Causes of death in children under 5 years old.[1,2]

reduction in deaths from 4.5 million to under 2 million.[7] This halving of death rates was due largely to focus on introduction of oral rehydration therapy (also known as oral rehydration salts, ORS), which was stated to be one of the most important medical advances of the 20th century.[8] When coupled with widespread efforts to educate caregivers, the impact was dramatic. However, other health issues have gained more public attention during the 1990s and 2000s with the result that today even the most optimistic estimates suggest that only 30–40% of children in developing countries receive ORS when required; and the fall in death rates from diarrhoeal diseases has halted.

Funding is a major issue. Despite the fact that neonatal and childhood deaths from diarrhoeal diseases account for nearly 20% of all deaths in children under 5 years of age, only 4.4% of global health funding is allocated to diarrhoeal disease research and development.[9] There is now an urgent need to shift attention and resources back to treating and preventing diarrhoea.

10.1.2 Geography of Diarrhoeal Diseases

Deaths from diarrhoeal diseases occur overwhelmingly in the developing world. In the more advanced countries diarrhoeal deaths have steadily dropped over the past century due to advances in the provision of clean water, sanitation, vaccines and antibiotics. Figure 10.2 gives a global overview of child death rates by region, together with major causes.[9]

According to UNICEF, for deaths in childhood from diarrhoeal diseases, India alone accounts for nearly one-quarter of the total, and Africa for nearly one-half (Figure 10.3).[1] Only 15 countries account for almost three-quarters of

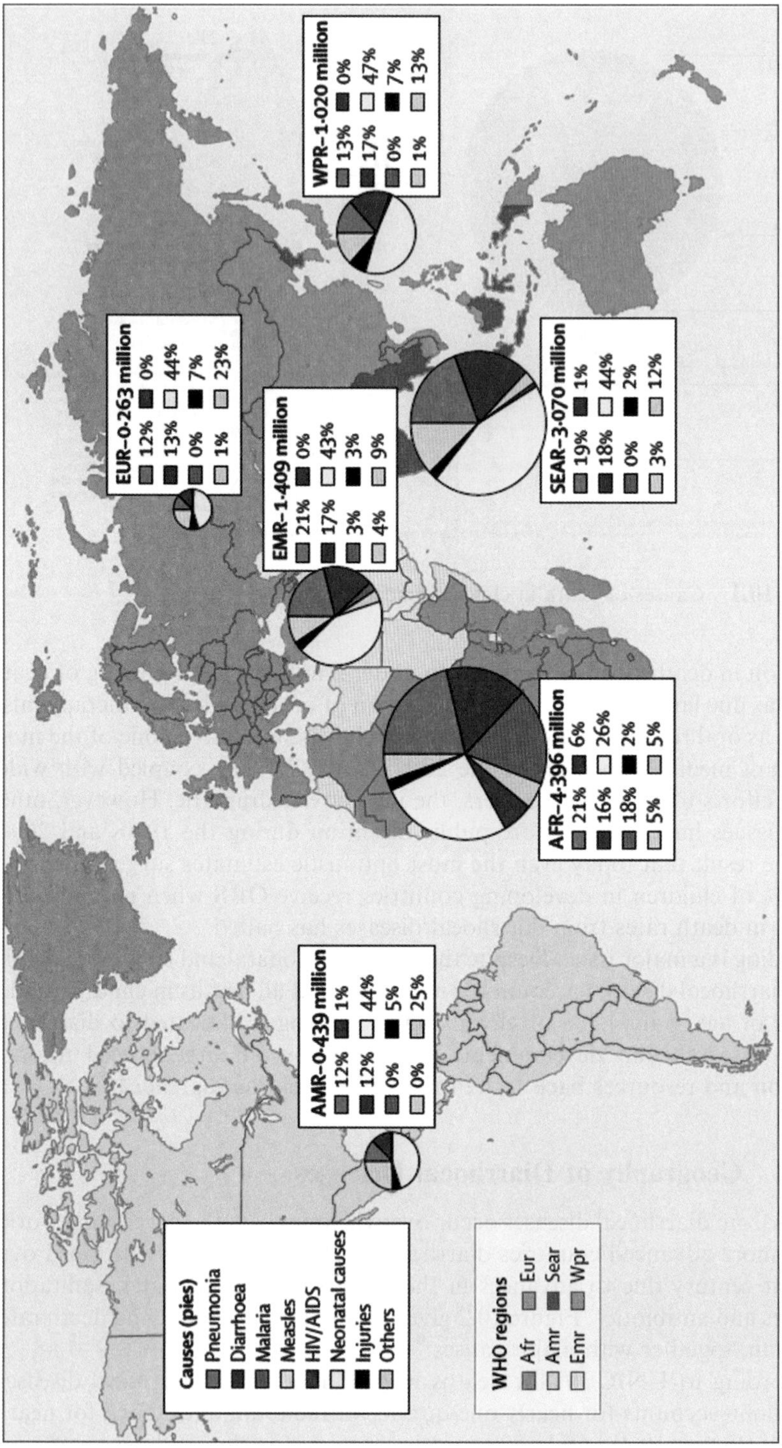

Figure 10.2 Global death rates of children under 5 years by continent, and major causes.[9]

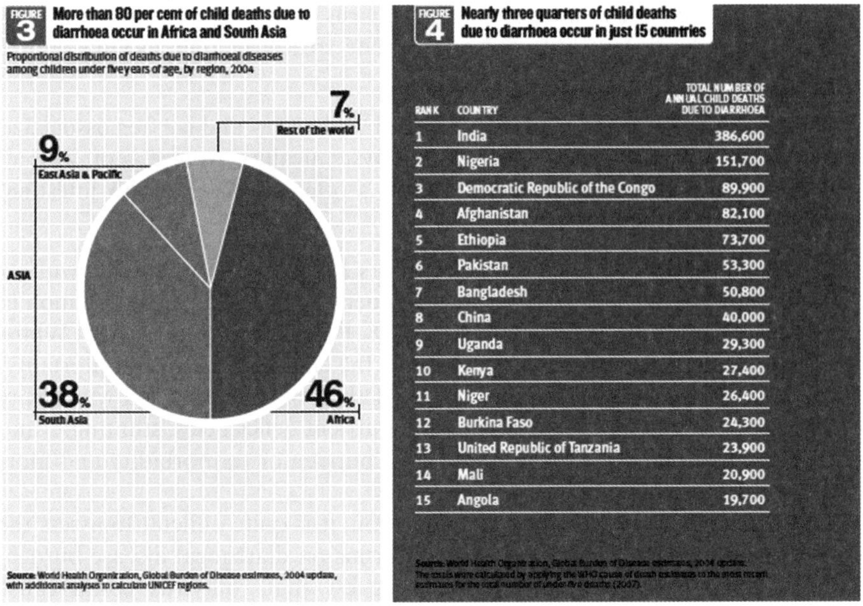

RANK	COUNTRY	TOTAL NUMBER OF ANNUAL CHILD DEATHS DUE TO DIARRHOEA
1	India	386,600
2	Nigeria	151,700
3	Democratic Republic of the Congo	89,900
4	Afghanistan	82,100
5	Ethiopia	73,700
6	Pakistan	53,300
7	Bangladesh	50,800
8	China	40,000
9	Uganda	29,300
10	Kenya	27,400
11	Niger	26,400
12	Burkina Faso	24,300
13	United Republic of Tanzania	23,900
14	Mali	20,900
15	Angola	19,700

Figure 10.3 Most diarrhoea deaths occur in just a few countries.[1,2]

diarrhoea deaths among children under five years of age – India, Nigeria, Congo, Afghanistan, Ethiopia, Pakistan, Bangladesh, China, Uganda, Kenya, Niger, Burkina Faso, Tanzania, Mali and Angola (Figure 10.3).[1]

10.1.3 Pathogenic Organisms Causing Diarrhoeal Diseases

Diarrhoeal diseases are more complicated to diagnose and treat than most diseases because there are many causative organisms, not just a single organism as is the case for e.g. AIDS, malaria or TB. Diarrhoea is caused by a number of gastrointestinal infections by viruses, bacteria or protozoa. However, just a small number of infectious organisms accounts for the vast majority of childhood diarrhoea in developing countries. Amongst the viruses, rotavirus causes about 40% of all hospital admissions and an estimated 30–40% of deaths.[10,11] Bacterial pathogens include species of *Escherichia coli* (*E. coli*), *Shigella* (dysentery), *Campylobacter, Salmonella* (typhoid fever) and *Vibrio cholerae. Cryptosporidium* and *Giardia lamblia* are major pathogenic protozoal causes of diarrhoea. *Entamoeba histolytica* is responsible for amoebic dysentery with an estimated 50 million cases per year and up to 100 000 deaths.

In clinical practice, symptoms are classified as follows:

Acute Watery Diarrhoea: This is associated with significant fluid loss and rapid dehydration, which can be rapidly lethal to infants in particular since water is a larger part of their body weight. Young children use relatively more water every day and their kidneys are less able to conserve water. Acute Watery

Diarrhoea is often due to infection with rotavirus, *E. coli* or cholera. Symptoms can last for several days but with appropriate treatment they usually resolve within 3–5 days.

Bloody Diarrhoea: Blood in faeces is called 'dysentery' and is often associated with infection by *Shigella*. This can be systemically invasive, not just an infection of the gastrointestinal tract, and is often associated with intestinal damage and nutritional losses.

Persistent Diarrhoea: If the condition lasts for more than 14 days, with or without blood in the faeces, it is defined as 'persistent'.[12] It is more common in under-nourished children or those with AIDS. Protozoa may be present. Whilst for Acute Watery Diarrhoea dehydration is the chief contributor to mortality, Persistent Diarrhoea is usually not acutely lethal, but is associated with many chronic adverse outcomes, including micronutrient deficiencies, stunting,[13] and cognitive impairment.[14] It is an important contributor to morbidity and mortality from other diseases.

Apart from this 'symptomatic' classification, it is quite difficult to assess the role of the many potential infectious organisms on diarrhoeal disease burden. For example, some current estimates suggest that rotavirus could cause between 450 000[15] to 700 000[16] diarrhoea deaths per year. Estimates of deaths caused by enterotoxigenic *E. coli* (ETEC) suggest well over 400 000 deaths per year in children under five.[17] In the most affected developing countries, the weak health infrastructure, limited formal diagnosis and virtually non-existent surveillance systems make it impossible to assess the true contribution of specific aetiologies to global diarrhoea morbidity and mortality. Moreover, there is a lack of suitable diagnostic tools to characterise infectious organisms in affected children whether in the home or hospital. This prevents the determination of the cause of many cases of diarrhoea. Improved diagnostic tools are needed to help health care providers make accurate diagnoses in order to prescribe the best available treatment.

Encouragingly, a multi-year international study is currently in progress which seeks to document the global burden of each infectious agent accurately. The study, called the Global Enteric Multi-center Study (GEMS),[18] also plans to test the efficacy of different interventions (e.g. ORS, vaccines, antibiotics). The study involves 8 sites on the two continents most affected by diarrhoeal diseases: 5 in Sub-Saharan Africa (Mozambique, Kenya, Gambia, Mali and Tanzania) and 3 in Asia (Bangladesh, India and Pakistan). When complete, the information should make a significant impact on effective therapy and vaccine design.

10.2 Prevention of Diarrhoeal Diseases

10.2.1 Hygiene, Sanitation and Public Health Policy

The level of faecal-oral pathogens in the environment is a major causative factor in diarrhoeal diseases. For this reason it is likely that clean water, effective sanitation and good personal hygiene play a dominant role in preventing disease transmission. It has been estimated that 88% of diarrhoeal deaths worldwide are attributable to unsafe water, inadequate sanitation and poor hygiene.[19]

Diarrhoea is more prevalent in the developing world, due in large part to the lack of safe drinking water, sanitation and hygiene, as well as poorer overall health and nutritional status. As of 2006, an estimated 2.5 billion people around the world were not using adequate sanitation facilities, and about 1 in 4 people in developing countries practiced open defecation.[1] Unsanitary conditions allow diarrhoea-causing pathogens to spread more easily. Simple remedies can be highly effective. Hand-washing with soap has been shown to reduce the incidence of diarrhoeal disease by over 40%, making it a highly cost-effective intervention.[1]

Improvement of unsanitary environments alone, however, will not be enough. Children with poor health and nutritional status are more vulnerable to serious infections like acute diarrhoea and suffer multiple episodes every year. At the same time, acute and prolonged diarrhoea seriously exacerbates poor health and malnutrition in children, creating a deadly cycle.

10.2.2 Breast-feeding and Micro-nutrient Supplementation

In developing countries, infants who are not breast-fed have a six-fold greater risk of death from infectious diseases during the first months of life, including diarrhoea.[20] Many cases of diarrhoea occur during weaning. Overall, death rates are highest during the first 12 months of life, and prolongation of breast-feeding confers very significant protection during this time. Currently, only 37% of infants in developing countries are exclusively breast-fed for the first 6 months.[21]

Vitamin A supplementation is another important preventative measure. Studies have shown reductions in infant death rates between 19% and 54%, with the majority of impact due to reductions in cases of diarrhoeal diseases and measles.[1]

Zinc is essential for normal development, yet many infants are zinc-deficient in developing countries. Supplementation trials have demonstrated a reduction in childhood diarrhoea. Zinc may also be used as a treatment (see below).[1]

10.2.3 Vaccines

Safe and effective vaccines are now available for prevention of infection by rotavirus, cholera, and typhoid. These vaccines alone could address approximately 25% of child deaths due to enteric and diarrhoeal diseases.

Rotavirus

Global rotavirus vaccine introduction has recently been recommended by the World Health Organization. Two products are available, Rotarix® from GlaxoSmithKline and RotaTeq® from Merck & Co. The GAVI Alliance (Global Alliance for Vaccines and Immunization), plans to roll out vaccine introduction in 42 GAVI-eligible countries. However the scale-up of use of rotavirus vaccines in the countries most in need remains uncertain due to the cost of vaccine and delivery programmes. At the moment only a few high- and middle-income countries provide for rotavirus vaccination in their routine immunization programmes. To ensure access to such vaccines in the developing world, the Bill and Melinda Gates Foundation is funding efforts by manufacturers in endemic countries to develop low-cost versions of these vaccines.[22]

Rotavirus vaccines are likely to find widespread use in developed countries too, since rotavirus infection is common in most countries of the world. Unusually amongst diarrhoeal diseases, good sanitation and hygiene do not prevent rotavirus, even in the wealthiest countries, though children rarely die in the wealthier countries because they receive rapid and effective rehydration therapy. Therefore, in both developed and developing countries, vaccination offers the best hope for preventing severe rotavirus illness. At the moment this disease cannot be treated effectively with drugs, and so prevention of rotavirus infections by vaccination remains an essential global strategy.

Because of the high incidence of rotavirus in wealthier countries, major vaccine manufacturers have had an incentive to develop vaccines. However, this is not the case for other diarrhoeal diseases, which have low incidence and death rates in wealthier countries. Therefore, development of the vaccines discussed below has often been dependent on government or philanthropic funding.

Cholera

An oral cholera vaccine has been available for some years (Dukoral,® a killed, whole-cell Plus toxin B subunit vaccine). However, outside the wealthier countries, only Vietnam has introduced the vaccine. Vietnam was able to do so by off-setting the prohibitive cost of the vaccine through access to a locally produced version.

To address the cost issue, the International Vaccine Institute (IVI) in Seoul has been aiming to develop and deploy a low-cost oral cholera vaccine. Two vaccines are in development. The first is based on the killed, whole-cell vaccine used in Vietnam. This product received a license in India in 2009 and is known as Shanchol™. However, because (like Dukoral) the approved killed, whole-cell vaccine requires two doses and provides only moderate levels of protection of the order 60–80%, IVI is also developing a second vaccine based on a live, attenuated Peru-15 strain which it is hoped will confer better protection after only a single dose.

Typhoid

The estimated global incidence of typhoid fever is approximately 21 million cases each year. Unless treated with antibiotics, typhoid fever can have a quite high case fatality rate. One estimate suggested that there were approximately 216 000 deaths from typhoid worldwide in the year 2000 and other estimates have suggested up to 600 000 typhoid-related deaths each year.[23]

Two vaccines are currently recommended by the WHO for the prevention of typhoid. One is a live, attenuated oral *S. typhi* (Ty21a) strain for oral immunization, and the other is an injectable subunit (Vi) polysaccharide vaccine. These were licensed in the late 1980s and early 1990s. Both are 50–80% protective. An older killed whole-cell vaccine is still used in countries where the newer preparations are not available, but this vaccine is no longer recommended for use by WHO, because it has a higher rate of side effects (mainly pain and inflammation at the site of the injection). There are extensive data documenting the safety, efficacy, and practicality of the Vi and Ty21a vaccines. However, they are not widely used in developing countries, despite WHO recommendation that typhoid vaccines be used for immunization of school-age children in areas where antibiotic-resistant typhoid is endemic.

These current licensed typhoid vaccines are potentially useful in school-age children but are not compatible with infant immunization: the unconjugated Vi vaccine is poorly immunogenic in infants, and the use of Ty21a in enteric-coated capsules is impractical.

Therefore, a second-generation conjugate vaccine that can be effective in younger age groups is being developed by IVI.[24] This vaccine is currently is expected to start Phase I/II clinical trials in 2010.[22]

Enterotoxigenic E. coli *(ETEC)*

ETEC is a leading bacterial cause of diarrhoea. No vaccine is currently available versus this organism though two are being assessed for early-stage clinical trials.[25] The lead candidate is ACE527 which is a live attenuated, whole-cell, oral vaccine, comprised of three ETEC strains. The second candidate under evaluation is an inactivated whole-cell vaccine, SBL 109.

Shigella

No vaccine is available *versus Shigella*. A candidate vaccine is currently at early-stage clinical trials.[26] CVD1208S is a live attenuated *Shigella* vaccine candidate under development by the University of Maryland, Baltimore. This oral vaccine is a multivalent vaccine designed to target five disease-causing strains of the bacteria.[27] Vaccination versus measles, an acute viral infection, is also important. Children may experience serious side effects from measles, including diarrhoea, which is one of the most common causes of death associated with measles worldwide.[1]

10.3 Treatment of Diarrhoeal Diseases

10.3.1 WHO Treatment Guidelines Summary

Current WHO and UNICEF diarrhoea management guidelines are set out in the Integrated Management of Childhood Illness handbook[28] and also in a WHO and UNICEF joint statement issued in 2004.[29] Diagnosis is based on clinical symptoms (extent of dehydration, the type of diarrhoea exhibited, whether blood is visible in the stool, and the duration of the diarrhoea episode). Identification of the responsible pathogen is not necessary before initiating treatment, even in high-income countries.

For treatment of Acute Watery Diarrhoea (often due to ETEC or cholera) WHO/UNICEF recommends the use of low-osmolarity ORS (LO-ORS, see below), home fluids and increased feeding. For non-weaned infants, continued breastfeeding is recommended. A 10–14 day course of zinc at a dose of 20 mg/day (or 10 mg/day for children under the age of six months) may be given. Antibiotics are recommended when bloody diarrhoea is present or cholera is suspected, but the WHO does not endorse the use of any other anti-diarrhoeal drugs. In particular, the use of anti-motility agents is contra-indicated in infants and children.

Despite these clear guidelines, many other therapies are used throughout developing countries. Some of these therapies may be inappropriate and even damaging.

10.3.2 Oral Rehydration Salts

Throughout history, the treatment of diarrhoea has taken various forms, including the use of oral coconut milk- and rice milk-based solutions. In the mid-1920s, a more effective treatment was developed, involving intravenous rehydration. However, hospitalization was necessary which prevented many children and adults from access to the treatment. In the late 1940s, it was realized that chronic diarrhoea leads to loss of critical electrolytes such as potassium and sodium. This paved the way for the development of more effective intravenous and oral treatments, culminating in the 1970s in development of the first generation of ORS.[30] Scientists and clinicians in Bangladesh played a key role in this advance. The dramatic reduction in child deaths from diarrhoeal diseases during the 1970s and 1980s was due in large part to this development of effective ORS.

More recently, an improved version known as low-osmolarity ORS has been devised which improves clinical outcomes. It was released in 2002. Stool output and vomiting are decreased in children by about 20% and 30%, respectively, when compared to children using the original ORS formula. Unscheduled intravenous therapy also declined by 33% among children with diarrhoea using this new formulation.[31]

Manufactured packets of ORS powder (to be dissolved in water) contain sodium chloride, glucose, potassium chloride, and trisodium nitrate. When manufactured ORS is not available, many countries advise use of 'recommended homemade fluids' made using readily available low-cost ingredients including salted rice water or chicken soup with salt. Potato and maize or other grain or root-crops may also be used. Once dehydration occurs, however, it is recommended to switch to an official ORS.

ORS is one of the most remarkable medical therapies: an estimated 50 million lives have been saved by its use.[32] It can be administered at home, and can be delivered to remote regions that do not have easy hospital access. As described above, a substitute can even be 'manufactured' at home using ingredients available to most homes. However, there are still barriers to success, and the coverage rate for ORS remains too low, probably in the region of 30–40% globally, with little improvement over the past decade. The newer version, low-osmolarity ORS has been slow to roll out over the past decade.[29] There are many cases of childhood diarrhoea and dehydration untreated by ORS. Many lives are being lost that could easily be saved if caregivers received education and access to ORS. Local manufacture is a key determinant of success. Some countries are manufacturing ORS locally, but other countries must make national policy changes to promote manufacturing, distribution and education.

However, while ORS helps counter total body fluid loss due to diarrhoea and promotes intestinal fluid absorption, it is ineffective in reducing stool output in Acute Watery Diarrhoea (in fact, stool output may increase as the child rehydrates) and it does not kill the pathogens responsible for the diarrhoea. Research continues on 'third-generation' ORS formulations, including the possibility of additional components such as zinc.[1] According to WHO, co-administration of zinc with ORS has shown significant benefits:

> "*... in reducing both the duration and severity of diarrhoea episodes as well as reducing stool volume and the need for advanced medical care. Children receiving zinc often have greater appetites and are more active during the diarrhoea episode; its use has also been associated with increased ORS uptake. The provision of zinc tablets by health workers may also reduce the demand from caregivers for other less effective drugs, such as antibiotics and anti-diarrhoeal medications, which should not be routinely administered.*[1]"

Zinc is reviewed in more detail in the following section.

10.3.3 Zinc

The research described above on addition of zinc to ORS is based on results from clinical trials with administered zinc, which have demonstrated efficacy both as a treatment and as a preventative. Zinc supplementation (10–20 mg per day until cessation of diarrhoea) significantly reduces the severity and duration of diarrhoea in children less than 5 years of age. Additional studies have shown that short-course supplementation with zinc (10–20 mg per day for 10–14 days) reduces the incidence of diarrhoea for 2– 3 months. This intriguing observation suggests that zinc may also have a preventative action.[33] Based on these studies, WHO recommends that zinc (10–20 mg/day) be given for 10–14 days to all children with diarrhoea.[31] In persistent diarrhoea, up to a 40% reduction in treatment failure and death have been demonstrated.[1]

As a result of these encouraging findings, large-scale diarrhoea treatment programmes using zinc have commenced in India, Pakistan and Mali. Initial results are reported to be very encouraging with efficacy levels appearing to surpass those reported from the initial controlled studies.[1]

10.3.4 Antibiotics

The World Health Organization recommends that antimicrobials including antibiotics should not be used routinely. The reason is that usually it is not possible to distinguish between episodes of diarrhoea that might respond to antibiotics (such as infections by *E. coli* and cholera) from those that will not (such as rotavirus and *Cryptosporidium*). However, there is a clear case for use of antibiotics when bloody diarrhoea (probably due to *Shigella*) is present or cholera is suspected.[28,29]

Nevertheless, pharmacies in many developing countries do dispense antibiotics, sometimes quite liberally. Inappropriate usage can lead to more-rapid development of resistance to the drug, making it less effective when really required.

Ciprofloxacin (**1**) is widely used to treat 'traveller's diarrhoea' due to *E. coli* infections. The same drug is also recommended by WHO for treatment of bloody diarrhoea (suspected Shigellosis) in adults or children, treatment being for three days with ciprofloxacin, or for five days with another oral antimicrobial to which most *Shigella* species in the area are sensitive.[31] Antimicrobial resistance is frequent and the pattern of resistance is unpredictable and so WHO recommends laboratory assessment of the sensitivity of local strains of *Shigella* whenever possible. Many well-known antimicrobials are ineffective for treatment of shigellosis, and WHO recommends that the following readily available drugs should never be given to treat bloody diarrhoea: metronidazole, streptomycin, tetracyclines, chloramphenicol, sulphonamides, amoxicillin, nitrofurans (e.g. nitrofurantoin, furazolidone), aminoglycosides (e.g. gentamicin, kanamycin), first- and second-generation cephalosporins (e.g. cephalexin, cefamandole), nalidixic acid.[31]

Cholera requires immediate treatment, particularly in infants and young children: the disease can cause death within hours from rapid and severe dehydration. Fluid replacement by intravenous and oral ORS is essential, and may be sufficient treatment for most patients. However, severe cases ideally require additional treatment with antibiotics. Several drugs are routinely used, though resistance is an ever-present issue. WHO recommends that all cases of suspected cholera with severe dehydration should receive an oral antibiotic known to be effective against strains of *V. cholerae* in the area. This reduces the total volume of stool passed, causes diarrhoea to stop within 48 hours, and shortens the period of faecal excretion of *V. cholerae*. The first dose should be given as soon as vomiting stops, which is usually 4–6 hours after starting rehydration therapy.

Antibiotics used against cholera – depending on resistance patterns in the area – include ciprofloxacin (**1**) and other fluoroquinolones, co-trimoxazole (**2**), doxycycline (**3**), chloramphenicol (**4**) and azithromycin (**5**). For children, furazolidone (**6**) has been an agent of choice, however resistance has been reported: ampicillin (**7**), erythromycin (**8**), and fluoroquinolones are potentially effective alternatives.

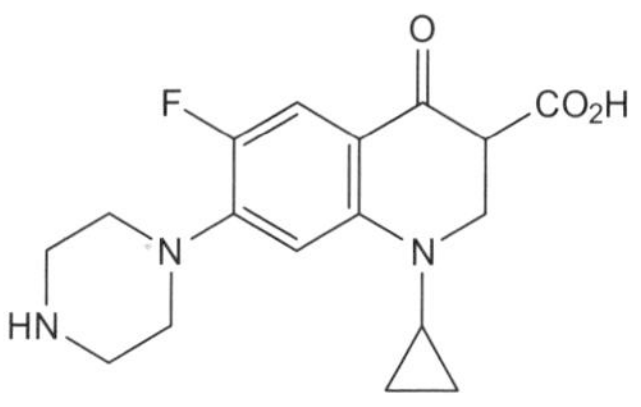

1

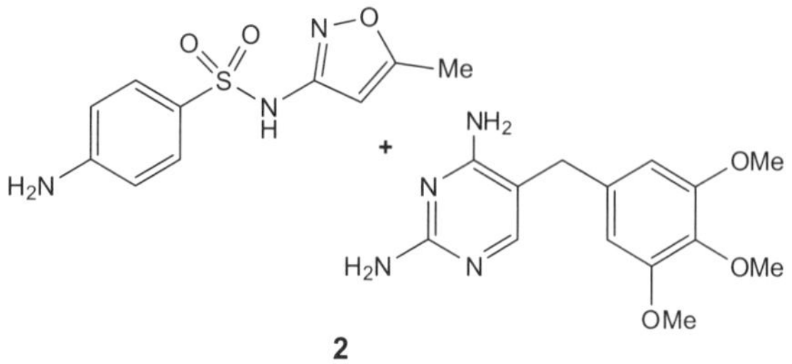

2

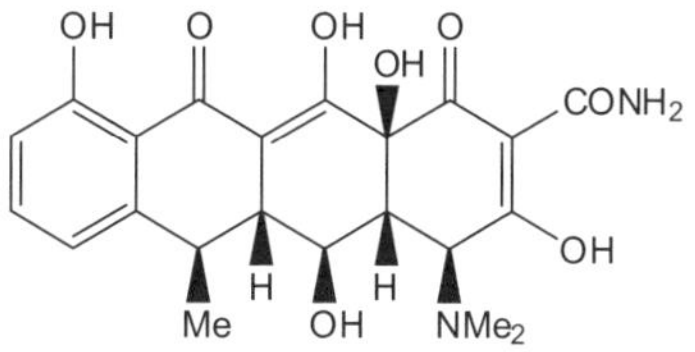

3

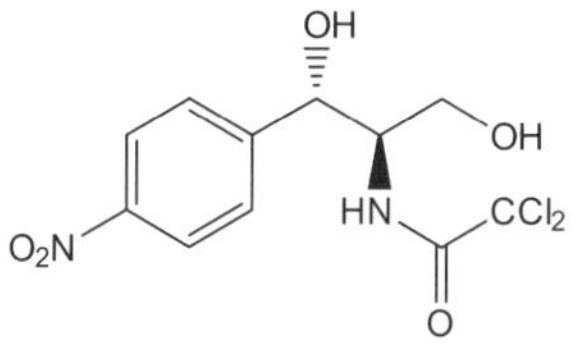

4

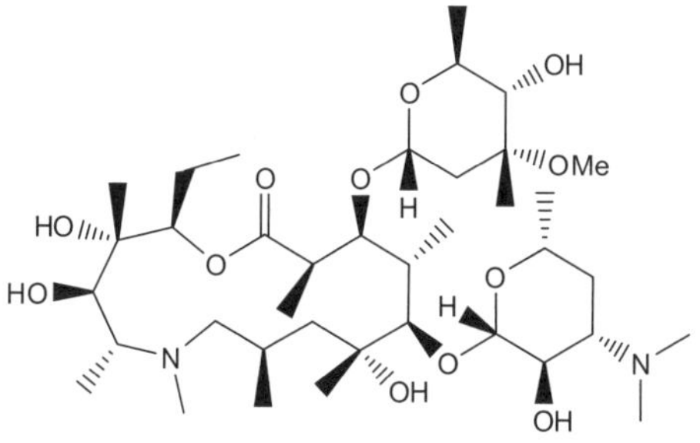

5

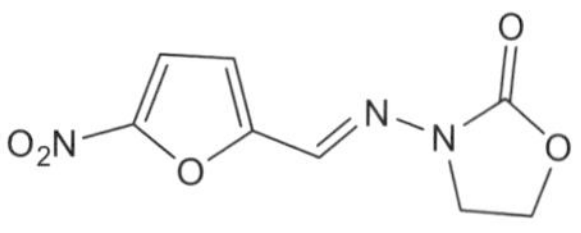

6

Rifaximin (**9**) is a nonsystemic, gastrointestinal-selective, oral antibiotic. It is a member of the rifamycin antimicrobial class, with a mechanism of action via RNA polymerase inhibition.[34] Rifaxamin is licensed by the US Food and Drug Administration to treat traveller's diarrhoea caused by non-invasive strains of *E. coli* in patients 12 years of age or older. It is not effective against *Campylobacter jejuni,* and there is no evidence of efficacy against *Shigella* species. Rifaximin is currently sold in the US, South America, Europe, and in India. The virtual lack of systemic absorption after oral administration provides very high local antibiotic concentrations in the gut, supporting efficacy against non-invasive species. Like its structural analogue rifampicin, rifaximin is associated with rapid development of bacterial resistance *in vitro*. However, the clinical resistance patterns to rifaxamin remain to be fully characterised.

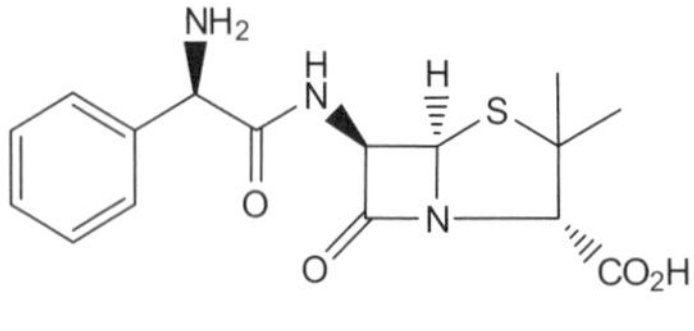

7

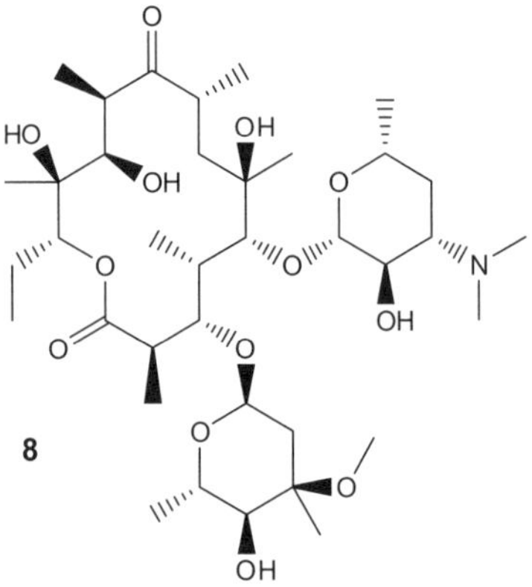

8

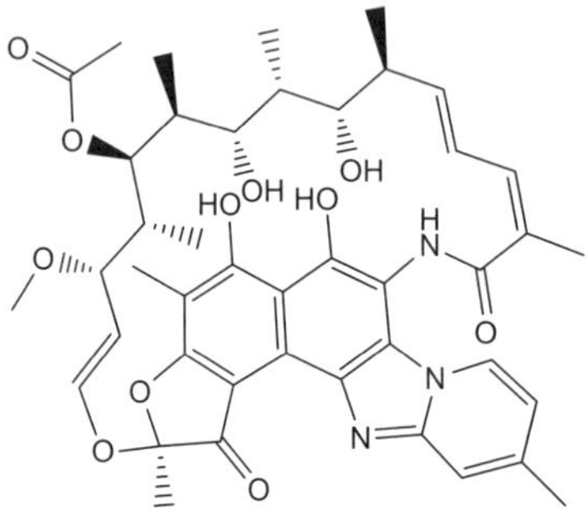

9

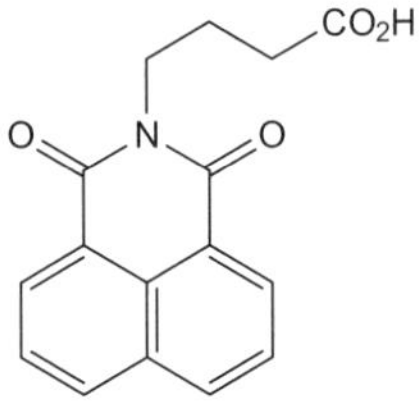

10

A number of other approaches to antibacterials have been investigated but not (yet) led to useful clinical drugs. For completeness, these will be summarised in the remainder of this section.

Virulence gene expression *in vivo* represents a potential target for antibiotic discovery. A high-throughput, phenotypic screen was used to identify a small molecule 4-[N-(1,8-naphthalimide)]-n-butyric acid (virstatin, **10**) that inhibits virulence regulation in *V. cholerae*.[35] By inhibiting the transcriptional regulator ToxT, virstatin prevents expression of two critical *V. cholerae* virulence factors, cholera toxin and the toxin coregulated pilus. Orogastric administration of virstatin protected infant mice from intestinal colonization by *V. cholerae*.

Bacteriophage therapy was utilised in the US until the discovery of penicillin, and throughout the Soviet Union until the 1990s. Bacteriophage therapy has been used as a single strain or a multi-strain 'cocktail' treatment for bacterial infections causing diarrhoea. Phage can be used alone or as an adjunct to antibiotic therapies. While phage are considered narrow-spectrum, it is possible to develop 'cocktails' for use against the dominant clinical strains in a given area. However, well-controlled clinical studies are lacking at the moment. Given the species-specific (even strain-specific) nature of phage, providing a broad coverage requires collection of live pathogenic specimens by a laboratory in the target geographical region where the phage therapy will be utilized. To address resistance, phage cocktail may need to be updated regularly, perhaps as often as every 6 months. Every new cocktail would currently need to be evaluated with clinical trials and re-registered with the regulatory agencies, which is very likely impractical. There is effort by the US Department of Defence to convince the FDA to allow delivery of phage cocktails in the same manner that flu vaccines are prepared; namely, every year, new phage are added to the cocktails and no new clinical trials are necessary. Phage-based medication, including preparations to treat intestinal infections that can be responsible for diarrhoea, is legal in all countries belonging to the Commonwealth of Independent States, and drug registration programmes have recently been underway in Central Asia, Latin America and Africa.[36]

In perhaps the only active discovery project currently underway with a specific focus on *Shigella*, small molecule inhibitors of *Shigella* tRNA guanine transglycosylase (TGT, EC 2.4.2.29) are under investigation. TGT is a tRNA-modifying enzyme common to nearly all organisms including humans. In bacteria, TGT catalyses the exchange of guanine in position 34 by preQ1, whereas eukaryotic TGT accelerates the replacement by queuine. This difference in substrate specificity offers the possibility for selective inhibition of the bacterial enzyme. TGT is essential for the pathogenic mechanism of Shigella. Low nanomolar lin-benzoguanine derivatives inhibitors have been designed as early lead structures,[37,38] though it is not clear whether any particular structure is being progressed.

10.3.5 Anti-protozoals

Protozoal infections lead to persistent diarrhoea lasting weeks or months if not treated. Causative organisms include *Giardia, Cryptosporidium*, and *Entamoeba* species (the latter causing 'ameobic dysentry').

Giardiasis occurs worldwide with a prevalence of 20–30% in developing countries. It also has a surprisingly high presence in developed countries. The Center for Disease Control and Prevention reports that in the USA, *Giardia* infects over 2.5 million people annually. There are multiple modes of transmission including person-to-person, water-borne, and venereal. Infection needs drug treatment since *Giardia* can persist and lead to severe malabsorption syndromes and weight loss. In all countries, in accordance with WHO guidelines, laboratory diagnosis of causative organism should ideally be obtained before initiating drug treatment, though in developing countries this may not always be possible. Metronidazole (**11**) clears the parasite with cure rates of approximately 85%. An alternative is the structurally related molecule tinidazole (**12**) which has similar efficacy and may be better tolerated. A more-recent introduction is nitazoxanide (**13**), from a different structural class. This drug is effective against metronidazole-resistant parasites and is approved by the FDA for treatment of diarrhoea caused by *Giardia* in children above age 1 year.

Nitazoxanide is a prodrug. Following oral administration of nitazoxanide, it is rapidly hydrolyzed to its active metabolite, tizoxanide (**14**), and the parent drug is not detected in plasma.[39]

The antiprotozoal activity of nitazoxanide is believed to be due to interference with the pyruvate: ferredoxin oxidoreductase (PFOR) enzyme-dependent electron transfer reaction which is essential to anaerobic energy metabolism. Studies have shown that the PFOR enzyme from *Giardia lamblia* directly reduces nitazoxanide by transfer of electrons in the absence of ferredoxin. The DNA-derived PFOR protein sequence of *Cryptosporidium parvum* appears to be similar to that of *Giardia lamblia*.[40] Interference with the PFOR enzyme-dependent electron transfer reaction may not be the only pathway by which nitazoxanide exhibits antiprotozoal activity.

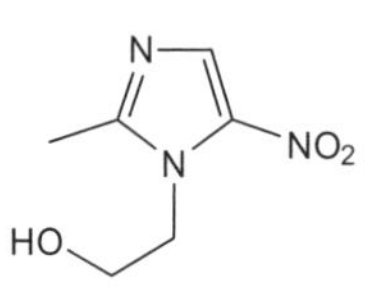

11

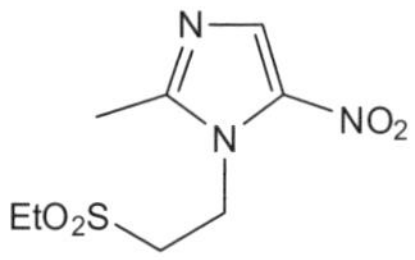

12

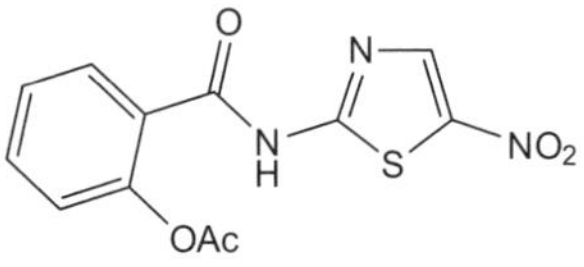

13

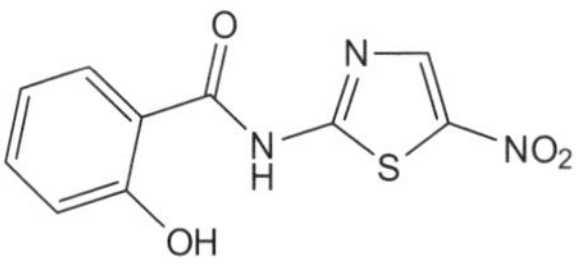

14

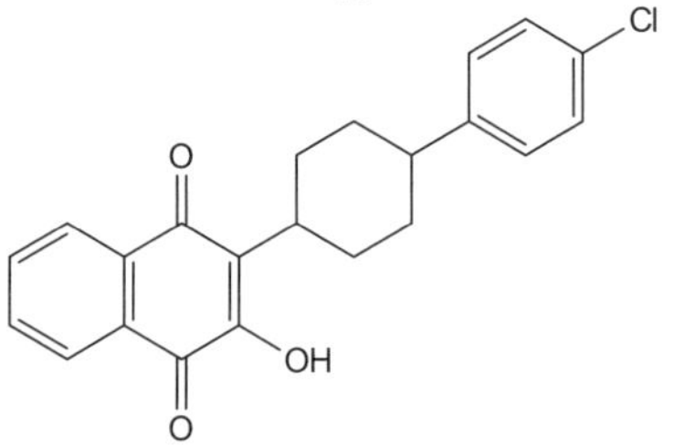

15

Cryptosporidiosis generally causes self-limiting diarrhoea in people with intact immune systems. However, in immunocompromised individuals, such as AIDS patients, the symptoms are particularly severe and often fatal. *Cryptosporidium* is the organism most commonly isolated in HIV positive patients presenting with diarrhoea. Treatment involves ORS and electrolyte correction.

Nitazoxanide (**13**) is approved by the FDA for treatment of diarrhoea caused by *Cryptosporidium* in children above age 1 year. Azithromycin (**5**) and atovaquone (**15**) are used but are not particularly effective.

Entamoeba species causing amoebic dysentery are initially managed by use of ORS. Ideally, no antimicrobial therapy should be administered until microbiological microscopy and culture studies have established the specific infection involved. However, when laboratory services are not available, it may be necessary to administer a combination of drugs, including an amoebicidal drug to kill the parasite and an antibiotic to treat any associated bacterial infection. Metronidazole (**11**) has been the drug of choice, with tinidazole (**12**) as an alternative. As mentioned above, nitazoxanide (**13**) is also effective and may be a pragmatic treatment for non-shigella/dysentery diarrhoea lasting 7 days or longer, if infectious aetiology is not clear.[33]

Persistent diarrhoea can be due to worm infections so it is interesting that nitazoxanide appears to have similar efficacy to benzimidazoles in treating Ascaris and Trichuris infections.[41]

10.3.6 Antisecretories

Although ORS has been tremendously successful in saving lives, the use of this treatment is considerably lower than warranted. One problem is that caregivers (usually young mothers who are often semi-literate or uneducated) observe that following ORS treatment the flow of diarrhoea may actually increase. Unfortunately they interpret this to mean the treatment is not working. To reduce fluid loss, unqualified doctors and mothers turn to the anti-motility agents such as loperamide (see section 10.3.8) which is not endorsed by the WHO guidelines and is contraindicated for use in children under age three years due to safety issues.

To address this issue, antisecretory agents have been developed to reduce fluid loss and stool output without impact on gut motility. A safe antisecretory drug that is successfully co-promoted with ORS and zinc should help to rehydrate a sick child. Moreover, it may actually help to increase the use of ORS. Effective, safe and cheap antisecretories may be the most urgent new therapeutic drug class needed for treatment of diarrhoeal diseases and have therefore received the most attention in recent years.

Currently, two antisecretory options are available for treatment of diarrhoeal diseases. These are racecadotril (**16**), a prodrug, whose active entity is thiorphan (**17**), and crofelemer (**18**–**21**), a mixture of natural products. Others are in earlier stages of development (see below).

Racecadotril (acetorphan, **16**) has shown efficacy in treatment of Acute Watery Diarrhoea in children.[42–45] It is claimed to be effective against diarrhoea associated with both bacterial infection and rotavirus infection. Convincing evidence for efficacy in cholera patients is lacking. It is available in some European countries, in Asia and in South America. However, some authors have questioned whether the drug confers sufficient benefit for widespread use in developing countries.[46,47] Racecadotril (chemical name acetorphan) is a prodrug which is hydrolyzed to an active metabolite, thiorphan (**17**) after oral dosing. Thiorphan is an enkephalinase inhibitor, and inhibition of this enzyme in the gastrointestinal tract is thought to prolong the natural antisecretory effects of endogenous enkephalins, with a reduction in water/electrolyte secretion into the intestinal lumen. Studies in animal models have demonstrated that the effects of racecadotril are antagonized by the opioid receptor antagonist nalaxone, confirming involvement of endogenous opioids in its action.[48]

In fact, racecadotril is a double prodrug: two pro-entities have to be removed by either metabolic means or *via* change in pH for the active entity to be released. Double prodrugs are relatively unusual: it can be quite difficult to achieve sequential release of the active entity reproducibly and the site of release of the two prodrug moieties could be critical. Release of active entity could in theory differ between populations according to nutritional status, gut content, gut microflora and gut pH. Could this account for the dispute about the effectiveness of the drug? There is insufficient evidence available to answer this question.

Crofelemer is under development for 4 diarrhoeal indications: acute infectious diarrhoea, paediatric diarrhoea, AIDS-related diarrhoea, and diarrhoea-predominant Irritable Bowel Syndrome.[49] Phase 2/3 studies have either been completed or are currently underway for these indications.

Crofelemer is a mixture of natural products, obtained from bark latex of the Amazonian tree, Croton lechleri. It is a purified heterogenous proanthocyanidin oligomer, with an average molecular weight of 2,200 Da. The basic monomers are mainly (+)-gallocatechin (**18**) and (−)-galloepicatechin (**19**), together with a smaller amount (+)-catechin (**20**) and (−)-epicatechin (**21**). The oligomer consists of linearly linked monomers (on average heptamers, ranging from penta- to 11-mers) of varying ratios.[50] This drug blocks the CFTR chloride channel (Cystic Fibrosis Transmembrane-conductance Regulator) in the intestinal lumen, reducing the flow of chloride ions and water into the gastrointestinal tract. It also demonstrates activity at calcium-activated chloride ion channels. Crofelemer is not absorbed systemically, it acts locally in the gut. Due to its mechanism of action, it is claimed the drug can treat Acute Watery Diarrhoea irrespective of aetiology.[51]

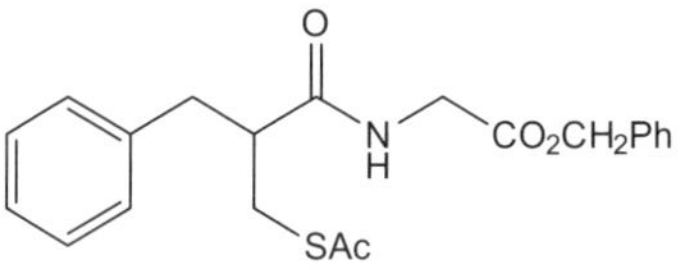

16

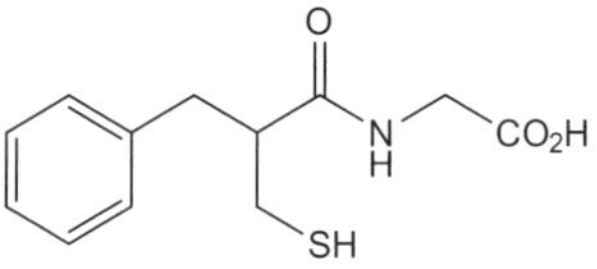

17

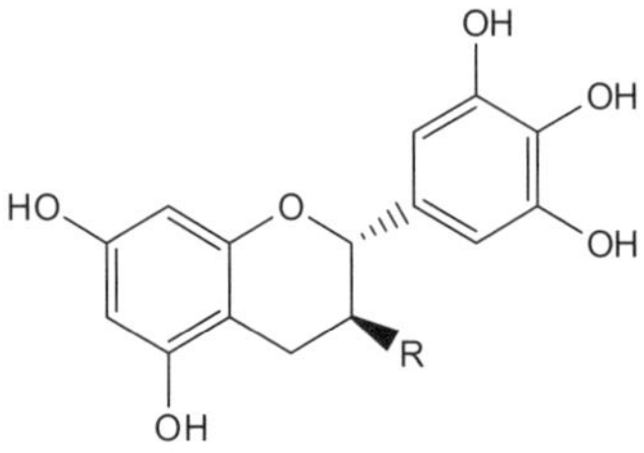

18 R = OH

19 R = OH

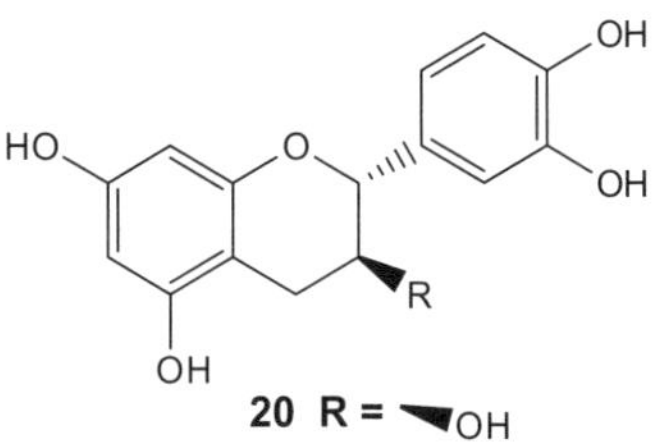

20 R = OH

21 R = OH

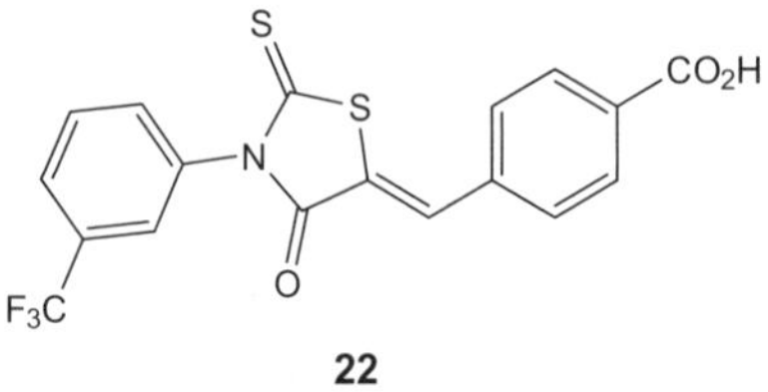

22

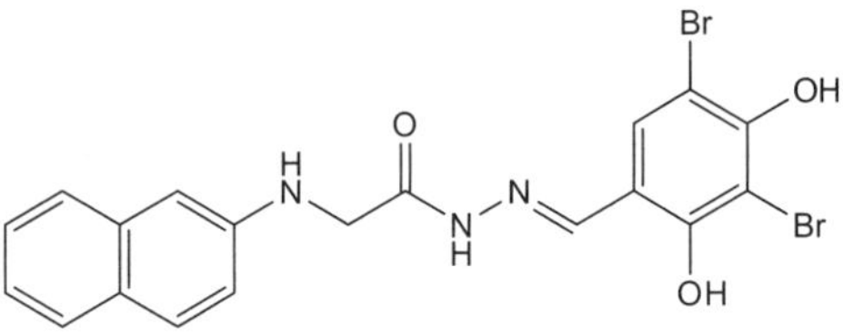

23

An advantage of this approach is that the CFTR chloride channel is a host target (rather than vector/pathogen-targeted), and therefore it is highly unlikely that resistance to drug action will arise. A number of groups are now pursuing host targets for this reason, as discussed below.

A search is underway for totally synthetic small molecule inhibitors acting by the same mechanism as crofelemer. Scientists at the University of California have published[52,53] structures of two lead series of inhibitors, e.g. the thiazolidinone (**22**) and also glycine acylhydrazone derivatives such as (**23**). However, these have not been progressed to clinical studies.

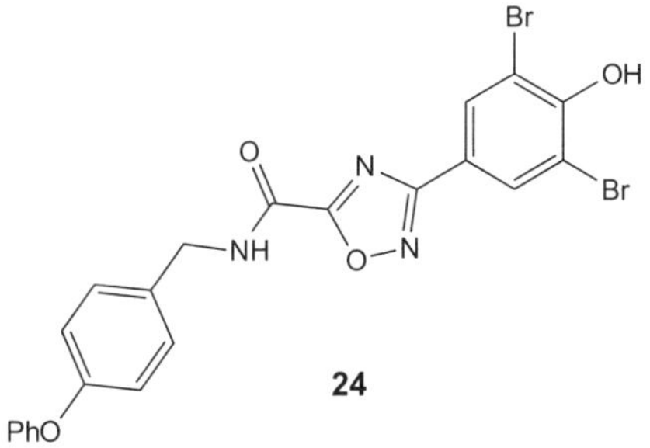

24

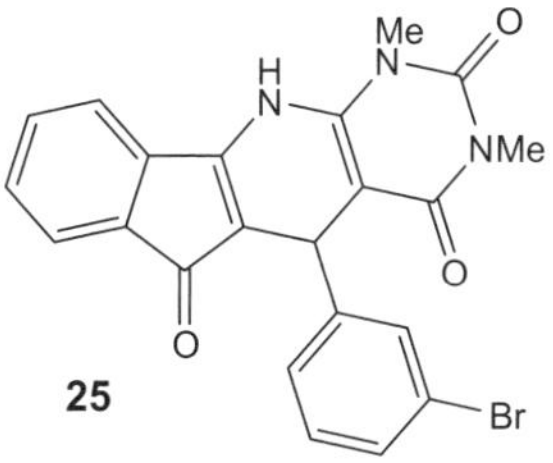

25

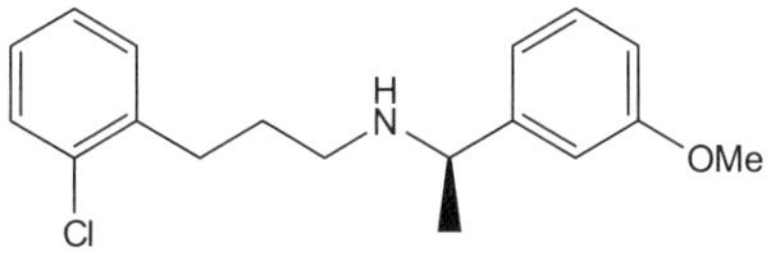

26

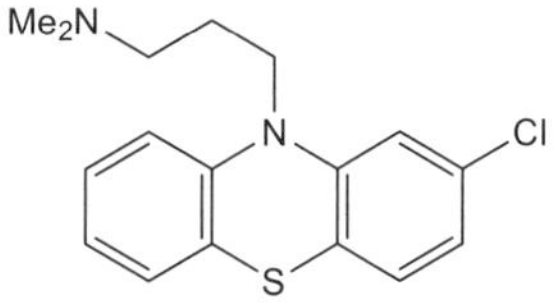

27

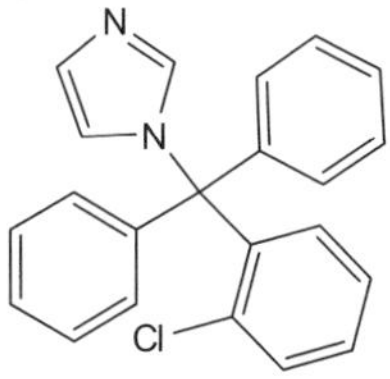

28

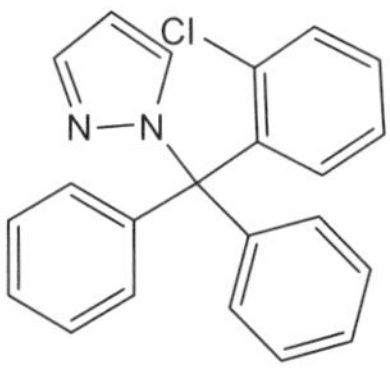

29

30

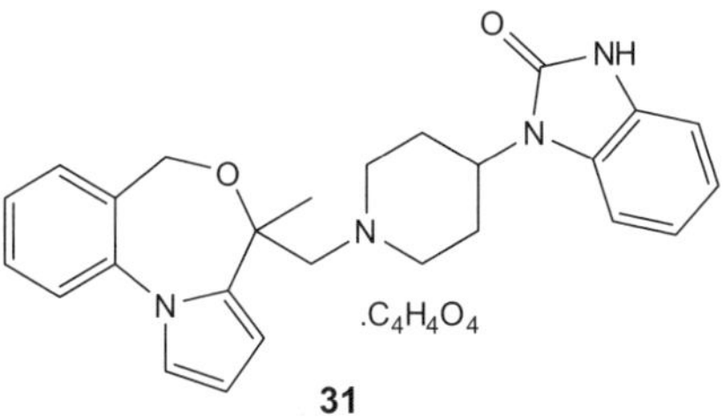

31

The Institute for One World Health also has small molecule inhibitors of the CFTR chloride channel under development, with lead compounds currently at the early clinical stage.[54] Structures were derived from the acylhydrazone series published by Verkman *et al.*,[53] by seeking 5- and 6-membered heterocyclic systems bioisosteric with acylhydrazones. IOWH032 (**24**) was stated to be the preferred compound. This compound entered Phase 1 clinical studies during calendar year 2011.

In addition to approaches focused directly on blockade of the CFTR-chloride channel, an alternative mechanism has been proposed, based on inhibition of guanyl cyclase.[55] Acute Watery Diarrhoea induced by infection with ETEC involves binding of stable toxin (STa) to its receptor on the intestinal brush border, guanylyl cyclase type C (GC-C). Intracellular cGMP is then elevated, inducing increase in chloride efflux and subsequent accumulation of fluid in the intestine. Screening of a compound library for inhibitors of GC-C led to identification of the pyridopyrimidine (**25**) and this compound was shown to suppress STa-stimulated cGMP accumulation by decreasing GC-C activation *in vitro*. The mechanism of inhibition appears to be complex and indirect, possibly associated with phospholipase C and tyrosine-specific phosphorylation. Compound **25** inhibited chloride-ion transport stimulated by activation of guanylyl or adenylyl cyclases and suppressed STa-induced fluid accumulation in an *in vivo* rabbit intestinal loop model. Thus, **25** may be a promising lead compound for optimization for treatment of diarrhoea and other diseases, however, the consequence of systemic inhibition of this essential system has not been reported.

In addition to the CFTR chloride channel, other chloride channels are being explored as potential targets for antisecretory drugs. There are at least five distinct classes of mammalian chloride channels, including the CFTR, CLC-type voltage-sensitive chloride channels, ligand-gated (GABA and glycine) chloride channels, volume-sensitive chloride channels, and calcium-activated chloride channels (CaCCs).[56]

The calcium-activated chloride channels have been proposed as potential targets for anti-diarrhoeal drugs.[56] It has been suggested that CaCCs in intestinal epithelial cells are involved in chloride and fluid secretion in secretory diarrhoeas caused by certain viruses, including rotavirus.[61] High-throughput screening has been pursued for initial lead molecules.[62]

Another host-target antisecretory mechanism under investigation is the calcium-sensing receptor (CaSR). Scientists at Harvard University have shown that activation of CaSR receptors in cells lining the intestine can stop fluid secretion.[63] Toxins released by cholera and ETEC stimulate the production of the cyclic nucleotides cAMP- and cGMP, leading to excess fluid secretion into the gut. CaSR receptor activation reverses this action. The Harvard group suggest that small molecule CaSR agonists may provide effective therapy for secretory diarrhoeas. This mechanism appears to be effective in both decreasing fluid/salt secretion and enhancing fluid/salt/nutrient absorption, so may be complimentary to ORS therapy. The small molecule CaSR agonist R-568 (**26**) provides a lead molecule.[63] A non-absorbed drug may be necessary since a key issue will be drug safety in view of the broad role of the calcium-sensing receptor throughout the body.

In addition to these current discoveries and clinical approaches to anti-secretory drugs, other approaches have been reported in the past which are included here for completeness even though most have not yielded useful therapies.

The anti-psychotic drug chlorpromazine (**27**) inhibits cholera toxin-stimulated intestinal fluid secretion in laboratory animals. Clinical trials also indicated some efficacy, particularly in severe cholera.[64,65] However, sedative effects at therapeutic doses limited interest in the drug.

The antifungal antibiotic, clotrimazole (**28**) also inhibits chloride secretion by human intestinal cells. The mechanism is *via* inhibition of the intermediate-conductance calcium-activated Ki channel, IKCa1. Clotrimazole exhibits efficacy in a mouse model of cholera.[66]

However, inhibition of cytochrome P450 enzymes by clotrimazole limits its therapeutic value. A rational design strategy was used to find a clotrimazole analog that selectively inhibits IKCa1 without blocking cytochrome P450 enzymes. A screen of 83 triarylmethanes showed that the pharmacophore for channel block is different from that required for cytochrome P450 inhibition. The 'IKCa1-pharmacophore' tolerated a pyrazole or tetrazole substituent, whereas cytochrome P450 inhibition requires the imidazole ring. The compound TRAM-34 (**29**) inhibits the cloned and the native IKCa1 channel in human T lymphocytes with a Kd of 20–25 nM and is 200- to 1,500-fold selective over other ion channels.[67] The channel residues involved in binding of TRAM-34 and related inhibitors have been identified by mutagenesis of channel residues.[68] However, these compounds have not been progressed into animal or human trials for treatment of diarrhoeal diseases.[69]

Studies have documented reduced frequency of diarrhoea in breast-fed children.[70] This effect has been attributed to a number of components including 1,2-linked fucosylated glycans, which function as soluble receptors that inhibit pathogens from adhering to their target receptors on the mucosal surface of the host gastrointestinal tract.[70] Another factor may be the antimicrobial action of human milk proteins such as lactoferrin and lysozyme.[71] Lactoferrin is a component of the innate immune response.[72] It has broad-spectrum antimicrobial activity against bacteria, fungi, viruses and protozoa resulting either from its ability to sequester iron or from a direct effect on microbial cell membranes.[73] Proteolysis of lactoferrin under acidic conditions, as would occur in the stomach or in the phagolysosomes of neutrophils, yields peptides called lactoferricins that have enhanced antimicrobial activity.[74] *In vitro*, recombinant human lactoferrin and lysozyme individually and in combination show antimicrobial activity against a wide spectrum of bacteria, viruses, parasites and fungi, including diarrhoea-associated organisms such as rotavirus, ETEC, *V. cholerae, Salmonella* and *Shigella*.[75] Clinical data are limited however, one study showed significant but small benefits of lactoferrin/lysozyme-supplemented ORS in paediatric acute secretory diarrhoea.[76]

Butyric acid – ('butyrate') – (**30**) is a short chain fatty acid produced in the colon as a result of microbial fermentation. Butyrate stimulates sodium absorption in colonic epithelia and inhibits chloride secretion induced by cAMP in these epithelia.[77] A key action of butyrate is that it induces expression in colonic epithelial cells of cathelicidin (LL-37), an endogenous antimicrobial small protein that kills pathogenic diarrhoea-causing bacteria. During infections by ETEC, cholera and shigella, expression of the antibacterial peptides LL-37 and also human β-defensin-1 is reduced or turned off.[78] The major virulence proteins of the pathogens are predominantly responsible for these effects, both *in vitro* and *in vivo*. Clinical trials versus shigellosis have demonstrated that butyrate treatment results in reduced illness.[79] Interest in these findings currently centres on use of mature green banana fruit which is rich in amylase-resistant starch that stimulates colonic production of short-chain fatty acids including butyrate. Green banana diet has been shown to reduce clinical severity in childhood shigellosis and could be a simple and useful adjunct for dietary management of this illness.[80]

Calmodulin is an important intermediate in the mucosal signal-transduction pathways that regulate intestinal NaCl transport. A selective inhibitor of calmodulin, zaldaride maleate, also known as CGS 9343B (**31**) was shown to be effective in two models of secretory diarrhoea. CGS 9343B was more potent in these models than morphine, loperamide and chlorpromazine though CGS 9343B did not exert any significant antimotility effects at antidiarrhoeal doses.[81] The compound was not further developed for use in diarrhoeal diseases however.

10.3.7 Antivirals

The main virus involved in diarrhoeal diseases in developing countries is rotavirus. This is treated by ORS and in future should be prevented in most countries by vaccination (see above). For those patients who unfortunately still become infected with rotavirus, and who respond poorly to ORS, options for drug treatment are very limited. Racecadotril (**16**) may have some efficacy though this is not well characterised and is disputed by some opinion-leaders, as discussed above.[46]

Nitazoxanide (**13**) has also been reported to have some clinical efficacy versus rotavirus.[82,83] However, it has not been developed and marketed for that indication.

10.3.8 Other drugs

Antimotility agents are often prescribed by pharmacists and physicians in developing countries. These include loperamide (**32**), diphenoxylate (**33**) and opioid-derivatives such as codeine (**34**), morphine (**35**) or an opium extract known as 'paragoric'. Use of antimotility agents is widespread by adults and in children too, despite the clear guidance from WHO that they should not be used in children. Of concern, anti-motility agents are often not promoted or prescribed in conjunction with ORS, and may in fact displace ORS use, hence these interventions may also result in a negative impact on rehydration of a sick patient presenting with AWD. Antimotility agents may have a role in treating some adult diarrhoeas, e.g. traveller's diarrhoea due to ETEC infections. But care must be taken over diagnosis; these agents should never be used by persons with fever or bloody diarrhoea, because they can increase the severity of disease by delaying clearance of causative organisms (specifically those diarrhoeal diseases caused by either *Shigella* or *Salmonella*).

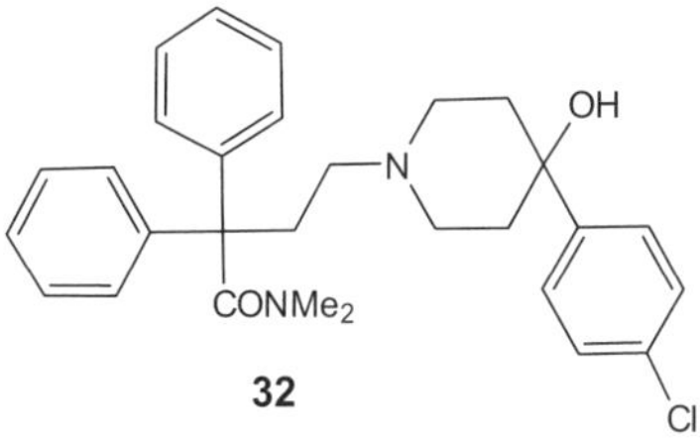

32

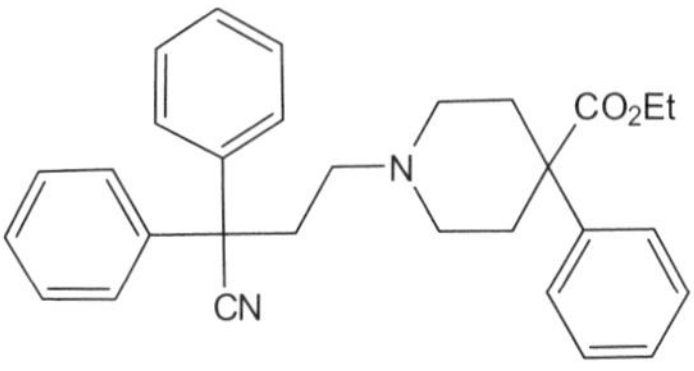

33

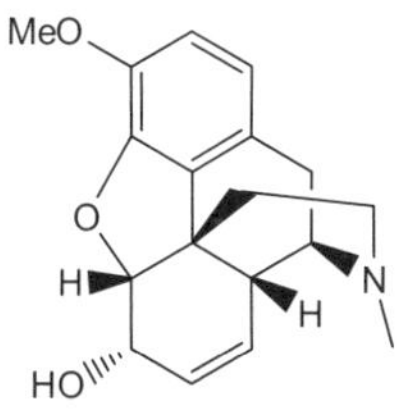

34

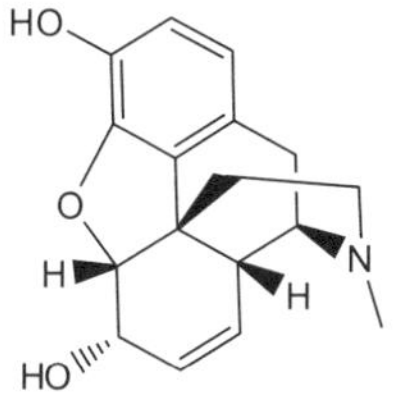

35

Probiotics have received increased attention for both prevention and treatment of several types of diarrhoea of different aetiologies. The mechanisms by which probiotics may prevent or treat diarrhoea include stimulation of the immune system, competition for binding sites on intestinal epithelial cells, and elaboration of bacteriocins. Organisms investigated for their medicinal uses include *Lactobacillus rhamnosus* GG, *L. reuteri*, certain strains of *L. casei, L. acidophilus, E. coli* strain Nissle 1917, and certain *Bifidobacteria* and *Enterococci* (*Enterococcus faecium* SF68) as well as the probiotic yeast *Saccharomyces boulardii*. Human trials have been relatively small and difficult to interpret, though a meta-analysis of randomised, controlled clinical trials of the efficacy of probiotics for the prevention of traveller's diarrhoea concluded that several probiotics (*Saccharomyces boulardii* and a mixture of *Lactobacillus acidophilus* and *Bifidobacterium bifidum*) had significant efficacy. In summary, although some encouraging data have been reported, large intervention studies and epidemiological investigations of long-term probiotic effects are required to fully validate this approach and show the most appropriate organisms or mixtures of organisms for particular diarrhoea aetiologies.[84]

10.4 Conclusions

Morbidity and mortality from diarrhoeal diseases was reduced dramatically in the 1970s and 1980s following introduction of ORS through an effective global campaign promoting their use. Deaths have slowly reduced further since then, probably in part due to increasing affluence across the globe and introduction of improved hygiene in home and public settings. However, the number of deaths remains high and the consensus view is that progress has effectively stalled over the past decade.

What opportunities are there to reduce morbidity and mortality in the short-to-medium term based on current knowledge and therapies? And what further research is required for the long term?

The immediate key actions are stated clearly in the 2009 report[1] from UNICEF and WHO titled *Diarrhoea: why children are still dying and what can be done*. The report includes a seven-point plan for comprehensive diarrhoea control, as follows:

Prevention package

- Rotavirus and measles vaccinations
- Promotion of early and exclusive breastfeeding and vitamin A supplementation
- Promotion of handwashing with soap
- Improved water quantity and quality, including treatment and safe storage of household water
- Promotion of community-wide sanitation

Treatment package

- Fluid replacement to prevent dehydration (ORS)
- Zinc supplements

Although not the primary focus of this review, hygiene and safe water are most important factors. If 88% of diarrhoeal deaths worldwide are attributable to unsafe water, inadequate sanitation and poor hygiene,[19] then it is in these areas that much attention must be focused.

Currently available therapies, in particular ORS and zinc for treatment of Acute Watery Diarrhoea, have tremendous potential to reduce the burden of death. Less than 40% of patients receive ORS and even fewer receive zinc. If these rates could be increased to 70–90% a quite dramatic reduction in death rates could be expected.

For rotavirus-induced diarrhoea it is essential that vaccines are made available for all children at risk around the world. The two available vaccines may have some impact but it is likely that financial subsidy may be required to support their widespread introduction in the poorest countries, or low-cost variants may be required also. Current efforts to produce these are encouraging but it could be another decade before low-cost options become widely available.

Looking to the longer term, beyond the seven-point plan from WHO-UNICEF, some additional observations can be made.

Zinc requires additional scientific focus, in particular the basic science behind its effect. Lack of understanding of its mechanism of action and lack of biomarkers has hindered design of trials, and it is unknown whether the current dose and duration of dosing is optimal. The 10-14 day course of zinc is difficult to adhere to for prophylaxis, and it is not consistent with the duration of ORS usage for treatment of diarrhoea.

In recent years *Shigella* has been well treated with available antibiotics. The incidence and severity of *Shigella* infections appears to have fallen substantially, for reasons not well understood. However, more recently there have been reports from South Asia of strains beginning to emerge with high levels of resistance to most current antibiotics.[85,86] There have also been reports from within the USA.[87] This situation needs careful monitoring and the research community may need to devise new antibiotics or combinations of antibiotics to halt the spread of resistant *Shigella*. The pharmaceutical industry has not acknowledged the need for discovery of new medicines versus Gram-negative bacteria such as *Shigella*, because of low medical need and commercial return in advanced countries, and has instead focused on Gram-positive bacteria such as MRSA. Therefore, current global investment in R&D for new antibiotics for Gram-negative bacteria such as *Shigella* is inadequate. This gap leaves the global health community in the precarious position of potentially having no replacements for current antibiotics for Gram-negative organisms such as *Shigella* (and *V. cholerae*), as resistance to front-line therapy continues to develop more widely.

The field of diarrhoeal diseases has lacked interest in combinations of antibiotics despite the fact drug combinations are a well-established approach in other diseases areas (double therapy for non-diarrhoeal antibacterial infections; triple therapy for AIDS, quadruple therapy for TB). If drug-resistant strains of *Shigella* do emerge the most rapid response could be to investigate antibiotic combinations.

In the longer term, there are several areas of research that appear worthy of support and focus. The clinical-stage and preclinical antisecretory approaches based on inhibition of the CFTR chloride channel discussed in section 10.3.6 hold promise of drugs which will stem fluid loss and reduce/prevent dehydration occurring. A parent's concern is to stop diarrhoea, not the dehydration due to diarrhoea, and so they tend towards therapies which do not treat the aspect of diarrhoea that is actually life-threatening to a child, the fluid loss. If antisecretory drugs prove to be compatible with ORS use, and preferably increase usage of ORS, then this combination would be a significant advance.

Calcium-activated chloride channels are thought to be important potential targets for therapeutics versus diarrhoeal diseases, but there are many gaps in basic scientific understanding of these channels. In addition, they have not been studied in detail in disease models. The suspected potential link between these channels and diarrhoea induced by rotavirus requires study as this could provide important therapeutic opportunities.

Finally, other drug discovery opportunities may be opened up by recent evidence, extending that discussed above, of how breast milk helps prevent diarrhoeal disease. Complex sugars constituting up to 21% human milk cannot be digested by babies and seem to have a purpose other than infant nutrition. They appear to be a major influence on the composition of the bacteria in the infant's gut, favouring 'protective bacteria': these bacteria prevent infectious bacteria from adhering to the infant's gut wall.[88] The protective bacterium is a subspecies of *Bifidobacterium longum*. It possesses genes that enable it to thrive on the indigestible component of milk, which are complex sugars derived from lactose, the principle component of breast milk. This new finding extends previous knowledge that the sugars themselves also serve as decoys for noxious bacteria that might attack the infant's intestines, as discussed earlier. Breast feeding remains the preferred means to deliver these sugars, but for mothers in whom this is not possible an exogenous substitute could be beneficial. These findings may open up drug discovery opportunities based on non-absorbed lactose derivatives.

References

1. UNICEF Diarrhea: *Why Children Are Still Dying And What Can Be Done*, UNICEF New York, 2009.
2. WHO: *The global burden of disease, 2004*. www.who.int/healthinfo/global_burden_disease/2004_report_update/en/index.html (accessed July 26, 2010).
3. D. You, T. Wardlaw, P. Salama and G. Jones, *Lancet*, 2009, **375**, 100.
4. A. D. Lopez, C. D. Mathers, M. Ezzati, D. T. Jamison and C. J. Murray, *Lancet*, 2006, **367**, 1747.
5. http://www.straitstimes.com/Breaking%2BNews/Tech%2Band%2BScience/Story/STIStory_448440.html (accessed July 30, 2010).
6. I. Rudan, S. El Arifeen, R. E. Black and H. Campbell, *Lancet Infect. Dis.*, 2007, **7**, 56.
7. C. Boschi-Pinto, C. F. Lanata and R. E. Black, in *International Maternal and Child Health*, ed. J. E. Ehiri and M. Meremikwu, Springer Publishing, Washington, DC, 2009, chapter 13, 225. Anon. *Lancet*, 1978, **312**, 300.
8. M. Moran, J. Guzman, A. L. Ropars, A. McDonald, T. Sturm, N. Jameson, L. Wu, S. Ryan and B. Omune, *Neglected disease research and development: how much are we really spending?*, The George Institute for International Health, 2009.
9. J. Bryce, C. Boschi-Pinto, K. Shibuya, R. E Black and the WHO Child Health Epidemiology Reference Group, *Lancet*, 2005, **365**, 1147.
10. C. J. Williams, A. Lobanov and R. G. Pebody, *Epidemiol. Infect.*, 2009, **137**, 607.
11. U. D. Parashar, C. J. Gibson, J. S. Bresee and R. I. Glass, *Emerging Infect. Dis.*, 2006, **12**, 304.
12. Z. A. Bhutta, E. A. Nelson, W. S. Lee, P. I. Tarr, R. Zablah, K. B. Phua, K. Lindley, D. Bass and A. Phillips, *J. Ped. Gastroent Nutr.*, 2008, **47**, 260.
13. A. A. Lima, S. R. Moore, M. S. Barboza Jr, A. M. Soares, M. A. Schleupner, R. D. Newman, C. L. Sears, J. P. Nataro, D. P. Fedorko, T. Wuhib, J. B. Schorling and R. L. Guerrant, *J. Infect. Dis.*, 2000, **18**, 1643.
14. D. S. Berkman, A. G. Lescano, R. H. Gilman, S. L. Lopem and M. M. Black, *Lancet*, 2002, **359**, 564.

15. WHO, External Review of Burden of Disease Attributable to Rotavirus, WHO, Geneva, 2005.
16. U. D. Parashar, C. J. Gibson, J. S. Bresee and R. I. Glass, *Emerging Inf. Dis.*, 2006, **12**, 304.
17. F. Qadri, ETEC Diarrhea Disease Burden – Global Perspectives with an Emphasis on Bangladesh, Global Vaccine Research Forum, Brazil, 2005.
18. http://medschool.umaryland.edu/GEMS/default.asp (accessed July 26, 2010).
19. R. E. Black, S. Morris and J. Bryce, *Lancet*, 2003, **361**, 2226.
20. http://www.unicef.org/nutrition/index_24824.html (accessed July 27, 2010).
21. WHO Collaborative Study Team on the Role of Breastfeeding on the Prevention of Infant Mortality and Bellagio Child Survival Study, *Lancet*, 2000, **355**, 451.
22. http://www.gatesfoundation.org/topics/Pages/diarrhea.aspx (accessed July 28, 2010).
23. http://www.ivi.int/program/tr_domi_typhoid.html (accessed July 28, 2010).
24. M M. Levine, *N. Engl. J. Med.*, 2009, **361**, 403.
25. http://www.path.org/projects/enteric_vaccine.php (accessed July 28, 2010).
26. K. L. Kotloff, J. K. Simon, M. F. Pasetti, M. B. Sztein, S. L. Wooden, S. Livio, J. P. Nataro, W. C. Blackwelder, E. M. Barry, W. Picking and M. M. Levine, *Human Vaccin.*, 2007, **3**, 268.
27. http://www.news-medical.net/news/2008/11/24/43367.aspx, (accessed July 28, 2010).
28. United Nations Children's Fund and World Health Organization, *Model IMCI Handbook: Integrated management of childhood illness*, WHO, Geneva, 2005, www.who.int/child_adolescent_health/documents/9241546441/en/index.html (accessed July 28, 2010).
29. United Nations Children's Fund and World Health Organization, *'WHO/UNICEF Joint Statement: Clinical management of acute diarrhoea'*, UNICEF, New York, 2004, www.afro.who.int/cah/documents/intervention/acute_diarrhoea_joint_statement.pdf (accessed July 28, 2010).
30. J. N. Ruxin, *Med. Hist.*, 1994, **38**, 363.
31. World Health Organization, *The Treatment of Diarrhoea: A manual for physicians and other senior health workers*, Geneva, 2005.
32. WHO, 2009, http://www.who.int/pmnch/media/membernews/2009/childhood_diarrhoea/en/index.html, (accessed July 25, 2010).
33. Z. A. Bhutta, R. E. Black, K. H. Brown, J. M. Gardner, S. Gore, A. Hidayat, F. Khatun, R. Martorell, N. X. Ninh, M. E. Penny, J. L. Rosado, S. K. Roy, M. Ruel, S. Sazawal and A. Shankar, *J. Pediatr.*, 1999, **135**, 689.
34. J. A. Adachi and H. L. DuPont, *Clin. Infect. Dis.*, 2006, **42**, 541.
35. D. T. Hung, E. A. Shakhnovich, E. Pierson and J. J. Mekalanos, *Science*, 2005, **310**, 671.
36. http://www.phageinternational.com/default.htm, (accessed July 27, 2010).
37. S. R. Hörtner, T. Ritschel, B. Stengl, C. Kramer, W. B. Schweizer, B. Wagner, M. Kansy, G. Klebe and F. Diederich, *Angew. Chem. Int. Ed. Engl.*, 2007, **46**, 8266.

38. P. C. Kohler, T. Ritschel, W. B. Schweizer, G. Klebe and F. Diederich, *Chemistry*, 2009, **15**, 10809.
39. L. M. Fox and L. D. Saravolatz, *Clin. Inf. Dis.*, 2005, **40**, 1173.
40. P. S. Hoffman, G. Sisson, M. A. Croxen, K. Welch, W. D. Harman, N. Cremades and M. G. Morash, *Antimicrob. Agents Chemother.*, 2007, **51**, 868.
41. C. White, *Expert Rev. Anti-infect Ther.*, 2004, **2**, 43.
42. E. salazar-lindo, J. Santisteban-Ponce, E. Chea-Woo and M. Gutierrez, *N. Engl. J. Med.*, 2000, **343**, 463.
43. J. P. Cézard, J. F. Duhamel, M. Meyer, I. Pharaon, M. Bellaiche, C. Maurage, J. L. Ginies, J. M. Vaillant, J. P. Girardet, T. Lamireau, A. Poujol, A. Morali, J. Sarles, J. P. Olives, C. Whately–Smith, S. Audrain and J. M. Lecomte, *Gastroenterol*, 2001, **120**, 799.
44. B. Cojocaru, N. Bocquet, S. Timsit, C. Wille, C. Boursiquot, F. Marcombes, D. Garel, N. Sannier and G. Chéron, *Arch. Pediatr.*, 2002, **9**, 774.
45. S. G. Rao, *J. Indian Med. Assoc.*, 2002, **100**, 530.
46. M. K. Bhan and S. Bhatnagar, *Indian Pediatr.*, 2004, **41**, 1204.
47. H. Szajewska, M. Ruszczyn'ski, A. Chmielewska and J. Wieczorek, *Aliment. Pharmacol. Ther.*, 2007, **26**, 807.
48. H. Marcais-Collado, G. Uchida, J. Costentin, J. C. Schwartz and J. M. Lecomte, *Eur. J. Pharmacol.*, 1987, **44**, 125.
49. http://www.napopharma.com/products/index.html, (accessed July 29, 2010).
50. http://www.faqs.org/patents/app/20090238901, (accessed July 29, 2010).
51. http://www.napopharma.com/products/mechanism.html, (accessed July 29, 2010).
52. C. Muanprasat, N. D. Sonawane, D. Salinas, A. Taddei, L. J. Galietta and A. S. Verkman, *J. Gen. Hysiol.*, 2004, **124**, 125.
53. N. D. Sonawane, J. Hu, C. Muanprasat and A. S. Verkman, *Faseb J.*, 2006, **20**, 130.
54. http://www.oneworldhealth.org/diarrheal_disease, accessed July 29, 2010; poster by E. de Hostos at Gordon Conference, New London, NH, USA, August 8-13, 2010; posters by K. Doyle at EFMC-ISMC 2010 conference, Brussels, Belgium, September 5-9, 2010.
55. A. Y. Kots, B.-K. Choi, M. E. Estrella-Jimenez, C. A. Warren, S. R. Gilbertson, R. L. Guerrant and F. Murad, *Proc. Natl. Acad. Sci.*, 2008, **105**, 8440.
56. R. De La Fuente, W. Namkung, A. Mills and A. S. Verkman, *Mol. Pharmacol.*, 2008, **73**, 758.
57. J. Eggermont, *Proc. Am. Thorac. Soc.*, 2004, **1**, 22.
58. M. J. Farthing, *Dig. Dis.*, 2006, **24**, 47.
59. K. Gyo mo rey, E. Garami, K. Galley, J. M. Rommens and C. E. Bear, *Pflugers Arch.*, 2001, **443**, S103.
60. C. Hartzell, I. Putzier and J. Arreola, *Ann. Rev. Physiol.*, 2005, **67**, 719.
61. M. Lorrot and M. Vasseur, *Virol. J.*, 2007, **4**, 31.
62. R. De La Fuente, W. Namkung, A. Mills and A. S. Verkman, *Mol. Pharmacol.*, 2008, **73**, 758.
63. J. Geibel, K. Sritharan, R. Geibel, P. Geibel, J. S. Persing, A. Seeger, T. K. Roepke, M. Deichstetter, C. Prinz, S. X. Cheng, D. Martin and S. C. Hebert, *Proc. Natl. Acad. Sci.*, 2006, **103**, 9390.

64. M. R. Islam, D. A. Sack, J. Holmgren, P. K. Bardhan and G. H. Rabbani, *Gastroenterol.*, 1982, **82**, 1335.
65. G. H. Rabbani, W. B. Greenough, J. Holmgren and B. Kirkwood, *Br. Med. J.*, 1982, **284**, 1361.
66. P. A. Rufo, D. Merlin, M. Riegler, M. H. Ferguson-Maltzman, B. L. Dickinson, C. Brugnara, S. L. Alper and W. I. Lencer, *J. Clin. Invest.*, 1997, **100**, 3111.
67. H. Wulff, M. J. Miller, W. Hansel, S. Grissmer, M. D. Cahalan and K. G. Chandy, *Proc. Nat. Acad. Sci.*, 2000, **97**, 8151.
68. H. Wulff, G. A. Gutman, M. D. Cahalan and K. G. Chandy, *J. Biol. Chem.*, 2001, **276**, 32040.
69. Personal communication from K. G. Chandy.
70. L. Morrow and J. M. Rangel, *Semin. Pediatr. Infect. Dis.*, 2004, **15**, 221.
71. B. Lonnerdal, *Am. J. Clin. Nutr.*, 1985, **42**, 1299.
72. P. Mohan and S. A. Abrams, *The Cochrane Library*, 2009, Issue 1.
73. L. Adlerova, A. Bartoskova and M. Faldyna, *Veterinarni Medicina.*, 2008, **53**, 457.
74. J. L. Gifford, H. N. Hunter and H. J. Vogel, *Cell Molec. Life Sci.*, 2005, **62**, 2588.
75. N. Zavaleta, D. Figueroa, J. Rivera, J. Sanchez, S. Alfaro and B. Lonnerdal, *J. Pediatr. Gasteroenterol. Nutr.*, 2007, **44**, 258.
76. Islam, L. Bandholtz, J. Nilsson, H. Wigzell, B. Christensson, B. Agerberth and G. H. Gudmundsson, *Nature Med.*, 2001, **7**, 180.
77. S. Vidyasagar and B. S. Ramakrishna, *J. Physiol.*, 2002, **539**, 163.
78. K. Chakraborty, S. Ghosh, H. Koley, A. K. Mukhopadhyay, T. Ramamurthy, D. R. Saha, D. Mukhopadhyay, S. Roychowdhury, T. Hamabata, Y. Takeda and S. Das, *Cell Microbiol.*, 2008, **10**, 2520.
79. R. Raqib, P. Sarker, P. Bergman, G. Ara, M. Lindh, D. A. Sack, K. M. Nasirul Islam, G. H. Gudmundsson, J. Andersson and B. Agerberth, *Proc. Natl. Acad. Sci.*, 2006, **103**, 8913.
80. G. H. Rabbani, S. Ahmed, I. Hossain, R. Islam, F. Marni, M. Akhtar and N. Majid, *Pediatr. Infect. Dis. J.*, 2009, **28**, 420.
81. J. E. Shook, T. F. Burks, J. W. Wasley and J. A. Norman, *J. Pharmacol. Exp. Ther.*, 1989, **251**, 247.
82. J-F. Rossignol, M. Abu-Zekry, A. Hussein and M G. Santoro, *Lancet*, 2006, **368**, 124.
83. C. G. Teran, C. N. Teran-Escalera and P. Villarroel, *Int. J. Infect. Dis.*, 2009, **13**, 518.
84. M. de Vrese and P. R. Marteau, *J. Nutr.*, 2007, **137**, 803S.
85. T. V. Nguyen, P. V. Le, C. H. Le and A. Weintraub, *Antimicrob. Agents Chemother.*, 2005, **49**, 816.
86. http://centre.icddrb.org/pub/publication.jsp?classificationID = 56&pubID = 5145, (accessed July 30, 2010).
87. http://www.shigellablog.com/tags/antibiotic-resistance/, (accessed July 30, 2010).
88. A. M. Zivkovic, J. B. German, C. B. Lebrilla and D. A. Mills, *Proc. Natl. Acad. Sci.*, 2010, published online before print August 2, 2010, doi: 10.1073/pnas.1000083107.

CHAPTER 11

Anthelmintic Discovery for Human Infections

TIMOTHY G. GEARY*[a] AND NOELLE GAUVRY[b]

[a] McGill University, Institute of Parasitology, 21,111 Lakeshore Road, Ste-Anne de Bellevue, H9X 3V9, Quebec, Canada; [b] Novartis Animal Health Inc., Chemistry and Bioactives, Schwarzwaldallee 215, CH-4002, Basel, Switzerland

11.1 Introduction and Background

Chemotherapy remains the only strategy available to treat helminth infections of humans. Helminth infections, or helminthiases, are remarkably prevalent in people, especially in regions of poverty. According to some estimates, as many as one-third of the human population are burdened by one or more of these parasites.[1–5] Although commonly grouped together, helminths are placed in two quite distinct phyla separated by hundreds of millions of years of evolution. These are the Nematoda, or roundworms, and Platyhelminths, or flatworms, a group that includes both trematodes (flukes) and cestodes (tapeworms). Although helminth infections are widespread in regions of poverty, these parasites generally do not cause acutely lethal diseases. Rather, they are agents of insidious but undeniable morbidity, typically resulting in disability, loss of productivity and delay or failure in development of physical and cognitive abilities.[6] These infections contribute to the cycle of poverty that complicates efforts to control infectious diseases and thus enable development in disadvantaged areas. The lack of acute lethality associated with helminthiases has limited investment in drug discovery targeted to them, even in

RSC Drug Discovery Series No. 14
Neglected Diseases and Drug Discovery
Edited by Michael J. Palmer and Timothy N. C. Wells

Published by the Royal Society of Chemistry, www.rsc.org

comparison with malaria, tuberculosis and HIV/AIDS. They are truly neglected tropical diseases (NTDs), which place an enormous burden on global health.

Because they are macroscopic, helminthiases were the first infectious diseases to be identified in human medicine and the first to receive chemotherapeutic attention, even if those early remedies were poorly efficacious and often toxic.[7] The modern era of chemotherapy for these diseases arose with the introduction of diethylcarbamazine (DEC) in the early 1950s for filariases, followed by the transition of thiabendazole from veterinary to human use in the 1960s.[7] The discovery and development of highly effective and generally safe anthelmintics in the 1970s and 1980s changed the landscape of parasite control in veterinary and human medicine, and led to the adoption of mass drug administration (MDA) programs to control and potentially eliminate some of the major human helminthiases. While notable gains have been made, especially in the control of filarial nematode infections and schistosomiasis, helminthiases remain far too common in developing regions and continue to impose an undue burden on human health and productivity.

11.2 Nematodes

Nematodes are among the most common infectious agents of humans, with estimates that ~1 billion people carry at least one species. These parasites primarily inhabit the gastrointestinal (GI) tract of humans, though filarial species reside in internal compartments (see below). The most important GI nematodes in prevalence and pathology are *Ascaris lumbricoides*, the hookworms *Ancylostoma duodenale* and *Necator americanus*, the whipworm *Trichuris trichiura*, and the threadworm *Strongyloides stercoralis*; other species are important in certain locations.[1–6,8–11] These species are globally distributed, with highest prevalence in tropical and sub-tropical countries. These are soil-transmitted helminths (STH), with the infectious larval stages present in soil contaminated by human faeces. Abundance and persistence of infectious stages in the environment is dictated by physical and cultural factors, including temperature, humidity, frequency and amount of rainfall, and sanitation infrastructure and habits. The infectious larvae can survive for years in the environment in optimal conditions and make control programs based solely on chemotherapy a very long-term prospect.

These parasites are characterized by a direct life cycle. Humans become infected by ingesting (or by skin exposure in the case of hookworms) infectious larvae in the environment.

For whipworms, the larvae remain in the intestinal tract, where they develop to the adult stage in the cecum. Development of *A. lumbricoides* includes a phase of migration through the lungs; larvae are coughed up and swallowed prior to ending up in the GI tract to complete development to the reproductive stage. Hookworm larvae penetrate the skin and migrate through the blood stream, exiting at the lungs before developing in the small intestine, similar to

Ascaris. A similar migratory phase characterizes infection with *Strongyloides* spp., although these parasites are unusual in that a free-living cycle occurs and the parasite may generate a pattern of autoinfection in immunosuppressed human hosts.[10,11] Eggs are produced by females and are shed in faeces; the cycle resumes when larval development in the environment provides new infective stages (the time required for development is species-specific and also varies with environmental factors).

Pathology of GI nematodes is associated with migrating larvae and the adult stages in the GI tract, although superinfection of immunosuppressed individuals by *Strongyloides* can result in multi-organ infection and carries a significant risk of mortality. Hookworms, which feed on host blood, cause anaemia, which can be quite severe in heavy infections. Whipworms may cause rectal prolapse in heavy infections, and large number of *Ascaris* may cause intestinal blockage. However, these infections tend to be chronic (adults may live for several years) and have less overt but nonetheless profound effects on host performance and development. This effect can be measured in the field when anthelmintics are used to eliminate most or all of the parasites, but the health burden imposed by these parasites is not always easy to quantify.[4–6]

Diagnosis of GI nematode infections is based on finding eggs in faecal samples. The morphology and size of the eggs allows species-level discrimination. These methods are not particularly sensitive or quantitative; they permit general epidemiological surveys and can be used to estimate the effects of anthelmintic treatment, but adult worm burdens are not necessarily correlated with the abundance of eggs in the faeces. Egg counts in faeces vary within and among days and regionally within a sample. Reading these tests requires low-technology equipment and settings, which is an advantage in resource-limited settings, but more quantitative tests that correlated with adult worm burdens would be valuable if consistent with use in the field in terms of cost and simplicity of operation.[12]

Chemotherapy remains the main method of control of GI nematode infections in humans, although sanitation and hygiene clearly have essential roles at the community level.[3,11,13–17] Anthelmintics are used to eliminate adult nematodes from the host; prophylaxis against infection is not included in current treatment strategies. The administration of anthelmintics for GI parasites can be done by a physician for patients in whom an infection has been diagnosed by the detection of nematode eggs in a faecal sample. The primary deployment of these drugs, however, is through MDA campaigns carried out in regions of high prevalence. These campaigns typically target children in elementary school, in whom worm burdens and egg output tend to be highest. Children in these campaigns receive one dose of an anthelmintic every 6–12 months regardless of diagnosis.[13,14,18,19]

By far the most commonly used drugs for GI nematodes are benzimidazole derivatives (Figure 11.1), including most notably mebendazole (**1**) and albendazole (**2**); flubendazole (**3**) is much less widely used.[13,14] The prototype drug in this class, thiabendazole (**4**), was introduced into veterinary practice in the 1960s and shortly thereafter was adopted for use in humans, but is no

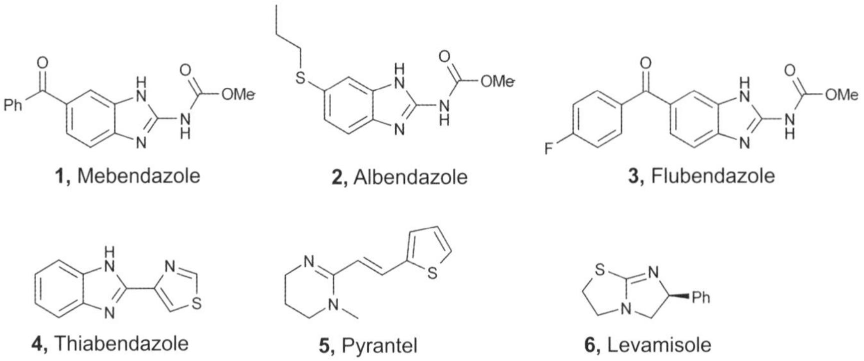

Figure 11.1 Current drugs against gastrointestinal nematodes.

longer manufactured for this indication. Benzimidazole anthelmintics disrupt the equilibrium between tubulin and microtubules.[20] Microtubules are essential structural components of all eukaryotic cells, and are dynamic polymers of a subunit composed of a dimer of α- and β-tubulin. Although tubulin sequences are very highly conserved across metazoan phyla, these drugs show ~10-fold selectivity for nematode tubulin compared to the host protein.[20] The safety profile of these drugs is very good, although concerns about teratogenicity have limited their use in pregnant women and women of child-bearing age. It should be noted that extensive deployment of these drugs in the field has not generated reports of birth defects in humans. The doses are not adjusted for weight; albendazole is administered as a single 400 mg tablet (given for 3–5 consecutive days for strongyloidiasis). The standard dose for mebendazole is a single 500 mg tablet, within a regimen of 100 mg daily for three days for hookworm and whipworm infections.

Efficacy of anthelmintics for human GI nematode infections is measured in terms of cure rate (CR: the percentage of patients with undetectable numbers of eggs in the faecal sample following treatment) or the egg reduction rate (ERR: the percentage difference in abundance of eggs per gram of faeces before and 1–2 weeks after treatment). Benzimidazoles are highly curative against *A. lumbricoides*, but are not usually 100% efficacious against hookworms, and in single doses, are rarely fully effective in eliminating whipworm infections.[11,12,14,21] A comprehensive analysis of trials of albendazole (**2**) and mebendazole (**1**) revealed that both drugs were equally efficacious against *A. lumbricoides* and equally inefficacious against whipworms in single-dose regimens, but that albendazole seems to be superior against hookworm infections.[21] Albendazole is available on the market in African countries at about 2 cents per dose, and GSK has recently announced a donation program for this drug for STH treatment in children.[19] This complements the existing mebendazole donation program from Johnson & Johnson.[18] These programs will increase exposure of parasite populations to what may be suboptimal efficacy drug pressure, raising concerns about the selection of benzimidazole-resistant parasites.[12,22]

Less commonly used drugs act as agonists of nicotinic acetylcholine receptors (nAChRs) in nematodes. Pyrantel (**5**) and levamisole (**6**), like benzimidazoles, are highly effective against *A. lumbricoides*, but are less active in single-dose regimens against hookworms and whipworms.[21] Pyrantel is much more commonly used in people, as the pamoate salt is very poorly bioavailable and host toxicity is minimized. It is administered at a standard single dose of 10 mg/kg. Levamisole is administered at a dose of 2.5 mg/kg (or as a single dose of 80 mg); it is less safe than pyrantel pamoate and is much less effective against hookworms and whipworms.[15,16,21] Nonetheless, these drugs offer an alternative mechanism of action that could be useful if resistance to benzimidazoles surface in human parasites.

Ivermectin (**7**), a macrocyclic lactone endectocide (Figure 11.2), is extensively used in veterinary medicine for the treatment of GI nematode infections in livestock, horses and companion animals. However, its primary use in humans is for tissue nematodes (see below). Ivermectin is only licensed for use against *Strongyloides* among the human GI nematodes (200 ug/kg oral doses for 1–2 days);[11] it is also used to treat ectoparasites, including lice and mites. Single 200 ug/kg doses are highly efficacious against ascariasis, but lack activity against one of the human hookworm species, *N. americanus*, for unknown reasons.[12] This lack of spectrum means that ivermectin is not suitable for use in humans for GI nematode infections that may include hookworms.[15,16,21]

11.2.1 Areas of Concern

1. A reliable standard operating procedure (SOP) for diagnosis of GI nematodes, less subject to operator error and providing an accurate estimate of adult worm burden, would offer tangible benefits for comparing efficacy of therapeutic regimens and new drugs. Reports of variable anthelmintic efficacy in field trials, especially against hookworms and whipworms, may be due to inadequate performance of tests designed to measure drug effects, sub-standard (non-GMP) drug supplies, or inherent differences in efficacy of the regimen or sensitivity of the parasite population. The first step toward resolving these variables is to develop and promulgate a validated SOP for routine use in efficacy studies.
2. Multiple-day regimens of benzimidazole and nAChR anthelmintics provide better efficacy than single-dose courses against hookworms and whipworms in monogastric animals, including humans. If single-dose regimens are to be maintained in MDA programs, strategies to enhance efficacy should be pursued. Such strategies could, in theory, include higher doses of benzimidazole anthelmintics or combinations of marketed anthelmintics.[12,23] Combinations of a benzimidazole and pyrantel are highly efficacious in single doses against the spectrum of GI nematodes in dogs[12] and may be of interest for human use as well.

3. Use of regimens that provide suboptimal efficacy, especially with benzimidazoles, may predispose parasite populations to resistance development. Experience in the veterinary arena has amply demonstrated the relative ease with which benzimidazole-resistant nematode populations may be selected, and that under-dosing (suboptimal efficacy) is a risk factor.[12] As MDA programs expand, attention must be paid to ensuring the continued efficacy of the therapies employed. There is already evidence that the relatively poor efficacy of benzimidazoles against whipworms may be associated with prevalence of a mutation found in benzimidazole-resistant trichostrongylid nematodes of ruminants.[22] Although benzimidazole resistance has not been confirmed in human parasites, ready selection for this trait in veterinary parasites in a variety of hosts under intense drug pressure suggests that concerns about the situation in humans are not unwarranted. This eventuality is a primary factor driving new anthelmintic discovery programs. Combination products that include agents with distinct mechanisms of action/resistance could be useful to retard the development of resistance; of particular interest in this regard are new drugs with new mechanisms.
4. A factor working against the introduction of new drugs for this indication is the very low cost of available anthelmintics for human use for GI nematodes, complicated further by broad-ranging donation programs for albendazole and mebendazole for use in children. In the veterinary arena, anthelmintic price is determined by market factors, including the perceived value of the product in comparison with competitor products. However, the availability of generic, low-cost products establishes a price ceiling even for new drugs. Market factors are not a primary determinant of cost-setting for human anthelmintics, as the primary distribution outlet is organized MDA campaigns in which the end-users do not pay for the treatment. In such a situation, cost-of-goods is balanced against efficacy and convenience. A less efficacious but cheaper (or free) product may be chosen for use, as the decision may be to clear some worms from many people instead of all worms from fewer people. This choice may predispose parasite populations to evolve resistance, threatening prospects for long-term control with available drugs. Establishing a minimally acceptable standard of efficacy for drugs used in MDA campaigns should be a priority for research to inform funders and disease-endemic countries about how to best obtain sustainable outcomes of worm control efforts.

Filarial nematodes, unlike species that infect the GI tract, are spread by vectors and reside as adults in internal tissue or compartments. Filariases of humans include lymphatic filariasis (LF) or elephantiasis,[24] caused by *Wuchereria bancrofti* and *Brugia malayi* (with rarer infections by *B. timori*), and river blindness or onchocerciasis,[25] caused by *Onchocerca volvulus*. These infections affect well over 100 000 000 people and have been the focus of intense MDA campaigns, which have markedly reduced transmission and pathology.

Filariae that cause LF reside as adults in the lymphatic system of the human host.[24,25] Parasite- and/or immune-mediated changes in lymph vessel function and lymph flow lead to oedema in affected tissues, accompanied eventually by swelling, secondary infection, and loss of function. Microfilariae (equivalent to advanced embryo or first-stage larvae) are released in large numbers from fertile females, circulate in the blood (typically found there only at night, when mosquitoes feed), and are acquired by mosquitoes in the genus *Aedes* during a blood meal. Development to the infectious L3 stage occurs in the mosquito prior to delivery to a new host during a subsequent blood meal. The adult stages, the host reaction to them, and secondary infections in affected oedematous areas are responsible for the pathology.

Onchocerciasis is transmitted by black flies in the genus *Simulium*, which breed in fast-flowing water. Adult *O. volvulus* parasites live in nodules under the skin and in deeper locations. Adult stages cause little overt pathology. Microfilariae released from fertile females inhabit the skin and cause little damage on their own; it is the host response to microfilariae which leads to skin pathology and blindness due to immune-mediate retinal damage.[25]

Diagnosis of both diseases was historically based on detection of microfilariae in either blood samples (for LF) or skin snips (onchocerciasis). Tests have been developed to detect circulating parasite antigens in LF, and this has proven to be a very useful method for mapping epidemiology of these infections.[26] The correlation of antigen level with worm burden has not been established, and the use of this parameter as a biomarker to compare chemotherapeutic agents or regimens on a quantitative basis remains to be established. Adult viability can be roughly monitored through ultrasonographic imaging of worms in lymph vessels ('worm nests'); the movement of living macrofilariae is termed the 'worm dance'.[27] Although useful, it is not clear if this assay is sufficiently quantitative to inform clinical trials of new drugs or improved regimens of existing agents.

Diagnostic tests beyond skin snips have yet to be validated for onchocerciasis.[28] Skin snips are painful and their routine use has been discouraged. Effects of drugs on microfilarial burdens still requires direct measures through snips, and adult viability is assessed by histological examination of worms in nodules removed surgically; although dead worms can be detected from live ones, this is not an easy analysis. A biomarker that was highly correlated with microfilarial or adult worm burdens would have considerable benefit for drug trials.[12]

Current chemotherapy for human filariases is based on the use of microfilaricidal drugs[15,16,27–35] (Figure 11.2), including diethylcarbamazine (DEC: **8**) and ivermectin (**7**). The precise mechanism of action of DEC remains obscure, but appears to require the host immune response.[12] DEC is used only for LF, as it causes severe side effects related to the immune-mediated killing of *O. volvulus* microfilariae in the skin (the so-called Mazzotti reaction). It provides a profound and long-lasting suppression of microfilariae and can be given on a yearly or twice-yearly schedule to prevent transmission through this action. MDA campaigns that employ DEC administer the drug along with

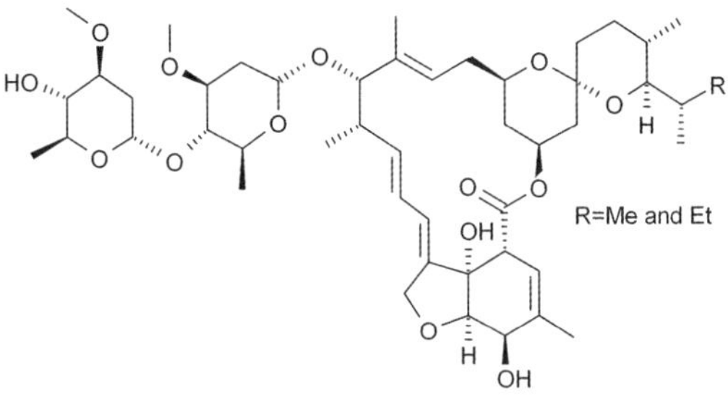

7, Ivermectin

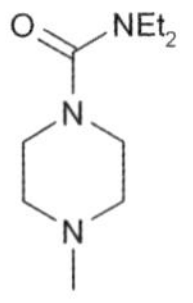

8, Diethylcarbamazine

Figure 11.2 Current drugs against filarial nematodes.

albendazole (**2**) (donated by GlaxoSmithKline for this purpose). The use of the combination provides some additional efficacy against LF, though the effect is not marked.[12,23] Instead, its activity against GI nematodes, which are frequently present as co-infections, may provide a boost to compliance and community acceptance of the MDA strategy.[12] While adulticidal activity of DEC can be demonstrated in some regimens, the consequent or co-incident immune response results in painful reactions to the dying worms, and this indication is not clinically pursued.

Macrocyclic lactone anthelmintics primarily act by opening glutamate-gated chloride channels in invertebrates, and the spectrum of action of these drugs coincides phylogenetically with the presence of such channels.[36] This class of ligand-gated ion channels is not found in mammals or other vertebrates. Although ivermectin is a potent activator of GABA-gated chloride channels, these proteins are only found in the mammalian CNS, and ivermectin is excluded from that compartment by the MDR-1 p-glycoprotein in the blood–brain barrier.[37] Ivermectin is used alone for onchocerciasis and is combined with albendazole for use in LF control in areas in which *O. volvulus* is present (to avoid DEC toxicity). As for DEC, yearly or twice-yearly dosing is sufficient to profoundly suppress microfilarial levels for months; the mechanism underlying this effect is not well understood for either drug.[12] The addition of albendazole does not seem to markedly enhance the antifilarial profile of ivermectin,[23] though it may be beneficial in other ways as noted above. Ivermectin does not have useful efficacy against adult filariae in humans, though repeated doses over multiple years may accelerate worm death.[28]

Although ivermectin is exceptionally safe for human use, worrisome severe adverse events, including fatalities, have been observed in patients co-infected with *O. volvulus* and another filarial parasite, *Loa loa*. The basis for this unanticipated drug-induced pathology is not understood.[38] Although the risk of a serious adverse event is correlated with the level of *L. loa* microfilariae in the blood,[39] it is not routinely possible to test people for this parameter prior to ivermectin administration. As a result, ivermectin distribution has been restricted in areas of Africa in which these two infections overlap.

The discovery that symbiotic bacteria in the genus *Wolbachia* are essential for fertility and viability in many species of filariae, including those that cause LF and onchocerciasis, opened a new avenue for chemotherapy.[40] Prolonged courses of several antibiotics, most prominently doxycycline (daily doses of 200–400 mg for 4–6 weeks) sterilize and kill adult filariids in animal models and in humans.[40–42] Efforts to find new antibiotics or dosage regimens that would shorten the duration of therapy to <2 weeks (a protocol more compatible with large-scale deployment in an MDA campaign) are underway.[43] Nonetheless, this breakthrough offers the immediate potential to treat people in at least some areas in which loiasis is found or in which possible ivermectin resistance has emerged.[41,42]

11.2.2 Areas of Concern for Filarial Nematodes

1. As long as resistance to the macrocyclic lactones does not appear, a new microfilaricide is not a priority. Ivermectin suppresses microfilarial survival and production for months, far longer than its ~35 h half-life of elimination from humans.[12] The basis for this effect is not understood, but it is difficult to envision a discovery strategy that could select for another, distinct kind of ultra-long-lasting compound in a high-throughput screen. However, concerns about ivermectin-resistant populations of *O. volvulus* have been raised,[44] and this phenotype has recently been reported in populations of the related filariid parasite, the dog heartworm *Dirofilaria immitis*.[45] If ivermectin resistance develops and spreads in human filariae, control strategies would require the immediate availability of a substitute compatible with the very limited dosing schedules favored for MDA operations; none is apparent.
2. A macrofilaricidal drug that can be given in one or two doses, with minimal adverse effects, would be a highly valuable addition to filariasis control programs.[12,15,33,34] As noted below, the absence of animal hosts other than humans for *W. bancrofti* and *O. volvulus* has been a significant impediment to macrofilaricide discovery.
3. A non-invasive biomarker temporally correlated with worm burdens (quickly and quantitatively responsive to changes) would be of great assistance in performing trials to compare new regimens of existing drugs and for evaluating candidate filaricides. Currently, trials require periods of 1–2 years for quantification of adulticidal efficacy, a serious impediment to dose optimization studies for registration.[12] A surrogate biomarker that was temporally and quantitatively correlated with worm burdens would improve the development to a considerable extent.

11.3 Trematodes

Trematode infections of humans are primarily due to several species in the genus *Schistosoma*, most prominently *S. mansoni, S. japonicum* (both of which reside as adults in the mesenteric vasculature around the intestinal tract) and *S. haematobium*, which reside as adults in the blood vessels around the bladder. Approximately 200 000 000 people are thought to be infected with schistosomes,[46,47] primarily in Africa. According to WHO/TDR, schistosomiasis is second only to malaria as a parasitic disease of serious public health importance in subtropical and tropical Africa. These parasites have as an intermediate host several species of aquatic snails. Humans are the definitive hosts. Schistosomiasis is a chronic disease; adult males and females *in copula* live for at least a decade in the bloodstream and produce hundreds to thousands of eggs per day. The host response to eggs results in their transit to the lumen of the gut or the bladder from whence they are shed into the environment. Eggs that reach a body of water following elimination hatch, and release larvae that infect susceptible snails. Following development and multiplication in the snail, infective cercariae are shed into the water to infect humans through contact with skin.

The host immune response also generates the granulomatous sequelae that leads to the pathology of the disease. Symptoms and consequences include fever, fatigue, hepatosplenomegaly, hematemesis, oesophageal varices, pulmonary hypertension, and hematuria and bladder cancer (the latter from *S. haematobium*). As with other human helminthiases, the consequences of chronic schistosomiasis are insidious and somewhat difficult to quantify, but are of undeniable severity in people who harbour this parasite.[48]

Schistosomiasis is a disease of chronic morbidity as opposed to acute mortality, especially since the advent of MDA campaigns with praziquantel (**9**, Figure 11.3). The precise mechanism of action of this drug remains enigmatic; there is evidence that it opens a type of voltage-gated Ca^{2+} channel in schistosome cell membranes,[49] but the situation is thought to be more complex.[50,51] Efficacy of the drug appears to also be dependent on the immune system of the host;[50,51] how this variable contributes to the attainment of efficacy in humans has not been fully resolved. Given in single doses of 40 or 60 mg/kg, praziquantel is efficacious but not routinely curative for existing infections. Reducing the number of adult worms in a human reduces the numbers of eggs shed, minimizing the development of serious pathological changes. As immunity to re-infection following chemotherapy is not profound, and exposure to contaminated water leads to frequent re-infection after chemotherapy, the current MDA programs have been very successful in reducing schistosome-associated disease and mortality, but have not led to broad-scale elimination of the parasites.[46,47]

Of some interest is the finding that the antischistosomal activity of the drug is largely restricted to the L-isomer, although the drug is only available as the diastereoisomer.[50,51] It is possible that a preparation containing only the active isomer would achieve efficacy with fewer issues of compliance and side effects and lower doses. From this perspective, the cost of goods of the pure L-isomer would have to be compared to the therapeutic and programmatic benefits of having the ability to widely distribute a cheap (0.10 USD/dose) and acceptable product. It is essential to keep in mind that choice of an antiparasitic drug for an MDA campaign is not consumer-driven, but instead is made by governmental or quasi-governmental agencies on the basis of relative cost, efficacy and compliance.

Diagnosis of schistosomiasis is made by detecting eggs in faecal or urine samples. There is an uncertain relationship between egg output and adult worm burden in humans, and the techniques available are more useful for discriminating infected from non-infected people rather than providing a precise estimate of the intensity of infection.

Recognizing that reliance on a single drug administered in MDA campaigns in a regimen that does not provide full efficacy in all patients is not an optimal strategy for eradication, there has been continuing interest, though at an insufficient level, in discovering new drugs for schistosomiasis.[52–55] In addition to incomplete efficacy and the threat of resistance development (which has not yet been confirmed in human populations),[46,50] praziquantel is not always well tolerated in humans; it is typically administered with fatty food to minimize gastrointestinal disturbances and improve compliance in MDA programs. Finally, praziquantel lacks activity against immature stages of the parasite in humans, compromising to some degree the benefits of annual or semi-annual treatments in areas of high transmission. The only other drug available for schistosomiasis is oxamniquine (**10**), which is manufactured in a very limited amount and has activity only against *S. mansoni*.[52–55] It is not recommended for use in pregnancy and is not without side effects.

Of increasing concern is a group of parasitic trematodes that are acquired primarily though food, including *Fasciola hepatica, Clonorchis sinensis* and parasites in the genera *Paragonimus* and *Opisthorchis*. Humans are infected by consuming unwashed vegetables, undercooked fish, or aquatic invertebrates; these parasites can reproduce in humans but are more commonly found in

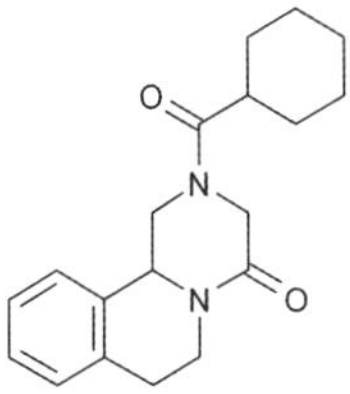

9, Praziquantel

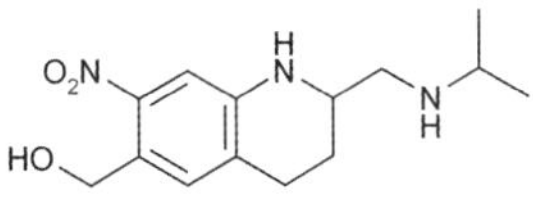

10, Oxamniquine

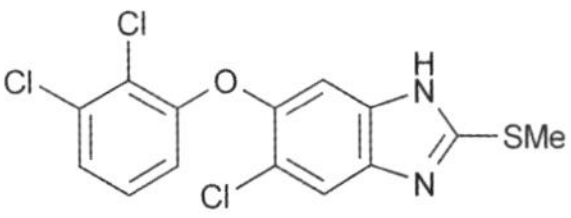

11, Triclabendazole

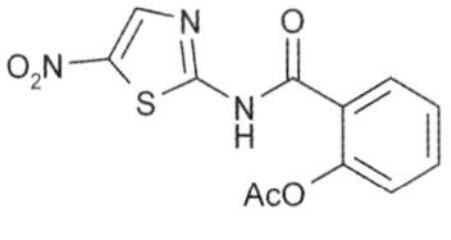

12, Nitazoxanide

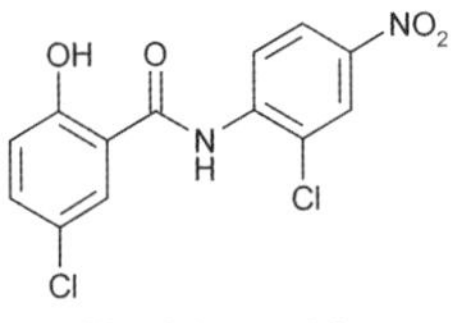

13, Niclosamide

Figure 11.3 Current drugs against trematodes and cestodes.

other mammals. Incidence of infection with each parasite can be as high as tens of millions people.[56] Some species reside in the liver, while others inhabit the lungs as adults. Pathology can range from severe hepatotoxicity due to *F. hepatica*, to liver cancer due to chronic infection with *C. sinensis* to pneumonia-like symptoms or chronic cough in lung infections. As the prevalence of the food-borne trematodiases was not appreciated until recently, relatively little effort has been made to identify optimal chemotherapies for them. Two drugs are currently in use: praziquantel (**9**) is effective against all food-borne trematodes except *F. hepatica*; it is given as a single dose of 40 mg/kg or as a dose of 25 mg/kg per day for 3 days.[56] The pharmacological basis for the inability of this broad-spectrum drug, which is highly active against most trematodes and cestodes, to control infections with *F. hepatica* is not understood.

The drug of choice in veterinary practice for fasciolosis in ruminants outside of the USA is triclabendazole[56] (**11**), a benzimidazole derivative which has no activity against nematodes. It is given as a single 10 mg/kg dose after a meal, but is registered in only a few countries for this application. Treatment can be repeated if needed. This drug may act through tubulin, like other benzimidazoles, but conclusive proof of this possibility has not been presented. Triclabendazole is not known to have useful activity against schistosomes, but is efficacious against paragonimiasis.[56] Extended courses of other drugs, including mebendazole and nitazoxanide (**12**), may have utility against *F. hepatica*, but offer no benefit over triclabendazole.[56] It is important to note that cases of triclabendazole-resistant *F. hepatica* are becoming of concern,[57] but have not been reported in humans.

11.3.1 Areas of Concern

1. Reliance on a single drug is unwise for intensive MDA campaigns. Although resistance to praziquantel has not been confirmed in humans, increasing drug pressure may give rise to resistant populations. Thus, discovering and developing viable alternatives to this drug should be a high priority. In addition, praziquantel lacks activity against immature worms and has some side effects that limit compliance in community-based MDA programs.
2. A cost-effective synthesis of the L-isomer of praziquantel could provide better compliance if its use reduced the incidence of side effects. However, the advantages would likely have to be substantial compared to the current product to justify the costs of registration and probable greater cost-of-goods of the pure L-isomer.
3. As for the nematodes, diagnostic methods available for trematode infections typically measure egg output as a surrogate for adult worm burdens. A biomarker that was highly correlated with adult worm numbers and was temporally responsive to changes in worm number would be a valuable tool for clinical trials of new drugs or regimens.

11.4 Cestodes

Human infections with tapeworms include those in which adult cestodes inhabit the GI tract and more serious infections in which larval tapeworms infect internal tissues (cysticercosis). Several cestode species can infect the human GI tract as adults, but pathology due to these infections is limited. Patients may request treatment when they notice segments being shed in the stool; praziquantel (**9**) is the routine treatment (5–10 mg/kg as a single dose).[58] Niclosamide (**13**) at a dose of 2 g is a second choice for treatment;[58] nitazoxanide (**12**) is also efficacious against intestinal cestodes.[59] Niclosamide is thought to uncouple oxidative phosphorylation in tapeworm mitochondria, an effect perhaps mediated by the mobile proton in the amide moiety. Niclosamide is poorly absorbed from the GI tract, which may make it safer for the host.

Treatment of cysticercosis is more challenging and much more important. Cysts containing the larvae of *Taenia solium* are thought to be present in as many as 50 000 000 people (and increasing in prevalence).[58] Humans are definitive hosts for the adult tapeworms, and infectious eggs are shed in tapeworm segments present in faeces. Typically, eggs are ingested by pigs exposed to untreated human faeces. The eggs develop into tissue cysts in the pigs, and human consumption of undercooked pork results in larval development into adult worms in the GI tract. Pathological problems develop when humans ingest *T. solium* eggs, which develop as tissue cysts in the new host. Of particular significance is neurocysticercosis, in which tapeworm cysts infect the CNS. Such infections are a leading cause of epilepsy in many developing countries with large swine populations and are increasingly found in countries with immigrant populations from disease-endemic countries.[58] Treatment of neurocysticercosis is complicated by the possibility of pathology due to host immune responses targeted to dead or dying worms, which may exacerbate symptoms such as seizures. Prolonged courses of albendazole (**2**) (400 mg twice daily for at least one week)[17] or praziquantel (**9**) (50–100 mg/kg per day in 3 divided doses for 30 days) are coupled with either dexamethasone or prednisone to limit immune-associated adverse events.[58]

Infections with larval stages of *Echinococcus spp.*, though much less common, are more life-threatening and more difficult to treat.[60] Parasites in this genus normally cycle between canids and intermediate hosts such as sheep. Humans become accidentally infected by acquiring eggs from household dogs that carry the adult tapeworms. Surgical removal is advocated for most cases, if possible; chronic treatment with albendazole (**2**) suppresses proliferation (and thus mortality) but is not fully lethal to the parasites.

Cysticercosis in humans in developed counties was largely eradicated by hygiene campaigns focused on thorough cooking of pork and on changes in swine production that minimized acquisition of parasites by pigs. The diminished profile of cestode infections has severely limited research on new drugs for this indication, and only a few are on the horizon.

11.4.1 Areas of Concern

1. The most prevalent problem in human cestode infections is cysticercosis involving *Taenia solium*, especially neurocysticercosis. Diagnosis of infection is often made after the development of epilepsy, often by imaging technology. Serological tests using cerebrospinal fluid or serum are available[61,62] but are not routinely applied for field surveys. It is not yet clear if chemotherapeutic intervention in pre-symptomatic cases can prevent the development of epilepsy, but more aggressive diagnostic efforts could enable such studies.
2. Efforts to diagnose and treat infected pigs could interrupt the life cycle in many areas (as was done in North America); of particular interest in this regard is the availability of a highly effective recombinant vaccine that can essentially prevent infection of pigs by *T. solium* eggs.[63] However, because the infection does not significantly reduce swine production, there is little economic incentive for poor farmers to vaccinate their herds in the interest of public health. The tapeworm vaccine represents the best (perhaps sole) example of a recombinant helminth vaccine with real potential to lead to disease eradication.

11.5 Late-stage Anthelmintic Leads

De novo discovery of new anthelmintics is primarily done in the context of the animal health industry (see below). Compounds arising from veterinary medicine have provided the vast majority of human anthelmintics at least since the introduction of thiabendazole. Such drugs, in addition to compounds approved for other indications in humans, can be considered as late-stage leads for possible registration for human anthelmintic use. Several recent reviews are available on such compounds,[15,16,31,52–55,63–69] which are therefore only briefly introduced here (Figure 11.4).

11.5.1 Emodepside

PF1022A **(14)** was patented some 20 years ago as an anthelmintic based on activity against *Ascaridia galli* in infected chickens.[70,71] The molecule is quite potent against a variety of GI nematodes in animal hosts, especially in monogastric animals; it is much less potent in ruminants.[72,73] It is worth noting that PF1022A was without acute toxicity in mice at quite high doses,[72,73] suggesting that it may have some potential for use in human GI nematodes. However, the (bis)-morpholino derivative emodepside **(15)** was chosen for development based on marked improvement in potency in livestock.[71,74] However, this drug was first marketed for use in companion animal GI nematode infections, as a topical treatment for cats and in an oral dosage form

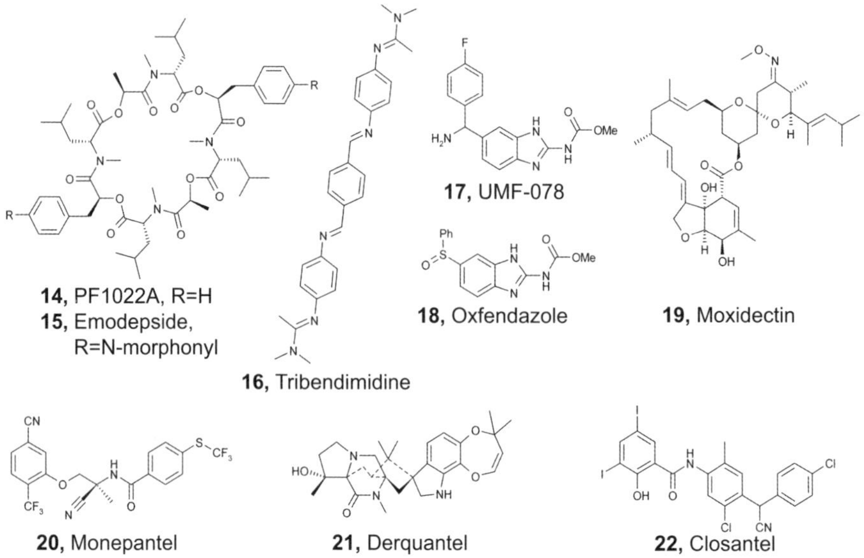

Figure 11.4 Late-stage leads against nematodes.

for dogs. Emodepside in single doses of 1–2 mg/kg (sold as a combination with praziquantel) is effective against ascarids, hookworms and whipworms in dogs.[75] Topical administration of ∼5 mg/kg to cats is effective against ascarids and hookworms.[76] These products suggest that emodepside could have an acceptable therapeutic index in humans and may exhibit the spectrum of action required for an anthelmintic for STH infections in humans. The possibility that PF1022A could be if use in this regard, given its lower cost of goods, loss of patent protection and reported potency and efficacy against relevant parasite species, should not be neglected.

The total synthesis of PF1022A has been reported,[71] but whether synthesis is compatible with an acceptable cost-of-goods is not clear. Many semi-synthetic derivatives have been reported.[71,74] None has evident advantages over the prototype or emodepside for consideration for evaluation in humans. The activity of these compounds against a variety of GI parasites in monogastric animals suggests that they could be usefully applied in human medicine. It is intriguing in this regard to consider whether cost savings with PF1022A, which is off-patent, could outweigh the potency disadvantage compared to emodepside.

These compounds, especially emodepside, also have intriguing antifilarial activity.[77] Of note is the exceptional potency of emodepside *in vitro* against adult *Onchocerca spp.*;[78] this compound clearly warrants evaluation in relevant animal models of onchocerciasis. It is interesting that potency was much lower against an LF parasite, *B. malayi*, in similar assays; the basis for this pharmacological discrepancy is unknown.

The target for emodepside in nematodes is a member of the Ca^{2+}- and voltage-gated K^{+} channel family.[79] In *C. elegans*, mutations in the *slo-1* gene generate high-level resistance to the drug,[80] supporting the assignment of the encoded channel as the primary drug target. A targeted discovery program based on this channel could have merit for the human situation; new templates with the same mechanism might provide improved pharmacological profiles and/or a lower cost-of-goods.

11.5.2 Tribendimidine

Tribendimidine (**16**) is under development as an anthelmintic for gastrointestinal nematodes and other helminths.[81] This drug is an agonist of nematode nAChRs in nematodes.[82] It is reported to be metabolized to deacylated amidantel, which also is a potent nematocide.[83] The drug exhibits efficacy against the range of human GI nematodes in the same range as benzimidazoles in single oral doses of 400 mg, with a good safety profile in field trials.[81] Intriguingly, it is also active against cestodes and a tissue trematode, *Opisthorchis viverrini*.[15,56,81] This spectrum is not typical of nicotinic agonists and illustrates that the anthelmintic pharmacology of this compound remains to be fully elucidated. If indeed the activity is solely due to actions at helminth nAChRs, concrete advantages in cost, safety or efficacy would have to be gained over pyrantel and levamisole to justify incorporation into therapeutic use.

11.5.3 Flubendazole

Flubendazole (**3**) is a potential macrofilaricide. Although registered for human use for gastrointestinal nematodes, as well as for veterinary anthelmintic use, the formulation used allows for very low oral bioavailability. When given parenterally, however, flubendazole is a highly active macrofilaricide in several models,[84] and was efficacious in a single trial against *O. volvulus* in humans.[85] Unfortunately, the study used intramuscular injection for delivery (5 weekly doses of 750 mg), which caused serious injection site reactions. Work on this drug as a macrofilaricide was subsequently halted. The formulation used was not disclosed, precluding a conclusion about whether the drug or the vehicle was responsible for the adverse events. Parenteral administration is thought to create a drug depot, providing extended parasite exposure to low drug concentrations.

In light of the apparent intolerability of parenteral flubendazole and its very low oral bioavailability in standard vehicles, chemical investment was dedicated to the production of more tolerable/bioavailable pro-drugs of flubendazole. Emerging from this effort was UMF-078 (**17**), which exhibited high macrofilaricidal activity in a single *i.m.* dose of 150 mg/kg in cows infected with the nodular parasite *Onchocerca ochengi*.[86] Pharmacokinetic studies suggest that this drug, while also orally bioavailable, is converted to flubendazole to a very small extent ($<10\%$) and exhibits neurotoxicity after chronic or multiple dosing in dogs and cattle.[86] Whether this effect is due to the parent drug or to flubendazole has not been determined; chronic dosing with albendazole for neurocysticercosis in humans has not generated reports of neurotoxicity,[17] suggesting that this is not a general characteristic of the benzimidazole anthelmintics. This pattern has, however, led to cessation of work with UMF-078.

More modern approaches to attaining oral bioavailability with pharmaceutically recalcitrant compounds may be of interest in this case. Flubendazole in a hydroxypropyl-β-cyclodextrin vehicle was effective against tissue cysts of the tapeworm *Echinococcus granulosus* in a mouse model following oral dosing.[87] It is not yet known if better oral bioavailability could provide exposure adequate for macrofilaricidal activity in a regimen that is compatible with field use. It is relevant to note that micronized flubendazole given in the diet for two weeks was markedly macrofilaricidal in rats infected with *Brugia pahangi* without apparent toxicity.[88] Further work on optimizing the formulation and regimen would seem to be warranted. Although flubendazole is licensed for human use, the systemic exposure needed for macrofilaricidal activity would demand a new registration dossier, including toxicological and manufacturing studies. It would also be necessary to determine if a macrofilaricidal regimen also kills microfilariae; to date, the drug has not shown microfilaricidal activity in macrofilaricidal regimens. Thus, flubendazole could be useful for end-stage elimination programs and in loaisis areas.

The commercial success of benzimidazoles in veterinary medicine led to the testing of a large number of analogs for anthelmintic activity. No published data show that any has advantages over flubendazole as a macrofilaricide. High-dose, prolonged therapy with albendazole[17] or oxfendazole (**18**)[56,89,90] controls some tissue helminths; whether either is macrofilaricidal in similar regimens has not been reported, but is worth investigating in appropriate animal models (see below).

11.5.4 Moxidectin

Prolonged exposure to macrocyclic lactone endectocides causes a slow loss of viability of *D. immitis* adults,[91] but conclusive evidence for high efficacy of ivermectin against adult filariae in humans has not been reported.[35] Ivermectin is generally not macrofilaricidal in the animal models used for antifilarial screening. Recent interest has focused on moxidectin (milbemycin) (**19**), another macrocyclic lactone, as a filaricide for onchocerciasis. Moxidectin has a much longer half-life than ivermectin[92] and is a poor substrate for P-glycoprotein drug transporters,[93] providing considerably longer exposure than is attainable with a single dose of IVM. Moxidectin is currently in field trials as a possible macrofilaricide in onchocerciasis based on this profile and unpublished studies showing efficacy against adult filariae in an animal model.[94] However, it is important to recognize that more frequent and intensive dosing of humans with ivermectin, partially mimicking the greater exposure to a macrocyclic lactone anthelmintic achieved with moxidectin, was not macrofilaricidal,[95] and that moxidectin was not macrofilaricidal in the cattle-*Onchocerca ochengi* model.[96] Results of nodule examinations in the human trials are expected in 2011. It is not clear if development of the drug to registration will continue if macrofilaricidal activity is not found.

However, resistance to ivermectin in *O. volvulus* may also be a driver for adoption of moxidectin. A phenotype has recently been reported which can be interpreted as 'early repopulation of skin by microfilariae'. Reappearance of microfilariae in skin after a dose of ivermectin is not normally marked until 6 months,[35] whereas people in some villages in Ghana exhibited microfilarial repopulation of skin within 3 months of dosing.[44] Macrocyclic lactone-resistant populations of *D. immitis* may have appeared in the USA.[45] In veterinary parasites, resistance to one drug in this class is associated with resistance to other macrocyclic lactones. However, it is possible that therapeutic doses of moxidectin may still provide extended suppression of microfilarial loads in skin even in ivermectin-resistant cases.

Severe adverse events in high-microfilaremia loaisis patients after therapeutic doses of ivermectin have compromised elimination onchocerciasis programs, as noted. The possibility that moxidectin would be safe in loaisis patients could also promote its registration for human use. However, experimental models that can reliably predict risk of such events have not been defined; deciding whether moxidectin could be used in such cases may require testing in co-infected patients, an unappealing prospect.

11.5.5 Monepantel

Monepantel (**20**) is the commercialized prototype of the new anthelmintic class of aminoacetonitrile derivatives, or AADs.[97,98] These drugs act on a type of nAChR that is insensitive to the classic cholinergic anthelmintics. Monepantel has an excellent safety profile and is highly efficacious against gastrointestinal and lung nematodes in ruminants.[99] Their activity against filariae has not yet been reported, but close analogs of this receptor are not known to be present in this clade.[100] Monepantel or another AAD may be of use against gastro-intestinal nematodes of humans, though the activity of these drugs against the kinds of worms that are most important in humans is not clear. If efficacy can be attained with a reasonable regimen, monepantel could represent a valuable addition to the list of available drugs, especially if benzimidazole resistance emerges.

11.5.6 Derquantel

Derquantel (**21**), 2-desoxoparaherquamide, is the licensed prototype of the paraherquamide/marcfortine class of anthelmintics.[101] These drugs act as antagonists at nAChRs.[102] These anthelmintics exhibit potent activity against trichostrongylid nematodes of ruminants.[103] The activity of derquantel against nematodes found in the human GI tract has not been reported, and its potential as a macrofilaricide is unclear. As for monepantel, the potential availability of a new anthelmintic, already licensed for veterinary use, bodes well for the future expansion of the chemotherapeutic options for the control of nematodes in humans. However, additional work is needed to determine if these new veterinary drugs will provide the spectrum of action, safety profile and cost-of-goods to be of use in MDA campaigns for human nematodiases.

11.5.7 *Bacillus Thuringiensis* (Bt) toxins

Proteins derived from isolates of the bacteria *Bacillus thuringiensis* have selective toxicity for invertebrates based on disruption of gut function,[104] and are widely used as insecticides in agriculture, both as products for application and in transgenic crops. Certain varieties of these proteins have potent activity against nematodes, including activity *in vivo*.[105] Acting through a novel mechanism and already (as a class) allowed in food products, it may be possible to develop a Bt toxin for use against human GI nematodes. Much more work will be required to verify spectrum against the major parasites of humans, but one could anticipate a favorable cost-of-goods and safety profile for a nematocidal Bt toxin in humans. Of interest in this regard is evidence for a synergistic interaction of levamisole or tribendimidine and a Bt toxin in a model nematode-host system.[106] This finding offers an intriguing way forward in the development of novel anthelmintic combinations for GI nematodes in humans.

11.5.8 Closantel

Closantel (**22**) is member of the salicylanilide class of veterinary anthelmintics. These drugs have a peculiar and somewhat limited spectrum of activity.[107] Closantel is used for the control of *Haemonchus contortus* in small ruminants and is also licensed for use against the liver fluke, *F. hepatica*. It is very highly protein-bound, which may account for its selective activity against the blood-feeding *H. contortus*. These drugs are thought to act at the level of helminth mitochondria, collapsing proton gradients and so poisoning ATP synthesis.[107]

A recent screen for chitinase inhibitors as leads for antifilarial drugs identified closantel as a potent inhibitor of this enzyme.[108] However, it was not very potent in assays of filarial moulting in culture and was not active in animal models of filariasis. Whether chitinase is a valid target for a new micro- or macrofilaricide remains to be proven. At this point, a new antifilarial drug based on the inhibition of moulting is of uncertain value, as it would have to maintain efficacious blood levels for long periods in order to intercept incoming larvae. Cuticle synthesis itself could be a viable target,[109] but too little is understood about the dynamics of cuticle synthesis and assembly in adult filariids to validate any of the enzymes which may play a role in this physiology.

11.5.9 Schistosomes

Praziquantel is manufactured as a mixture of two stereoisomers, only one of which is a potent schistosomicide. Attention has been devoted to developing a production scheme that would generate only the bioactive L-isomer.[50,51] However, a major factor enabling the spread of MDA campaigns for schistosomiasis control has been the exceptionally low price of praziquantel (0.10 USD per dose). The requirement for a very low cost-of-goods limits options for the manufacture of L-praziquantel through a process that would provide an inexpensive product. Some additional work has also been devoted to the synthesis of new praziquantel analogs that might have improved activity against juvenile stages,[110] but none appears to offer clear benefits over the approved drug.

New potential leads for schistosomicides have emerged from the antimalarial area. Artemisinin derivatives (**23a–c**), including the synthetic trioxolanes (**24a–b**), show significant activity against these helminth parasites, including in humans[15,53,54,111] (Figure 11.5). Haemin or iron, in the blood or in the parasite, possibly derived from the breakdown of ingested haemoglobin, has been implicated in the action of these drugs in both schistosomes and malaria parasites, although a precise understanding of their mechanisms remains to be derived. However, a major and so far unresolved concern relates to the adoption of essential and last-line antimalarial drugs for use against a different parasite in areas in which they co-occur. It will be necessary to show that such drugs can be used safely in a way that prevents the development of resistance of malaria parasites to these drugs. As noted above, cost-of-goods issues also favor the continuing use of praziquantel. Whether the goals of malaria and schistosome control can be integrated to allow registration and use of artemisinin derivates for the latter indication remains an open question. Certainly, it is valuable to have an already approved drug available in case praziquantel resistance does arise.

Mefloquine (**25**), a structurally unrelated antimalarial drug, also has some activity in human schistosomiasis,[111] although the antimalarial dose regimen does not produce high efficacy on its own. Combination of mefloquine and artesunate was highly efficacious and is active against juvenile worms.[111] Although the combination could allay concerns about the development of artemisinin resistance in co-exposed malaria parasite populations, it should be noted that administration of these drugs in the antimalarial regimen (over 3 days) is not obviously compatible with MDA use, and there are concerns

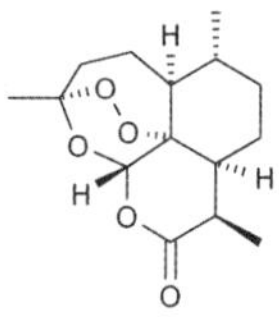

23a, Artemisinin

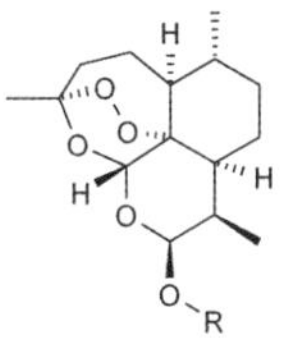

23b, Artemether, R=Me
23c, Artesunate, R=C(=O)(CH2)2CO2H

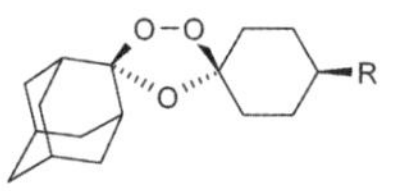

24a, OZ78, R=CH2CO2H
24b, OZ288, R=Ph(4-OH)

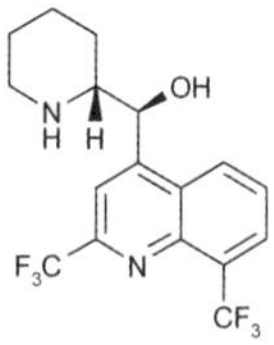

25, Mefloquine

Figure 11.5 Late-stage leads against trematodes and cestodes.

about the side effect profile of mefloquine, especially with regard to chronic CNS toxicity;[112] concerns are greater for the higher doses used in curative therapy compared to those used for prophylaxis. Nonetheless, and as noted above, it is reassuring to know effective alternatives are available should praziquantel resistance arise. While these drugs do not appear to be ideally suited for deployment in MDA campaigns, they could be used to cure praziquantel-resistant cases and to limit the spread of such parasites from the area of origin, if it is detected in time. There is still a need for a praziquantel substitute with better single-dose efficacy against immature worms and a low cost-of-goods. It should be noted that mefloquine is also effective against other flukes in animal models,[113] but again, it is not clear that it is worth registration of the drug for those indications unless resistance problems surface with praziquantel.

11.5.10 Cestodes

Nitazoxanide (**12**), originally patented as a cestocide,[59] is approved for use against some GI nematodes and protozoa;[114,115] it is also active in single doses of 25–50 mg/kg against intestinal tapeworm infections in humans.[59] It also has some activity in animal models of echinococcosis,[116] but its potential for the treatment of larval tapeworm infections (cysticercosis) in humans has not been demonstrated. Interest in the thiazolide template for this indication is evident from the publication of screening results.[116]

Although albendazole is routinely used for echinococcosis therapy in humans,[17] there is evidence in animal models that oxfendazole may be more efficacious against systemic helminths, including echinococcosis,[89,90] based on trials in infected sheep. Flubendazole (**3**) is efficacious against tissue echinococcosis in mice in a novel hydroxypropyl β-cyclodextrin vehicle.[87] The basis for variation in anthelmintic spectrum among the benzimidazoles is not completely understood, but must reflect differences in pharmacokinetic behavior or intrinsic affinity for tubulin in different helminth species among them.

11.6 New Anthelmintic Leads

Early-stage anthelmintic leads arising from focused discovery programs are primarily disclosed in the patent literature. Development of such drugs to registration is a long-term process which requires a substantial investment in multiple areas of pharmaceutical research, such as formulation, manufacturing, clinical pharmacology and toxicology. Outside the animal health industry, an investment strategy to support these costly and time-consuming studies is not readily apparent, though attention is being paid to novel partnership strategies that could propel drugs through development to registration for human use; these include potential partnerships that involve animal health companies.[66–69,117–119] However, some compounds and series warrant mention as interesting leads.

11.6.1 Monepantel Analogs

Since the discovery of the AAD class and the introduction of monepantel into the veterinary arena by Novartis AH, interest in this first new fully synthetic anthelmintic class in 25 years has grown. Pfizer claimed a closely related series bearing the peculiar pentafluorosulfanyl (SF_5) substituent.[120] For example, compound (**26a**) shows a minimum effective concentration (MEC) of 1 ug/ml in an *in vitro* assay against *Haemonchus contortus* L3 larvae (Figure 11.6). Although *in vivo* activity is not described, it can be anticipated, considering that SF_5 has been reported several times as a bioisostere of the trifluoromethyl (CF_3) group.

Replacement of monepantel phenoxy group by bicyclic cores has been reported by both Pfizer[121] and Merial.[122–124] Claimed activity against several nematode species orally in sheep was exemplified by compound (**26b**), with high efficacy at a dose of 1.5 mg/kg. It remains to be seen if any of the new compounds is better suited for use in humans than monepantel.

11.6.2 Closantel Analogs

Benzo[d]isoxazolyl benzamide derivatives have been described by Janssen.[125] These compounds can be assigned to the salicylanilide class and categorized as fused closantel analogues. Compound (**27**), for example, exhibits *in vivo* efficacy against a multi-resistant *H. contortus* strain orally in a jird model at a dose of 5 mg. Potential advantages over closantel have not been reported for veterinary or human medical indications.

11.6.3 Aminocyclohexanol Derivatives

Intervet has reported a new squeletal formula with anthelmintic activity. No information regarding mode of action has been revealed. However, similar *in vitro* MEC values were reported against a sensitive and a resistant (to

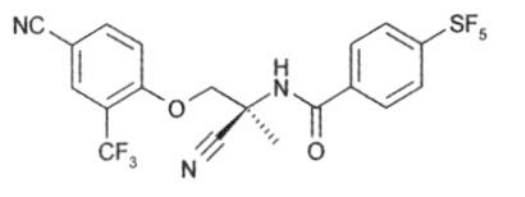

26a, Monepantel analogue

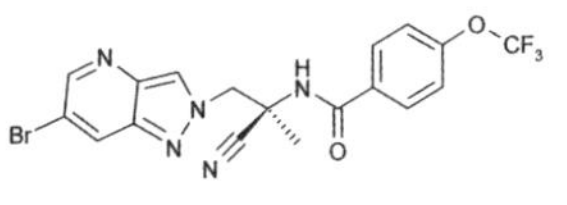

26b, AAD analogue

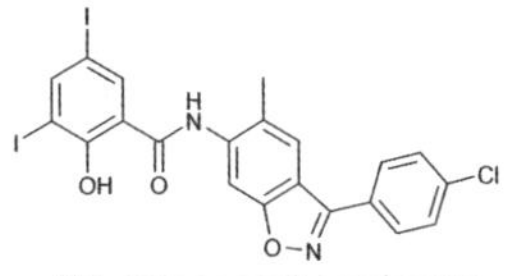

27, Closantel analogue

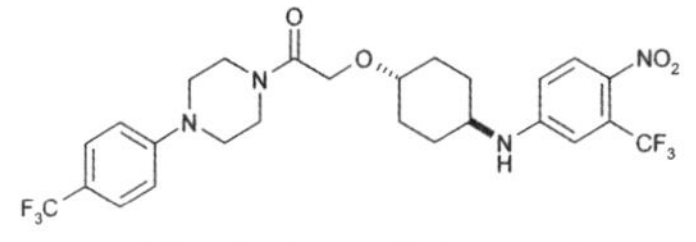

28a, Aminocyclohexanol derivative

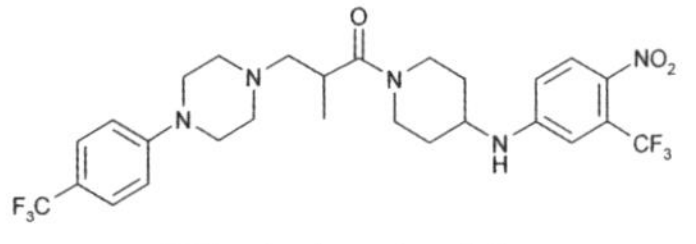

28b, Aminopiperidine derivative

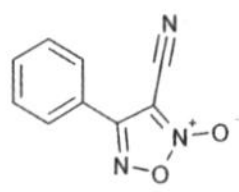

29, Substituted Furoxan

Figure 11.6 Early-stage anthelmintic leads.

benzimidazoles and ivermectin) strain of *H. contortus*. The general structure encompasses a phenylpiperazine subunit linked by various chains to an aryl moiety. Preferred chains contain a 4-aminocyclohexanol[126,127]- like moiety as in compound (**28a**), for which a 95% worm reduction of *H. contortus* orally in sheep at a dose of 5 mg/kg is described.

Another series in this class, containing a piperidine ring[128] in the linker, has been reported, but seems less potent than the previous class. For example, compound (**28b**) achieves only a 74% worm reduction against *H. contortus* orally in sheep at a dose of 10 mg/kg. Activity in either case against the types of nematode parasites important for humans has not been revealed.

11.6.4 Oxadiazole N-oxide Derivatives

The only report of *in vivo* anthelmintic efficacy outside the veterinary field involves the identification of oxadiazole *N*-oxide derivatives as new leads for the control of schistosomiasis.[129,130] The primary target of this chemotype is the parasite redox protein thioredoxin-glutathione reductase (TGR).[131] The activity of these compounds is associated with their ability to release nitric oxide (NO). Furoxan (**29**) exhibits *in vivo* efficacy against *S. mansoni* in a mouse model after five consecutive daily intraperitoneal administrations at 10 mg/kg. Latest efforts seem to be devoted to the search for analogues with better properties for oral delivery.[132]

11.7 Drug Discovery and Development: Pathways and Problems

Outside the animal health industry, investment in the discovery of new anthelmintics for people has been sporadic and sadly limited, compared to the magnitude of human helminthiases. As noted, with the exception of DEC (**8**), all drugs used for human helminth infections were derived from animal health discovery programs. Considering the origins of the advanced leads discussed above, this trend is likely to continue. Indeed, resistance to all available anthelmintics in the veterinary sector has re-energized discovery research in this area.[66–68,97,101,119,133] Discovery programs in this area benefit from ready access to target parasites for whole-organism screening, several convenient laboratory models for target parasite species, and facile testing in target host species; these advantages do not fully extend to discovery of drugs for human parasites.

The discovery stream for anthelmintics has historically focused on whole-organism assays, employing either target parasites in culture (especially larval or immature adult stages), or model organisms, such as the free-living nematode *C. elegans*.[134,135] Translation of actives found against whole organisms in culture to activity in animal models has been exceptionally difficult, and very few new *in vivo* actives with genuine therapeutic potential have been identified in our current screening systems; prospects and challenges in some of these areas have been recently reviewed.[136–138]

A particular emphasis had been placed on the use of *C. elegans* as a convenient surrogate for parasitic species for whole organism screening for novel nematocides; despite significant investment in this process, especially in the animal healthy industry, no novel leads were developed to the market based on this strategy.[139] This lack of success could be interpreted to mean that 'anthelmintic chemical space' is very sparsely populated. An alternative explanation has recently surfaced; it has been shown that bioaccumulation of many kinds of even drug-like synthetic molecules into intact *C. elegans* is markedly and unexpectedly limited.[140] This phenomenon does not seem to be the case for adult stages of parasitic species[141] and may reflect the challenges a free-living species encounters in the environment;[142] this phenomenon may also extend to free-living larval stages of parasitic species which develop in the environment. One conclusion is that our screening strategy has been unable to identify a significant and unknown proportion of anthelmintic space. Although recent examples of successes in whole-organism screening approaches are available[97,101] and work continues in this area,[143] alternative approaches are still of considerable interest.

Mechanism- or target-based screens have not been an abundant source of new anthelmintic leads (although see [131,132]), but they have not been routinely employed for discovery in this area. A considerable amount of thought has gone into their development for these indications.[119,135,144–146] The major benefit of such screens is that they avoid the problem of prioritizing the large number of hits identified in typical whole-organism anthelmintic assays in culture. Additional chemical investment can be devoted to hits with a valuable mechanism of action with some confidence that this investment will pay off in either a true lead or rejection of the target as a discovery tool (if a ligand for it can be shown to achieve 'active' *in situ* concentrations in an infected animal model without apparent consequences for the parasite).

Identification of suitable targets in helminths has been hampered by a dearth of basic research on the physiology and biochemistry of these organisms.[135] Few targets have been chemically validated by interaction with known anthelmintics. Strategies to identify and prioritize potential targets based on bioinformatics and RNA interference data, particularly from *C. elegans*, have been advanced and are likely to have merit.[147] However, we lack the ability to conduct functional genomics studies, including those revolving around RNA interference,[148] in target species and target stages of almost all helminths. In addition, RNA interference only identifies targets for antagonists or inhibitors, whereas most anthelmintics are agonists or channel openers. Our inability to conduct such studies in parasites impedes authentic biological validation of a target prior to the acquisition of chemical matter that acts on it. We are also unable at this time to systematically identify key proteins that mediate viability of a parasite in a host, which may not be present or phenotypically evident in a model, non-parasitic organism. Great benefit would be realized if robust genetic systems were available in parasitic helminths; this may require development of methods that enable egg-to-egg cultivation of parasites. This goal has existed for decades, but little effort is currently expended on it.

Similarly, genetic screens in model organisms, such as *C. elegans*, which associate gain-of-function mutations with a deleterious phenotype could expand the list of candidate targets for drug discovery, but the technical challenges involved will not be simple to overcome.

As noted, pathways for pursuing screen-derived hits against parasites of unique importance in humans are somewhat rudimentary in comparison to those for important veterinary parasites, despite admirable efforts to standardize and streamline them. Whole-organism assays employing the parasites of human relevance can be accomplished for some species (e.g., *Schistosoma mansoni*), but not others (e.g., *O. volvulus, Wuchereria bancrofti*). Animal models that accurately replicate the biology of human lymphatic filariasis and onchocerciasis are not routinely available at low cost in laboratory settings. This means that final validation may require human testing, which can be a daunting prospect for a drug that will have minimal economic rewards. It is challenging to acquire funding to optimize animal models for drug screening in a hypothesis-driven grants world, but additional investment in this area is likely to be needed to translate hits into leads for neglected tropical diseases caused by helminths.

11.8 Conclusions

Helminth infections continue to exact a toll on humans living in resource-poor environments. Infrastructure improvements that led to the virtual eradication of these infections in developed countries are not yet attainable on a global scale. In the interim, chemotherapeutic solutions are needed to minimize the considerable morbidity and losses in productivity and development associated with helminthiases. Historically, anthelmintics used in humans have been adopted from veterinary medicine, and have not been optimized *via* medicinal chemistry or delivery systems for application to humans, especially for mass drug administration campaigns. We need better systems for discovering new anthelmintic leads for human parasites and development pathways to facilitate their registration. Continued ties to the animal health industry can provide an economic pull-through to enable development, but additional models of drug development must evolve to sustain these efforts. In the short term, several candidate molecules for GI nematodes are evident. New antischistosomal drugs and drugs with macrofilaricidal activity remain some distance from identification, registration and introduction.

References

1. S. Brooker, *Int. J. Parasitol.*, 2010, **40**, 1137.
2. S. Brooker, P. J. Hotez and D. A. P. Bundy, *PLoS Negl. Trop. Dis.*, 2010, **4**, e779.
3. P. J. Hotez, A. Fenwick, L. Savioli and D. H. Molyneux, *D. H. Lancet*, 2009, **373**, 1570.

4. P. J. Hotez, P. J. Brindley, J. M. Bethony, C. H. King, E. J. Pearce and J. Jacobson, *J. Clin. Invest.*, 2008, **118**, 1311.
5. J. Bethony, S. Brooker, M. Albonico, S. M. Geiger, A. Loukas, D. Diemart and P. J. Hotez, *Lancet*, 2006, **367**, 1521.
6. C. H. King, *Adv. Parasitol.*, 2010, **73**, 51.
7. W. C. Campbell, in *Chemotherapy of Parasitic Diseases*, ed. W.C. Campbell, R. S. Rew, Plenum Press, NY, 1986, p. 3.
8. I. Coombs and D. W. T. Crompton, *A Guide to Human Helminths*, Taylor & Francis, New York, 1991, p. 196.
9. D. W. T. Crompton and L. Savioli, *Handbook of Helminthiasis for Public Health*, CRC Taylor & Francis, Boca Raton, FL, 2007, p. 362.
10. M. E. Viney and J. B. Lok in *WormBook*, ed. The *C. elegans* Research Community, doi/10.1895/wormbook.1.141.1, http://www.wormbook.org, 2007 (accessed 13 April 2011).
11. M. Satoh and A. Kokaze, *Exp. Opin. Pharmacother.*, 2004, **5**, 2293.
12. T. G. Geary, K. Woo, J. S. McCarthy, C. D. Mackenzie, J. Horton, R. K. Prichard, N. R. de Silva, P. L. Olliaro, J. K. Lazdins-Helds, D. A. Engels and D. A. Bundy, *Int. J. Parasitol.*, 2010, **40**, 1.
13. D. W. T. Crompton and L. Savioli, *Handbook of Helminthiasis for Public Health*, CRC Taylor & Francis, Boca Raton, FL, 2007, p. 362.
14. M. Albonico, A. Montresor, D.W.T. Crompton and L. Savioli, *Adv. Parasitol.*, 2006, **61**, 311.
15. J. Keiser and J. Utzinger, *Adv. Parasitol.*, 2010, **73**, 197.
16. E. van den Enden, *Exp. Opin. Pharmacother.*, 2009, **10**, 435.
17. J. Horton, *Curr. Opin. Infect. Dis.*, 2002, **15**, 599.
18. http://www.childrenwithoutworms.org/ (accessed January 2011).
19. http://endtheneglect.org/2010/10/gsk-announces-expansion-of-albendazole-donation/ (accessed January 2011).
20. E. Lacey, *Int. J. Parasitol.*, 1988, **18**, 885.
21. J. Keiser and J. Utzinger, *J. Amer. Med. Assoc.*, 2008, **299**, 1937.
22. A. Diawara, L. J. Drake, R. R. Suswillo, J. Kihara, D. A. Bundy, M. E. Scott, C. Halpenny, J. R. Stothard and R. K. Prichard, *PLoS Negl. Trop. Dis.*, 2009, **3**, e397.
23. A. Olsen, *Trans. Royal Soc. Trop. Med. Hyg.*, 2007, **101**, 747.
24. T. B. Nutman, ed., *Lymphatic Filariasis*, Imperial College Press, London, 2000, p. 292.
25. M. J. Taylor, A. Hoerauf and M. Bockarie, *Lancet*, 2010, **376**, 1175.
26. D. H. Molyneux, *Trans. Royal Soc. Trop. Med. Hyg.*, 2009, **103**, 338.
27. E. A. Ottesen, *Adv. Parasitol.*, 2006, **61**, 395.
28. B. Boatin, F. Richards and F. , *Adv. Parasitol.*, 2007, **61**, 349.
29. J. O. Gyapong, V. Kumaraswami, G. Biswas and E. A. Ottesen, *Exp. Opin. Pharmacother.*, 2005, **6**, 179.
30. E. A. Ottesen, P. J. Hooper, M. Bradley and G. Biswas, *PLoS-Negl. Trop. Dis.*, 2008, **2**, e317.
31. A. Müllner, A. Helfer, D. Kotlyar, J. Oswald and T. Efferth, *Curr. Med. Chem.*, 2011, **18**, 767.

32. B. Thylefors, *Ann. Trop. Med. Parasitol.*, 2008, **102** (Suppl 1), 39.
33. M. J. Bockarie and R. M. Deb, *Curr. Opin. Infect. Dis.*, 2010, **23**, 617.
34. E. W. Cupp, M. Sauerbrey and F. Richards, *Acta Trop.*, in press.
35. M. G. Basañez, S. D. Pion, E. Boakes, J. A. Filipe, T. S. Churcher and M. Boussinesq, *Lancet Infect. Dis.*, 2008, **8**, 310.
36. T. G. Geary and Y. Moreno, *Curr. Pharmaceut. Biotech.*, in press.
37. K. L. Mealey, *Vet. Parasitol.*, 2008, **158**, 215.
38. C. Bourguinat, J. Kamgno, M. Boussinesq, C. Mackenzie, R. Prichard and T. Geary, *Am. J. Trop. Med. Hyg.*, 2010, **83**, 282.
39. M. Boussinesq, J. Gardon, N. Gardon-Wendel and J. P. Chippaux, *Filaria J.*, 2008, **2** (Suppl 1), S4.
40. A. Hoerauf, *Curr. Opin. Infect. Dis.*, 2008, **21**, 673.
41. S. Wanji, N. Tendongfor, N. Nji, M. Esum, J. N. Che, A. Nkwescheu, F. Alassa, G. Kamnang, P. A. Enyong, M. J. Taylor, A. Hoerauf and D. W. Taylor, *Parasit. Vectors*, 2009, **2**, 39.
42. J. D. Turner, N. Tendongfor, M. Esum, K. L. Johnston, R. S. Langley, L. Ford, B. Faragher, S. Specht, S. Mand, A. Hoerauf, P. Enyong, S. Wanji and M. J. Taylor, *PLoS Negl. Trop. Dis.*, 2010, **13**, e660.
43. http://www.a-wol.net/index.htm (accessed 13 April 2011).
44. M. Y. Osei-Atweneboana, J. K. Eng, D. A. Boakye, J. O. Gyapong and R. K. Prichard, *Lancet*, 2007, **369**, 2021.
45. C. Bourguinat, K. Keller, B. Blagburn, R. Schenker, T. G. Geary and R. K. Prichard, *Vet. Parasitol.*, 2011, **176**, 374.
46. A. Fenwick and J. P. Webster, *Curr. Opin. Infect. Dis.*, 2006, **19**, 577.
47. A. Fenwick, D. Rollinson and V. Southgate, *Adv. Parasitol.*, 2006, **61**, 567.
48. C. H. King, *Acta Trop.*, 2010, **113**, 95.
49. M. C. Jeziorski and R. M. Greenberg, *Int. J. Parasitol.*, 2006, **36**, 625.
50. M. J. Doenhoff, D. Cioli and J. Utzinger, *Curr. Opin. Infect. Dis.*, 2008, **21**, 659.
51. A. Dömling and K. Khoury, *Chem. Med. Chem.*, 2010, **5**, 1420.
52. C. Caffrey, *Curr. Opin. Chem. Biol.*, 2007, **11**, 433.
53. J. Keiser and J. Utzinger, *Curr. Opin. Infect. Dis.*, 2007, **20**, 605.
54. R. Abdul-Ghani, N. Loufty, A. El Sahn and A. Hassan, *A. Parasitol., Res.*, 2009, **104**, 955.
55. C. R. Caffrey, D. L. Williams, M. H. Todd, D. L. Nelson, J. Keiser and J. Utzinger, in *Antiparasitic and Antibacterial Drug Discovery: From Molecular Targets to Drug Candidates*, ed. P.M. Selzer, Wiley-Blackhall, Weinheim, 2009, p. 301.
56. J. Keiser and J. Utzinger, *Clin. Microbiol. Rev.*, 2009, **22**, 466.
57. G. P. Brennan, I. Fairweather, A. Trudgett, E. Hoey, M. McCoy, M. McConville, M. Meaney, M. Robinson, N. McFerran, L. Ryan, C. Lanusse, L. Mottier, L. Alvarez, H. Solana, G. Virkel and P. M. Brophy, *Exp. Mol. Pathol.*, 2007, **82**, 104.
58. A. L. Willingham III and D. Engels, *Adv. Parasitol.*, 2006, **61**, 509.

59. J. F. Rossignol and H. Maisonneuve, *Am. J. Trop. Med. Hyg.*, 1984, **33**, 511.
60. P. S. Craig and E. Larrieu, *Adv. Parasitol.*, 2006, **61**, 443.
61. C. M. Coyle and H. B. Tanowitz, *Interdisciplin. Persp. Infect. Dis.*, 2009, article ID 180742, doi: 10.115/2009/180742.
62. S. Handali, M. Klarman, A. N. Gaspard, X. F. Dong, R. LaBorde, J. Noh, Y.-M. Lee, S. Rodriguez, A. E. Gonzalez, H. H. Garcia, R. H. Gilman, V. C. W. Tsang, P. P. Wilkins and P. P. , *Clin. Vacc. Immunol.*, 2010, **17**, 631.
63. C. R. Caffrey and D. Steverding, *Exp. Opin. Drug Disc.*, 2008, **3**, 173.
64. R. P. Tripathi, D. Katiyar, N. Dwivedi, B. K. Singh and J. Pandey, *Curr. Med. Chem.*, 2006, **13**, 3319.
65. P. K. Singh, A. Ajay, S. Kushwaha, R. P. Tripathi and S. Misra-Bhattacharya, *Fut. Med. Chem.*, 2010, **2**, 251.
66. D. J. Woods and T. M. Williams, *Inv. Neurosci.*, 2007, **7**, 245.
67. D. J. Woods, C. Lauret and T. G. Geary, *Expert Opin. Drug. Disc.*, 2007, **2** (Suppl 1), S25.
68. T. G. Geary, D. J. Woods, T. Williams and S. Nwaka, in *Drug Discovery in Infectious Diseases*, ed. Selzer PM, Wiley-VCH, Weinheim, De, 1–16, (2009).
69. S. Nwaka and A. Hudson, *Nature Drug Disc.*, 2006, **5**, 941.
70. J. Scherkenback, J. Jeschke, A. Harder and A., *Curr. Top. Med. Chem.*, 2002, **2**, 759.
71. A. Harder, H.-P. Schmitt-Wrede, J. Krücken, P. Marinovski, F. Wunderlich, J. Willson, K. Amliwala, L. Holden-Dye and R. Walker, *Int. J. Antimicrob. Agents*, 2003, **22**, 318.
72. G. A. Conder, S. S. Johnson, D. S. Nowakowski, T. E. Blake, F. E. Dutton, S. J. Nelson, E. M. Thomas, J. P. Davis and D. P. Thompson, *J. Antibiot.*, 1995, **48**, 820.
73. G. von Samson-Himmelstjerna, A. Harder, N. C. Sangster and G. C. Coles, *Parasitology*, 2005, **130**, 343.
74. P. Jeschke, K. Iinuma, A. Harder, M. Schindler and T. Mirakami, *Parasitol. Res.*, 2005, **97** (Suppl 1), S11.
75. G. Altreuther, I. Radeloff, C. LeSueur, A. Schimmel, K. J. Krieger and K. J., *Parasitol. Res.*, 2009, **105** (Suppl 1), S23.
76. G. Altreuther, J. Buch, S. D. Charles, W. L. Davis, K. J. Krieger, I. Radeloff and I., *Parasitol. Res.*, 2005, **97** (Suppl 1), S58.
77. H. Zahner, A. Taubert, A. Harder and G. von Samson-Himmelstjerna, *Int. J. Parasitol.*, 2001, **31**, 1515.
78. S. Townson, A. Freeman, A. Harris and A. Harder, *Am. J. Trop. Med. Hyg.*, 2005, **73** (Suppl 1), 93.
79. A. Harder, K. Bull, M. Guest, L. Holden-Dye and R. Walker, in *Antiparasitic and Antibacterial Drug Discovery: From Molecular Targets to Drug Candidates*, ed. P. M. Selzer, Wiley-Blackhall, Weinheim, 2009, p. 339.

80. M. Guest, K. Bull, R. J. Walker, K. Amliwala, V. O'Connor, A. Harder, L. Holden-Dye and N. A. Hopper, *Int. J. Parasitol.*, 2007, **37**, 1577.
81. S.-H. Xiao, W. Hui-Ming, M. Tanner, J. Utzinger and W. Chong, *Acta Trop.*, 2005, **94**, 1.
82. Y. Hu, S.-H. Xiao and R. F. Aroian, *PLoS-Negl. Trop. Dis.*, 2009, **3**, e499.
83. J. Xue, S.-H. Xiao, L. L. Xu and H. Q. Qiang, *Parasitol. Res.*, 2010, **106**, 775.
84. C. D. Mackenzie and T. G. Geary, *Expert Rev. Anti-infect. Ther.*, 2011, **9**, 497.
85. A. Dominguez-Vasquez, H. R. Taylor, B. M. Greene, A. M. Ruvalcaba-Macias, A. R. Rivas-Alcala, R. P. Murphy and F. Beltran-Hernandez, *F. Lancet*, 1983, **1(8317)**, 139.
86. B. M. deC Bronsvoort, B. L. Makepeace, A. Renz, V. N. Tanya, L. Fleckenstein, D. Ekale and A. J. Trees, *Parasit. Vect.*, 2008, **1**, 18.
87. L. Ceballos, M. Elissondo, S. S. Bruni, G. Denegri, L. Alvarez and C. Lanusse, *Parasitol. Int.*, 2009, **58**, 354.
88. I. Van Kerckhoven and V. Kumar, *Trans. Royal Soc. Trop. Med. Hyg.*, 1988, **82**, 890.
89. C. M. Gavidia, A. E. Gonzalez, L. Lopera, C. Jayashi, R. Angelats, E. A. Barron, B. Ninaquispe, L. Villarreal, H. H. Garcia, M. R. Versategui, R. H. Gilman and R. H., *Am. J. Trop. Med. Hyg.*, 2009, **80**, 367.
90. C. M. Gavidia, A. E. Gonzalez, E. A. Barron, B. Ninaquispe, M. Llamosas, M. R. Verastegui, C. Robinson and R. H. Gilman, *PLoS-Negl. Trop. Dis.*, 2010, **4**, e616.
91. J. W. McCall, *Vet. Parasitol.*, 2005, **133**, 197.
92. M. M. Cotreau, S. Warren, J. L. Ryan, L. Fleckenstein, S. R. Vanapalli, K. R. Brown, D. Rock, C. Y. Chen and U. S. Schwertschlag, *J. Clin. Pharmacol.*, 2003, **43**, 1108.
93. A. Lespine, S. Martin, J. Dupuy, A. Roulet, T. Pineau, S. Orlowski and M. Alvinerie, *Eur. J. Pharm. Sci.*, 2007, **30**, 84.
94. World Health Organization/Tropical Diseases Research, Annual Report 2009: Drug Development and Evaluation for Helminths and Other Neglected Tropical Diseases. http://apps.who.int/tdr/publications/about-tdr/annual-reports/bl6-annual-report/pdf/bl6-annual-report-2009.pdf (accessed 2 February 2010).
95. B. O. L. Duke, M. C. Pacqué, B. Muñoz, B. M. Greene and H. R. Taylor, *Bull. WHO*, 1991, **69**, 163.
96. B. M. de C Bronsvoort, A. Renz, V. Tchakouté, V. N. Tanya, D. Ekale and A. J. Trees, *Filaria J.*, 2005, **4**, 4.
97. R. Kaminsky, P. Ducray, M. Jung, R. Clover, L. Rufener, J. Bouvier, S. S. Weber, A. Wenger, S. Wieland-Berghausen, T. Goebel, N. Gauvry, F. Pautrat, T. Skripsky, O. Froelich, C. Komoin-Oka, B. Westlund, A. Sluder and P. Mäser, *Nature*, 2008, **452**, 176.
98. P. Ducray, N. Gauvry, F. Pautrat, T. Goebel, J. Fruechtel, Y. Desaules, S. Schorderet Weber, J. Bouvier, T. Wagner, O. Froelich and R. Kaminsky, *Bioorg. Med. Chem. Lett.*, 2008, **18**, 2935.

99. B. C. Hosking, R. Kaminsky, H. Sager, P. F. Rolfe and W. Seewald, *Parasitol. Res.*, 2010, **106**, 529.
100. L. Rufener, J. Keiser, R. Kaminsky, P. Mäser and D. Nilsson, *PLoS Path.*, 2010, **6**, e1001091.
101. B. H. Lee, M. F. Clothier, F. E. Dutton, S. J. Nelson, S. S. Johnson, D. P. Thompson, T. G. Geary, H. D. Whaley, C. L. Haber, V. P. Marshall, G. I. Kornis, P. L. McNally, J. I. Cialdella, D. G. Martin, J. W. Bowman, C. A. Baker, E. M. Coscarelli, S. J. Alexander-Bowman, J. P. Davis, E. W. Zinser, V. Wiley, M. F. Lipton and M. A. Mauragis, *Curr. Top. Med. Chem.*, 2002, **2**, 779.
102. E. W. Zinser, M. L. Wolfe, S. J. Alexander-Bowman, E. M. Thomas, V. E. Groppi, J. P. Davis, D. P. Thompson and T. G. Geary, *J. Vet. Pharmacol. Therap.*, 2002, **25**, 241.
103. P. R. Little, A. Hodges, T. G. Watson, J. A. Seed and S. J. Maeder, *NZ Vet. J.*, 2010, **58**, 121.
104. K. van Frankenhuysen, *J. Invert. Pathol.*, 2009, **101**, 1.
105. Y. Hu, S. B. Georghiou, A. J. Kelleher and R. V. Aroian, *PLoS-Negl. Trop. Dis.*, 2010, **4**, e614.
106. Y. Hu, E. G. Platzer, A. Bellier and R. V. Aroian, *Proc. Natl. Acad. Sci. U.S.A*, 2010, **107**, 5955.
107. G. E. Swan, *J. South African Vet. Assoc.*, 1999, **70**, 61.
108. C. Gloeckner, A. L. Garner, F. Mersha, Y. Oksov, N. Tricoche, L. M. Eubanks, S. Lustigman, G. F. Kaufmann and K. D. Janda, *Proc. Natl. Acad. Sci.*, 2010, **107**, 3424.
109. D. P. Thompson, T. G. Geary, in *Molecular Medical Parasitology*, ed. J. J. Marr and R. Komuniecki, Academic Press, Oxford, UK, 2003, p. 297.
110. Y. Dong, J. Chollet, M. Vargas, N. R. Mansour, Q. Bickle, Y. Alnouti, J. Huang, J. Keiser and J. L. Vennerstrom, *Bioorg. Med. Chem. Lett.*, 2010, **20**, 2481.
111. J. Keiser, N. A. N'Guessan, K. D. Adoubryn, K. D. Silué, P. Vounatsou, C. Hatz, J. Utzinger and E. K. N'Goran, *Clin. Infeci. Dis.*, 2010, **50**, 1205.
112. C. R. Meier, K. Wilcock and S. S. Jick, *Drug Safety*, 2004, **27**, 203.
113. S.-H. Xiao, J. Xue, X. Li-li, Y.-N. Zhang and H.-Q. Qiang, *Parasitol. Res.*, 2010, **107**, 1391.
114. L. M. Fox and L. D. Saravolatz, *Clin. Inf. Dis.*, 2005, **40**, 1173.
115. V. R. Anderson and M. P. Curram, *Drugs*, 2007, **67**, 1947.
116. B. Stadelmann, S. Scholl, J. Müller and A. Hemphill, *Antimicrob. Agents Chemother.*, 2010, **65**, 512.
117. A. Hudson and S. Nwaka, *Expert Opin. Drug Disc.*, 2007, **2(Suppl. 1)**, S3.
118. S. Nwaka, B. Ramirez, R. Brun, L. Maes, F. Douglas and R. R. Ridley, *PLoS-Negl. Trop. Dis.*, 2009, **3**, e440.
119. T. G. Geary, D. J. Woods, T. Williams and S. Nwaka, in *Antiparasitic and Antibacterial Drug Discovery: From Molecular Targets to Drug Candidates*, ed. P.M. Selzer, Wiley-Blackhall, Weinheim, 2009, p. 3.

120. S. N. Comlay, J. C. Hannam, W. Howson, C. Lauret and Y. A. Sabnis, Antiparasitic agents, WO2008096231, Aug. 14, 2008.
121. S. P. Gibson and C. Lauret, Antiparasitic agents, US20080200540, Aug. 21, 2008.
122. M. D. Soll, L. P. Le Hir De Fallois, S. K. Huber, H. I. Lee, D. E. Wilkinson and R. T. Jacobs, Aryloazol-2-yl cyanoethylamino compounds, method of making them and method of using thereof, WO2008144275, Nov. 27, 2008.
123. L. P. Le Hir De Fallois, H. I. Lee and R. P. Timmons, Thioamide compounds, method of making them and method of using thereof, WO2010048191, Apr. 29, 2010.
124. M. D. Soll, L. P. Le Hir De Fallois, S. K. Huber H. I. Lee, D. E. Wilkinson R. T. Jacobs and B. C. Beck, Enantiomerically enriched aryloazol-2-yl cyanoethylamino parasiticidal compounds, WO2010056999, May 20, 2010.
125. J. Heeres and J. P. Lewi, Anthelmintic benzo[d]isoxazolyl benzamide derivatives, WO2008152081, Dec. 18, 2008.
126. C. P. A. Chassaing, J. Schroeder, T. S. Ilg, M. Uphoff and T. Meyer, Anthelmintic agents and their use, WO2009077527, Jun. 25, 2009.
127. C. P. A. Chassaing and T. Meyer, Anthelmintic agents and their use, WO2010115688, Oct. 14, 2010.
128. C. P. A. Chassaing and T. Meyer, Anthelmintic agents and their use, WO2010146083, Dec. 23, 2010.
129. D. L. Williams and A. A. Sayed, Treatment for the control of schistosomiasis, WO2009076265, Jun. 18, 2009.
130. C. J. Thomas, D. J. Maloney, G. R. Bantukallu, A. A. Sayed, A. Simeonov and D. L. Williams, Oxadiazole-2-oxides as anti-schistosomal agents, WO2010019772, Feb. 18, 2010.
131. A. A. Sayed, A. Simeonov, C. J. Thomas, J. Inglese, C. P. Austin and D. L. Williams, *Nature Med.*, 2008, **14**, 407.
132. G. Rai, A. A. Sayed, W. A. Lea, H. F. Lueke, H. Chakrapani, S. Prast-Nielsen, A. Jadhav, W. Leister, M. Shen, J. Inglese, C. P. Austin, L. Keefer, E. S. J. Arner, A. Simeonov, D. J. Maloney, D. L. Williams and C. J. Thomas, *J. Med. Chem.*, 2009, **52**, 6474.
133. C. Chassaing and H. Sekljic in *Antiparasitic and Antibacterial Drug Discovery: From Molecular Targets to Drug Candidates*, ed. P. M. Selzer, Wiley-Blackhall, Weinheim, 2009, p. 117.
134. D. P. Thompson, R. D. Klein and T. G. Geary, *Parasitol.*, 1996, **113**, S217.
135. T. G. Geary, D. P. Thompson and R. D. Klein, *Int. J. Parasitol.*, 1999, **29**, 105.
136. S. Townson, B. Ramirez, F. Fakorede, M-A. Mouries and S. Nwaka, *Expert Opin. Drug. Discov.*, 2007, **2 (Suppl 1)**, S63.
137. B. Ramirez, Q. Bickle, Y. Yousif, F. Fakorede, M-A. Mouries and S. Nwaka, *Expert Opin. Drug Discov.*, 2007, **2 (Suppl 1)**, S53.
138. J. Keiser, *Parasitol.*, 2010, **137**, 589.
139. T. G. Geary and D. P. Thompson, *Vet. Parasitol.*, 2001, **101**, 371.

140. A. R. Burns, I. M. Wallace, J. Wildenhain, M. Tyers, G. Giaever, G. D. Bader, C. Nislow, S. R. Cutler and P. J. Roy, *Nature Chem. Biol.*, 2010, **6**, 549.
141. N. F. H. Ho, S. M. Sims, T. J. Vidmar, J. S. Day, C. L. Barsuhn, E. M. Thomas, T. G. Geary and D. P. Thompson, *J. Pharmaceut. Sci.*, 1994, **83**, 1052.
142. E. Ruiz-Lancheros, C. Viau, T. N. Walter, A. Francis and T. G. Geary, *Int. J. Parasitol.*, 2011, **41**, 455.
143. T. Meyer, J. Schröder, M. Uphoff, S. Noack, A. R. Heckeroth, M. Gassel, P. Rohrwild and T. Ilg, in *Antiparasitic and Antibacterial Drug Discovery: From Molecular Targets to Drug Candidates*, ed. P. M. Selzer, Wiley-Blackhall, Weinheim, 2009, p. 357.
144. J. P. McCarter, *Trends Parasitol.*, 2004, **20**, 462.
145. S. Kumar, K. Chaudhary and J. M. Foster, *et al. PLoS One*, 2007, **11**, e1189.
146. G. J. Crowther, D. Shanmugam, S. J. Carmona, M. A. Doyle, C. Hertz-Fowler, M. Berriman, S. Nwaka, S. A. Ralph, D. S. Roos, W. C. Van Voorhis and F. Agüero, *PLoS-Negl. Trop. Dis.*, 2010, **4**, e804.
147. C. A. Behm, M. M. Bendig, J. P. McCarter and A. E. Sluder, *Trends Parasitol.*, 2005, **21**, 97.
148. D. P. Knox, P. Geldhof, A. Visser and C. Britton, *Trends Parasitol.*, 2007, **23**, 105.

CHAPTER 12

Managing the HIV Epidemic in the Developing World – Progress and Challenges

ELNA VAN DER RYST,[a] MICHAEL J PALMER[a] AND CLOETE VAN VUUREN[b]

[a] Pfizer Global Research and Development, Ramsgate Road, Sandwich, CT13 9NJ, United Kingdom; [b] Faculty of Health Sciences, University of the Free State, Bloemfontein, 9301, South Africa

12.1 The HIV Epidemic

In 1981 the Centers for Disease Control and Prevention's *Mortality and Morbidity Weekly Report* described 5 cases of *Pneumocystis carinii* (now *jirovecii*) pneumonia occurring in previously healthy men in Los Angeles.[1] This was of interest as *P. jirovecii* infection does not normally occur in immunocompetent individuals.[2] These initial reports were followed by several more, as well as reports of an increase in other immune deficiency-associated conditions such as Kaposi's sarcoma, mucosal candidiasis, disseminated cytomegalovirus and peri-anal herpes simplex virus infections.

In 1983, a group from the Pasteur Institute in Paris isolated a retrovirus from lymph node tissues of an AIDS patient.[3] The virus was initially called lymphadenopathy-associated virus, but subsequently renamed human immunodeficiency virus type 1 (HIV-1). In 1986 a second retrovirus associated with AIDS (HIV-2) was isolated from patients in West Africa.[4]

RSC Drug Discovery Series No. 14
Neglected Diseases and Drug Discovery
Edited by Michael J. Palmer and Timothy N. C. Wells

Published by the Royal Society of Chemistry, www.rsc.org

It soon became clear that this new syndrome, now called acquired immune deficiency syndrome (AIDS), was not limited only to homosexual men and intravenous drug users, but also affected haemophiliacs, transfusion recipients, sexual partners of people at risk and babies born to infected mothers.[5] Furthermore, it was apparent that HIV was already well established in East and Central Africa by the early 1980s.[6] In fact, there is clear evidence that hundreds of people in Africa may already have been infected as early as the 1960s. Stored blood samples from an American malaria research project carried out in the Congo in 1959 provide proof for one example of early HIV infection.[7]

12.1.1 HIV Transmission

Both HIV-1 and –2 are transmitted via the following routes: (i) sexual contact, (ii) parenteral inoculation or transfusion of blood and blood products, and (iii) perinatal transmission.[5]

The virus is transmitted through both homo- and heterosexual contact. The risk of transmission is related to the number of sexual partners, type of sexual practice, presence of other sexually transmitted infections, especially ulcerative genital disease, the HIV prevalence in the area and the viral load of the transmitting partner.[8–11]

Whole blood, cellular blood components, plasma and clotting factors have all been implicated in the transmission of HIV-1 infection, but since the introduction of extensive antibody and nucleic acid testing of donated blood and blood components, transmission through this route has dropped to extremely low levels in most of the world.[12,13] However, a significant risk still exists in some lower income countries with high HIV prevalence, especially in sub-Saharan Africa.[14]

Among injecting drug users, HIV is transmitted through sharing of contaminated equipment, especially needles. A recent study estimated that there are approximately 16 million injecting drug users worldwide, with the largest numbers found in China, the USA, and Russia. It was estimated that worldwide about 3 million injecting drug users are HIV infected.[15] Injecting drug use is therefore a major factor in the global HIV epidemic, including in developing countries.

Transmission of HIV from mother to child (MTCT) can occur *in utero*, during delivery, or through breast-feeding; and is responsible for the majority of paediatric HIV-1 infections.[16] The relative importance of the three routes of transmission has not been clearly defined, but several factors including high maternal HIV viral load and mixed breast/formula feeding increase the risk of transmission.[17,18] In 2008, an estimated 1600 children worldwide became infected with HIV-1 each day, 90% of whom live in sub-Saharan Africa, where vertically-acquired HIV-1 disease remains a major contributor to child mortality.[15]

12.1.2 The Global Spread of HIV Infection

Since the recognition of AIDS in 1981, and the subsequent isolation of HIV-1 in 1983, the number of HIV-1 infections has increased rapidly, resulting in a

Table 12.1 Estimates of number of people living with HIV, prevalence, new infections and AIDS deaths at the end of 2008.[19]

	People living with HIV	*Prevalence*[a]	*New infections*	*AIDS deaths*
Sub-Saharan Africa	22.4 million	5.2%	1.9 million	1.4 million
Middle East and North Africa	310 000	0.2%	35 000	20 000
South and South-East Asia	3.8 million	0.3%	280 000	270 000
East Asia	850 000	<0.1%	75 000	59 000
Latin America	2.0 million	0.6%	170000	77 000
Caribbean	240 000	1%	20000	12 000
Eastern Europe and Central Asia	1.5 million	0.7%	110000	87 000
Western and Central Europe	850 000	0.3%	30000	13 000
North America	1.4 million	0.6%	55000	25 000
Oceania	59 000	0.3%	3900	2 000
Total	33.4 million	0.8%	2.7 million	2.0 million

[a]Adult prevalence.

global pandemic with enormous humanitarian and financial implications.[19] In contrast, HIV-2 is mainly confined to West Africa and India.[20–23]

Epidemiological data indicate that, globally, the spread of HIV appears to have peaked in 1996, when 3.5 million new HIV infections occurred. In 2008, the estimated number of new HIV infections (2.7 million) was approximately 30% lower than at the epidemic's peak 12 years earlier in 1996. The number of people living with HIV worldwide was estimated to be 33.4 million in 2008 (Table 12.1).[19]

Assessment of regional epidemiology data indicates that the developing world carries the vast majority of the HIV burden. Importantly, it is estimated that the prevalence of HIV-1 infection in adults in sub-Saharan Africa is 5.2% with an estimated 22.4 million people living with HIV in this region (Table 12.1). By the end of 2008, sub-Saharan Africa accounted for 67% of HIV infections worldwide, 68% of new HIV infections among adults and 91% of new HIV infections among children. The region also accounted for 72% of the world's AIDS-related deaths in 2008. The epidemic continues to have an enormous impact on individuals, communities, businesses, public services and national economies in the region.[19]

There are several reasons for the high rates of HIV transmission in sub-Saharan Africa. These include early sexual debut, intergenerational sex, transactional sex, high partner turnover, concurrent sexual partnerships, substance abuse and lack of knowledge of HIV status. One of the most important issues is the high rate of sexual concurrency.[24] The introduction of HIV into sexual networks leads to rapid spread of infection throughout the network, due to the high viral load during sero-conversion. In societies with the same number of lifetime sexual partners but low sexual concurrency rates, the spread of HIV in sexual networks is much slower.[25,26]

Gains in life expectancy attained through improvement in living conditions and management of other diseases have been largely wiped out in many countries.[19,27] More than 14.1 million children in sub-Saharan Africa were estimated to have lost one or both parents to AIDS. It is estimated that 390 000 children were infected with HIV-1 in this region in 2008, the vast majority as a result of MTCT.[19] Although sub-Saharan Africa is the worst affected region in the developing world, large numbers of infections occur in the rest of the developing world too (Table 12.1).[19]

12.1.3 HIV-1 Structure and Variability

HIV-1 belongs to the *Lentivirus* genus of the family *Retroviridae*. This group of retroviruses can infect a broad range of animals including monkeys (simian immunodeficiency virus), cats (feline immunodeficiency virus), sheep (Visna/Maedi virus), horses (equine infectious anaemia virus) and goats (caprine encephalitis-arthritis virus).[28]

The virion is a spherical particle about 110 nm in diameter and consists of a lipid bilayer membrane surrounding a conical nucleocapsid. The inner core of the viral particle contains the double stranded ribonucleic acid (RNA) genome in association with reverse transcriptase and the nucleocapsid protein and is surrounded by a 5 nm capsid shell. The lipid bilayer envelope contains approximately 70 knobs which are 9-10 nm long with an ovoid distal end that is linked to the lipid membrane by a stalk. Each knob is thought to contain four heterodimers of the envelope glycoprotein. Each heterodimer consists of a heavily glycosylated surface unit (gp120) that interacts with the transmembrane unit (gp41) through noncovalent bonds (Figure 12.1).[29,30]

The HIV-1 genome is about 9.5 kilobases in length. It contains three major structural genes: (i) *gag*, encoding the matrix (p17), capsid (p24) and nucleocapsid (p9) proteins; (ii) *pol*, encoding the viral enzymes reverse transcriptase (p66), RNAse H (p51), protease (p11) and integrase (p32); and (iii) *env*, encoding the external surface envelope (gp120) and the transmembrane (gp41) proteins.[29] While the *gag-pol* regions of HIV-1 are relatively conserved, *env* contains several hypervariable regions. HIV-1 isolates have been classified into three groups, the M (major), O and N groups.[31,32] The M group is sub-divided into at least 9 subtypes (designated A through H and J), on the basis of sequence homologies in the *env* and *gag* genes.[33] Although certain subtypes are found preferentially in certain regions, there does not appear to be a strict localisation of subtypes to precise geographical areas.[34,35]

12.1.4 Pathogenesis and Clinical Manifestations of HIV Infection

In contrast to most other infectious agents, there is a long period of clinical latency between infection with HIV and the development of clinical symptoms (and eventually AIDS). Up to 70% of infected individuals have an acute sero-conversion illness at the period of maximum viral replication, which is followed by an asymptomatic period that can last from a few months to more than 10 years.[36]

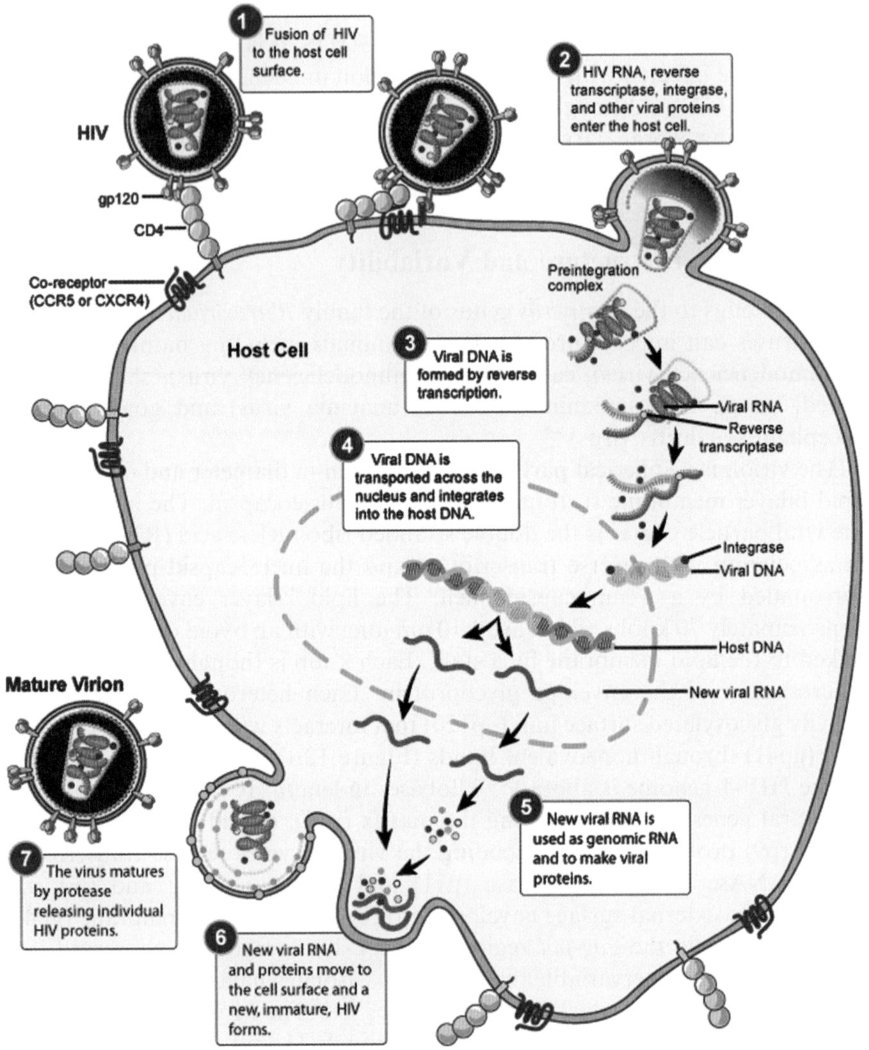

Figure 12.1 HIV-1 structure and replication. Schematic representation of HIV-1 structure and replication cycle. Reprinted with gratitude from the National Institute for Allergy and Infectious Diseases website (http://www.niaid.nih.gov/topics/HIVAIDS/Understanding/Biology/Pages/hivReplicationCycle.aspx).[30]

During the asymptomatic period the virus continues to replicate at high levels and there is a gradual decrease in $CD4^+$ T-cell numbers over time.[37] The asymptomatic period is eventually followed by the development of clinical symptoms including weight loss, chronic diarrhoea, fevers and opportunistic infections; and eventually overt AIDS develops. The CDC classification for HIV infection defines the spectrum of disease as ranging from asymptomatic disease to AIDS. According to these criteria, AIDS is defined as a CD4 count of

<200 cells/mm^3 and/or the presence of certain conditions known to be associated with advanced HIV-related immuno-suppression (CDC category C conditions).[38] Although opportunistic infections and AIDS-related cancers are classically associated with HIV, the virus is also implicated in a range of other conditions including autoimmune disorders, chronic liver disease, non-AIDS-related malignancies and cardiovascular disease. The great majority of HIV-infected people, especially in the absence of any therapeutic interventions, will eventually die from AIDS, but a small percentage (<5%) will survive for more than 15 years without any evidence of immunological deterioration.[39,40]

The degree of immunodeficiency associated with HIV-1 infection correlates closely with plasma CD4 cell counts. Additionally, the rate at which immunosuppression develops also closely reflects the levels of HIV-1 RNA in plasma, such that the higher the HIV-1 viral load, the greater the loss of circulating CD4 cells per year. Initially, it was thought that the slow plasma CD4 cell depletion rate was reflective of the total body CD4 pool and was driven directly by virus replication.[37] However, more recent data indicate that this relationship may be more complex than initially thought. In fact, it is clear that with HIV-1 infection destruction of the vast majority of memory CD4 cells occurs in the gastrointestinal tract in the first few weeks after infection. The depleted mucosal CD4 cells are not completely replaced and the host remains deficient in memory CD4 cells. The depleted mucosal immune barrier cannot adequately control invading organisms, which can lead to a state of generalized activation of the immune system.[41,42] The immune activation may also be stimulated by co-infection with other viruses including HCV, EBV and CMV.[43–45] Over time, this can lead to chronic inflammation, fibrosis and irreversible scarring of the gut-associated lymphoid tissue.[46] It is therefore important not to see HIV infection purely as a condition of immunosuppression, but rather as a condition of immune dysregulation.

12.1.4.1 HIV in the Developing World

In the developing world the epidemic is largely driven by heterosexual transmission, with females infected at a younger age and representing 60% of infected people. This, coupled with inadequate access to, or uptake of, interventions to prevent MTCT, also leads to a high number of infected children. The clinical manifestations of HIV in the developing world are complicated by the interaction of HIV-associated immune-deficiency with high rates of endemic infectious agents and malnutrition associated with poverty. Furthermore, in sub-Saharan Africa, the HIV epidemic occurs in a region where malaria and tuberculosis have a major impact on childhood, maternal and overall mortality.

Tuberculosis is the most important opportunistic infection complicating HIV infection in developing countries, and may present at any stage in the course of HIV-associated disease. Importantly, the severe immune-deficiency associated with HIV can result in atypical presentations of tuberculosis, which complicate diagnosis. The prevalence of latent tuberculosis infection may be as high as 80% in certain high-transmission settings.[47,48] The risk of developing active tuberculosis is significantly increased with HIV infection, even in patients with relatively high CD4 counts.[49] It is estimated that 70% of smear positive cases of tuberculosis are also HIV positive. Patients who start on anti-TB treatment have a high rate of late-stage adverse effects, deterioration, death and progression of their HIV.[50] Finally, treatment of HIV infection may "unmask" tuberculosis infection, especially in patients with low CD4 counts, resulting in immune reconstitution inflammatory syndrome which contributes to HIV-associated morbidity in these patients.[51,52]

Malaria and HIV both occur in tropical and subtropical regions of the world and are especially concentrated in sub-Sahara Africa where they contribute to childhood and maternal mortality. Malaria increases plasma HIV RNA concentrations with subsequent higher MTCT rates. Anaemia associated with malaria and that needs transfusion in areas with unsafe blood supply puts patients at risk for acquiring HIV. Malaria treatment and prophylaxis are also less effective in HIV-infected persons, while the pharmacodynamics and pharmacokinetics of antiretroviral drugs in patients with malaria is not well studied.[53]

Sub-Saharan Africa also carries the highest burden of cryptococcal meningitis.[54] A low baseline CD4 count has a strong association with the development of cryptococcal meningitis.[55] Amphothericin B, the preferred first-line treatment, is often not available in resource-poor settings, complicating the management of this condition.

Other conditions commonly associated with HIV infection in the developing world are bacterial septicaemia, caused by non-typhoid *Salmonella* spp. and pneumococci, as well as parasitic conditions including cryptosporidiosis, microsporidiosis, isosporiasis, which contribute to HIV-associated chronic diarrhoea, and cerebral toxoplasmosis.[56]

12.2 HIV-1 Replication and Development of Antiretroviral Drugs

In spite of the evidence that chronic immune activation plays an important role in HIV-infection, the amount of virus in the plasma (HIV RNA copies) is strongly associated with clinical outcome in HIV-infected individuals, and complete control of virus replication has been firmly established as the ultimate aim of treatment.[57,58] Consequently, development of drugs that decrease the amount of circulating virus through suppression of virus replication has been the main focus of treatment research to date.

The first breakthrough in the development of antiretroviral drugs came shortly after the discovery of HIV, when it was demonstrated that zidovudine, a nucleoside analogue initially evaluated as a cancer treatment, was able to suppress HIV replication in cell culture by inhibiting reverse transcriptase and subsequent transcription of proviral DNA by the enzyme.[59–61] This was followed by a clinical study which demonstrated that zidovudine administration decreased mortality and the frequency of opportunistic infections compared to placebo in patients with AIDS or AIDS-related complex. The impact of zidovudine treatment was so dramatic that the study was terminated early, and patients randomised to placebo offered the opportunity to receive zidovudine in an open-label study.[62]

The discovery of the anti-HIV activity of zidovudine and the demonstration of its clinical efficacy provided the first proof of concept that the replication of HIV and the clinical course of HIV-associated disease could be influenced by treatment. This established the foundation for antiretroviral drug discovery research and provided a platform for remarkable progress. In 2010, only 25 years later, there were 25 approved single-agent antiretroviral drugs from 6 mechanistic drug classes, acting on different stages of the viral replication cycle, as well as several fixed-dose combination products (Table 12.2).[60,63] These drugs can be combined to form multiple combination regimens to achieve control of viral replication even in patients with advanced disease and multi-drug resistant virus. There are also several investigational antiretroviral drugs in late-stage development (Table 12.3).[64–72] Of these, dolutegravir, a novel integrase inhibitor, shows particular promise, especially as it retains activity against raltegravir resistant virus strains.

12.2.1 HIV-1 Entry and Inhibitors of Virus Entry

The first step in the HIV-1 replication cycle is entry into the host cell (Figure 12.1). This process is initiated by the specific binding of viral gp120 to CD4, the primary cellular receptor for HIV-1. This is followed by binding to a human chemokine receptor (CCR5 or CXCR4) that acts as an essential co-receptor for HIV-1 infection.[73] The binding of gp120 to CD4 causes a conformational change in gp120 that exposes the bridging sheet and forms a co-receptor binding site.[74] Once this has occurred, co-receptor binding triggers conformational changes in gp41, which drives the remaining steps in fusion of the viral membrane with that of the host cell.[75]

12.2.1.1 CCR5 Antagonists

The chemokine receptors most commonly utilised by HIV-1 *in vivo* are CCR5 and/or CXCR4.[73] The ability of gp120 to bind to either one or both receptors defines the tropism of the virus and HIV-1 strains are categorised as R5 (CCR5-tropic), X4 (CXCR4-tropic) or R5/X4 (strains using both CCR5 and CXCR4; also referred to as dual-tropic).[76]

The potential of CCR5 as an antiretroviral target was recognised following the observation that individuals with a mutation in their CCR5 gene showed resistance to HIV-1 infection.[77–79] Approximately 1% of the Caucasian population is homozygous for a natural 32-base-pair deletion in the CCR5 gene (CCR5-D32), which results in complete absence of CCR5 expression on the cell surface. Furthermore, HIV-infected individuals who are heterozygous for the CCR5-D32 allele show delayed disease progression.[77,78] Confidence in the safety of a CCR5 antagonist was based on the observation that individuals who are CCR5-D32 homozygous are apparently healthy.[79]

Table 12.2 Currently marketed antiretroviral drugs approved for use by the United States Food and Drug Administration.[60]

Generic name	*Year of registration*
Nucleoside/nucleotide reverse transcriptase inhibitors	
Zidovudine (ZDV/AZT)	1987
Didanosine (ddI)	1991
Stavudine (d4T)	1994
Lamivudine (3TC)	1995
Abacavir (ABC)	1998
Tenofovir (TDF)	2001
Emtricitabine (FTC)	2003
Non-nucleoside reverse transcriptase inhibitors	
Nevirapine (NVP)	1996
Delavirdine (DLV)	1997
Efavirenz (EFV)	1998
Etravirine (ETR)[a]	2008
Protease inhibitors	
Saquinavir (SQV)	1995
Ritonavir (RTV)	1996
Indinavir (IDV)	1996
Nelfinavir (NFV)	1997
Amprenavir (APV)	1999
Lopinavir/ritonavir (LPV/r)	2000
Atazanavir (ATV)	2003
Fosamprenavir (FPV)	2003
Tipranavir (TPV)[a]	2005
Darunavir (DRV)	2006
Integrase inhibitors	
Raltegravir (RGV)	2007
Entry inhibitors	
Enfuvirtide (T-20)[a]	2003
Maraviroc (MVC)	2007
Fixed dose combinations	
Lamivudine/zidovudine	1997
Abacavir/zidovudine/lamivudine	2000
Abacavir/lamivudine	2004
Tenofovir/emtricitabine	2004
Efavirenz/tenofovir/emtrictabine	2006
Nevirapine/stavudine/lamivudine[b]	

[a]Only approved for use in treatment experienced patients.
[b]Triommune is available and used some countries in the developing world, but is not FDA approved.[143]

Although the identification of CCR5 as a putative anti-HIV-1 target prompted widespread discovery activities across academia and industry, only one CCR5 antagonist, maraviroc, has been approved to date. Maraviroc is the result of medicinal chemistry optimization of a lead molecule defined

Table 12.3 Investigational antiretroviral drugs currently in phase 2b/3 development.

Drug class/generic name	*Potential advantages*	*Potential disadvantages*
Nucleoside reverse transcriptase inhibitors		
Elvucitabine[64] 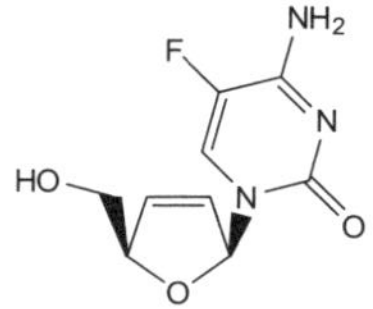	• Low-dose once a day	• No clear advantage over lamuvidine or emtricitabine with regards to efficacy or resistance profile
Non-nucleoside reverse transcriptase inhibitors		
*Rilpivirine[65,66] 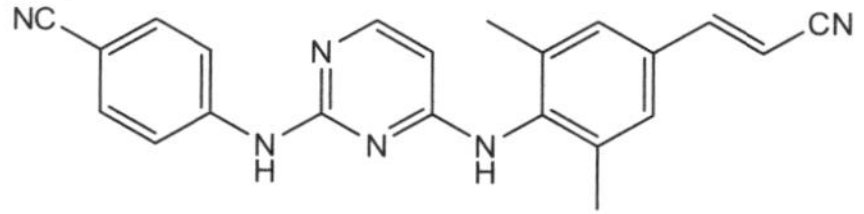	• Low-dose once-daily dosing • Potential for long-acting injectable formulation • Initial data demonstrate tolerability may be improved compared to efavirenz • Potential for higher barrier to resistance	• Drug-drug interactions with CYP3A4 inducers and inhibitors and oral contraceptives • Cannot be used with proton pump inhibitors • Potential for QTc prolongation at high exposures
Lersivirine[67–69] 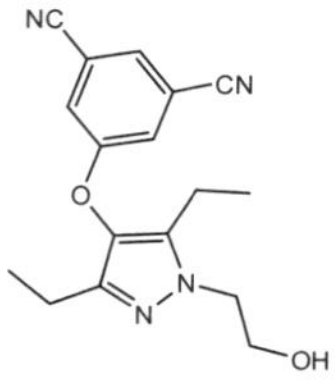	• Resistance profile is different from other NNRTIs • Lower potential for drug-drug interactions	• High dose • Nausea seen at higher doses in early clinical studies

Table 12.3 (*Continued*)

Drug class/generic name	*Potential advantages*	*Potential disadvantages*
Integrase Inhibitors		
Elvitegravir[70,71] 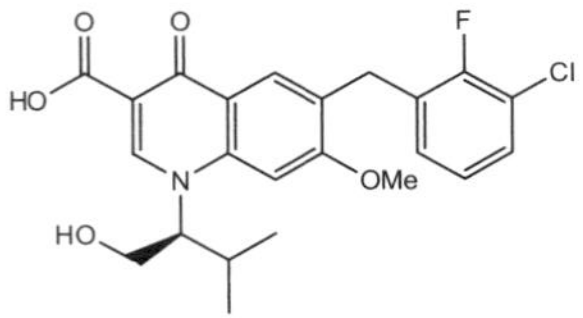	• Once-daily dosing	• Requires pharmacokinetic boosting with ritonavir • Higher propensity for drug-drug interactions
Dolutegravir[72] 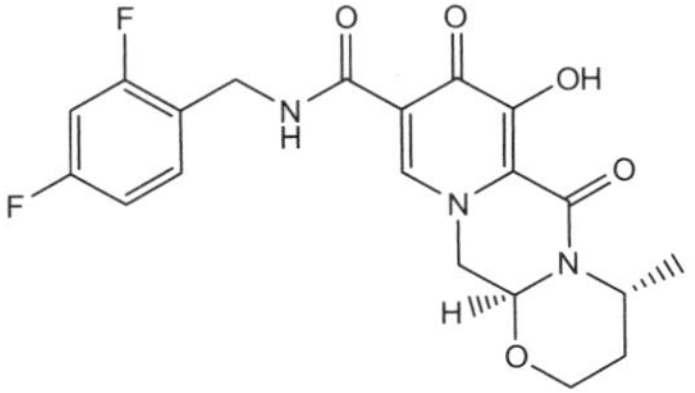	• Once-daily dosing • Low dose • Active against raltegravir resistant virus	

*Rilpivirine was approved by FOA for the treatment of HIV in treatment-naïve patients in May 2011.

through high-throughput screening and is effective against R5 HIV strains (including strains resistant to other drug classes) in the low nanomolar range (Figure 12.2 Panel A).[80] It has been demonstrated to be active in both treatment-naïve and treatment-experienced patients infected with R5 HIV, with an excellent safety and tolerability profile.[81–83] Initial concerns regarding the potential for emergence of X4 virus strains in patients treated with CCR5 antagonists have been mitigated by data demonstrating that there are no adverse clinical consequences of failure with X4 virus in maraviroc-treated patients.[82,84] The unique mechanism of action of maraviroc means there is no cross-resistance to other classes of antiretroviral agents, making this a valuable option even in patients with drug resistant HIV strains. Furthermore, the excellent safety and tolerability profile, extracellular mode of action, and the fact that it prevents the cell from getting infected through prevention of entry, makes this an attractive candidate for both oral and topical prevention strategies.[85] The use of maraviroc is complicated by the fact that it is extensively metabolized via the CYP3A4 pathway and requires dose adjustment when co-administered with potent CYP3A4 inhibitors or inducers. However, it has no effect on the metabolism of co-administered drugs.[86]

The biggest disadvantage of maraviroc is that it is only active against R5 virus, thus a virus tropism test is required prior to treatment with maraviroc.[87] The cost, limited availability and long turnaround times of some of these assays is a significant barrier to the use of maraviroc, especially in resource limited settings. The development of cheaper, more rapid and widely available tropism tests suitable for use in smaller less sophisticated laboratories will be crucial to expand access to this compound in resource-limited settings.

12.2.1.2 Fusion Inhibitors

Compounds that disrupt gp41-mediated membrane fusion (fusion inhibitors) were the first entry inhibitors to be identified for the treatment of HIV infection.[88] This discovery was triggered by biochemical and crystallization studies of gp41 peptides identified in epitope-mapping experiments designed to identify targets for vaccine development. These compounds prevent the formation of the six-helix bundle by competing for binding to the HR1 and HR2 domains on gp41, and have potent antiviral activity.[89]

Enfuvirtide (T20), is a linear, 36 amino acid synthetic peptide with a sequence identical to part of the HR2 region of gp41 (Figure 12.2, Panel B).[90] It competes for binding to HR1 and is the only agent from this class which is approved for the treatment of HIV infection.[88] It was approved by the FDA for treatment-experienced, HIV-infected patients in 2003 following the demonstration of excellent efficacy and safety in clinical trials in patients with advanced disease and limited treatment options.[91–93] Similar to maraviroc, enfuvirtide has a unique mechanism of action and retains activity against viruses resistant to agents from other antiretroviral drug classes.[94] However, despite the undisputed utility of this compound in the treatment of HIV, its

A

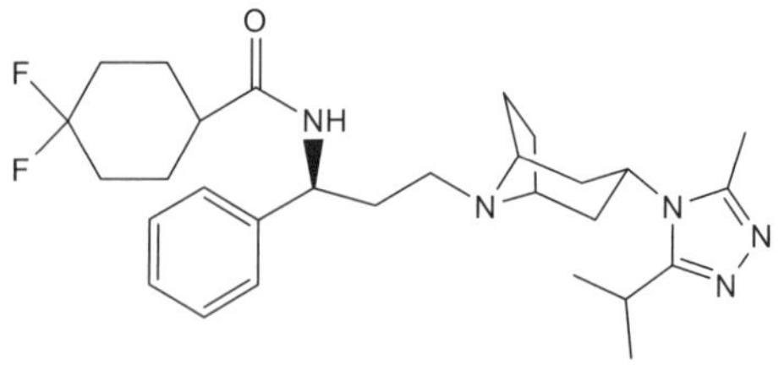

B

Ac-Tyr-Thr-Ser-Leu-Ile-His-Ser-Leu-Ile-Glu-Glu-Ser-Gln-Asn-Gln-Gln-Glu-Lys-Asn
-Glu-Gln-Glu-Leu-Leu-Glu-Leu-Asp-Lys-Trp-Ala-Ser-Leu-Trp-Asn-Trp-Phe-NH2

Figure 12.2 Inhibitors of HIV entry. Chemical structure of the CCR5 antagonist maraviroc (Panel A) and the peptide sequence of enfuvirtide, an HIV fusion inhibitor (Panel B).

use has been limited by need for administration by subcutaneous injection twice daily and the high cost of the drug. This is likely to be an insurmountable barrier to the widespread use of this compound in the developing world.

12.2.2 Reverse Transcription and Reverse Transcriptase Inhibitors

Following fusion, the virus particle is partly uncoated and the core enters the cellular cytoplasm where the viral RNA is copied to form double-stranded complementary DNA (Figure 12.1). This process is mediated by the viral reverse transcriptase (RT) enzyme. RT mediates three essential activities in viral replication: (i) RNA dependent DNA polymerase activity (reverse transcription), (ii) RNase H activity (cleavage of the genomic RNA from RNA/DNA hybrids), and (iii) DNA-dependent DNA polymerase activity (synthesis of the second strand of proviral DNA).[95] Two antiretroviral drug classes, nucleoside/nucleotide RT inhibitors (NRTIs) and non-nucleoside RT inhibitors (NNRTIs), target this stage in the HIV-1 replication cycle.

12.2.2.1 Nucleoside/Nucleotide RT Inhibitors

This was the first antiretroviral drug class to be discovered and there are currently 7 NRTIs on the market for the treatment of HIV infection, as well as several fixed-dose dual NRTI combinations (Table 12.2). Additionally, two triple-class fixed-dose combinations containing a dual NRTI backbone are available. They are analogues of endogenous 2'-deoxy-nucleosides and nucleotides (Figure 12.3). In this form they are "prodrugs" and require modification by host cell kinases and phosphotransferases to form deoxynucleoside triphosphate (dNTP) analogues, which compete with endogenous dNTPs for incorporation into cDNA by the HIV RT enzyme.[64,65] They are structurally diverse and are metabolized to analogues of all four natural dNTPs used during

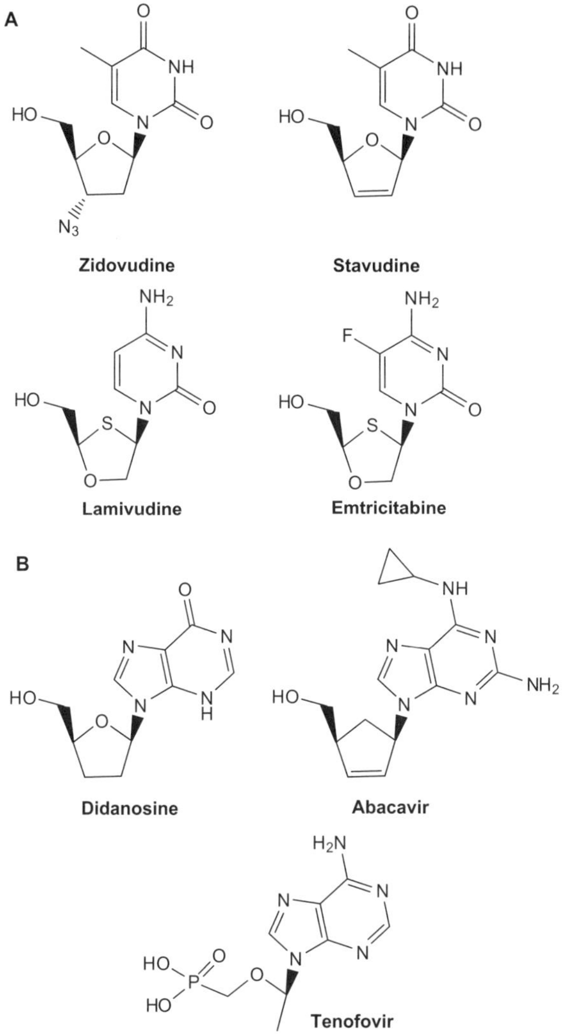

Figure 12.3 Nucleoside/nucleotide reverse transcriptase inhibitors. Chemical structures of the approved nucleoside/nucleotide reverse transcriptase inhibitors. Panel A: Pyrimidine analogues. Panel B: Purine analogues

DNA synthesis. All are obligate chain-terminators of DNA elongation thus preventing proviral DNA synthesis.[64]

The NRTIs remain the backbone of most current antiretroviral regimens and are combined with an anchor drug from one of the other drug classes to form preferred first-line regimens.[87,96] The continuing success of NRTIs is in part due to intracellular accumulation and prolonged retention of active metabolites of some NRTIs; this allows for once-daily dosing, more forgiving pharmacokinetics, and can buffer the pharmacokinetic fluctuations in levels of other drugs in the regimen.[64] The synergistic antiviral activity with other drug classes, especially NNRTIs, may also contribute to their success in preferred first-line combination regimens.[64,97] The unique metabolism of NRTIs results in less frequent and smaller magnitude drug-drug interactions compared with other antiretroviral drug classes. Many NRTIs are available as generic formulations, or are made available to low-income countries at cost price; they thus play a key role in the management of HIV in resource poor settings.[64]

The key disadvantage of NRTIs is that they may compete with each other for intracellular phosphorylation and metabolism (for example, zidovudine and stavudine), which may result in less than additive antiviral activity. Other interactions, such as that seen between tenofovir and didanosine, may result in increased levels of one of the components with increased rates of adverse events.[64,87] The propensity for some of the NRTIs (especially the thymidine analogues zidovudine and stavudine) to affect mitochondrial DNA polymerase γ may result in significant long-term toxicities.[64] This has resulted in a shift away from the use of these agents, but they are still frequently used in resource-poor settings.[96] Although all NRTIs have different resistance profiles, activity of all are reduced by the presence of thymidine analogue mutations (TAMs). The degree of reduction in susceptibility correlates with the number of TAMs present.[64]

12.2.2.2 Non-nucleoside RT Inhibitors

In contrast to the NRTIs, compounds from the NNRTI class inhibit RT by binding to the enzyme in a hydrophobic pocket located close to its catalytic site. Although the NNRTIs are chemically diverse, they all bind to the same site on

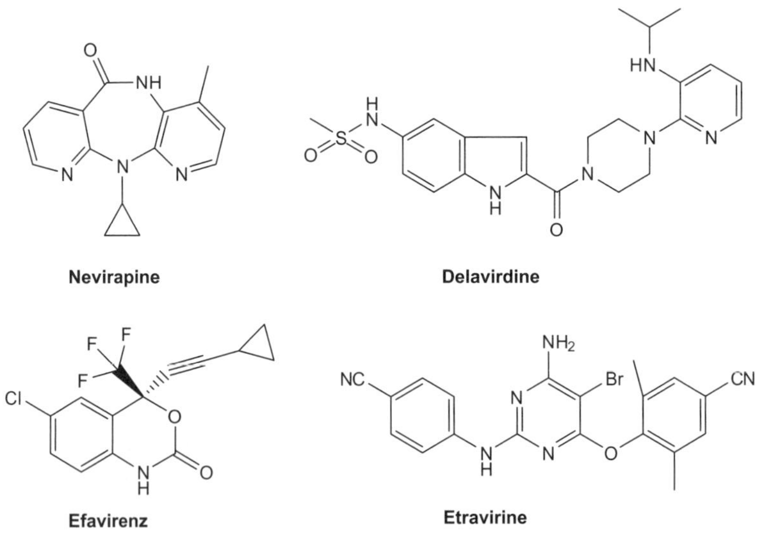

Figure 12.4 Non-nucleoside reverse transcriptase inhibitors. Chemical structures of the approved non-nucleoside reverse transcriptase inhibitors.

RT (albeit in slightly different ways for the second-generation compounds) (Figure 12.4). This interaction induces conformational changes in the RT enzyme that impacts its catalytic activity.[65] The exact mechanism through which the catalytic activity of RT is inhibited by this class of compounds remains unclear.

Three first-generation NNRTIs were approved by the FDA in the late 1990s. The first of these was nevirapine in 1996, followed by delavirdine and efavirenz.[65] Although delavirdine is seldom used today, efavirenz, and nevirapine to a lesser extent, continue to form the cornerstone of triple combination regimens for initial antiviral therapy.[65,87,96]

Efavirenz has demonstrated safety and durable efficacy in multiple controlled clinical trials and remains the most commonly used anchor drug in first-line antiretroviral regimens.[98] One of the key advantages of efavirenz is the long half-life, enabling once-daily administration. It is also co-formulated with two NRTIs (tenofovir and emtricitabine) to form a once-daily fixed dose combination regimen. This lowers the pill burden and simplifies adherence.[98] The ease of use, proven efficacy, forgiveness and availability of generic formulations makes this a good treatment option in developing countries. The key disadvantages of efavirenz are the low genetic barrier to resistance, the propensity for drug-drug interactions and the potential for teratogenicity, which raises concerns regarding its use in females who may become pregnant. A single point mutation in the HIV pol gene (K103N) results in high-level resistance to not only efavirenz, but also to other first generation NNRTIs, limiting further use of this class following efavirenz first-line failure.[65] Efavirenz is both a substrate and inducer of several P450 enzymes (CYP2B6, CYP3A4, CYP2A6, CYP2C9 and CYP2C19), resulting in multiple drug-drug interactions.[98] It reduces the plasma concentrations of several antiretroviral drugs, including the protease inhibitors and maraviroc.[98,99]

Nevirapine has similar efficacy to efavirenz and is an alternative NNRTI option in females who may want to become pregnant or in patients who cannot tolerate efavirenz. However, the high risk of hepatotoxicity in patients who start therapy at higher CD4 counts means that treatment has to be carefully monitored, especially early on.[87] It is a CYP3A4 substrate and inducer and is subject to drug-drug interactions with other drugs affecting, or affected by, this metabolic pathway. One of the most difficult interactions to manage in the developing world is that with rifampicin, one of the cornerstone drugs used in the treatment of tuberculosis and itself a potent CYP3A4 inducer. This complicates the use of nevirapine, and to a lesser extent efavirenz, in areas with high rates of TB and HIV co-infection.[100,101] There are several generic formulations of nevirapine available, and it is available as a twice-daily fixed-dose combination regimen with lamivudine and stavudine (Triomune™).

Etravirine is the only registered second-generation NNRTI. Unlike other NNRTIs, etravirine requires multiple mutations for the development of resistance, and it retains activity against viruses with NNRTI resistance-associated mutations.[102] The results from the clinical studies in treatment experienced patients with extensive NRTI and PI resistance, and documented NNRTI resistance, have demonstrated efficacy of etravirine, thus confirming the activity of etravirine against these virus strains.[103,104] Etravirine is a substrate and an inducer of CYP3A4 and an inhibitor of CYP2C9 and CYP2C19 and is subject to numerous drug-drug interactions. This significantly complicates its use and prevents co-administration with PIs.[87,102]

12.2.3 Integration of Proviral DNA and Integrase Inhibitors

The integrase enzyme combines with viral DNA and cellular cofactors to form a pre-integration complex (Figure 12.1). The pre-integration complex enters the host cell nucleus to begin strand transfer. Following completion of this process, any gaps remaining between viral and host cell DNA are repaired by host cell enzymes.[105]

Integrase is a unique viral enzyme and as such was recognised as a promising target for intervention very early on. However, the development of integrase inhibitors proved to be extremely challenging. Early approaches focussing on using ribozymes and small molecules with DNA-binding activity; triple helix forming oligonucleotides and aptamers demonstrated significant risk of toxicities and none proceeded into the clinic.[70] The development of *in vitro* assays for the analysis of integrase activity, including high-throughput screening assays in the 1990s provided an opportunity for rapid progress in finding safe and potent small molecule inhibitors of HIV integrase.[70] This led to the discovery of raltegravir, a potent and specific inhibitor of HIV integrase (Figure 12.5). Raltegravir, the only registered integrase inhibitor to date, targets the strand transfer step of viral DNA integration, through inhibition of the binding of the pre-integration complex to host cell DNA, thereby halting the integration process.[71]

Raltegravir demonstrated excellent efficacy in clinical trials in both treatment experienced and treatment naïve HIV infected patients and was first approved by the FDA in 2007.[106–108] Raltegravir also has an excellent safety profile, and as it is not metabolised by CYP450, it has limited drug-drug interaction potential.[71,106,107] However, raltegravir has a low genetic barrier to resistance, with a single-point mutation resulting in high-level resistance in some cases.[109] Furthermore, cross-resistance of raltegravir-resistant virus strains with elvitegravir (currently in advanced development) has been demonstrated.[110]

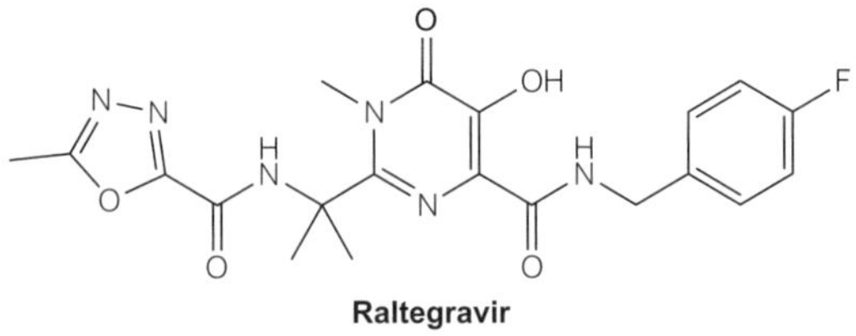

Figure 12.5 Integrase inhibitors. Chemical structure of the integrase inhibitor, raltegravir.

12.2.4 Production and Maturation of Progeny Virions and Inhibitors of Viral Protease

The next step is transcription of HIV-1 proteins from the integrated proviral cDNA (Figure 12.1). Full length HIV transcripts serve three purposes; (i) genomic RNA for progeny virions that are assembled at the plasma membrane, (ii) mRNA for translation of Gag and Gag-Pol polyproteins in the cytoplasm, and (iii) precursors for alternatively spliced mRNAs that are translated in the cytoplasm to produce Env glycoproteins and accessory proteins. The ratio of spliced to full-length RNA is controlled by the regulator of viral gene expression (Rev). These viral RNAs are then transported out of the nucleus and translated into viral proteins.[95] The viral particle is then assembled at the plasma membrane and the newly assembled virus pushes out (or buds) from the host cell, acquiring its lipid bilayer membrane. Following budding, the final maturation process takes place through proteolytic cleaving of the Gag and Gag-Pol precursor polyproteins by the viral protease enzyme.[111] The progeny HIV virions are now able to infect new cells.

Detailed knowledge of the structure of the HIV protease enzyme has allowed the development of protease inhibitors (PIs). They bind the viral protease with high affinity, but in such a way that they occupy more space than the natural substrate. All of the currently available PIs apart from tipranavir, are competitive peptidomimetic inhibitors thus mimicking the natural substrate of the viral protease (Figure 12.6).[112] The PIs were the second class of antiretroviral drugs to be discovered, and the seminal finding that treatment of HIV with a highly active regimen consisting of a 3-drug combination has a durable impact on viral replication and a profound impact on mortality, stemmed from their discovery. Since then PIs have continued to play a key role in HIV therapy, especially in treatment-experienced patients.

The key advantage of PI-based antiretroviral therapy is that it has a high genetic barrier to resistance and for the development of high-level resistance multiple mutations are required.[87,112] Furthermore, virologic failure with one PI does not normally result in class-resistance, which allows sequential treatment with different PIs.[113] Unlike efavirenz, they are not associated with teratogenic effects and can thus be used in females who want to become pregnant.[87,113] The major disadvantage is that they require pharmacokinetic boosting with ritonavir for optimal efficacy in most cases. Also, in addition to being CYP3A4 substrates, most PIs are both inhibitors and inducers of the CYP3A4 pathway and can result in significant increases in concentrations of co-administered drugs metabolised via this pathway.[113] This is extremely complex to manage when co-administered with treatment for tuberculosis, especially as very limited drug-drug interaction data exists. Boosted PIs are contra-indicated for use with rifampicin, severely complicating their use in co-infected patients.[96] Many PIs have a relatively high pill burden, as well as significant gastro-intestinal adverse effects and long-term metabolic complications, which can impact on adherence. Finally, PIs are relatively expensive, making them less attractive options for use in first-line regimens in low- and middle-income countries.

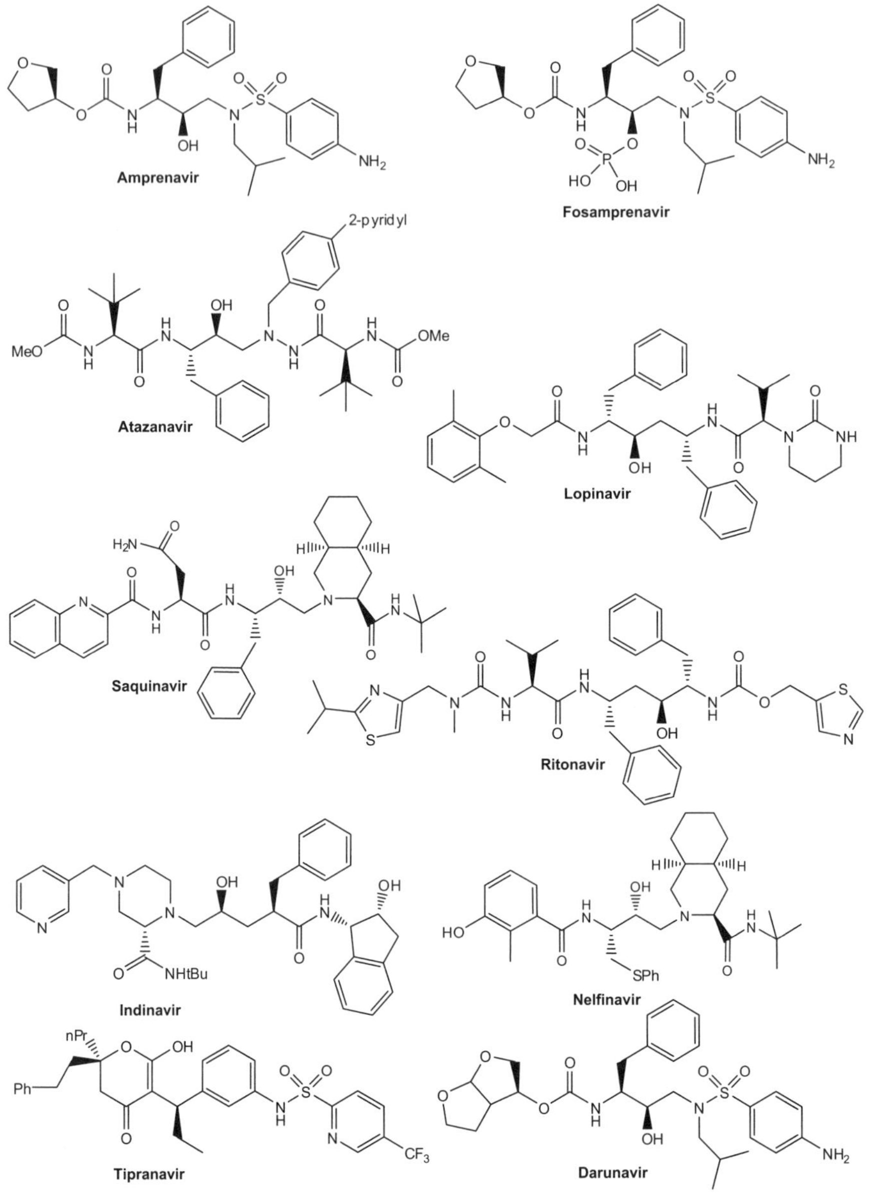

Figure 12.6 Protease inhibitors. Chemical structures of the approved, protease inhibitors.

12.2.5 Ongoing Challenges – Managing Adverse Effects and Drug Resistance

In spite of the huge progress made in the treatment of HIV there are several remaining challenges. These include the intriguing possibility of virus eradication, and management of chronic immune activation. However, the most important short-term challenges remain those of addressing adverse effects and overcoming resistance.

Antiretroviral drugs are associated with a number of both short- and long-term adverse effects summarised in Table 12.4[87,114] These adverse effects have

Table 12.4 Important adverse effects associated with antiretroviral drugs.[87,114]

Drug	*Adverse effect*
Nucleoside reverse trancriptase inhibitors	
Short-term adverse effects	
Abacavir	Hypersensitivity reactions
Didanosine	Pancreatitis Lactic acidosis with hepatic steatosis
Stavudine	Pancreatitis Lactic acidosis with hepatic steatosis Ascending neuromuscular weakness
Zidovudine	Bone marrow suppression, Gastro-intestinal upset, Lactic acidosis with hepatic steatosis
Long-term adverse effects	
Didanosine	Peripheral neuropathy
Stavudine	Mitochondrial toxicity (lipoatrophy and dyslipidaemia) Peripheral neuropathy
Tenofovir	Renal insufficiency Decreased bone mineral density
Zidovudine	Dyslipidemia Mitochondrial toxicity (lipoatrophy and myopathy)
Non-nucleoside reverse transcriptase inhibitors	
Short term adverse effects	
Delavirdine	Hepatotoxicity Rash (with rare cases of Stevens Johnson syndrome) Headache
Efavirenz	Central nervous system and neuropsychiatric effects Rash (with rare cases of Stevens Johnson syndrome) Hepatotoxicity
Etravirine	Rash/hypersensitivity reactions (include organ failure) Nausea
Nevirapine	Rash Stevens Johnson syndrome Severe hepatotoxicity including hepatic failure[a]
Long-term adverse effects	
Efavirenz	Potential for teratogenicity Dyslipidemia
Protease inhibitors	
Short-term adverse effects	
All PIs	Gastro-intestinal adverse effects (less common for ATV and DRV) Possible risk of increased bleeding in haemophilia patients

Table 12.4 (*Continued*)

Drug	*Adverse effect*
Atazanavir	Hyperbilirubinemia and jaundice Prolonged PR-interval with first degree heart block
Darunavir	Rash (including Stevens Johnson syndrome and erythema multiforme) Hepatoxicity Headache
Fosamprenavir	Rash Hepatoxicity Headache
Indinavir	Hyperbilirubinemia and jaundice Headache, blurred vision, dizziness Asthenia Rash Metallic taste Alopecia Thrombocytopenia Hemolytic anaemia
Lopinavir/ritonavir	Hepatotoxicity PR interval prolongation QTc interval prolongation and *torsade de pointes*
Nelfinavir	Hepatotoxicity
Ritonavir	Paresthesias (circumoral and extremities) Hepatotoxicity Abnormal taste
Saquinavir	Hepatotoxicity Headache
Tipranavir	Hepatotoxicity (especially in patients with underlying liver disease) Rash Rare cases of intracranial bleeding
Long-term adverse effects	
All PIs	Metabolic changes (hyperglycemia, hyperlipidemia, insulin resistance and fat maldistribution) Increased cardiovascular risk
IDV, ATV, FPV	Nephrolithiasis
Integrase inhibitors	
Raltegravir	Nausea Headache Asthenia and fatigue
Fusion inhibitors	
Enfuvirtide	Injection site reactions Bacterial pneumonia Hypersensitivity reactions
CCR5 antagonists	
Maraviroc	Postural hypotension and dizziness Rash Hepatotoxicity

[a]More common in females with CD4 count >250 cells/mm^3 and males with CD4 count >400 cells/mm^3.

an impact on tolerability and play a major role in determining adherence to antiretroviral therapy. High-level adherence is one of the key factors driving success of treatment.[115] Furthermore, as HIV is now a chronic manageable condition, long-term adverse effects are becoming an important source of morbidity and mortality in HIV-infected patients.[114] It is critical to consider the impact of potential adverse events on all components when deciding on an appropriate regimen for any patient, including the possibility of additive effects and the potential impact or interaction with underlying conditions the patient may have, or drugs used for the treatment of these conditions. Thus, one of the key challenges in the development of new antiretroviral drugs is the discovery and development of drugs that are well tolerated and safe for long-term use.

Resistance to antiretroviral drugs is driven by the inherent genetic diversity of HIV. The reverse transcriptase enzyme is error-prone and lacks proofreading activity, resulting in frequent strand transfers during reverse transcription, leading to multiple recombination events.[116,117] Together with the high rates of virus replication, this results in any one individual in a swarm or quasispecies of different but genetically related viral variants instead of a unique viral genotype. Models of virus replication predict that any single mutant and many double mutants can be generated daily through viral replication.[118] Suboptimal regimens, either through reduced potency, resistance to one or more agent(s) in the combination, or lack of adherence, may select for mutants with a fitness advantage in the presence of therapy. Persistent viraemia in the presence of antiretroviral therapy leads to further accumulation of mutations, which further increases the degree of resistance.[118] Antiretroviral drug resistance remains a major obstacle for the adequate management of HIV infection. The development of new drugs that work against different stages of the HIV lifecycle and retain activity even against multidrug-resistant virus strains therefore remains a key priority.

12.3 Current State of the Art in the Management of HIV-1 Infection

Antiretroviral therapy for treatment of human immunodeficiency virus type 1 (HIV-1) infection has improved steadily since the advent of potent combination therapy in 1996. Analysis of 18 trials that included more than 5000 participants with viral load monitoring showed a significant association between a decrease in plasma viraemia and improved clinical outcome.[119] The ultimate goal of therapy today is complete suppression of plasma HIV RNA levels to below limits of assay detection (<50 copies/ml for most assays).[87] New drugs that offer new mechanisms of action, improvements in potency, and activity even against multidrug-resistant viruses, dosing convenience and tolerability, means that this is an achievable goal, even in patients with very advanced disease and/ or those carrying multidrug-resistant virus strains.

There are multiple guidelines for the treatment of HIV, and individual countries or regions often have their own guidelines. However, these guidelines are very similar and have evolved over time to base recommendations primarily on evidence from well-conducted randomized controlled trials and to keep up with new data on antiretroviral therapies and their use.

All guidelines agree that any adult with symptomatic HIV infection and/or a CD4 count of <200 cells/mm^3 should receive antiretroviral treatment (ART). However, most current guidelines, including those from WHO, agree that antiretroviral therapy should be started when the CD4 count declines to <350 cells/mm^3, while the United States Department of Health and Human Services (DHHS) guidelines also recommend treatment for patients with CD4 count >350 and less than 500 cells/mm^3.[87,96] The DHHS recommendation is based on the results of several large studies that have demonstrated a reduction in both AIDS and non-AIDS related complications, as well as all-cause mortality when ART was initiated above 350 cells/mm^3, resulting in increasing calls for even earlier initiation of therapy.[120,121] However, the potential benefit of earlier initiation of therapy should be weighed against the increased risk of toxicity and impact on quality of life, as well as cost.[87]

The choice of the initial regimen for a patient should include consideration of possible transmitted resistance (a baseline resistance test should ideally be done), pill burden, dosing schedule, tolerability profile, co-morbid conditions like cardiovascular or renal disease and co-infections like hepatitis B and C, as well as, the anticipated long-term toxicity profile of the ART regimen.[122] The most commonly recommended initial regimens are an NNRTI, ritonavir boosted PI or an integrase inhibitor combined with a dual nucleoside/nucleotide combination. Of note, stavudine, which is still commonly used in many developing countries as a component of first-line regimens (due to low cost) is not now recommended for initial therapy due to concerns regarding high rates of serious toxicities, especially lactic acidosis, pancreatitis and peripheral neuropathy.[87,96]

Given our current understanding of viral dynamics during treatment, it is expected that most first-line antiretroviral regimens should be able to suppress the virus for a very long period of time, assuming that an optimal regimen is selected and assuming that the patient can adhere to that regimen indefinitely. However, antiretroviral treatment failure is not uncommon and as it increases the risk of HIV disease progression it should be addressed aggressively.[87] There is not complete agreement on when to switch treatments due to virologic failure, but it is generally accepted that patients with low-level viraemia (HIV-1 RNA <1000 copies/ml) could be closely followed, while efforts are made to ensure optimal adherence and adequate plasma drug levels. For patients with persistent viraemia (HIV-1 RNA >1000 copies/ml), resistance testing should be performed to guide further treatment decisions.[87] When a patient is switched to a new regimen, it is critical to obtain a resistance test, a careful history of the patient's antiretroviral treatment history, a review of previous resistance testing results, and a review of adherence patterns. This is to ensure selection of at least two, but preferably three fully active agents that the patient is able to tolerate to provide an optimal regimen for the patient.[87,122]

12.3.1 Management of HIV Infection in Paediatric Patients

Highly active combination antiretroviral regimens including at least three drugs are associated with enhanced survival, reduction in opportunistic infections and other complications of HIV infection, improved growth and neurocognitive function, improved quality of life, and a reduction in mortality in children.[123,124]

Factors to consider in selecting regimens for treatment of HIV-infected children are similar to that of adults, but in addition, the availability of appropriate (and palatable) drug formulations, pharmacokinetic information on appropriate dosing in the child's age group, and potential previous exposure as part of MTCT must be considered.[125] It is important to take into account growth and development of the patient as this can significantly affect drug absorption and disposition, and thus may change over time.[126] Furthermore, immature renal function, developing hepatic enzyme activity (particularly for cytoctrome P450 involved in the metabolism of several ARVs) and differences in drug absorption lead to variations in systemic exposure of ARVs among children of different ages.[127] It is therefore important that children on antiretroviral therapy are monitored very regularly (both for efficacy and toxicity of their regimen) and that their treatment is adjusted to cope with their changing physiology and needs. All of these factors present considerable challenges and require highly trained health care workers.

12.3.2 Prevention of Mother to Child Transmission

In the developed world there has been dramatic progress in reducing MTCT of HIV. Early identification of HIV infection in pregnant women through routine antenatal HIV testing; immediate assessment of HIV-infected pregnant women to assess whether they need treatment, and provision of antiretroviral treatment or drug prophylaxis regimens has substantially reduced the risk of infection among infants during pregnancy and delivery. When combined with elective caesarean delivery and complete avoidance of breastfeeding, these interventions have reduced the risk of HIV transmission to 1–2%.[128]

When considering the prevention of MTCT, it is critical to firstly consider the optimal treatment of the mother. Females who meet the criteria for treatment should be managed as indicated by the relevant treatment guidelines with consideration for selecting the most appropriate drug regimen for pregnancy. Optimal MTCT prevention strategies aim to reduce perinatal transmission through several mechanisms, including lowering maternal viral load and pre- and post-exposure prophylaxis of the infant.

The DHHS perinatal guidelines recommend a combination of antepartum, intrapartum and infant prophylaxis, as well as ceasarean section for females with HIV-1 RNA $>$1000 copies/ml at week 38.[129] It is recommended that breastfeeding is avoided where feasible. However, in resource-limited settings this is often not feasible because of cost, lack of an alternative food supply and no access to clean water. Thus, there is a critical need to identify strategies to address breastfeeding-associated HIV transmission. Exclusive breastfeeding has been shown in observational studies to lower the risk of postnatal transmission, compared with mixed feeding, but does not eliminate risk.[130]

The current WHO guidelines recommend treatment of all pregnant women with a CD4 count <350 cells/mm^3, using a triple antiretroviral combination regimen. For females who do not qualify for treatment, two options, both of which should start at 14 weeks gestation, or as soon as possible thereafter, are recommended. The first option is twice-daily zidovudine for the mother and infant prophylaxis with either zidovudine or nevirapine for six weeks after birth if the infant is not breastfeeding. If the infant is breastfeeding, daily nevirapine infant prophylaxis should be continued until one week after the end of the breastfeeding period. The second option is a three-drug prophylactic regimen for the mother taken during pregnancy and throughout the breastfeeding period, as well as infant prophylaxis for six weeks after birth, whether or not the infant is breastfeeding.[131]

12.4 Universal Access to Antiretroviral Drugs – What are the Challenges?

Access to, and outcomes of, HIV treatment are intricately related to economic, political and geographic factors. The United Nations Millennium Development Goals included two key goals related to HIV, namely to have halted and begun to reverse the spread of HIV by 2015, and universal access to treatment for all HIV-infected individuals who need it by 2010. Although significant progress has been made towards the goal of universal access to treatment, with >4 million people with HIV in low- and middle-income countries having received antiretroviral treatment at the end of 2008; $>50\%$ of the 9.5 million people in low- to middle-income countries needing HIV treatment still did not have access to it at the end of 2008. Encouragingly, the greatest progress was seen in sub-Saharan Africa, the region hardest hit by the HIV epidemic. The number of children receiving antiretroviral therapy has also increased significantly, but still falls far short with only 38% of those in need receiving therapy. Approximately 45% of pregnant women with HIV received treatment to prevent MTCT.[132]

Significant decreases in the prices of the most commonly used antiretroviral drugs have contributed to wider availability. The cost of most first-line regimens decreased by 10–40% between 2006 and 2008, but later regimens for people who have either failed, or cannot tolerate first-line regimens, remain expensive.[132] At the end of 2009, the median cost of the most commonly used second-line regimen was $853 per person per year in low income countries, compared to $137 for that of the most widely used first-line regimens.[133] Overall, access to treatment services is still falling far short of need and there is fear that the global economic crisis may impact on sustainability of existing service levels.[132]

12.4.1 Key Challenges for HIV Treatment in the Developing World

The clinical management of the HIV-infected individual is complex, and requires well-trained health care providers and adequate facilities, in addition to drug supplies. Furthermore, optimal management requires monitoring of HIV-RNA levels and CD4 count at regular intervals, as well as laboratory safety testing to monitor for common adverse effects. In patients who fail therapy, resistance testing is recommended to allow for optimal management and selection of the next regimen.

Many developing countries are faced with a very high prevalence of HIV infection in the context of inadequate resources including facilities, health care workers and funding. It is not only the demand for services that impacts on the health care system, but also the high prevalence of HIV infection in health care workers. Two surveys conducted in South Africa found that 13.7% and 16.3% of nurses were HIV positive.[134] Outside of research settings, staff are often not supported and become overwhelmed by numbers, having to care for very ill patients in ill-equipped environments. They are often infected themselves and are dealing with sickness and death in their own households and families. This leads to despair and burnout and to health care workers leaving the system. Task-shifting to allow nurses to prescribe and monitor antiretroviral therapy to overcome the shortage of physicians has been investigated in several trials.[135]

There are multiple factors affecting the prognosis of patients receiving treatment in resource-poor settings. Despite the recently updated WHO treatment guidelines which suggest a CD4 count threshold of <350 cells/mm^3 for the initiation of antiretroviral therapy in otherwise asymptomatic patients, many countries still use a CD4 count threshold of <200 cells/mm^3 to limit the demand on overstretched health care systems.[136] Early treatment offers several advantages including better immune restoration, less drug resistance, less adverse events, more durable and robust viral suppression, decreased rates of active tuberculosis and secondary cases of TB infection. However, these advantages need to be balanced with the cost of treating higher numbers of patients.

Patients in developing countries in general present with much more advanced disease. The median CD4 count at initiation of antiretroviral therapy in South Africa is as low as 87 cells/mm^3 and patients often first present with opportunistic infections.[137] There are multiple reasons for this, including stigma and the inefficiencies of the health care system. Up to 40% of patients are lost to care prior to initiation of treatment due to the necessity of multiple visits to do CD4 counts, return for results (which often are not yet available), and to exclude tuberculosis prior to the initiation of therapy. Cheap and robust point-of-care viral load and CD4 testing, serum biochemistry and tuberculosis diagnostics for staging and monitoring will ensure that the number of visits are reduced and the interval between testing and being provided with a result is significantly shortened. It is important that such tests are easy to use, portable and heat stable with high throughput, as well as affordable.[138]

Based on the data from the DART study that compared clinical monitoring of adverse events and treatment failure to laboratory safety and CD4 count monitoring, and where the 5-year survival rate in both arms was more than 75%, it has been suggested that clinical monitoring only could be used in HIV care. This would make it easier to decentralize care to make it more accessible and the saving on laboratory investigations could be used to fund a significant number of patients' antiretroviral drugs.[139] However, to achieve long-term survival of HIV-infected patients and prevent the emergence of circulating multidrug-resistant HIV strains, it is critical to do regular viral load monitoring to detect early virologic failure.[118] The sheer number of people on treatment in the developing world where patients are often only clinically monitored without regular viral load testing will eventually lead to widespread circulation of drug-resistant virus strains and transmitted resistance; this will result in newly diagnosed patients not responding to first-line treatment regimens.[140] However, most sub-Saharan African countries do not have the resources to do viral load monitoring. Even in South Africa where viral load monitoring is part of the public sector programme, the National Health Laboratory Services struggle to keep up with the demand. The development of cost-effective qualitative viral load measurement technologies which can differentiate between viral loads more or less than 500 copies/ml, and can be used under field conditions, to ensure that patients are not left on failing regimens for prolonged times, are critical to enable the long-term success of treatment and to reduce the risk of transmitted resistance. The use of single or dual drug regimens in MTCT may also contribute to this risk.[16]

The WHO currently requires monitoring for transmitted drug resistance among recently infected patients. If resistance rates in recently infected treatment-naïve individuals exceeds 15%, all patients should have drug resistance genotyping before selection of the initial antiretroviral regimen.[141] However, current genotypic resistance assays are labour intensive and too expensive for widespread implementation in developing countries. There is a need for cost-effective genotyping that can be done in less sophisticated laboratories. Additionally, current resistance algorithms are mainly derived from subtype B virus isolates. Algorithms that include data from non-subtype B HIV that accounts for the majority of infections in the developing world, must be developed. One such an example is a public database for the analyses of subtype C HIV strains developed by the SATuRN network.[142]

12.4.2 Optimisation of Antiretroviral Drugs for Developing Countries

For development of any new antiretroviral drug it is critical not only to consider the cost of goods, but also ease of use as that will determine the potential "total cost" to use the drug. In addition to being well tolerated and having once-a-day dosing with a "forgiving" pharmacokinetic profile, and a low pill burden, an ideal antiretroviral regimen for use in the developing world should require minimal safety monitoring, have no drug-drug interactions with other commonly used drugs in these settings, be safe in pregnancy, require no special storage, and have a high barrier to resistance. This is a significant hurdle and unlikely to be achieved in the immediate future.

However, consideration should also be given as to how currently available antiretroviral drugs could be optimised for use in these settings. Genetic background, concomitant diseases, circulating HIV subtypes, and nutritional status in the developing world often differ significantly from those in countries where the parent drugs were originally evaluated. There is thus a need for pharmacokinetic and pharmacodynamic data on all drugs or drug combinations when introduced into resource-poor settings.[143] Of critical importance is the extensive evaluation and complete understanding of the drug-drug interactions with drugs used in the treatment of other prevalent conditions such as malaria and tuberculosis.

Fixed-dose combination regimens to decrease the pill burden will improve adherence, reduce costs and improve long-term success. Of the currently available combinations, Atripla® (efavirenz/tenofovir/emtricabine) is the most widely used, but it may have teratogenic potential and is therefore not recommended for use in the first trimester of pregnancy by most current treatment guidelines.[87,129,131] Triomune® (nevirapine/lamivudine/stavudine) is also commonly used in the developing world, but mitochondrial toxicity due to stavudine and the interaction between rifampicin and nevirapine means that this is not an ideal combination.[144] Therefore, more research to develop other fixed-dose combination regimens that are safe, effective, easy to use and affordable, is critical. Furthermore, the vast majority of paediatric patients with HIV infection are found in resource-poor countries. Data on bioavailability, pharmacokinetics, drug-drug interactions, efficacy and safety of antiretroviral drugs in children in these settings is very limited and it is critical that these data are obtained in order to optimise the treatment of children with HIV in the developing world.

12.5 Antiretroviral Drugs and Prevention of HIV-1 Infection – Future Directions

In addition to treatment of patients already infected by HIV, there is the intriguing possibility of using antiretroviral drugs in a prophylactic fashion, i.e. to prevent infection with HIV. The utility of prophylactic therapy has been demonstrated for many infections, including surgical, gastrointestinal, upper respiratory, malaria and meningococcal infections. Furthermore, as discussed above, provision of antiretroviral therapy for pregnant women to prevent MTCT of HIV has resulted in a dramatic reduction in the number of children infected via this route.

12.5.1 Pre-exposure Prophylaxis Using Oral Antiretroviral Therapy

Pre-exposure prophylaxis (PrEP) has been proposed as a method to prevent HIV infection in very high-risk populations, such as commercial sex workers.[145] In support of this, a randomised, blinded, placebo-controlled randomized trial in African women at high risk of HIV infection confirmed the safety of tenofovir once daily in this population.[146] Clinical trials are currently evaluating whether daily oral tenofovir with or without emtricitabine provides additional protection compared to standard prevention care, including HIV testing, counselling, condoms, and management of sexually transmitted infections.[147] Drugs with a long half-life that can be administered with injectable contraceptives once a month will be ideal for pre-exposure prophylaxis in at risk populations.

However, establishing antiretroviral therapy for widespread use to prevent infection in healthy uninfected people is likely to be challenging. Firstly, in the absence of a surrogate marker for efficacy, the incidence of new HIV infections is the only potential endpoint; this means that prevention trials must be large, enrolling many thousands of participants who are followed up for several years at great expense. Secondly, the burden of proof for demonstrating safety (and an acceptable risk/benefit) of an oral antiretroviral drug or drug combination in a healthy uninfected population will be huge. Thirdly, in the case of failure of the intervention to prevent infection, development of resistance to the drugs could occur, especially if suboptimal mono or dual therapy regimens are used in PrEP. Finally, the cost of providing the drugs and monitoring for efficacy and safety in large population is likely to be prohibitive.

12.5.2 Microbicides

The vast majority of HIV-1 infections are transmitted sexually. Although condoms are highly effective in preventing HIV-1 infection they are often not used. Reasons for this include individuals not being able to negotiate their use, condoms not being available, or couples choosing not to use them. Topical (vaginal and rectal) HIV microbicides may offer an important alternative means of HIV-1 prevention in these situations.[148] Furthermore, it could provide a female-controlled method of HIV prevention in paternalistic societies where females may have limited bargaining power to negotiate safe sex. An effective microbicide could thus have a significant impact on public health, especially in areas of the developing world with high transmission rates.

However, there are considerable challenges to the development of a microbicide, including; (i) the need to ensure that the microbicide product is active in the vaginal and/or rectal environments, including during and after sexual intercourse, (ii) the need for an effective formulation that is acceptable to the target population, (iii) the need to protect the normal vaginal microenvironment from harm, (iv) the need to protect from both cell-free and cell-associated innoculae and (v) the need to understand the degree of systemic absorption of the product from the vaginal or rectal mucosae. Finally, as these products will be used in both the developed and developing world, low cost, ease of application and excellent stability characteristics are essential. Ideally, any preparation should include compounds that inactivate virus in the seminal and/or vaginal fluid, or that prevent establishment of infection (for example, by preventing virus entry).[148,149]

In order to overcome these challenges, concerted, internationally coordinated efforts will be needed. In 2002, the International Partnership for Microbicides (IPM) was established with the aim of accelerating the development and accessibility of microbicides to prevent HIV-1 transmission in women. By screening multiple compounds, evaluating optimal combinations, designing optimal formulations, establishing manufacturing capacity, developing trial sites and conducting clinical trials, the efficiency of all efforts to develop and deliver safe and effective microbicides should be improved. IPM has been granted licenses to develop several established and experimental antiretroviral compounds as microbicides. Their position as an independent organization should also facilitate the development of microbicides containing combinations of antiretroviral products.

The first generation of microbicides included agents that were primarily surfactants like nonoxynol-9, products that enhanced vaginal defences through mechanisms such as maintenance of vaginal acidity, or sulphated polysaccharides that disrupt CD4 binding of the virus. Five products (Nonoxynol-9, Carraguard, cellulose sulfate, SAVVY, and PRO2000) have been tested and failed to demonstrate efficacy in phase 3 clinical trials. Some have failed due to lack of efficacy, but more worryingly for some compounds, an increased risk of infection in women using the product was observed. This was attributed to damage to the vaginal epithelium and breaching of natural vaginal defences.[150–154]

Following these disappointing results, the focus of microbicide research has now largely moved towards products containing antiretroviral drugs. The recent demonstration of an important proof of principle, that topical antiretroviral drugs could be an effective tool for prevention of HIV-1 infection through the results of the CAPRISA 0004 trial, has increased the sense of urgency in this area of HIV research. This double-blind, randomised, controlled study assessed effectiveness and safety of a 1% vaginal gel formulation of tenofovir for the prevention of HIV infection in women over a 30-month period. The results demonstrated that the tenofovir gel reduced HIV incidence, compared to placebo, by an estimated 39% overall and by 54% in women with high gel adherence.[155]

The presence of circulating drug-resistant viral strains is likely to increase as more people are treated, and it is likely that compounds containing combinations of antiretroviral drugs will be needed. However, the development of combination formulations that are active in the vaginal or rectal milieu may present a significant challenge. Some progress has been made, and recently it has been demonstrated that a maraviroc and dapivirine combination could be effectively incorporated within a single matrix-type silicon vaginal ring device and provide sustained release.[156]

12.5.3 Potential of Large Scale Treatment Programmes to Reduce Transmission

Prevention strategies currently are mainly focussed on the uninfected population. However, the treatment of infected persons may potentially have the biggest impact as: (i) transmission only occurs from infected persons, (ii) the amount of virus in the plasma and other body fluids is the most important determinant of the risk of transmission for all modes of transmission, (iii) antiretroviral therapy is extremely effective in reducing viral load, (iv) success in prevention of MTCT has demonstrated proof of concept, and (v) there is evidence of reduced transmission within discordant heterosexual couples when the index partner is on effective antiretroviral therapy.[157]

Several studies have assessed the effect of antiretroviral therapy on transmission of HIV.[158,159] However, the most compelling support for this strategy came from a recently published study by Granich and colleagues at the WHO. They used mathematical modelling to explore the effect of testing all people older than 15 years, starting them on antiretroviral therapy as soon as they were found to be HIV positive, and taking into account the reproductive ratio of the infection and the long-term dynamics of the HIV epidemic. The results demonstrated that this strategy could reduce HIV incidence and mortality to less than one case per 1000 people per year within 10 years of full implementation, and reduce the prevalence of HIV to less than 1% within 50 years. Furthermore, while the cost to implement this strategy would initially be very high (up to 3 times the cost of the current strategy), within approximately 20 years it would cost the same and then decrease to be significantly less costly than the current strategy of treating only individuals at highest risk of progression.[160]

This approach to HIV prevention is unique in that it addresses the needs of both infected and uninfected individuals through targeting viral load which is the key factor driving both disease progression and transmission.[157] There are significant challenges to such an approach including operational and financial feasibility; ethical and human rights challenges, and the potential for non-adherence, drug resistance and toxicity leading to treatment failure and failure of the strategy. However, as progress in HIV vaccine research remains slow and other HIV prevention strategies have demonstrated limited success, further research in this area is urgently needed.[157]

12.6 HIV Vaccine Development – Progress and Challenges

In spite of the advances in antiretroviral drug therapy, HIV/AIDS remains a challenge globally, especially in developing countries. Efforts to control the pandemic through education and behaviour modification have had only limited success and the need for a vaccine remains desperate. However, this is a formidable challenge. HIV differs from other infectious diseases where successful preventative vaccines have been developed. Most importantly, there is no evidence of natural recovery from HIV and immune correlates of protection are not known. Scientists have even raised the question of whether any vaccine against HIV could possibly be effective in preventing infection. It is clear that the development of such a vaccine presents the greatest challenge vaccine researchers have ever had to face.

12.6.1 Requirements for Vaccine-induced Immune Responses

HIV persists in the host despite a vigorous immune response. The cell-mediated immune response includes natural killer cells and cytotoxic T-lymphocytes (CTL) targeted to cells expressing HIV antigens, as well as non-lytic suppression of HIV by CD8 cells through the secretion of chemokines.[161,162] Most HIV-infected individuals also eventually develop a neutralising antibody response.[163] Why these antiviral immune responses fail to clear the virus remains unknown.

An effective vaccine against HIV will most likely need to induce both a humoral and a cell-mediated immune response. It is important that the virus should be rapidly neutralised, and ideally be prevented from multiplying at the portal of entry, which would require the induction of high levels of neutralising antibodies. The detection of HIV-1-specific T-helper cells in HIV-1 seronegative sex partners of seropositive individuals has led to the hypothesis that the immune system might be able to successfully clear a low-dose HIV-1 infection *via* a cell-mediated immune response.[164,165] This implies that a cell-mediated immune response, particularly CTL, which are able to recognise and destroy virus-infected cells, may also be critical in immune protection from HIV infection. A single vaccine is unlikely to be able to elicit both a strong neutralising antibody response and a strong CTL response, and it will probably be necessary to use two vaccines in a prime/boost combination. Finally, it is critical that an HIV vaccine induces an immune response that will protect from infection at mucosal surfaces.

An HIV-1 vaccine will also have to elicit immune responses that recognise viruses from multiple subtypes. Several studies in the chimpanzee model have shown that inter-subtype cross-protection might be difficult to achieve.[166,167] Evidence for dual HIV-1 infections with HIV-1 strains from the same and different subtypes in humans also exists.[168] However, the identification of cross-neutralising activity in sera from some patients, as well as the demonstration of broadly neutralising activity of monoclonal antibodies directed to discontinuous epitopes in gp120 or gp41, indicate that this may not be an insurmountable barrier.[169,170]

12.6.2 Candidate Vaccine Approaches

The classic vaccine approaches that have proven most successful for viral diseases thus far, namely live-attenuated virus strains and killed virus vaccines, are unlikely to result in a successful HIV vaccine. Studies using an attenuated *nef*-deleted SIV strain have demonstrated efficacy in the macaque SIV infection model.[171,172] However, the safety concerns when using a live attenuated retrovirus vaccine are daunting. These include reversion or recombination to form pathogenic strains, transmission of the attenuated virus in the population, development of disease in immuno-compromised individuals, and integration of HIV into host DNA which could result in life-long persistence and insertional mutagenesis.[173,174] Whole-killed HIV preparations have not shown efficacy in the nonhuman primate model, and the risk of inadequate inactivation of the vaccine virus presents a significant concern.[174,175]

HIV vaccine design has thus relied almost exclusively on gene expression technologies, using parts of the virus either in protein or genetic form to create antigens that cannot reassemble into functional HIV. These include live recombinant vector expression HIV genes, protein subunit vaccines, synthetic peptides, virus-like particles and DNA vaccines.[174]

12.6.3 Progress to Date

Although studies of HIV vaccines in animal models have contributed valuable information, trials in human volunteers at risk of HIV infection is critical to assess the true value of any candidate HIV vaccine. However, there are many complex ethical concerns with clinical efficacy trials of HIV vaccines in humans, including the issue of true informed consent, potential for coercion, increased risk behaviour due to confidence in protective effect of vaccine, protection of confidentiality, and that the patient will test positive for antibodies to HIV.

In spite of these challenges, multiple HIV vaccine candidates have been tested in more than one hundred clinical trials, including several large-scale efficacy studies. Two phase 3 studies of AIDSVAX, a recombinant gp120 subunit vaccine, have failed to protect volunteers from infection, apparently because the vaccine did not induce broadly neutralising antibodies.[176] A recombinant adenovirus vector vaccine (MRKAd5 HIV-1 *gag*/*pol*/*nef* trivalent vaccine), has been evaluated in two efficacy trials. The STEP trial was conducted in North America, South America, the Caribbean, and Australia. A second study (Phambili) was conducted in South Africa. Both trials were terminated ahead of schedule when data from the STEP trial showed that the vaccine failed to prevent HIV infection and failed to lower virus levels in vaccinated volunteers who became infected.[177] Furthermore, there was a trend towards a greater number of new infections among vaccinated individuals. Exploratory analyses demonstrated an increased risk of HIV-1 infection in adenovirus type 5 seropositive and uncircumcised men who received the vaccine. The mechanism of this apparent increased susceptibility to infection is not currently understood.[178]

Most recently, results of a study evaluating a prime boost vaccination strategy consisting of four priming doses of a recombinant canarypox vector vaccine (ALVAC-HIV) plus two booster injections of a recombinant glycoprotein 120 subunit vaccine (AIDSVAX B/E) in Thailand, demonstrated a modest trend towards the prevention of HIV-1 infection among the vaccine recipients, with a vaccine efficacy of 26.4%. In a modified intention-to-treat analysis (excluding 7 subjects who were found to have been HIV infected at baseline), the vaccine efficacy was 31.2%. Vaccination did not affect the degree of viraemia or the CD4+ T-cell count in subjects in whom HIV-1 infection was subsequently diagnosed.[179]

Although these trials have not yet resulted in an approved vaccine, they have led to a better understanding of the complexities of virus biology and identification of possible targets against which improved and novel technologies can be used to design an effective prophylactic vaccine.[180] Importantly, these studies have demonstrated that large-scale HIV vaccine studies are feasible and paved the way for optimum study set-up.

12.7 Conclusions

The discovery and development of multiple drugs to treat HIV infection, thereby turning an almost universally fatal infection into a chronic manageable condition, less than 30 years after the discovery of the causative virus, is undoubtedly one of the greatest achievements of modern medical science. The multitude of researchers who contributed to this can rightly be very proud of the positive impact their work has had on the lives of millions of people.

However, there is no room for complacency, as much remains to be done in the fields of both prevention and treatment of HIV. Key ongoing needs are new drugs to treat drug-resistant virus, management of long-term complications of antiretroviral therapy, management of HIV infection in the context of ageing, and the treatment of HIV-induced chronic immune activation and its consequences. Additional requirements are that new drugs have oral once-daily dosing pharmacokinetics, minimal drug-drug interactions and be low cost. Current antiretroviral therapy effectively suppresses but does not eradicate HIV-1 infection. Much still needs to be done to understand the processes driving viral latency and reservoirs in order to enable treatments that will lead to virus eradication.

In spite of the undisputed progress towards the goal of universal access to antiretroviral therapy, large numbers of patients in the developing world do not yet have access to even basic antiretroviral therapy or are suboptimally managed. Further research on the optimal and most cost-effective strategies for managing HIV infection in large numbers of patients in resource-limited settings is critical. More also needs to be done to understand both the side effects of antiretroviral drugs in the context of other infectious disease in the developing world, and the potential of drug-drug interactions with drugs used for their treatment.

Further research on the potential use of early widespread antiretroviral therapy of infected persons as a prevention mechanism, and ways to overcome the key challenges for this approach, is urgently needed. Following the recent success of the CAPRISA 0004 microbicide trial, it is critical that momentum is maintained in this area. Finally, a vaccine remains elusive and progress has been painfully slow to date.

The complexity of these challenges means that it would be impossible for any one organisation or company to address them, and international collaboration through public-private partnerships will be critical to drive progress. It is up to those engaged in research to continue to lead the way to resolving these remaining issues.

References

1. Centers for Disease Control (CDC), *Morb. Mortal. Wkly. Rep.*, 1981, **30** 305.
2. C. F. Thomas Jr and A. H. Limper, *N. Engl. J. Med.*, 2004, **350**, 2487.
3. F. Barrè-Sinoussi, J. C. Chermann, F. Rey, M. T. Nugeyre, S. Chamaret, J. Gruest, C. Dauguet, C. Axler-Blin, F. Brun-Vezinet, C. Rouzioux, W. Rozenbaum and l. Montagnier, *Science*, 1983, **220**, 868.
4. F. Clavel, D. Guetard, F. Brun-Vezinet, S. Chamaret, M. A. Rey, M. O. Santos-Ferreira, A. G. Laurent, C. C. Dauguet, C. Katlama, C. Rouzioux, D. Klatzmann, J. L. Champalimaud and L. Montagnier, *Science*, 1986, **233**, 343.
5. M. S. Hirsch, J. Curran in *Fields Virology* (Third Edition) ed. B. N. Fields, D. M. Knipe and P. M. Howley, Lippincott-Raven Publishers, Philadelphia, 1996, p. 1953.
6. D. Serwadda, R. D. Mugerwa, N. K. Sewankambo, A. Lwegaba, J. W. Carswell, G. B. Kirya, A. C. Bayley, R. G. Downing, R. S. Tedder, S. A. Clayden, R. A. Weiss and A. G. Dalgleish, *Lancet*, 1985, **2**, 849.
7. A. J. Nahmias, J. Weiss, X. Yao, F. Lee, R. Kodsi, M. Schanfield, T. Matthews, D. Bolognesi, D. Durack, A. Motulsky, P. Kanki and M. Essex, *Lancet*, 1986, **1**, 1279.
8. R. M. Anderson and R. M. May, *Nature*, 1988, **313**, 514.
9. F. A. Plummer, J. N. Simonsen, D. W. Cameron, J. O. Ndinya-Achola, J. K. Kreiss, M. N. Gakinya, P. Waiyaki, M. Cheang, P. Piot, A. R. Ronald and E. N. Ngugi, *J. Infect. Dis.*, 1991, **163**, 233.
10. S. Cu-Uvin, A. M. Caliendo, S. Reinert, A. Chang, C. Juliano-Remollino, T. P. Flanigan, K. H. Mayer and C. C. Carpenter, *AIDS*, 2000, **14**, 415.
11. R. Granich, S. Crowley, M. Vitoria, C. Smyth, J. G. Kahn, R. Bennett, Y. R. Lo, Y. Souteyrand and B. Williams, *Curr. Opin. HIV AIDS*, 2010, **5**, 298.
12. J. P. Allain, S. L. Stramer, A. B. Carneiro-Proietti, M. L. Martins, S. N. Lopes da Silva, M. Ribeiro, F. A. Proietti and H. W. Reesink, *Biologicals*, 2009, **37**, 71.
13. M. Vermeulen, N. Lelie, W. Sykes, R. Crookes, J. Swanevelder, L. Gaggia, M. Le Roux, E. Kuun, S. Gulube and R. Reddy, *Transfusion*, 2009, **49**, 1115.
14. S. Jayaraman, Z. Chalabi, P. Perel, C. Guerriero and I. Roberts, *Transfusion*, 2010, **50**, 433.

15. B. M. Mathers, L. Degenhardt, B. Phillips, L. Wiessing, M. Hickman, S. A. Strathdee, A. Wodak, S. Panda, M. Tyndall, A. Toufik and R. P. Mattick, *Lancet*, 2008, **372**, 1733.
16. R. Becquet, D. K. Ekouevi, E. Arrive, J. S. A. Stringer, N. Meda, M.-L. Chaix, J-M. Treluyer, V. Leroy, C. Rouzioux, S. Blanche and F. Dabis, *Clin. Infect. Dis.*, 2009, **49**, 1936.
17. G. Scarlatti, *AIDS Rev.*, 2004, **6**, 67.
18. G. E. Gray and H. Saloojee, *N. Engl. J Med.*, 2008, **359**, 189.
19. UNAIDS, *AIDS epidemic update*, 2009, UNAIDS, Geneva. Available at: http://data.unaids.org/pub/Report/2009/JC1700_Epi_Update_2009_en.pdf, accessed 19 October 2010.
20. U. Dietrich, J. K. Maniar and H. Rübsamen-Waigmann, *Trends Microbiol.*, 1995, **3**, 17.
21. T. I. De Silva, M. Cotten and S. L. Rowland-Jones, *Trends Microbiol.*, 2008, **16**, 588.
22. R. Kannangai, S. C. Nair, G. Sridharan, S. Prasannakumar and D. Daniel, *Indian J. Med. Microbiol.*, 2010, **28**, 111.
23. C. Tienen, M. S. van der Loeff, S. M. Zaman, T. Vincent, R. Sarge-Njie, L. Peterson, A. Leligdowicz, A. Jaye, S. Rowland-Jones, P. Aaby and H. Whittle, *J. Acquir. Immune Defic. Syndr.*, 2010, **53**, 640.
24. DoH South Africa, *HIV & AIDS and STI Strategic Plan for South Africa 2007-2011*. Available at http://www.doh.gov.za/docs/misc/stratplan-f.html, accessed 19 October 2010.
25. M. N. Lurie, B. G. Williams, K. Zuma, D. Mkaya-Mwamburi, G. P. Garnett, M. D. Sweat, J. Gittelsohn and S. S. Karim, *AIDS*, 2003, **17**, 2245.
26. M. Morris, A. E. Kurth, D. T. Hamilton, J. Moody and S. Wakefield, *Am. J. Public. Health.*, 2009, **99**, 1023.
27. A. Whiteside and J. Smith, *Glob. Health.*, 2009, **5**, 15.
28. J. M. Coffin in *The Retroviridae,* ed. J. A. Levy, Plenum Press, New York, 1992, p. 19.
29. P. A. Luciw in *Fields Virology* (Third Edition), ed. B. N. Fields, D. M. Knipe and P. M. Howley, Lippincott-Raven Publishers, Philadelphia, 1996, p. 1881.
30. National Institute of Allergy and Infectious Diseases, HIV replication cycle, United States Department of Health and Human Services, Bethesda, MD, 2010, Available at: http://www.niaid.nih.gov/topics/HIVAIDS/Understanding/Biology/Pages/hivReplicationCycle.aspx, accessed 25 November 2010.
31. P. Charneau, A. M. Borman, C. Quillent, D. Guetard, S. Chamaret, J. Cohen, G. Remy, L. Montagnier and F. Clavel, *Virology*, 1994, **205**, 247.
32. F. Simon, P. Mauclere, P. Roques, I. Loussert-Ajaka, M. C. Muller-Trutwin, S. Saragosti, M. C. Georges-Courbot, F. Barre-Sinoussi and F. Brun-Vezinet, *Nature Med.*, 1998, **4**, 1032.
33. B. Korber, C. Kuiken, B. Foley, B. Hahn, F. E. McCutchan, J. W. Mellors, J. Sodroski, *Human retroviruses and AIDS: A compilation and*

analysis of nucleic acid and amino acid sequences, Los Alamos National Laboratory, Los Alamos, New Mexico, 1998.
34. F. Gao, S. G. Morrisson, D. L. Robertson, C. L. Thornton, S. Craig, G. Karlsson, J. Sodroski, M. Morgado, B. Calvao-Castro and H. von Briesen, *J. Virol.*, 1996, **70**, 1651.
35. D. L. Robertson, P. Sharp, F. E. McCutchan and B. H. Hahn, *Nature*, 1995, **374**, 124.
36. T. Niu, D. S. Stein and S. Schnittman, *J. Infect. Dis.*, 1993, **168**, 1490.
37. D. D. Ho, A. U. Neumann, A. S. Perelson, W. Chen, J. M. Leonard and M. Markowitz, *Nature*, 1995, **373**, 123.
38. Centers for Disease Control and Prevention, *MMWR Recomm. Rep.*, 1992, **41**, 1. Available at www.cdc.gov/mmwr/preview/mmwrhtml/00018871.htm, accessed 19 October 2010.
39. M. Mendila, H. Heiken, S. Becker, M. Stoll, A. Kemper, R. Jacobs and R. E. Schmidt, *Eur. J. Med. Res.*, 1999, **4**, 417.
40. W. B. Dyer, J. J. Zaunders, F. F. Yuan, B. Wang, J. C. Learmont, A. F. Geczy, N. K. Saksena, D. A. McPhee, P. R. Gorry PR and J. S. Sullivan, *Retrovirology*, 2008, **5**, 112.
41. J. M. Brenchley, T. W. Schacker, L. E. Ruff, D. A. Price, J. H. Taylor, G. J. Beilman, P. L. Nguyen, A. Khoruts, M. Larson, A. T. Haase and D. C. Douek, *J. Exp. Med.*, 2004, **200**, 749.
42. S. M. Smith, *Retrovirology*, 2006, **3**, 60.
43. A. Kovacs, L. Al-Harthi, S. Christensen, W. Mack, M. Cohen and A. Landay, *J. Infect. Dis.*, 2008, **197**, 1402.
44. J. M. Doisne, A. Urrutia, C. Lacabaratz-Porret, C. Goujard, L. Meyer, M-L. Chaix, M. Sinet and A. Venet, *J. Immunol.*, 2004, **173**, 2410.
45. L. Papagno, C. A. Spina, A. Marchant, M. Salio, N. Rufer, S. Little, T. Dong, G. Chesney, A. Waters, P. Easterbrook, P. R. Dunbar, D. Shepherd, V. Cerundolo, V. Emery, P. Griffiths, C. Conlon, A. J. McMichael, D. D. Richman, S. L. Rowland-Jones and V. Appay, *PLoS. Biol.*, 2004, **2**, 173.
46. T. W. Schacker, C. Reilly, G. J. Beilman, J. Taylor, D. Skarda, D. Krason, M. Larson and A. T. Haase, *AIDS*, 2005, **19**, 2169.
47. R. Wood, H. Liang, H. Wu, K. Middelkoop, T. Oni, M. X. Rangaka, R. J. Wilkinson, L. G. Bekker and S. D. Lawn, *Int. J. Tuberc. Lung. Dis.*, 2010, **14**, 406.
48. Y. Hanifa, A. D. Grant, J. Lewis, E. L. Corbett, K. Fielding and G. Churchyard, *Int. J. Tuberc. Lung. Dis.*, 2009, **13**, 39.
49. S. D. Lawn, L. Myer, D. Edwards, L. G. Bekker and R. Wood, *AIDS*, 2009, **23**, 1717.
50. D. J. Pepper, S. Marais, R. J. Wilkinson, F. Bhaijee, G. Maartens, H. McIlleron, V. De Azevedo, H. Cox, C. McDermid, S. Sokhela, J. Patel and G. Meintjes, *BMC. Infect. Dis.*, 2010, **10**, 83.
51. Y. C. Manabe, R. Breen, T. Perti, E. Girardi and T. R. Sterling, *J. Infect. Dis.*, 2009, **199**, 437.
52. U. G. Lalloo and S. Pillay, *Curr. HIV/AIDS. Rep.*, 2008, **5**, 132.
53. V. Idemyor, *HIV. Clin. Trials*, 2007, **8**, 246.

54. B. J. Park, K. A. Wannemuehler, B. J. Marston, N. Govender, P. G. Pappas and T. M. Chiller, *AIDS*, 2009, **23**, 525.
55. J. N. Jarvis, T. S. Harrison, E. L. Corbett, R. Wood and S. D. Lawn, *AIDS*, 2010, **24**, 612.
56. A. S. Grant and K. M. De Cock, *B. M. J.*, 2001, **322**, 1475.
57. J. W. Mellors, C. R. Rinaldo, P. Gupta, R. M. White, J. A. Todd and L. A. Kingsley, *Science*, 1996, **272**, 1167.
58. J. W. Mellors, A. Muñoz, J. V. Giorgi, J. B. Margolick, C. J. Tassoni, P. Gupta, L. A. Kingsley, J. A. Todd, A. J. Saah, R. Detels, J. P. Phair and C. R. Rinaldo, *Ann. Intern. Med.*, 1997, **126**, 946.
59. H. Mitsuya, K. J. Weinhold, P. A. Furman, M. H. St Clair, S. N. Lehrman, R. C. Gallo, D. Bolognesi, D. W. Barry and S. Broder, *Proc. Natl. Acad. Sci. U.S.A.*, 1985, **82**, 7096.
60. S. Broder, *Antivir. Res.*, 2010, **85**, 1.
61. P. A. Furman, J. A. Fyfe, M. H. St Clair, K. Weinhold, J. L. Rideout, G. A. Freeman, S. N. Lehrman, D. P. Bolognesi, S. Broder, H. Mitsuya and D. W. Barry, *Proc. Natl. Acad. Sci. U.S.A.*, 1986, **83**, 8333.
62. M. A. Fischl, D. D. Richman, M. H. Grieco, M. S. Gottlieb, P. A. Volberding, O. L. Laskin, J. M. Leedom, J. E. Groopman, D. Mildvan, R. T. Schooley, G. G. Jackson, D. T. Durak and D. King, *N. Engl. J. Med.*, 1987, **317**, 185.
63. J. A. Este and T. Cihlar, *Antivir. Res.*, 2010, **85**, 25–33.
64. T. Cihlar and A. S. Ray, *Antivir. Res.*, 2010, **85**, 39.
65. M. P. De Bethune, *Antivir. Res.*, 2010, **85**, 75.
66. C. A. Hughes, L. Robinson, A. Tseng and R. D. MacArthur, *Expert. Opin. Pharmacother.*, 2009, **10**, 2445.
67. R. Corbau, J. Mori, C. Phillips, L. Fishburn, A. Martin, C. Mowbray, W. Panton, C. Smith-Burchnell, A. Thornberry, H. Ringrose, T. Knöchel, S. Irving, M. Westby, A. Wood and M. Perros, *Antimicrob. Agents Chemother.*, 2010, **54**, 4451.
68. M. Vourvahis, M. Gleave, A. N. Nedderman, R. Hyland, I. Gardner, M. Howard, S. Kempshall, C. Collins and R. LaBadie, *Drug. Metab. Dispos.*, 2010, **38**, 789.
69. G. Fätkenheuer, S. Staszewski, A. Plettenburg, F. Hackman, G. Layton, L. McFadyen, J. Davis and T. M. Jenkins, *AIDS*, 2009, **23**, 2115.
70. D. J. McColl and X. X. Chen, *Antivir. Res.*, 2010, **85**, 101.
71. J. J. Schafer and K. E. Squires, *Ann. Pharmacother.*, 2010, **44**, 145.
72. L. Vandekerckhove, *Curr. Opin. Investig., Drugs*, 2010, **11**, 203.
73. Y. Feng, C. C. Broder, P. E. Kennedy and E. A. Berger, *Science*, 1996, **272**, 872.
74. P. D. Kwong, R. Wyatt, J. Robinson, R. W. Sweet, J. Sodroski and W. A. Hendrickson, *Nature*, 1998, **393**, 648.
75. D. C. Chan and P. S. Kim, *Cell*, 1998, **93**, 681.
76. E. A. Berger, R. W. Doms, E. M. Fenyö, B. T. Korber, D. R. Littman, J. P. Moore, Q. J. Sattentau, H. Schuitemaker, J. Sodroski and R. A. Weiss, *Nature*, 1998, **391**, 240.

77. M. Dean, M. Carrington, C. Winkler, G. A. Huttley, M. W. Smith, R. Allikmets, J. J. Goedert, S. P. Buchbinder, E. Vittinghoff, E. Gomperts, S. Donfield, D. Vlahov, R. Kaslow, A. Saah, C. Rinaldo, R. Detels and S. J. O'Brien, *Science*, 1996, **273**, 1856.
78. Y. Huang, W. A. Paxton, S. M. Wolinsky, A. U. Neumann, L. Zhang, T. He, S. Kang, D. Ceradini, Z. Jin, K. Yazdanbakhsh, K. Kunstman, D. Erickson, E. Dragon, N. R. Landau, J. Phair, D. D. Ho and R. A. Koup, *Nature. Med.*, 1996, **2**, 1240.
79. M. Samson, F. Libert, B. J. Doranz, J. Rucker, C. Liesnard, C. M. Farber, S. Saragosti, C. Lapoumeroulie, J. Cognaux, C. Forceille, G. Muyldermans, C. Verhofstede, G. Burtonboy, M. Georges, T. Imai, S. Rana, Y. Yi, R. J. Smyth, R. G. Collman, R. W. Doms, G. Vassart and M. Parmentier, *Nature*, 1996, **382**, 722.
80. P. Dorr, M. Westby, S. Dobbs, P. Griffin, B. Irvine, M. Macartney, J. Mori, G. Rickett, C. Smith-Burchnell, C. Napier, R. Webster, D. Armour, D. Price, B. Stammen, A. Wood and M. Perros, *Antimicrob. Agents. Chemother.*, 2005, **49**, 4721.
81. D. A. Cooper, J. Heera, J. Goodrich, M. Tawadrous, M. Saag, E. DeJesus, N. Clumeck, S. Walmsley, N. Ting, E. Coakley, J. D. Reeves, G. Reyes-Teran, M. Westby, E. Van Der Ryst, P. Ive, L. Mohapi, H. Mingrone, A. Horban, F. Hackman, J. Sullivan and H. Mayer, *J. Infect. Dis.*, 2010, **201**, 803.
82. G. Fatkenheuer, M. Nelson, A. Lazzarin, I. Konourina, A. I. Hoepelman, H. Lampiris, B. Hirschel, P. Tebas, F. Raffi, B. Trottier, N. Bellos, M. Saag, D. A. Cooper, M. Westby, M. Tawadrous, J. F. Sullivan, C. Ridgway, M. W. Dunne, S. Felstead, H. Mayer and E. van der Ryst, *N. Engl. J. Med.*, 2008, **359**, 1442.
83. R. M. Gulick, J. Lalezari, J. Goodrich, N. Clumeck, E. DeJesus, A. Horban, J. Nadler, B. Clotet, A. Karlsson, M. Wohlfeiler, J. B. Montana, M. McHale, J. Sullivan, C. Ridgway, S. Felstead, M. W. Dunne, E. van der Ryst and H. Mayer, *N. Engl. J. Med.*, 2008, **359**, 1429.
84. E. van der Ryst and M. Westby, presented at the 47th Interscience Conference on Antimicrobial Agents and Chemotherapy (ICAAC) 2007, Chicago, USA.
85. M. Westby and E. van der Ryst, *Antivir. Chem. Chemother.*, 2010, **20**, 179.
86. S. Abel, D. J. Back and M. Vourvahis, *Antivir Ther.*, 2009, **14**, 607.
87. Panel on Antiretroviral Guidelines for Adults and Adolescents. Department of Health and Human Services, 2011, pp. 1–1174 Available at http://www.aidsinfo.nih.gov/ContentFiles/AdultandAdolescentGL.pdf, accessed 26 January 2011.
88. J. C. Tilton and R. W. Doms, *Antivir. Res.*, 2010, **85**, 91.
89. C. Wild, J. W. Dubay, T. Greenwell, T. Baird Jr, T. G. Oas, C. McDanal, E. Hunter and T. Matthews, *Proc. Natl. Acad. Sci. U.S.A.*, 1994, **91**, 12676.
90. C. Wild, T. Greenwell and T. Matthews, *AIDS. Res. Hum. Retrovir.*, 1993, **9**, 1051.

91. J. M. Kilby, J. P. Lalezari, J. J. Eron, M. Carlson, C. Cohen, R. C. Arduino, J. C. Goodgame, J. E. Gallant, P. Volberding, R. L. Murphy, F. Valentine, M. S. Saag, E. L. Nelson, P. R. Sista and A. Dusek, *AIDS. Res. Hum. Retroviruses*, 2002, **18**, 685.
92. J. Lalezari, K. Henry, M. O'Hearn, J. S. Montaner, P. J. Piliero, B. Trottier, S. Walmsley, C. Cohen, D. R. Kuritzkes, J. J. Eron, J. Chung, R. DeMasi, L. Donatacci, C. Drobnes, J. Delehanty and M. Salgo, *N. Engl. J. Med.*, 2003, **348**, 2175.
93. A. Lazzarin, B. Clotet, D. Cooper, J. Reynes, K. Arasteh, M. Nelson, C. Katlama, H. J. Stellbrink, J. F. Delfraissy, J. Lange, L. Huson, R. DeMasi, C. Wat, J. Delehanty, C. Drobnes and M. Salgo, *N. Engl. J. Med.*, 2003, **348**, 2186.
94. A. Lazzarin, *Expert Opin. Pharamacother.,* , 2005, **6** 453.
95. J. M. Coffin, in *Fields Virology* (Third Edition), ed. B. N. Fields, D. M. Knipe and P. M. Howley, Lippincott-Raven Publishers, Philadelphia, 1996, p. 1953.
96. WHO, *Antiretroviral therapy for HIV infection in adults and adolescents - Recommendations for a public health approach*, Geneva, World Health Organization, 2010. Available at http://whqlibdoc.who.int/publications/2010/9789241599764_eng.pdf, accessed 26 January 2011.
97. J. Y. Feng, J. K. Ly, F. Myrick, D. Goodman, K. L. White, E. S. Svarovskaia, K. Borroto-Esoda and M. D. Miller, *Retrovirology*, 2009, **6**, 44.
98. N. Y. Rakhmanina and J. N. van der Anker, *Expert Opin. Drug Metab. Toxicol.*, 2010, **6**, 95.
99. S. Abel, T. M. Jenkins, L. A. Whitlock, C. E. Ridgway and G. J. Muirhead, *Br. J. Clin. Pharmacol.*, 2008, **65**, 38.
100. A. Boulle, G. Van Cutsem, K. Cohen, K. Hilderbrand, S. Mathee, M. Abrahams, E. Goemaere, D. Coetzee and G. Maartens, *JAMA*, 2008, **300**, 530.
101. A. Kwara, G. Ramachandran and S. Swaminathan, *Expert Opin. Drug. Metab. Toxicol.*, 2010, **6**, 55.
102. P. Yeni, A. Mills, M. Peeters, J. Vingerhoets, T. N. Kakuda, G. De Smedt and B. Woodfall, *Curr. HIV. Res.*, 2010, **8**, 564.
103. A. Lazzarin, T. Campbell, B. Clotet, M. Johnson, C. Katlama, A. Moll, W. Towner, B. Trottier, M. Peeters, J. Vingerhoets, G. de Smedt, B. Baeten, G. Beets, R. Sinha and B. Woodfall, *Lancet*, 2007, **370**, 39.
104. J. V. Madruga, P. Cahn, B. Grinsztejn, R. Haubrich, J. Lalezari, A. Mills, G. Pialoux, T. Wilkin, M. Peeters, J. Vingerhoets, G. de Smedt, L. Leopold, R. Trefiglio and B. Woodfall, *Lancet*, 2007, **370**, 29.
105. T. K. Chiu and D. R. Davies, *Curr. Top. Med. Chem.*, 2004, **4**, 965.
106. D. A. Cooper, R. T. Steigbigel, J. Gatell, J. Rockstroh, C. Katlama, P. Yeni, A. Lazzarin, B. Clotet, P. Kumar, J. E. Eron, M. Schechter, M. Markowitz, M. R. Loutfy, J. L. Lennox, J. Zhao, J. Chen, D. M. Ryan, R. R. Rhodes, J. A. Killar, L. R. Gilde, K. K. Strohmaier, A. R. Meibohm, M. D. Miller, D. J. Hazuda, M. L. Nessly, M. J. DiNubile, R. D. Isaacs, H. Teppler and B. Y. Nguyen, *N. Engl. J. Med.*, 2008, **359**, 355.

107. R. T. Steigbigel, D. A. Cooper, P. N. Kumar, J. E. Eron, M. Schechter, M. Markowitz, M. R. Loutfy, J. L. Lennox, J. M. Gatell, J. K. Rockstroh, C. Katlama, P. Yeni, A. Lazzarin, B. Clotet, J. Zhao, J. Chen, D. M. Ryan, R. R. Rhodes, J. A. Killar, L. R. Gilde, K. M. Strohmaier, A. R. Neibohm, M. D. Miller, D. J. Hazuda, M. L. Nessly, M. J. DiNubile, R. D. Isaacs, B. Y. Nguyen and H. Teppler, *N. Eng. J. Med.*, 2008, **359**, 339.
108. J. L. Lennox, E. DeJesus, A. Lazzarin, R. B. Pollard, J. V. Madruga, D. S. Berger, J. Zhao, X. Xu, A. Williams-Diaz, A. J. Rodgers, R. J. Barnard, M. D. Miller, M. J. DiNubile, B. Y. Nguyen, R. Leavitt and P. Sklar, *Lancet*, 2009, **374**, 796.
109. D. J. Hazuda, M. D. Miller, B. Y. Nguyen and J. Zhao, *J. Antiviral. Ther.*, 2007, **12**, S10.
110. W. G. Powderly, *J. Antimicrob. Chemother.*, 2010, **65**, 2485.
111. L. De Luca, S. Ferro, R. Gitto, M. L. Barreca, S. Agnello, F. Christ, Z. Debyser and A. Chimirri, *Bioorg. Med. Chem.*, 2010, **18**, 7515.
112. A. M. J. Wensing, N. M. van Marseveen and M. Nijhuis, *Antivir. Res.*, 2010, **85**, 59.
113. L. Cuzin, C. Allavena, P. Morlat and P. Dellamonica, *AIDS. Rev.*, 2008, **10**, 205.
114. T. Hawkins, *Antivir. Res.*, 2010, **85**, 201.
115. A. d'Arminio Monforte, A. C. Lepri, G. Rezza, P. Pezzotti, A. Antinori, A. Phillips, G. Angarano, V. Colangeli, A. De Luca, G. Ippolito, L. Caggese, F. Soscia, G. Filice, F. Gritti, P. Narciso, U. Tirelli and M. Moroni, *AIDS*, 2000, **14**, 499.
116. J. M. Coffin, *Science*, 1995, **267**, 483.
117. E. Domingo and J. J. Holland, *Annu. Rev. Microbiol.*, 1997, **51**, 151.
118. R. Paredes and B. Clotet, *Antivir. Res.*, 2010, **85**, 245.
119. J. S. Murray, M. R. Elashoff, L. C. Iacono-Connors, T. A. Cvetkovich and K. A. Struble, *AIDS*, 1999, **13**, 797.
120. The Strategies for Management of Antiretroviral Therapy (SMART) Study Group, *N. Engl. J. Med.*, 2006, **355**, 2283.
121. M. Kitahata, S. Gange and A. Abraham, *N. Engl. J. Med.*, 2009, **360**, 1815.
122. A. R. Zolopa, *Antivir. Res.*, 2010, **85**, 241.
123. D. S. Storm, M. G. Boland, S. L. Gortmaker, Y. He, J. Skurnick, L. Howland and J. M. Oleske, *Pediatrics*, 2005, **115**, e173.
124. R. M. Viani, M. R. Araneta, J. G. Deville and S. A. Spector, *Clin. Infect. Dis.*, 2004, **39**, 725.
125. Panel on Antiretroviral Therapy and Medical Management of HIV-Infected Children, *Department of Health and Human Services*, 2010, p. 1. Available at http://aidsinfo.nih.gov/ContentFiles/PediatricGuidelines.pdf, accessed 26 January 2011.
126. M. Penazzato, D. Donà, P.-S. Wool, O. Rampon and C. Giaquinto, *Antivir. Res.*, 2010, **85**, 266.
127. G. L. Kearns, S. M. Abdel-Rahman, S. W. Alander, D. L. Blowey, J. S. Leeder and R. E. Kauffman, *N. Engl. J. Med.*, 2003, **349**, 1157.

128. C. L. Townsend, M. Cortina-Boria, C. S. Peckham, A. de Ruiter, H. Lyall and P. A. Tookey, *AIDS*, 2008, **22**, 973.
129. Panel on Treatment of HIV-Infected Pregnant Women and Prevention of Perinatal Transmission, *Department of Health and Human Services*, 2010, p. 1. Available at http://aidsinfo.nih.gov/ContentFiles/PerinatalGL.pdf, accessed 26 January 2011.
130. L. M. Mofenson, *Clin. Infect. Dis.*, 2010, **50 (S 3)**, S130.
131. WHO, *Antiretroviral drugs for treating pregnant women and preventing HIV infections in infants*, Geneva, World Health Organization, 2010. Available at http://whqlibdoc.who.int/publications/2010/9789241599818_eng.pdf, accessed 26 January 2011.
132. UNAIDS, *Towards Universal Access: Scaling up Priority HIV/AIDS Interventions in the Health Sector—Progress Report, 2009*, 2009, UNAIDS, Geneva. Available at www.who.int/hiv/pub/2009progressreport/en/index.html, accessed 26 January 2011.
133. J. Zarocostas, *BMJ*, 2010, **341**, c5393.
134. DoH South Africa, HIV & AIDS and STI Strategic Plan for South Africa 2007-2011. Available at http://www.doh.gov.za/docs/misc/stratplan-f.html.
135. M. Callaghan, N. Ford and H. Schneider, *Hum. Resour. Health*, 2010, **8**, 8.
136. DoH South Africa, *The South African Antiretroviral Treatment Guidelines*, 2010 Available at http://www.sanac.org.za/documents/2010%20ART%20Guideline-Short.pdf, accessed 26 January 2011.
137. I. M. Sanne, D. Westreich, A. P. Macphail, D. Rubel, P. Majuba and A. Van Rie, *J. Int. AIDS. Soc.*, 2009, **12**, 38.
138. I. Jani, N. Sitoe, E. Alfai, P. Chongo, J. Lehe, B. Rocha, J. Quevedo, T. Peter, presented at the International AIDS conference, Vienna, 2010.
139. DART Trial Team, *Lancet*, 2010, **375**, 123.
140. M. S. Hirsch, H. F. Günthardt, J. M. Schaprio, F. Brun-Vézinet, B. Clotet, S. M. Hammer, V. A. Johnson, D. R. Kuritzkes, J. W. Mellors, D. Pillay, P. G. Yeni, D. M. Jacobson and D. D. Richman, *Clin. Infect. Dis.*, 2008, **47**, 266.
141. WHO, Guidelines for HIV Diagnosis and Monitoring of Antiretroviral Therapy, Geneva, World Health Organization, 2010. Available at http://www.searo.who.int/LinkFiles/Publications_SEA-HLM-382.pdf, accessed 26 January 2011.
142. T. De Oliveira, R. W. Shafer and C. Seebregts, *Nature*, 2010, **464**, 673.
143. H. McIlleron, presented at the 17th Conference of Retroviruses and Opportunistic Infections, 2010.
144. M. V. Garcia, D. Mukeba-Tshialala, D. Vaira and M. Moutschen, *Rev. Med. Liege.*, 2009, **64**, 32.
145. M. Youle and M. Wainberg, *AIDS*, 2003, **17**, 937.
146. L. Peterson, D. Taylor, R. Roddy, G. Belai, P. Phillips, K. Nanda, R. Grant, E. E. Clarke, A. S. Doh, R. Ridzon, H. S. Jaffe and W. Cates, *PLoS. Clin. Trials.*, 2007, **2**, e27.

147. R. M. Grant, *Clin. Infect. Dis.*, 2010, **50(S3)**, S96.
148. I. McGowan, *Biologicals*, 2006, **34**, 241.
149. R. W. Jr. Buckheit, K. M. Watson, K. M. Morrow and A. S. Ham, *Antivir. Res.*, 2010, **85**, 142.
150. S. McCormack, G. Ramjee, A. Kamali, H. Rees, A. M. Crook, M. Gafos, U. Jentsch, R. Pool, M. Chisembele, S. Kapiga, R. Mutemwa, A. Vallely, T. Palanee, Y. Sookrajh, C. J. Lacey, J. Darbyshire, H. Grosskurth, A. Profy, A. Nunn, R. Hayes and J. Weber, *Lancet*, 2010, **367**, 1329.
151. L. Van Damme, R. Govinden, F. M. Mirembe, F. Guédou, S. Solomon, M. L. Becker, B. S. Pradeep, A. K. Krishnan, M. Alary, B. Pande, G. Ramjee, J. Deese, T. Crucitti and D. Taylor, *N. Engl. J. Med.*, 2008, **359**, 463.
152. L. Van Damme, G. Ramjee, M. Alary, M. Vuylsteke, V. Chandeying, H. Rees, P. Sirivongrangson, L. Mukenge-Tshibaka, V. Ettiègne-Traoré, C. Uaheowitchai, S. S. Karim, B. Mâsse, J. Perriëns and M. Laga, *Lancet*, 2002, **360**, 971.
153. S. Skoler-Karpoff, G. Ramjee, K. Ahmed, L. Altini, M. G. Plagianos, B. Friedland, S. Govender, A. De Kock, N. Cassim, T. Palanee, G. Dozier, R. Maguire and P. Lahteenmaki, *Lancet*, 2008, **372**, 1977.
154. P. J. Feldblum, A. Adeiga, R. Bakare, S. Wevill, A. Lendvay, F. Obadaki, M. O. Olayemi, L. Wang, K. Nanda and W. Rountree, *PLoS One*, 2008, **3**, e1474.
155. Q. A. Karim, S. S. Karim, J. A. Frohlich, A. C. Grobler, C. Baxter, L. E. Mansoor, A. B. Kharsany, S. Sibeko, K. P. Mlisana, Z. Omar, T. N. Gengiah, S. Maarschalk, N. Arulappan, M. Mlotshwa, L. Morris and D. Taylor, *Science*, 2010, **329**, 1168.
156. A. Faheem, M. McBride, K. Malcolm, D. Woolfson, M. Sparks presented at the 16th Conference on Retroviruses and Opportunistic Infections, 2009, Poster 1069.
157. K. M. De Cock, S. P. Crowley, Y-R. Lo, R. M. Granich and B. G. Williams, *Bull. World. Health. Organ.*, 2009, **87**, 488.
158. B. Auvert, S. Males, A. Puren, D. Taljaard, M. Carael and B. Williams, *J. Acquir. Immune. Defic. Syndr.*, 2004, **36**, 613.
159. V. D. Lima, K. Johnston, R. S. Hogg, A. R. Levy, R. Harrigan, A. Anema and J. S. Montaner, *J. Infect. Dis.*, 2008, **198**, 59.
160. R. M. Granich, C. F. Gilks, C. Dye, K. M. De Cock and B. G. Williams, *Lancet*, 2009, **373**, 48.
161. C. M. Walker, D. J. Moody, D. P. Stites and J. A. Levy, *Science*, 1986, **234**, 1563.
162. C. Mackewicz, E. Barker and J. A. Levy, *Science*, 1996, **274**, 1393.
163. I. Pellegrin, E. Legrand, D. Neau, P. Bonot, B. Masquelier, J-L. Pellegrin, J-M. Ragnaud, M. Bernard and H. J. A. Fleury, *J. Acquir. Immune. Defic. Syndr.*, 1996, **11**, 438.
164. S. Mazzoli, D. Trabattoni, S. Lo Caputo, S. Piconi, C. Ble, F. Meacci, S. Ruzzante, A. Salvi, F. Semplici, R. Longhi, M. L. Fusi, M. Tofani,

M. Biasin, M. L. Villa, F. Mazotto and M. Clerici, *Nature Med.*, 1997, **3**, 1250.
165. S. Rowland-Jones, J. Sutton, K. Anyoshi, T. Dong, F. Gotch, S. McAdam, D. Whitby, S. Sabally, A. Gallimore and T. Corrah, *Nature Med.*, 1995, **1**, 59.
166. P. N. Fultz, L. Yue, Q. Wei and M. Girard, *J. Virol.*, 1997, **71**, 7990.
167. M. Girard, L. Yue, F. Barré-Sinoussi, E. van der Ryst, B. Meignier, E. Muchmore and P. N. Fultz, *J. Virol.*, 1996, **70**, 8229.
168. A. W. Artenstein, T. C. VanCott, J. R. Mascola, J. K. Carr, P. A. Hegerich, J. Gaywee, E. Sanders-Buell, M. L. Robb, D. E. Dayhoff, S. Thitivichianlert, S. Nitayaphan, J. G. McNeill, D. L. Birx, R. A. Michael, D. S. Burke and F. E. McCutchan, *J. Infect. Dis.*, 1995, **171**, 805.
169. J. P. Moore, Y. Cao, J. Leu, L. Qin, B. Korber and D. D. Ho, *J. Virol.*, 1996, **70**, 427.
170. A. Trkola, A. B. Pomales, H. Yuan, B. Korber, R. J. Maddon, G. P. Allaway, H. Katinger, C. F. Barbas 3rd, D. R. Burton, D. D. Ho and J. M. Moore, *J. Virol.*, 1995, **69**, 6609.
171. M. P. Cranage, A. M. Whatmore, S. A. Sharpe, N. Cook, N. Polyanskaya, S. Leech, J. D. Smith, E. W. Rud, E. J. Dennis and G. A. Hall, *Virology*, 1997, **229**, 143.
172. D. Daniel, F. Kirschoff, S. C. Czajak, P. K. Seghal and R. C. Desrosiers, *Science*, 1992, **258**, 1938.
173. R. M. Ruprecht, T. W. Baba and M. F. Greene, *Lancet*, 1995, **346**, 177.
174. E. van der Ryst, *Oral. Dis.*, 2002, **8 (S2)**, 21.
175. M. Neidrig, J. P. Gregerson, P. N. Fultz, M. Broker, S. Mehdi and J. Hilfenhaus, *Vaccine*, 1993, **11**, 67.
176. M. I. Johnston and A. S. Fauci, *N. Engl. J. Med.*, 2007, **356**, 2073.
177. M. I. Johnston and A. S. Fauci, *N. Engl. J. Med.*, 2008, **359**, 888.
178. S. P. Buchbinder, D. V. Mehrotra, A. Duerr, D. W. Fitzgerald, R. Mogg, D. Li, P. B. Gilbert, J. R. Lama, M. Marmor, C. Del Rio, M. J. McElrath, D. R. Casimiro, K. M. Gottesdiener, J. A. Chodakewitz, L. Corey and M. N. Robertson, *Lancet*, 2008, **372**, 1881.
179. S. Rerks-Ngarm, P. Pitisuttithum, S. Nitayaphan, J. Kaewkungwal, J. Chiu, R. Paris, N. Premsri, C. Namwat, M. de Souza, E. Adams, M. Benenson, S. Gurunathan, J. Tartaglia, J. G. McNeil, D. P. Francis, D. Stablein, D. L. Birx, S. Chunsuttiwat, C. Khamboonruang, P. Thongcharoen, M. L. Robb, N. L. Michael, P. Kunasol and J. H. Kim, *N. Engl. J. Med.*, 2009, **361**, 2209.
180. G. P. Bansal, A. Malaspina and J. Flores, *Curr. Opin. Mol. Ther.*, 2010, **12**, 39.

CHAPTER 13

Drug Discovery for Lower Respiratory Tract Infections

J CARL CRAFT, MD

318 Camelot Lane, Libertyville, Illinois 60048, USA

13.1 Introduction

Unlike most infectious diseases that mainly affect the developing world, lower respiratory tract infections (LRTI) are just as important to the developed world. They are the leading cause of death associated with infectious diseases in the world, accounting for nearly 4 million deaths worldwide per year and significant morbidity.[1] The vast majority of cases that result in death are due to pneumonia, which itself is the leading cause of death in children less than five years of age.[2] Lower respiratory tract infections can be divided into six groups: community-acquired pneumonia, hospital-acquired or nosocomia pneumonia, aspiration pneumonia, lung abscess, acute bronchitis, and chronic bronchitis. Tuberculosis can also be considered as a Lower Respiratory Tract Infection, but this is covered separately in Chapter 8. Treatment of these indications requires all types of antimicrobials: antibacterials, antifungals and antivirals. As with all infectious agents, resistance to therapy eventually develops and new agents will be needed. This should be a reason for continually increasing investment in the research and development of new therapeutics, but in fact the opposite is happening.[3]

Before the discovery of antibiotics, there was little the physician could do for patients other than provide bed rest and good nursing care to keep the patient comfortable. Lewis Thomas comments on the first uses of antibiotics in his

RSC Drug Discovery Series No. 14
Neglected Diseases and Drug Discovery
Edited by Michael J. Palmer and Timothy N. C. Wells

Published by the Royal Society of Chemistry, www.rsc.org

book, *The Youngest Science*. "I remember the astonishment when the first cases of pneumococcal and streptococcal septicemia were treated in Boston in 1937. The phenomenon was almost beyond belief. Here were moribund patients, who would surely have died without treatment, improving in their appearance within a matter of hours of being given the medicine and feeling entirely well within the next day or so ... we became convinced, overnight, that nothing lay beyond reach for the future. Medicine was off and running."[4]

Antibiotics have made a significant change in the lives of people with respiratory tract infections. Prior to antibiotics, patients with community-acquired pneumonia had a mortality rate of over 35%, whereas now with antibiotics it is only 10%.[5,6] In patients with hospital-acquired pneumonia, mortality has dropped from approximately 60% to 30%.[6] Antibiotics lowered the US infection death rate by approximately 220 per 100,000 from 1938 to 1953.[7] In the developing world this improvement has been less dramatic: the cost of medicines is a concern, but also the cost of the health care system to deliver them, and even an approximate diagnosis of the infectious agent has always limited their use. Now, there is a risk that the trend is reversing, and many of the gains of the last fifty years have been lost due to the development of resistance. As of 2002, more than two million people succumbed to hospital-acquired infections per year in the U.S with 99 000 deaths.[8,9] Resistant infections prolong the hospital stays by 24% and increase costs by 29%, compared with susceptible infections, and the cost of antibiotic resistance in the US is estimated at US dollars 21–34 billion per year.[10–12] The epidemic of antibiotic resistance should have been considered a call to action for the research and development community. We understand that antibiotics lose efficacy over time and must be continually replaced, so why aren't we doing the research and development to replace them? The main reasons for this lack of progress are: first, concerns over the economics – being able to achieve a return on investment; and, second the uncertainty about the changing landscape of the regulatory process.

13.1.1 The Economics of Antibiotics: Getting a Return on Investment

The discovery of anti-infective medicines has been historically easier than in many other therapeutic areas. Nature has done the work over millions of years as part of microbial evolution. Scientists only had to identify the molecules nature had selected, and confirm activity using *in vitro* testing (in test tubes) and *in vivo* testing (in animal models). There is little intrinsic clinical risk around mechanism of action: if the Phase I studies show the compound has good pharmacokinetics, good tolerability and no toxicity issues, then the compound should be active in patients. The "target validation" risk in development is much lower than other therapeutic areas, making anti-infectives somewhat cheaper to develop. Against this, there are many complications. Clinical trials for respiratory tract infections must allow for seasonal variations in the timing and severity of the respiratory infections, and so may take several seasons to complete, and require studies in both hemispheres. The problem is that the average cost of drug development has grown dramatically from US dollars 100 million in 1979 to an estimated US dollars 1.3 billion in 2005. The pharmaceutical industry has therefore focused on medicines which potentially produce a billion dollars of revenue a year.[13,14] This is difficult to achieve in anti-infectives: most anti-infective therapy is only taken for a short duration and the patient is cured, therefore the total amount of therapy consumed is less than in chronic disease. Although companies have been able to justify premium prices, particularly for intravenous antibiotics, the financial valuation of a life saved from Lower Respiratory Tract Infection has not reached the same heights as for oncology products. This lack of a clear financial return has resulted in large pharmaceutical companies being hesitant to invest in the research needed to develop new agents and particularly new classes of agent. Small biotech companies have taken up the challenges, often as a result of technology spun out from the pharmaceutical companies, but they too suffer from inadequate funding to complete the pivotal clinical studies. In the developing world things are no rosier, here comparative prices with Western markets adds an additional limit on availability. This need not be the case: antibiotics could be purchased on the cost-plus a small margin seen in HIV, tuberculosis and malaria, then substantially cheaper molecules could be available. However, the research and development still needs to be paid for.

13.1.2 Regulatory Uncertainty for Antibiotic Trials

Many companies developing antibiotics are concerned that the regulatory pathway to the market is getting harder, with increasing need to show clearer evidence of safety than was needed in the past. This has deterred new players from starting in antibiotic discovery. This caution has been called "the Ketek effect": and the regulatory agencies such as the US Food and Drug Administration appear to have become increasingly cautious. The agency came under criticism, and even a congressional investigation, over its handling of the semisynthetic derivative of erythromycin, Ketek, which was approved, and later turned out to cause multiple cases of hepatic failure and even some deaths. The other concern has been around shifting endpoints required to demonstrate efficacy.[15] The simple question is: what is the best way to determine clinically if a new medicine is effective? There have been multiple public workshops with the United States Food and Drug Administration (US FDA) to try to address this question, and the apparently shifting ground may have also discouraged investment in antibiotics.

In the past, many anti-infective agents were approved based on clinical data showing that the patient's condition improved, combined with microbiological confirmation that the pathogen had been eliminated. In many lower respiratory tract infections, and especially community-acquired pneumonia, this is not easy to do, since it is difficult to isolate the infecting pathogen. In most cases, clinical trial endpoints therefore were simply based around clinical cure: does the patient get better?[16–21] Newer agents were then generally compared against older agents. However, the clinical trend in the past has been to try to show that the new medicine was similar to the old one (non-inferiority), rather than statistically demonstrated as better (superiority). The flaw in this was often that the data on the efficacy of the old medicine were weak. This built a system where eventually the true efficacy of modern antimicrobial therapy was called into question, and the FDA became concerned about the adequacy of clinical trial data. Placebo controlled trials would seem to be the way forwards, but for community-acquired pneumonia, historical data from the 1930s and 1940s showed a significant mortality for untreated patients, making this unethical. The agency therefore has moved from what used to be purely clinical endpoints to quantitative ones. The primary endpoint used to be, does the fever go away at day 3? However, this now has to be defined by a whole range of criteria: the patient's temperature must be $\leq 37.8\,^{\circ}C$, the heart rate returning to normal (≤ 100 beats/min), with normal systolic blood pressure of ≥ 90 mm Hg, a normal intake of food and drink, and stable mental status. These all add considerably to the cost of the clinical study and, although more rigorous, arguably increase the chance of the study producing a false negative result.

In the case of sinusitis, acute bronchitis and acute exacerbations of chronic bronchitis (where many of the most profitable medicines are sold), the infections were suggested by some experts to be self-limited, cured by the patient's immune system and so not to require antibiotics. This is at odds with the data: a meta-analysis of treatment of sinusitis infections against placebo showed that infections resolved 35% faster if treated.[22] However, there is now an expectation that new agents in this condition will have to demonstrate statistical superiority – perfectly possible, but requiring much larger and more expensive trials. This so far has meant that no company has been willing to take this risk.

In addition to changes in the clinical endpoints, the US Food and Drug Administration has asked for more rigorous determination of the non-inferiority margins, which will be used in clinical trials. This makes it difficult to estimate how large the studies should be, since there are limited data for these new endpoints.[23,24] This uncertainty over statistical methodology brings the additional fear that the criteria for approval will change over the course of a development program, and this is another reason for the decline of development of new agents for lower respiratory tract infections.

13.2 Lower Respiratory Tract Infections Indications

13.2.1 Community-acquired Pneumonia

Pneumonia is defined as respiratory disease characterized by inflammation of the lung parenchyma (excluding the bronchi) with congestion caused by viruses or bacteria or irritants. Community-acquired Pneumonia is the leading cause of death associated with lower respiratory tract infections, with up to 2.4 million deaths annually, mostly in the African, Southeast Asian and the Eastern Mediterranean. It is particularly dangerous for neonates and children: causing one-third of newborn infant deaths, and representing as much as twice as many child deaths than malaria and HIV infection in these same regions.[25] These infections include bacterial infections, and so could be prevented with vaccines. Up to one million children could be saved with effective vaccination against *Streptococcus pneumoniae*.[26–29] Mortality from pneumonia generally decreases with age in children, but increases in the elderly. In the United Kingdom, pneumonia incidence increases from 0.6% of the adult population (18–39 age group) to 7.5% for those over 75 years. People with increased risk factors, such as alcoholism, or immune system diseases are also more likely to have repeated episodes of pneumonia. Lung disease and air pollution from charcoal and petroleum fuels in the developing world also increase the risk from the disease. The organisms associated with community-acquired pneumonia are listed below in Table 13.1.

Table 13.1 Organisms associated with community-acquired pneumonia.

Age	*Organisms*	*Resistance Problem?*
Birth to 3 weeks[30–34]	Group B streptococci	No
	Listeria monocytogenes	No
	Gram-negative bacilli	Yes
	Cytomegalovirus	No
3 weeks to 3 months[30–34]	*Streptococcus pneumoniae*	Penicillin, Macrolides
	Respiratory syncytial virus	No
	Parainfluenzae viruses	No
	Metapneumoviruses	No
	Bordetella pertussis	No
	Staphylococcus aureus	MRSA, VISA
	Chlamydia trachomatis	No
4 months to 4 years[30–34]	Streptococcus pneumoniae	Penicillin, Macrolides
	Respiratory syncytial virus	No
	Parainfluenzae viruses	No
	Influenza viruses	Yes
	Adenovirus	No
	Rhinovirus	No
	Metapneumovirus	No
	Mycoplasma pneumoniae	Yes
	Group A streptococcus	No
5 to 15 years[30–34]	*Streptococcus pneumoniae*	Penicillin, Macrolides
	Mycoplasma pneumoniae	Yes
	Chlamydia pneumoniae	No
Adults	*Streptococcus pneumoniae*	Penicillin, Macrolides
	Hemophilus influenzae	Ampicillin, Chloramphenical
	Chlamydia pneumoniae	No
	Respiratory syncytial virus	No
	Adenovirus	No
	Influenza viruses	Yes
	Metapneumovirus	No
	Parainfluenzae viruses	No

13.2.2 Hospital-acquired (Nosocomial) Pneumonia

Hospital-Acquired Pneumonia is defined as any pneumonia developing after at least 48–72 hours of being admitted to a hospital. Again, it has increased morbidity and mortality in infants, young children, and the elderly. Patients with underlying lung disease, immunosuppression or recovering surgery, and those on ventilation are particularly at risk.[35–37]

The disease is frequently polymicrobial,[38–47] with a predominance of gram-negative bacilli.[38,40–42,48–52] *Staphylococcus aureus* (especially the methicillin-resistant strains)[39,43,50,53–55] and other gram-positive cocci, such as *Streptococcus pneumoniae*[39,50] are significant causes.[56] Viral infection can lead to hospital-acquired pneumonia: influenza viruses transmitted by patients breathing the contaminated aerosols produced by others in the hospital, and respiratory syncitial virus infections occur from touching the conjunctivae or nasal mucosa with contaminated hands.[35–37] The organisms associated with hospital-acquired pneumonia are shown in Table 13.2.

Table 13.2 Organisms associated with hospital-acquired pneumonia.

Type Organism	*Organism*	*Frequency/References*
Bacterial pneumonia		
	Gram-negative bacilli	50%[38,40–42,48–52]
	Pseudomonas aeruginosa	
	Enterobacter sp.	
	Klebsiella pneumonia	
	Escherichia coli	
	Serratia marcescens	
	Proteus sp.	
	Staphylococcus aureus (MRSA)	19%[39,43,50,53–55]
	MRSA type	
	Streptococcus pneumoniae	[39,50,56]
	Haemophilus spp.	6%[37,38,42,43,50,57]
	Legionella	
Intensive care units		
	Staphylococcus aureus	17.4%
	P. aeruginosa	17.4%
	Klebsiella pneumoniae and Enterobacter spp.	18.1%
	Haemophilus influenzae[35]	4.9%[37,38,42,43,50,57]
Viral pneumonias		10–20% of infections
	Influenza virus	
	Respiratory syncytial virus	
	Cytomegalovirus	Immunocompromised

13.2.3 Aspiration Pneumonia

Aspiration pneumonia develops following the entrance of oral or gastric secretions into the lung.[58–61] There are three separate causes: aspiration of oropharyngeal bacteria, particulate matter, and acidified gastric contents.[62] The outcome depends on the acidity of the aspirate and the amount of bacterial contamination. Gastric contents are generally sterile, but this is not true in all patients or in those using antacids.[59–63] Regardless of the initial bacterial load, bacterial superinfection may occur after the acid-mediated injury.[59,61,64] 5–15% of the 4.5 million cases of community-acquired pneumonia each year in the US are due to aspiration pneumonia.[60,65] The mortality associated with aspiration pneumonia is similar to that of community-acquired pneumonia: approximately 1% in the outpatient setting and up to 25% in those requiring hospitalization.[65] A wide range of bacteria are implicated in aspiration pneumonia: amongst the aerobic *Haemophilus influenza, Staphylococcus aureas, Streptococcus pneumonae, Psuedomonas aeruginosa* and *Escherichia coli.*[58,59] Anaerobic causes include *Peptostreptococcus, Bacterioides, Fusobacterium* and *Prevotella* species.

13.2.4 Chronic Lung Infections: Abscess, Empyema, Bronchiectasis

Lung abscess is not a common form of lower respiratory tract infection but has significant morbidity and mortality. Following necrosis of the pulmonary tissue, cavities develop, which fill with necrotic debris and fluid caused by the infecting organism.[66] Lung abscess frequently follows aspiration, and poor oral hygiene is a risk factor. Abscesses are associated with a large number of infectious agents, including anaerobic and aerobic bacteria, fungi (aspergillus, blasomcyes, coccidioides, cryptococcus and pneumocystis species amongst others), mycobacteria and parasites such as *Entamoeba histolytica* and *Echinococcus* species.[67–72]

13.2.5 Acute Bronchitis

Although acute bronchitis is the most common lower respiratory tract infection, it is not associated with significant mortality. It is an inflammation of the large bronchi caused by airway irritants, allergens and viruses.[73] Characteristic symptoms include cough, sputum production, shortness of breath and wheezing. Treatment is typically symptomatic. Most infectious acute bronchitis is caused by viruses, and the FDA considers it a self-limited disease.[74,75] No new drugs have been registered for this indication and this would require a placebo controlled trial. Potential causative organisms include respiratory syncytial virus, rhinovirus and influenza virus as well as atypical bacteria such as *Mycoplasma pneumoniae* and *Chlamydia pneumoniae*.

13.2.6 Chronic Bronchitis Including Acute Bacterial Exacerbations of Chronic Bronchitis

Chronic bronchitis is a chronic inflammation of the bronchi of the lungs,[76,77] causing a persistent cough that produces sputum and mucus, for at least three months in two consecutive years.[78] Acute exacerbations of chronic bronchitis are when breathing becomes much more difficult because of further airway narrowing and increased secretion of thicker mucus. These exacerbations are an important cause of mortality linked to lower respiratory tract infections.[79] Many patients respond to symptomatic treatment such as cough suppressants, inhaled bronchodilators, corticosteroids, theophylline, and oxygen, but some will require antibiotics to see improvement. Improvement is only temporary and the underlying lung function will deteriorate over time. The role of bacterial infection in acute exacerbations of chronic bronchitis is controversial.[80–83] Respiratory viruses such as influenza, parainfluenzae, rhinovirus, coronavirus, adenovirus and respiratory syncytical virus are associated with 30% of exacerbations. Atypical bacteria, mostly *Chlamydia pneumoniae*, and *Mycoplasma pneumoniae* are implicated in around 10% of exacerbations.[84–89] The role of bacteria has become better defined recently, with immune responses suggesting up to 80% of exacerbations have a bacterial component.[90–92] There are many noninfectious factors in chronic bronchitis, such as tobacco smoking, occupational inhalation of noxious agents, air pollution and allergies.

13.3 Anti-infective Drug Research and Development

Ancient man had a need for something to treat infections. The ancient Greeks used natural substances such as ground onion, woods like myrrh, and wine and honey to treat or prevent wound infections. The Chinese used moldy tofu and the Egyptians used moldy bread.[93–95] The rational search for cures for infections started in the era of Paul Ehrlich, who coined the term chemotherapy. In 1932 Prontosil rubrum, a red dye, was shown to be effective in treating mice, it was discovered to be metabolized to sulfanilamide as the active agent and by 1933, the first humans were treated, with miraculous results.[96]

The first true antibiotic, penicillin was discovered in 1929, but it was not used to treat patients until much later.[97,98] A steady stream of antibiotics has followed over the last fifty years. Early antibiotic research was about discovering the molecules already produced as a means of survival in a crowded microbiological world. Over 150 different antimicrobials exist today and the majority are antibiotics originating from nature, or molecules based on concepts uncovered from natural antibiotics.[99]

There is a real need for new anti-infective agents, replacing those which can no longer be used because of resistance. In the developed world, patients survive longer and their care becomes more complicated, and new anti-infectives are needed to treat new opportunistic pathogens. Doubtless, the pharmaceutical industry should be investing in the research and development of new classes of antibiotics, but this will require financial support to be economically viable. Such medicines could also benefit patients in the developing world, where resistance is often only a few years behind.

13.3.1 Classes of Antibiotics Important in Lower Respiratory Tract Infections

13.3.1.1 β-Lactam Antibiotics: Penicillins, Cephalosporins, Carbacephems, Carbapenems, Monobactums, and β-Lactamase Inhibitors

Penicillin was the first antibiotic discovered, isolated from *Penicillin chrysogenum*[100] and one of the most useful for treating Lower respiratory tract infections. New generations of β-lactams have broader spectrum, including

Gram-negative bacteria such as *Pseudomonas aeruginosa*. Penicillin remained active against *Streptococcus pneumoniae* for over 40 years, until resistance became widespread in the 1990s.[101–104] Penicillins were also developed which were more active against *Staphylococcus aureus,* but methicillin-resistant *Staphylococcus aureus* (MRSA) was first seen in Europe in 1961 and arrived in the USA 1968 and Australia in 1970.[105]

Cephalosporins were isolated from a *Cephalosporium acremonium* from an Italian sewer in 1948.[106] Penicillins have a 6-aminopenicillanic acid core, whereas cephalosporins have a 7-aminopenicillanic acid core. Carbapenem antibiotics were originally developed from thienamycin, isolated from *Streptomyces cattleya.*[103] They are highly active against most β-lactamases. Synthetic carbacephems are similar, with a carbon substituted for the sulfur at the 1 position.[107] Monobactams are monocyclic beta-lactams isolated from the purple bacteria *Chromobacterium violaceum.*[108,109] The only registered monobactam is aztreonam, active against Gram-negative bacteria.

Mechanism of action. The bacterial cell wall is a complex lattice composed of glycan chains of two alternating sugars, N-acetylglucosamine and N-acetylmuramic acid. The N-acetylmuramic acids are linked to a pentapeptide that ends with two unusual D-alanine residues. These peptides are cross-linked by a D,D-transpeptidase which is the main site of action of penicillin, and is known as a penicillin binding protein or PBP. The structure of penicillin is such that it mimics the D-alanine dimer, and is easily accommodated in the enzyme active site, covalently modifying the active site. Each bacterium has a number of penicillin binding proteins, which are numbered according to their molecular weight, with PBP-1 being the largest.[110,111] Inhibition of the different PBPs may cause lysis (PBP-1), delayed lysis (PBP 2) or the development of long, filamentous forms of the bacterium (PBP 3). β-lactams also have a nonlytic killing mechanism involving holin-like proteins in the bacterial membrane that collapse the membrane potential.[112]

Mechanism of bacterial resistance. This is most commonly the generation of β-lactamases, and bacteria have developed hundreds of different β-lactamases in response to the many different β-lactams.[113–116] Carbapenems are one of the antibiotics of last resort, since they retain activity against most β-lactamases. However, carbapenem resistance has been detected in Gram-negative bacteria, caused by the enzyme, NDM-1. A second resistance mechanism is the mutation of the penicillin binding proteins, preventing drug binding. This is the mechanism which lowers penicillin activity against *Streptococcus pneumoniae*, compromising one of the most important agents in the fight against lower respiratory tract infection.[110,111] Thirdly, Gram-negative organisms developed changes in porin structure preventing β-lactams reaching their site of activity and also developed efflux pumps.[117]

Next-generation compound research and development has given us a large armamentarium of broad-spectrum β-lactams and β-lactamase inhibitors. Recent developments have only been minor improvements and further progress will be very difficult: there are no new antibiotics to combat carbapenem resistance.[118,119] Table 13.3 shows the β-lactams that are in clinical development.

Table 13.3 β-Lactams in development.

Product Name	*Sponsor*	*Class/Mechanism*	*Development Status*
Doribax® S4661 Doripenem 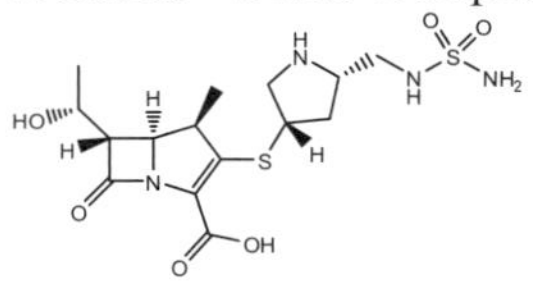	Shionogi USA Florham Park, NJ	Carbapenem for intravenous use	Launched March 2008
CXA-201 CXA-101/tazobactam 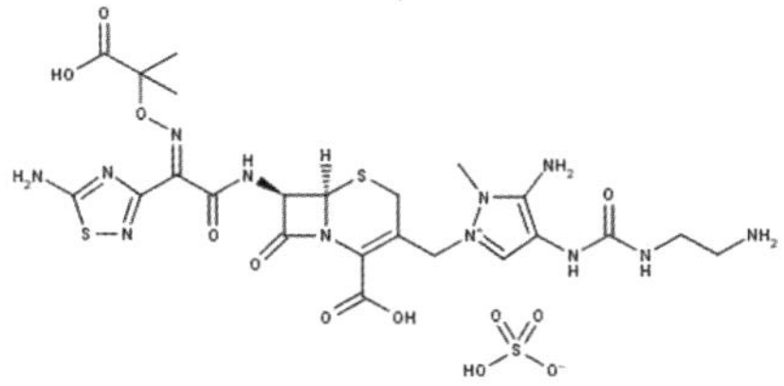	Cubist Pharmaceuticals Lexington, MA	Novel cephalosporin β-lactam CXA-101 with the β-lactamase inhibitor tazobactam	Phase II
NXL104/Ceftazidime 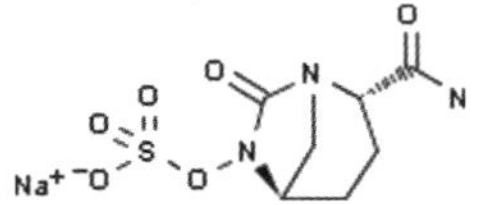	Novexel (subsidiary of AstraZeneca)	Combination β-lactamase enhancer NXL 104 with injectable cephalosporin, ceftazidime	Phase II
MP-601205 No structure reported	Mpex Pharmaceuticals San Diego, CA	Carbopenem β-lactam Bacterial efflux pump inhibitor	Phase I, but no progress reported since 2008
BAL 30072 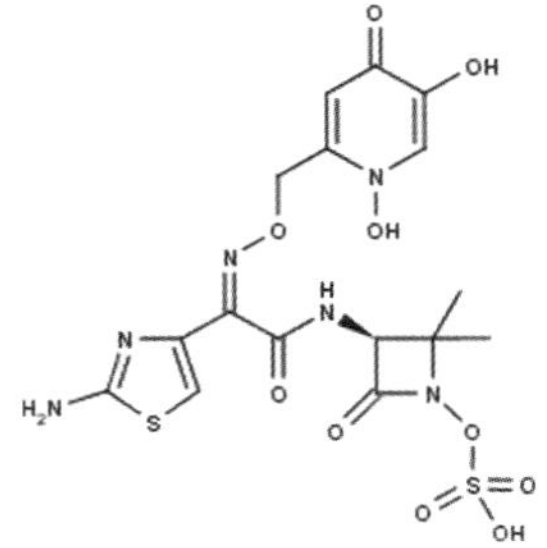	Basilea Pharmaceutica	Monobactam	Phase I

Table 13.3 (*Continued*)

Product Name	*Sponsor*	*Class/Mechanism*	*Development Status*
sulopenem (CP-70429) 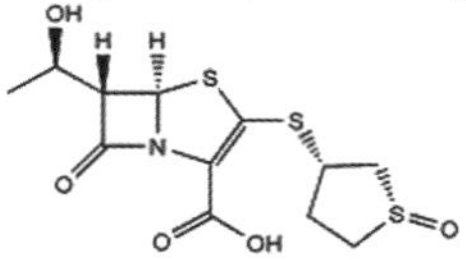	Pfizer New York, NY	Penem β-lactam	Discontinued Sept 2010
BLI-489 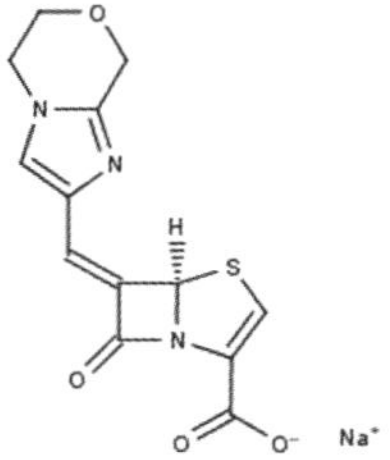	Pfizer New York, NY	Combination of a β-lactam and a penem β-lactamase inhibitor	Phase I Assumed to be discontinued

13.3.1.2 Tetracyclines

The tetracyclines are broad-spectrum antibiotics produced by strains of *Streptomyces*, genus of *Actinobacteria*. They are a subclass of polyketides with a octahydrotetracene-2-carboxamide skeleton.[120] The first tetracyclines were discovered in the late 1940s. Since then, several generations of compounds have been made by modification of the basic scaffold.

Mechanism of action: tetracyclines are protein synthesis inhibitors and work by binding to the 16S ribosomal RNA part of the 30S subunit of microbial ribosomes. They block the attachment of charged aminoacyl-tRNAs preventing the introduction of new amino acids into the growing peptide chain. They are bacteristatic since binding is reversible.[120]

Mechanism of resistance: there are three mechanisms which result in resistance to tetracycline. The first is that the bacteria acquire an efflux pump, which actively transports out the tetracycline, and stops the inhibitory concentration building up. This is usually plasmid-controlled, the plasmids allowing the rapid shuttling of resistance to other bacteria.[121–123] The transporter is a member of the major facilitator superfamily (MFS), which exports the tetracycline as a divalent metal-cation complex.[124] The second method is that the ribosome mutates to overcome the tetracycline action: usually in a soluble protein encoded by the *tetM* and *tetQ* genes which share homology with the GTPases participating in protein synthesis.[125] The third mechanism is rare and involves a cytoplasmic protein that chemically modifies tetracycline.[126]

Next-generation compounds: tetracyclines have been very useful for treating lower respiratory tract infections particularly atypical pneumonia. Research continues with more significant changes to the base structure and side changes, mainly carried out by small companies like Paratek Pharmaceuticals, Inc. The main problems with this class still include teeth staining, photosensitivity and inactivation by calcium from food and antacids.[120] Table 13.4 summarizes the tetracyclines in development.

Table 13.4 Tetracyclines in development.

Product Name	*Sponsor*	*Class/ Mechanism*	*Development Status*
Omadacycline 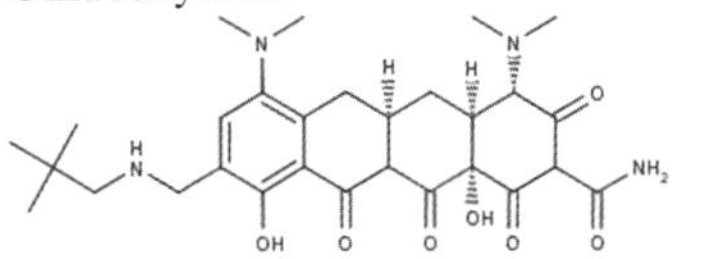	Novartis Basel, Paratek Pharmaceuticals Boston, MA	Protein synthesis inhibitor	Phase III
TP-434: intravenous 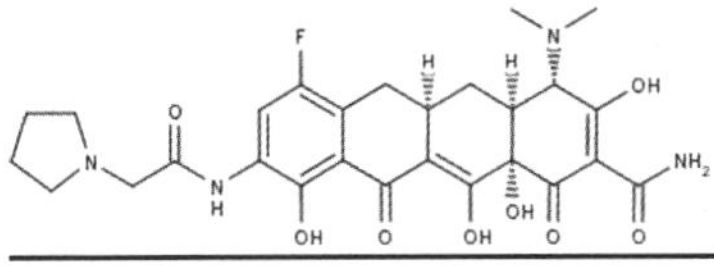	Tetraphase Pharmaceuticals Watertown MA	Protein synthesis inhibitor	Phase I, potential for oral formulation

13.3.1.3 The Macrolide–Lincosamide–Streptogramin B Class

Erythromycin was isolated in 1952 from *Streptomyces erythreus*. Erythromycin and clarithromycin are macrolides with 14-member lactone rings with a cladinose and desoamine sugar attached. Azithromycin is the only 15-membered ring antibiotic, and spiramycin, (used for treating toxoplasmosis) has a 16-membered ring. Ketolides are based on the clarithromycin ring with the cladinose sugar removed and replaced by bulky side chains. The lincosamides are structurally distinct, but have a similar mechanism of action, and so are discussed together. They are based on lincomycin which was isolated from *Streptomyces lincolnensis* in 1962. Closely related clindamycin is the only marketed lincosamide, where the 7(*S*)-chloro- is replaced by a 7(*R*)-hydroxyl.[127]

The streptogramin antibiotic family are cyclic hexa- or hepta-depsipeptide natural products isolated from various members of the genus of bacteria *Streptomyces*.[128,129] The amino acid composition of streptogramin B consists of 3-hydroxypicolinic acid, L-threonine, D-aminobutyric acid, L-proline, 4-N,N-(dimethylamino)-L-phenylalanine, 4-oxo-L-pipecolic acid and phenylglycine. Semi-synthetic, water-soluble derivatives of pristinamycin IA (B type streptogramin) and pristinamycin IIA have given rise to quinupristin and dalfopristin the FDA approved combination drug Synercid.[130]

Mechanism of action: the macrolide/lincosamide/streptogramin B class are structurally different but they all work by binding to the 23S rRNA molecule in the 50 S ribosomal subunit and block the path of the growing peptides as they exit from the ribosome.[131,132] They each have a different way of interacting with the peptidyl-tRNA on the ribosome. Macrolides inhibit protein synthesis by stimulating dissociation of the peptidyl-tRNA molecule from the ribosome during elongation without reaching the peptidyl transferase center. This results in chain termination and a reversible stoppage of protein synthesis.[133,132] Lincosamides block protein synthesis and do reach the peptidyl transferase center in a position between the macrolide binding site and the chloramphenicol binding site.[131] Pristinamycin IA (a streptogramin) blocks dissociation of peptidyl-tRNAs in a way more similar to macrolides. Newer macrolides such as telithromycin, cethromycin, and solitromycin are active against resistant organisms. The macro-lactone rings of solitromycin, telithromycin and erythromycin all occupy the same binding position.[135] Solithromycin has better anchoring to the ribosome binding site from additional hydrogen bonding interactions of aminophenyl at G745. Proximity of solithromycin's fluorine modification provides better binding to base pair C2611-G2057. In addition the lower atomic displacement parameter for solithromycin compared with telilithromycin may account for its improved activity. Mammalian ribosomes have larger 60S subunits, and the differences in sequences presumably accounts for the safety and tolerability for these classes. These binding sites are also targeted by chloramphenicol, macrolides, quinupristin-dalfopristin and linezolid.[131]

Mechanism of Resistance: the first mechanism of macrolide resistance described was post-transcriptional modification of the 23S rRNA by the adenine-N^6 methyltransferase.[133,136] These enzymes modified the ribosome by adding one or two methyl groups to a single adenine A2058 in the 23S rRNA. Since then, many adenine-N^6-methyltransferases have been described. These genes have been designated erythromycin ribosome methylation (*erm*). Both macrolide inducible and the more common constitutive *erm* have been identified, and plasmids carrying multiple *erm* genes have also been found.[137,136] Adenine methylation reduces the affinity for all three classes of antibiotics, resulting in resistance against macrolides, lincosamides, and streptogramin B antibiotics (MLS_B) but not to ketolides.

Additional mutations have resulted in two different ribosomal proteins on the 50S ribosome, L4 and L22. The phenotypes associated with these mutations are ML (macrolide lincomycin) and MSB (macrolide Streptogramin B), which can be confused with *erm* and *mef*. The ever-increasing number of *erm* gene types have been grouped as *erm*(A) and *erm*(B).[133,137] These mutations are in the same location as the adenine that gets methylated by *ermA* ("A2058"), and are adenine to guanine changes at position 2058, 2059, or 2611. The resistance phenotype increases with increasing numbers of mutated ribosomes. *Streptococcus pneumoniae* has four copies of ribosomal RNA and at least two of the four ribosomes must include the mutation to show the *erm* phenotype.[138–146] Some alternative mechanisms confer resistance to only one or two of the class. These include efflux proteins *mef*(A), *mef*(E) and *lmr*(A), which transport antibiotic out of the cell,[133] Other mechanisms include Streptogramin B hydrolysis, stretogramin acetyltransferase (related to chloramphenicol acetyltransferase), erythromycin esterification genes, macrolide phosphotransferase genes and specific lincomycin nucleotidyltransferases genes.[133] The new ketolides such as solithromycin from Cempra and Enanta's EDP-322 are not susceptible to these resistance mechanisms.[141]

Safety and tolerability: the macrolide/lincosamide/streptogramin B class is generally well tolerated. Erythromycin was limited by gastrointestinal tolerability due to acid instability, but this was resolved in the semi-synthetic macrolides clarithromycin and azithromycin. There are isolated reports of QTc prolongation with erythromycin, most frequently associated with intravenous 'push' where the plasma concentrations are ten times higher than with the oral dose. In addition, interactions with drugs which cause QTc prolongation, or are potent inhibitors of CYP3A4 are a concern.[142]

Telithromcin is an erythromycin derivative, where the cladinose sugar is replaced by a ketone, hence the name ketolide. It causes specific adverse events, believed to be due to the pyridine-imidazole moiety not shared by older macrolides. These adverse events include blurred vision, rapid progression of exacerbation of myasthenia gravis and most importantly liver toxicity, which ultimately led to the withdrawal of the drug. These are due to inhibition of various nicotinic acid acetylcholine receptor subtypes. *In vitro* telithromycin shows a significantly greater inhibition of nicotinic receptors in peripheral ganglia ($\alpha3\beta4$ and $\alpha7$) than other macrolides. These receptor subtypes are responsible for visual accommodation, suggesting a mechanism for the visual disturbances. Antagonism of the $\alpha7$ subtype of the vagus nerve in the liver increases hepatic inflammation through cytokine release and this blockade may be partly responsible for the rare cases of liver failure seen with this drug.[143] Both lincomycin and clindamycin have a clear association with pseudomembranous colitis. They are active against the anaerobic bacteria of the colon, facilitating the growth and colonization with *Clostridium difficile*.[144]

Next-generation compounds. When the hepatic side effect of telithromyin caused the drug to be withdrawn, and even became a topic of concern for the US Congress, development of the next generation of the macrolide/lincosamide–streptogramin B went on hold. With a fuller understanding of the mechanistic basis of these side effects, this area of drug discovery should go forward. New agents in development are shown in Table 13.5, and this class is

Table 13.5 Macrolide/lincosamide/streptogramin B antibiotics in development.

Product Name	*Sponsor*	*Class/ Mechanism*	*Development Status*
CEM-101 (solithromycin) oral, intravenous 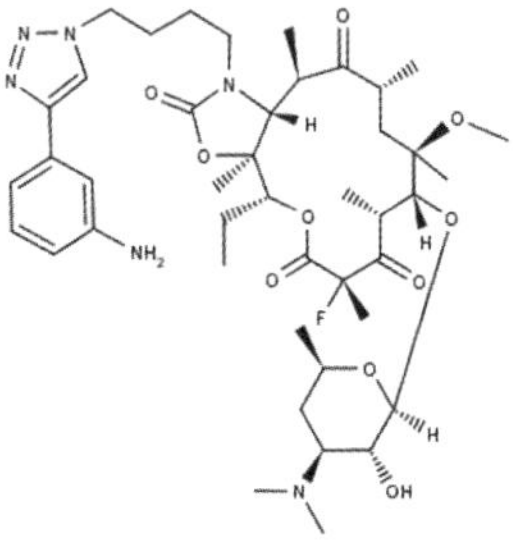	Cempra Pharmaceutical Chapel Hill, NC, Optimer Pharmaceuticals San Diego, CA	Fluroketolide inhibits protein synthesis by binding to the 50 S ribosomal subunit	Phase II (oral formulation), most potent macrolide antibiotic known
Fidaxomicin/tiacumicin 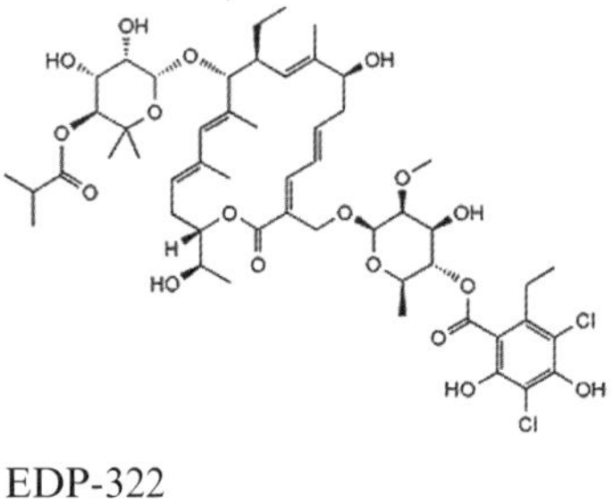	Optimer Pharmaceuticals San Diego CA Astellas Pharma Tokyo	RNA polymerase?	Pre-registration
EDP-322 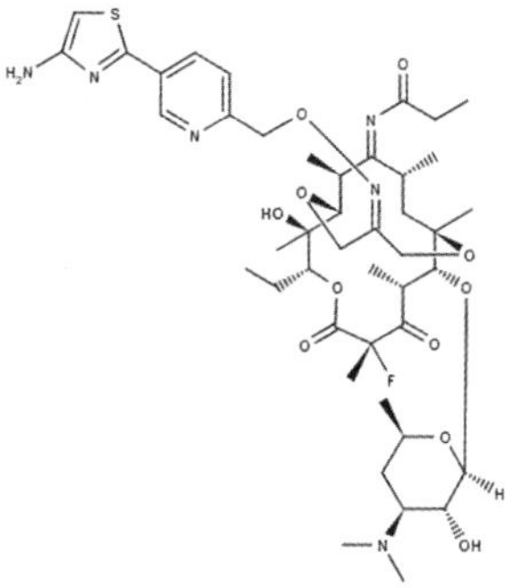	Enanta Pharmaceuticals Watertown, MA	Ketolide inhibiting protein synthesis by binding to the 50 S ribosomal subunit	Phase I

Table 13.5 (*Continued*)

Product Name	*Sponsor*	*Class/ Mechanism*	*Development Status*
Modithromycin EDP-420 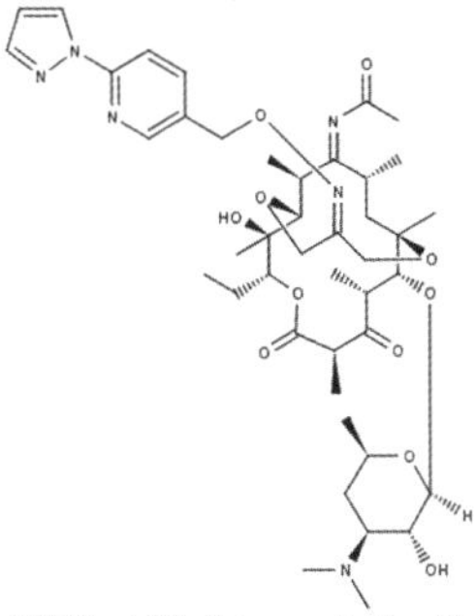	Enanta Pharmaceuticals Watertown, MA Shionogi Osaka Japan	Ketolide inhibits protein synthesis by binding to the 50 S ribosomal subunit	Phase II
NXL 103 Linopristin (top) and Flopristin (bottom) 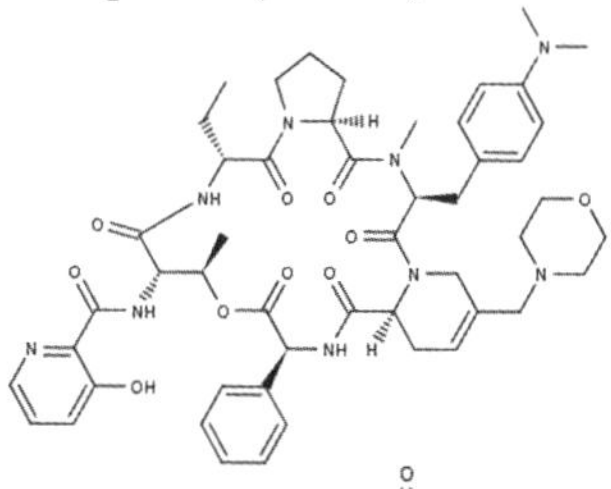 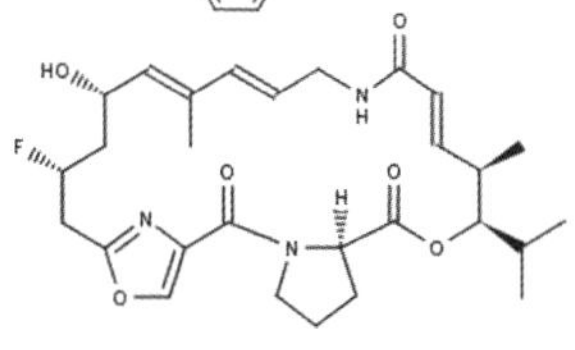	Novexel, AstraZeneca Wilmington, DE	Streptogramin Combination: antibiotics linopristin and flopristin, inhibiting bacterial ribosomes	Phase II
Restanza™ (Cethromycin) 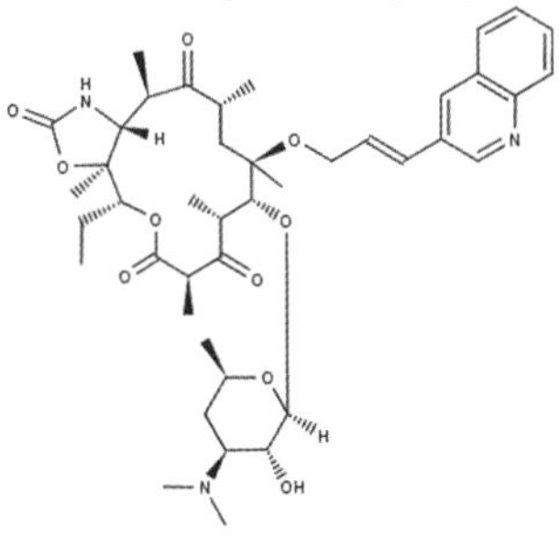	Abbott Laboratories Advanced Life Sciences, Woodridge, IL	Ketolide; Inhibits protein synthesis by binding to the 50 S ribosomal subunit	Phase III oral, phase I intravenous, did not get FDA approval

very important to the treatment of lower respiratory tract infections. They cover the most important organisms that cause community-acquired pneumonia including *Streptococcus pneumoniae* and the atypical organisms. There appears to be room for modification of these structures with large side chains and removal of the desoamine sugar. There are several discovery programs working on this class, led by Enanta and Cempra: biotechnology companies who have developed a strategic interest in the area.

13.3.1.4 Aminoglycosides

The aminoglycosides are derived from either *Streptomyces* or *Micromonospora*. Aminoglycosides derived from *Streptomyces* genus are named with the suffix *-mycin*, whereas those that are derived from *Micromonospora* are named with the suffix *-micin*. They are organic bases composed of amino sugars and a common inositol residue. The main application of aminoglycosides in treatment of lower respiratory tract infection was for hospitalized patients with Gram-negative infections.[145] Streptomycin was also the first effective drug for the treatment of tuberculosis. The following agents are available today: amikacin, arbekacin, gentamicin, kanamycin, neomycin, netilmicin, paromomycin, rhodostreptomycin, streptomycin, tobramycin, and apramycin.[146] Due to resistance, they are not used as much today, but are still used for the treatment of cystic fibrosis including by inhalation.

Mechanism of action: aminoglycosides are protein synthesis inhibitors. They bind irreversibly to the bacterial 30S ribosomal subunit and some may also bind to the 50S subunit. Disruption of the proofreading process leads to increased error rates in the protein synthesis and premature termination. Inhibition of ribosomal translocation occurs when the peptidyl-tRNA moves from the A-site to the P-site. Aminoglycosides are also reported to disrupt the integrity of bacterial cell membrane.[147]

Mechanism of resistance: aminoglycoside resistance results from three mechanisms: reduction in uptake or decreased cell permeability, mutations in the ribosomal binding site, and production of aminoglycoside-modifying enzymes. Reduced uptake or decreased cell permeability may provide resistance to all aminoglycosides.[147–150] Alteration of ribosome binding sites is common in streptomycin resistance, but cross-resistance with the other aminoglycoside is rare since they have multiple binding sites.[147–150] Enzymatic modification is the most common type of resistance, encoded on plasmids and transposons, and resulting in high-level resistance. There are three types of aminoglycoside modifying enzymes:[147–150] N-Acetyltransferases (AAC), catalyzing acetyl CoA-dependent acetylation of the amino group, O-Adenyltransferases (ANT) – catalyzing the ATP-dependent adenylation of the hydroxyl group, and O-Phosphotransferases (APH) catalyzing the ATP-dependent phosphorylation of a hydroxyl group

Next-generation compounds: there has been very little research to find new members of this class, with most activity being on development of new inhaled formulations. These have been useful for the treatment of patients with cystic fibrosis. See Table 13.6 for present aminoglycoside pipeline. Durante-Mangoni asked the rhetorical question "Do we still need the aminoglycosides?" He may be right.[151]

Table 13.6 Aminoglycoside and glycopeptide antibiotics in development.

Product Name	*Sponsor*	*Class/Mechanism*	*Development Status*
TIP™ tobramycin dry-powder 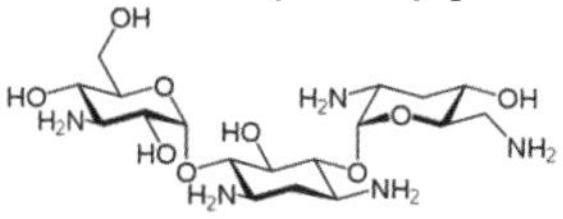	Novartis Pharmaceuticals East Hanover, NJ	Fluoro quinolone DNA Gyrase Inhibitor of Topoisomerase II	Phase III
GS9310/11 tobramycin/fosfomycin inhalation. Fosfomycin: 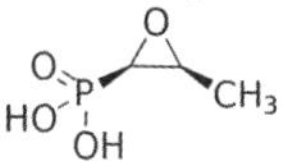	Gilead Sciences Forster City, CA	Aminoglycoside binds to the 50 S and 30 S subunits preventing complexation Fosfomycin inhibits cell wall formation biogenesis inhibiting UDP-acetyl glucosamine enolpyruvyl transferase (MurA)	Phase II
Arikace™ (amikacin sustained) release 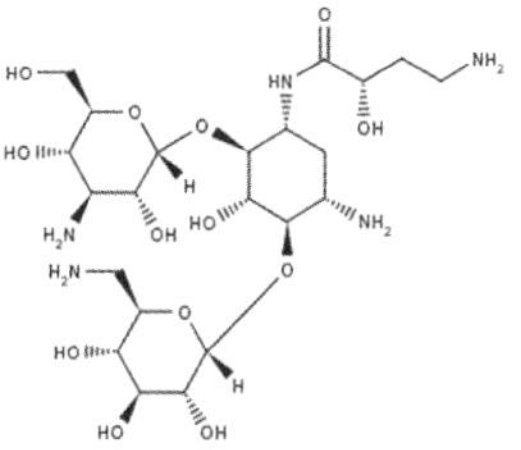	Transave Monmouth Junction, NJ	Aminoglycoside 30 S ribosomal protein inhibitor	Phase I
Oritavancin 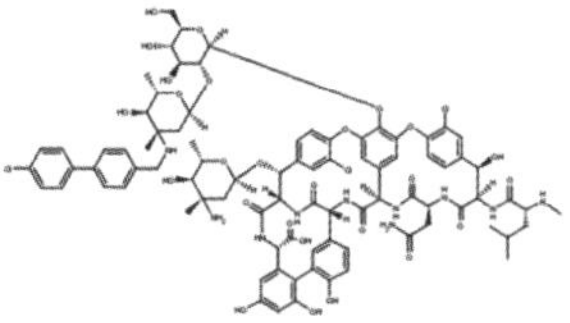	The Medicine Company Parsippan, NJ	Glycopeptide. Disrupts the cell membrane of gram positive bacteria	Phase III
TD-1792 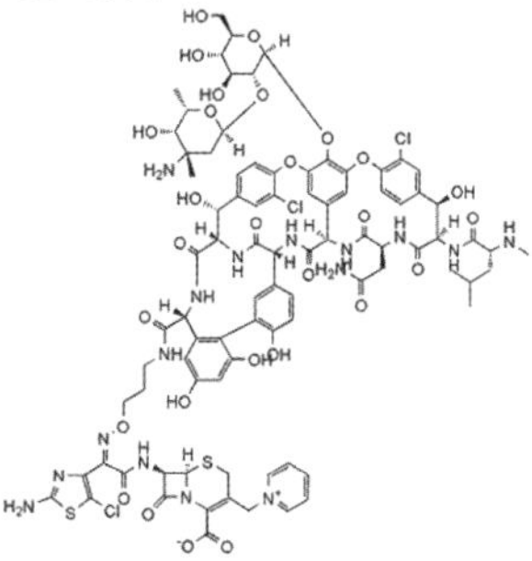	Theravance South San Francisco, CA	Glycopeptide Disrupts the cell membrane of gram positive bacteria; conjugated with a beta lactam	Phase II

Table 13.6 (*Continued*)

Product Name	*Sponsor*	*Class/Mechanism*	*Development Status*
Vibativ™ Televancin 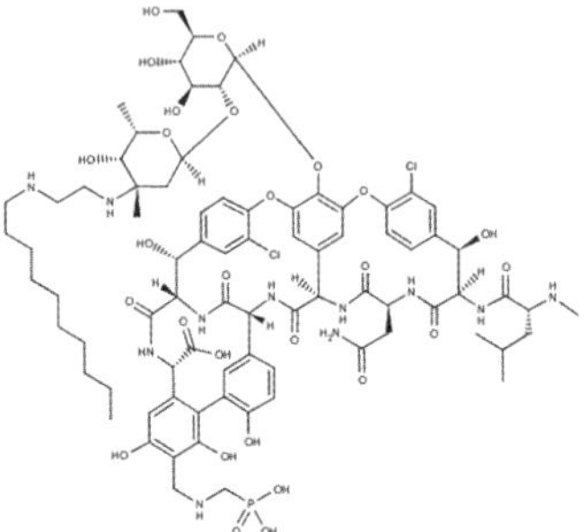	Atellas Pharma US Deerfield, IL Theravance South San Francisco	Glycopeptide: Disrupts the cell membrane of gram positive bacteria	Launched for skin infection, in pre-registration for pneumonia

13.3.1.5 Glycopeptides

Vancomycin was the first glycopeptide antibiotic to be isolated in 1953, produced by *Amycolatopsis orientalis* identified in a soil sample collected in Borneo.[152,153] Glycopeptide antibiotics are glycosylated cyclic or polycyclic non-ribosomal peptides and the class includes vancomycin, teicoplanin, telavancin, bleomycin, ramoplanin, and decaplanin.[154] Their main use is for treating Gram-positive organisms, particularly methicilin-resistant *Staphlococcus aureus* (MRSA). Their use in lower respiratory tract infection is limited to severe infections with MRSA or in a combination with another agent with Gram-negative coverage.

Mechanism of action: glycopeptides inhibit bacterial cell wall synthesis by a mechanism similar to the β-lactams. They prevent incorporation of N-acetylmuramic acid (NAM) and N-acetylglucosamine (NAG) peptide subunits by forming hydrogen bonds blocking formation of the peptidoglycan matrix. This disrupts the major structural component of the Gram-positive cell wall and leads to lysis.[155]

Resistance: for many years, Vancomycin was the agent of choice for MRSA but resistance is a growing problem.[156] In 1987 vancomycin-resistant enterococcus was reported.[157] Since then, vancomycin resistance has increased and now includes vancomycin-intermediate *Staphylococcus aureus* (VISA), vancomycin-resistant *Staphylococcus aureus* (VRSA), and vancomycin-resistant *Clostridium difficile*.[158–160]

Mechanism of resistance: the main mechanism of resistance is alteration of the terminal amino acid residues of the NAM/NAG-peptide subunits, decreasing the importance of the hydrogen bonding interaction, and lowering the affinity 1000-fold.[156] The *VanA* mutation effects inducible resistance to both vancomycin and teicoplanin; *VanB* is lower-level resistance inducible by vancomycin and remains susceptible to teicoplanin; *VanC* produces constitutive resistance only to vancomycin but is the least important clinically.

The glycopeptides have had a place in lower respiratory tract infections, particularly as methicilin resistant *Staphylococcus aureus* has become more important.[161] They are also used in the treatment of *Clostridium difficile*.[160] Glycopeptides are well tolerated: the main adverse event is nephrotoxicity, which can be easily managed.[162]

Next-generation compounds: research has continued on several new glycopeptides including oritavancin and dalbavancin, which have improved pharmacokinetics and activity against vancomycin-resistant bacteria. Table 13.6 shows the glycopeptides antibiotics currently in development.[163]

13.3.1.6 Lipopeptide Daptomycin

Daptomycin is a novel lipopeptide, isolated from cultures of *Streptomyces roseosporus*, active against Gram-positive pathogens.[164,165] It is a bactericidal cyclic anionic lipopeptide which perturbs the bacterial cell membrane, a mode of action different from most other antibiotics except polymyxin and gramicidin. Daptomycin (Figure 13.1) inserts into the cytoplasmic membrane of Gram-positive bacteria in a calcium-dependent fashion. The ion-channel so formed allows the efflux of potassium, depolarizing the cell membrane, leading to cell

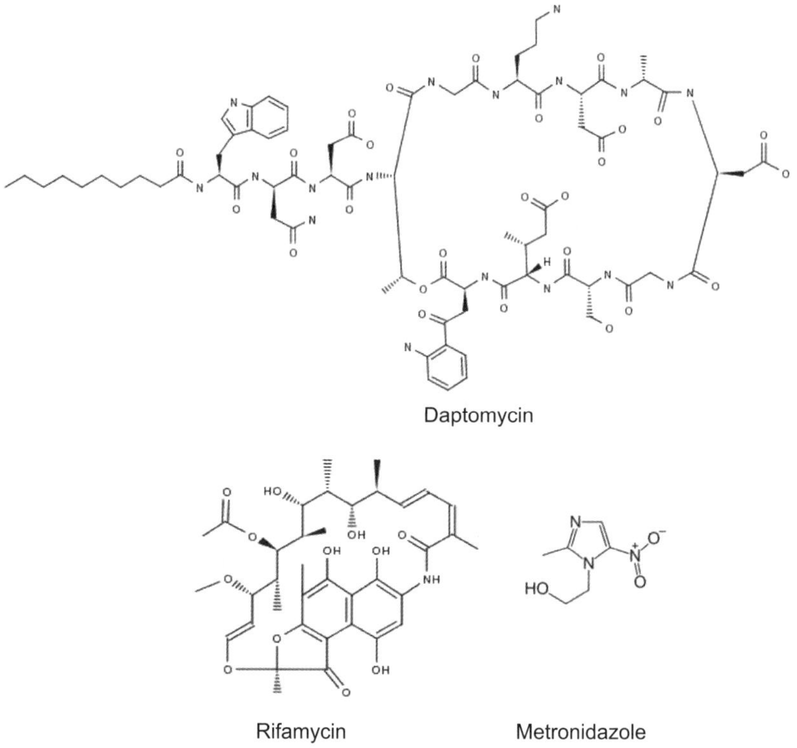

Figure 13.1 Daptomycin, rifamycin and metronidazole.

death. Gram-negative bacteria are not affected because of their lipopolysaccharide layer. Although daptomycin has been approved to treat Gram-positive infections, it was not approved for the treatment of community-acquired pneumonia (CAP)[166](Dap3). The lack of efficacy in community-acquired pneumonia is considered to be due to inhibition by pulmonary surfactants.[167] Research into this class may be fruitful in the future.

13.3.1.7 The Ansamycins and Rifamycins

The ansamycins are a family of antibiotics active against many Gram-positive and some Gram-negative bacteria. This class of antibiotics are products of various *Actinomycetes* with a carbon framework arises from the polyketide pathway *via* a polyketide synthase but can be synthesized chemically. Rifamycin (Figure 13.1), produced by *Streptomyces mediterranei* is a medically important ansamycin.[168,164] The rifamycin group also includes the derivatives rifampicin, rifabutin and rifapentine. They are important because of their use for the treatment tuberculosis and leprosy. They are also active against the serious lower respiratory tract infections caused by Gram-positive bacteria such as *Staphylococcus aureus* or *Streptococcus pyogenes*. Rifamycin is a potent bactericidal antibiotic against both free-living and biofilm-associated Gram-positive bacteria.[170]

Mechanism of action: the mechanism of activity of the rifamycins is inhibition of DNA-dependent RNA synthesis.[171] This is due to their high affinity for prokaryotic RNA polymerase. Rifamycin blocks synthesis of the growing mRNA.[172,173]

Mechanism of resistance: use of rifamycin monotherapy is limited by the rapid emergence of rifamycin-resistant organisms,[174–180] which are largely due to mutations in the RNA polymerase. Rifamycin is mainly used as part of a multidrug treatment regimen, where the partner prevents emergence of resistance, primarily in tuberculosis therapy.

13.3.1.8 Metronidazole

The class of nitroimidazole antibiotics have activity against anaerobic bacterial and protozoa infections. Metronidazole (Figure 13.1), tinidazole and nimorazole are the market examples of this class. The nitroimidazoles have an important place in respiratory tract infections for the treatment of empyema, pneumonia, aspiration pneumonia, and lung abscess including lung abscess due to *Entamoeba histolytica*.

Mechanism of action: nitroimidazoles are taken up by anaerobic bacteria and activated by reduction in a reaction with reduced ferredoxin. The reduction of the nitro group results in activated intermediates, which then covalently modify DNA causing strand breaks and lead to cell death. Metronidazole can only work when the target organism is in a low redox potential environment.[178] After more than 40 years metronidazole has continued to be effective, but an increase in the incidence of metronidazole resistance among *Bacteroides* spp. is raising concern.[179,180]

Mechanism of resistance: resistance to nitroimidazoles can be due to several mechanisms including decreased uptake or efflux, mutation of the biological target on the chromosomal and plasmid-borne *nimR* genes, increased oxygen scavenging capabilities (SOD/catalase/peroxidase), and enhanced activity of DNA repair enzyme.[181,182]

Research into new nitroimidazoles has been limited by concerns over carcinogenity since metronidazole has been shown to promote tumour formation in experimental animals, although clinically it is known to be safe in humans.

13.3.1.9 Pleuromutilins

The pleuromutilin are derivatives of the naturally occurring pleuromutilin produced by an edible mushroom *Pleurotus mutilus*. They have a unique type of protein synthesis inhibition, which involves inhibition of bacterial protein synthesis by binding to domain V of 23S rRNA blocking peptide formation directly interfering with substrate binding.[183,184] Pleuromutilins have no target-specific cross-resistance to other antibacterials.[191] Unfortunately, mutations in the genes encoding 23S rRNA have led to reduced susceptibility to tiamulin, a topical pleuromutilin used in veterinary practice.[186]

There are ongoing activities in research and development. Retapamulin (SB-275833), sold by GlaxoSmithKline, was the first pleuromutilin to be developed specifically for human topical use and is 8-fold more potent than the veterinary product tiamulin (Figure 13.2).[187] The program produced other molecules

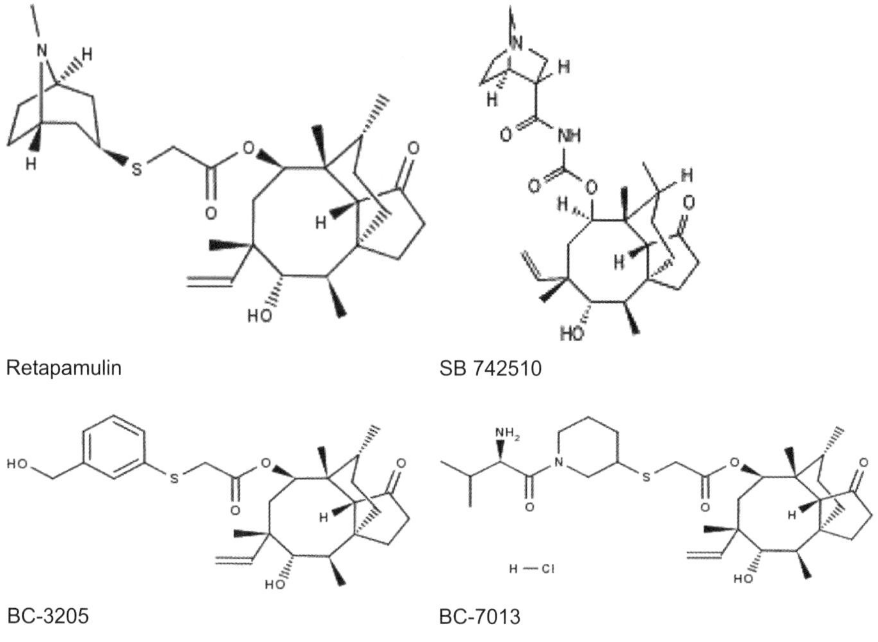

Figure 13.2 Pleuromutilins.

including SB-742510 which went into phase I studies, but was not developed further. Another topical pleuromutilin, BC-7013, is being developed by Nabriva (Vienna, Austria, a spin-off from Sandoz/Novartis) and is in Phase II trials. BC-3205 and 3781 are the first generation of pleuromutilins with excellent systemic bioavailability, and are being considered for lung infection. It is active against gram-positive pathogens and atypicals as well as some Gram-negative pathogens including multidrug-resistant pathogens; methicillin-resistant *Staphylococcus aureus* (MRSA), *Streptococcus pneumoniae*, and vancomycin-resistant *Enterococcus faecium*. Both oral and IV administration formulations are being developed. The structure of BC-3781 has not been disclosed.

GlaxoSmithKline are working to develop an oral pleuromutilin for the treatment of respiratory tract infections and also working with the Global Alliance against Tuberculosis to develop pleuromutilins for tuberculosis. New agents for lower respiratory tract infections are certainly needed, but there are risks in developing the same class of agents for topical and systemic use. The veterinary use of tiamulin has already reduced its effectiveness.[186]

13.3.1.10 Chloramphenicol and Thiamphenicol

Chloramphenicol was originally isolated from *Streptomyces venezuelae* and is the only naturally occurring antibiotic containing a nitrobenzene group. Chloroamphenicol was first used in clinical practice in 1949. It was also the first antibiotic to be manufactured synthetically on a large scale.[188] Chloramphenicol (Figure 13.3) was initially widely used but safety concerns resulted in a marked decrease in its use. One concern is bone marrow suppression, a direct toxic effect of the drug, but it is reversible. The more serious and often fatal adverse effect is aplastic anemia, which is idiosyncratic, rare and unpredictable. Intravenous chloramphenicol is associated with gray baby syndrome.[189,190] Due to immature liver enzymes like UDP-glucuronyl transferase in the newborn, there is an accumulation of unmetabolized chloramphenicol in the body, causing hypotension and cyanosis.[191] Its use in the developing world is largely driven by its low cost: a 250 mg tablet costs between 1.2 and 2.2 cents in Africa.

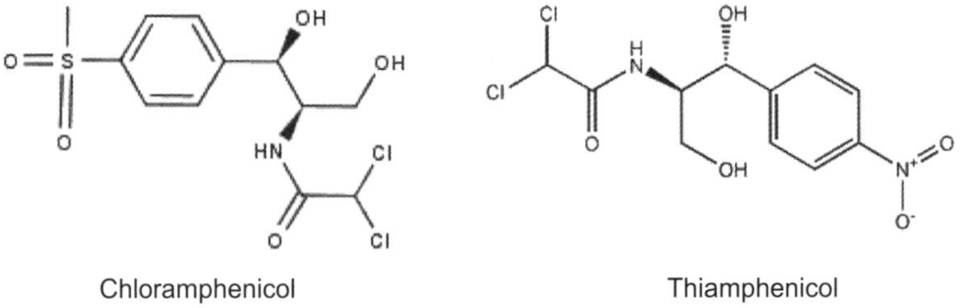

Figure 13.3 Chloramphenicol and thiamphenicol.

Mechanism of action: chloramphenicol and thiampenicol are broad-spectrum bacteriostatic antimicrobials that act by inhibiting the peptidyl transferase activity of the bacterial ribosome, binding to A2451 and A2452 residues in the 23S rRNA of the 50S ribosomal subunit.[192,193]

Mechanism of resistance: the first mechanism of resistance are modifications which greatly reduce membrane permeability, the second is mutation of the 50S ribosomal subunit to prevent binding of the drug to its target, and the third is elaboration of chloramphenicol acetyltransferase.[194]

No clinical development of any further chloramphenicol analogs is ongoing, reflecting the relatively limited possibilities for structural variation in this class.

13.3.2 Target-based Synthetic Antimicrobials Important to Lower Respiratory Tract Infections

13.3.2.1 Folate Synthesis: Sulfonamides and Diaminopyrimidines

Mechanism of action: sulfonamides were the first synthetic antimicrobial agents that were used as antimicrobial drugs, starting the antimicrobial revolution in medicine.[195] Prontosil was the first sulfonamide developed in 1932. Its anitbacterial effect was only seen *in vivo*, since it was a prodrug of sulfanilamide the active agent. These compounds work by inhibiting the enzyme dihydopterate synthetase, an essential enzyme in folate biosynthesis. Many thousands of molecules containing the sulfanilamide structure have been synthesised since its discovery, yielding improved effectiveness and decreased toxicity.[195] Folic acid metabolism is essential for bacterial growth, however mammals do not obtain folic from their diets and are not affected by these agents.[196] Use of sulfonamides has been limited by allergic reactions including delayed hypersensitivity reaction and occasionally more serious immune complications such as Stevens-Johnson syndrome, toxic epidermal necrolysis, agranulocytosis, hemolytic anemia, thrombocytopenia, and fulminant hepatic necrosis.[197–201]

Diaminopyrimidines work by inhibiting the enzyme dihydrofolate reductase, also blocking the essential folic acid biosynthesis. Optimisation of this scaffold led to the development of trimethoprim as an antibacterial agent. As montherapy, trimethoprim use is limited, but is used in combination with sulfamethoxazole, known as co-trimoxazole. This has become one of the most successful drugs in therapy of lower respiratory tract infections and is cheap (a 400/80 mg tablet is available in Africa for less than 1 cent).

Mechanism of resistance: resistance to sulfonamides may be due to overproduction of the substrate p-aminobenzoic acid, and is plasmid mediated. Chromosomally mediated resistance due to mutation of dihydopterate synthetase is rare. Resistance to the combination of trimethoprim and sulfamethoxazole is increasing due to several mechanisms including decreased permeability, efflux pumps and mutations in dihydrofolate reductase. Because of this resistance the combination is no longer frequently used for respiratory infections.

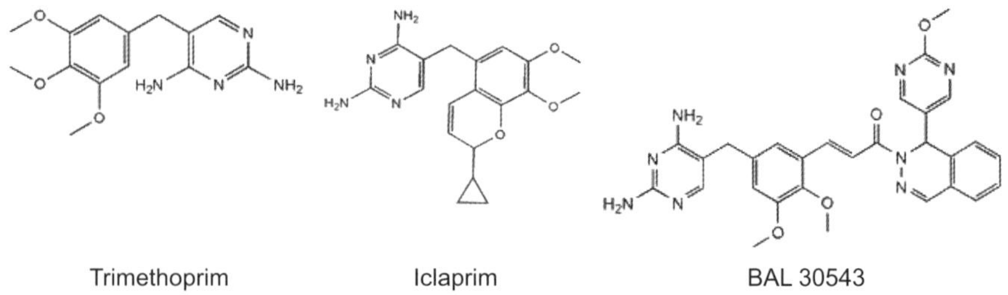

Figure 13.4 Dihydrofolate reductase inhibitors.

Next-generation molecules. Much of the development of DHFR inhibitors has focused on inhibiting the human enzyme, however there are a few antibacterials. Their design has been aided by the availability of the enzyme's three-dimensional structure of the enzyme. Iclaprim is in phase II development as an intravenous formulation for pneumonia by Evolva, and Basilea has related compounds in discovery (Figure 13.4).

13.3.2.2 Topoisomerase Inhibitors: Quinolones and Fluoroquinolones

The quinolones are a family of synthetic broad-spectrum antibiotics which were identified in a distillate from chloroquine synthesis.[201–204] In 1962, nalidixic acid was the first quinolone to be introduced for the treatment of urinary tract infections. It was followed by pipemidic acid, oxolinic acid, and cinoxacin which were introduced in the 1970s, but were only marginal improvements.[205] Since then, more than 10 000 analogs have been synthesized, more than thirty received marketing approval, including for veterinary use, but only a handful are still available.[206] Many quinolones were removed from the market due to toxicity: nalidixic acid, cinoxacin, clinafloxacin, enoxacin, gatifloxacin, grepafloxacin, temafloxacin, and trovafloxacin. Most of the others were removed because of a lack of market share. Ciprofloxacin, levofloxacin and moxifloxacin are the most frequently used quinolones now.

Mechanism of action: quinolones interfere with DNA replication, by inhibiting the bacterial topoisomerase type II enzymes, including DNA gyrase and topoisomerase IV enzymes, thereby inhibiting DNA replication and transcription.[207] Quinolones also concentrate intra-cellularly which accounts for their activity against intracellular pathogens such as *Legionella pneumophila, Mycoplasma pneumoniae,* and *Chlamydia pneumoniae*. The target for most Gram-negative bacteria is DNA gyrase and for Gram-positive bacteria the target is topoisomerase IV. In eukaryotic cells quinolones have been shown to damage mitochondrial DNA.[208,209]

Quinolones are heterocycles with a bicyclic core structure.[210] A carboxylic acid function at position 3 and a bulky substituent on the bicyclic core mainly at positions 1 and 7 and/or 8. Twenty years after the original discovery of nalidixic acid, the addition of a fluorine at position 6 of the 4-quinolone and the replacement of the 7-methyl side-chain of nalidixic acid with a piperazine group were introduced. The addition of the C6 fluorine atom is required for the antibacterial activity of this class.[211]

The fluoroquinolones have become a major part of the treatment of lower respiratory tract infections particularly in the outpatient setting.[212,213] At present fluoroquinolones are recommended for the treatment of community-acquired pneumonia because of the high level of resistance seen to penicillins and macrolides.[214] They are frequently prescribed to treat bronchitis even though only 5–10% are due to bacterial infection and should resolve without treatment.[215]

Adverse effects: fluoroquinolones are generally well tolerated but can be associated with serious adverse effects and some have been removed from the market because of toxicity.[216,217] These adverse events include neurological and psychiatric effects such as tremor, confusion, anxiety, insomnia, agitation, and, in severe cases, psychosis. Concerns about cardiovascular safety are underlined by QTc interval prolongation and cardiac arrhythmias (*torsade de pointes*). Other issues include tendon and articular issues, hypoglycemia, and rarely hepatic toxicity.[218–225] Children and the elderly are at increased risk of adverse reactions and adverse events may occur during, as well as after, completed therapy.[219,220,226] *Clostridium difficile*-associated diarrhea is more frequently seen with fluoroquinolone therapy than other antibiotics.[227–229] Drug interactions are a concern for fluoroquinolones, which are mainly metabolized by CYP1A2, leading to enhanced toxicity with, for example, theophylline.

Mechanism of resistance: resistance to quinolones can evolve rapidly, even during a course of treatment. Numerous pathogens, including *Staphylococcus aureus*, enterococci, and *Streptococcus pyogenes* now exhibit resistance worldwide. There are three major mechanisms of resistance: efflux pumps, plasmid-mediated resistance genes producing proteins that bind to DNA gyrase, and mutations of DNA gyrase or topoisomerase IV.[230,231]

Quinolone research has cooled because of concerns over the safety of this class, and little discovery work is currently being done. Table 13.7 shows the present extent of research and development. The majority of research is directed at modification of formulations to forms which can be inhaled, are useful in cystic fibrosis, and the development of new agents to combat tuberculosis.

13.3.2.3 Oxazolidinones

The oxazolidinones were originally known for their role as monoamine oxidase inhibitors. The antibacterial activity was first described in 1978 for treatment of plant infections by DuPont. The activity against clinically important bacteria was discovered six years later, and Linezolid was approved in 2010. It is worth underlining that these were the first new class of antibiotics to be brought to the market in decades.

Table 13.7 Quinolones in development.

Product Name	*Sponsor*	*Class/Mechanism*	*Development Status*
Prulifloxacin 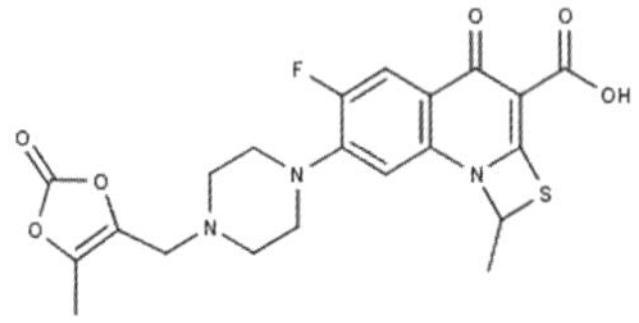	Nippon Shinyaku Kyoto Japan	Fluoro quinolone Inhibits DNA gyrase and topoisomerase II	Launched
ARD-3100, 3150 (liposomal ciprofloxacin) 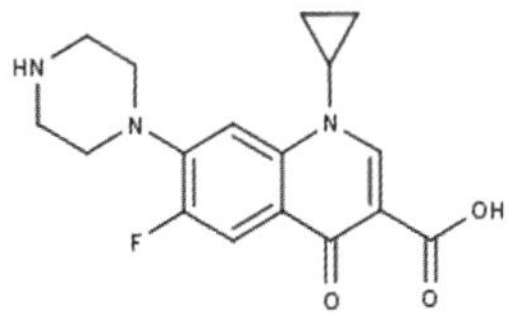	Aradigm Hayward, CA	Fluoroquinolone Inhibits DNA gyrase and topoisomerase II	Phase II
MP-376 (levofloxacin inhalation) 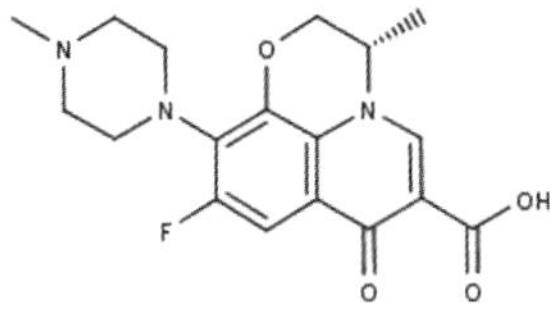	Mpex Pharmaceuticals San Diego, CA	Fluoroquinolone Inhibits DNA gyrase topoisomerase II	Phase II
Finafloxacin 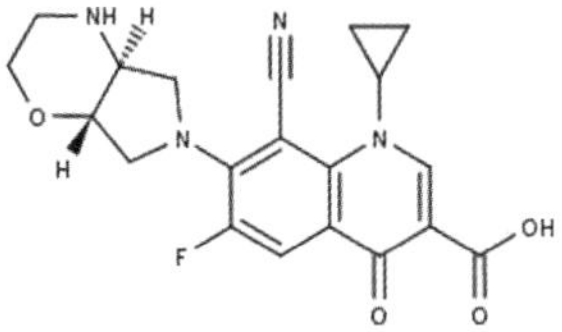	Merlion Pharmaceuticals Singapore	Fluoroquinolone Unique acid activated activity	Phase II for Urinary Tract Infections
DX-619 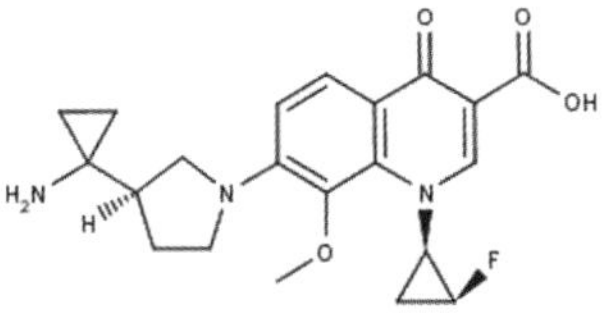	Daiichi Seiyaku Kyoto Japan	Des-fluoro (6) quinolone Inhibits DNA gyrase topoisomerase II	Phase I for multidrug resistant infections, but appears to be discontinued
GSK-945237 No structure available	GlaxoSmithKline	Inhibits DNA gyrase and topoisomerase II	Phase I

Mechanism of action: oxazolidinones are protein synthesis inhibitors with a unique mechanism of action, preventing initiation of synthesis. They affect the ribosomal peptidyltransferase center and modify tRNA positioning. Linezolid has bacteriostatic activity against methicillin-resistant *Staphylococcus aureus*, vancomycin-resistant enterococci, and penicillin- and cephalosporin-resistant *Streptococcus pneumoniae*. Their bioavailabilities are excellent, with comparable pharmacokinetics between oral and intravenous formulations.

Linezolid is the only presently marketed oxazolidinone with an acceptable risk/benefit ratio.[232] However, there are large numbers of side effects including reversible myelosuppression, optic neuropathy, an irreversible peripheral neuropathy, and serotonin syndrome. These are only seen with prolonged therapy, and so treatment for longer than two weeks is discouraged, and the drug is under a 'Black Triangle' warning from the UK regulatory agency, meaning that there is extremely close pharmacovigilance. Further modification of the oxazolidinone nucleus may yield agents with even greater potency. Torezolid and radezolid (RX-1741) are presently in Phase II-III clinical development.[232] Table 13.8 shows the oxazolidinones in development, many of which are being specifically targeted towards tuberculosis infections.

13.3.3 Antifungals

13.3.3.1 Imidazole, Triazole, and Thiazole Antifungals

The azole antifungals work by inhibiting the enzyme lanosterol 14 α-demethylase, which converts lanosterol to ergosterol. Depletion of ergosterol in the fungal membrane disrupts the structure of the membrane leading to inhibition of fungal growth.[233] The azoles are generally well tolerated, but in rare cases can cause liver damage and anaphylaxis. These are mainly due to interactions with and the cytochrome CYP3A4 and P-glycoprotein.[233]

13.3.3.2 Echinocandins

Echinocandins are broad-spectrum antifungals, inhibiting the synthesis of glucan in the cell wall, *via* noncompetitive inhibition of the enzyme 1,3-β glucan synthase.[234,235] These drugs can only be give parenterally. They may be used for lower respiratory tract fungal infections in immunocompromised patients. The echinocandins are lipopeptides, consisting of large cyclic (hexa)peptides linked to a long-chain fatty acid. They were first isolated from *Papularia sphaerosperma* and semisynthetic analogs of the echinocandins gave rise to the present day antifungals caspofungin and later micafungin and anidulafungin were also approved. They are well tolerated, but are embryotoxic and cannot be used during pregnancy.[236] Table 13.9 summarizes present antifungal research and development.

Table 13.8 Oxazolidinone antibiotics in development.

Product Name	*Sponsor*	*Class/ Mechanism*	*Development Status*
Linezolid 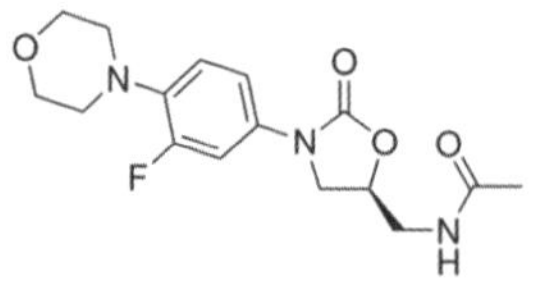	Pfizer New York	Oxazolidinone. Binds 50 S ribosomal subunit	Launched (2000)
Radezolid (RX1741) 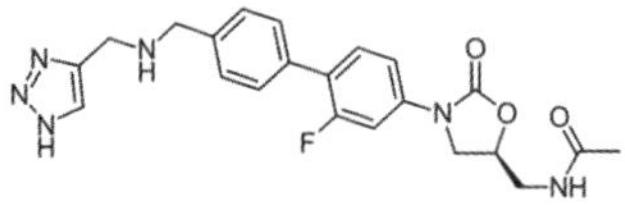	Rib-X Pharmaceuticals San Diego, CA	Oxazolidinone. Binds 50 S ribosomal subunit	Phase II for skin infections and community acquired pneumonia
Torezolid phosphate (TR-701 oral, intravenous) 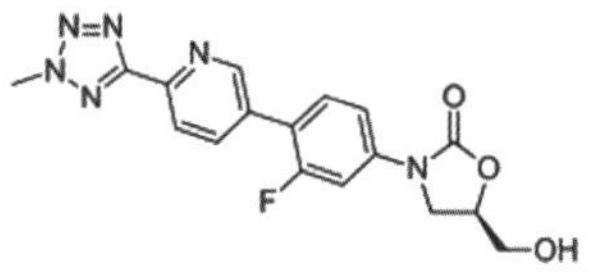	Trius Therapeutics San Diego, CA	Oxazolidinone. Binds 50 S ribosomal subunit	Phase III for skin infections Phase I for Gram-positive infections, shows high lung exposure
PF-2341272 PNU-100480 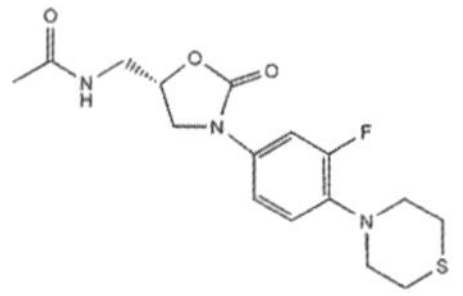	Pfizer New York, NY	Oxazolidinone. Binds 50 S ribosomal subunit	Phase I Targeting Tuberculosis
AZD5847 Structure not available	AstraZeneca Wilmington, DE	Oxazolidinone Binds 50 S ribosomal subunit	Phase I Targeting Tuberculosis

13.3.4 Antivirals

Influenza infection is one of the most important causes of lower respiratory tract infection. Influenza can affect approximately 20% of the world's population during any season. Vaccination is the primary strategy for controlling infection, but drugs are still needed. Influenza viruses have two surface glycoproteins, a hemagglutinin and a neuraminidase, which are the antigens that define each strain of influenza. By varying these molecules, the virus evades the host immune response. This necessitates a new vaccine each season, but also drugs for influenza control.[237,238]

Table 13.9 Anti-fungals in development.

Product Name	*Sponsor*	*Class/Mechanism*	*Development Status*
Eraxis™ (Anidulafungin)/Vfend® (Voriconazole) combination 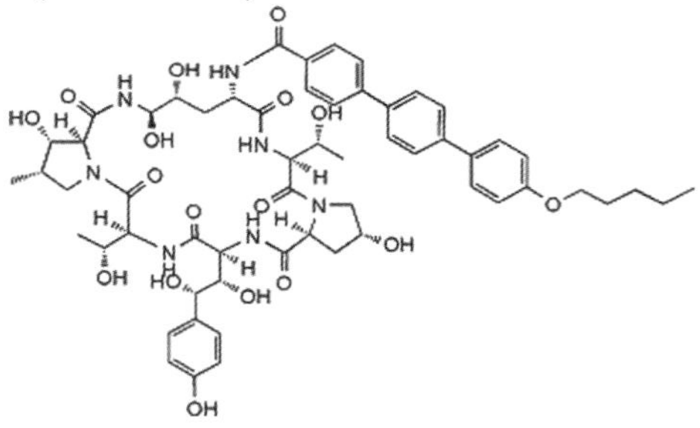 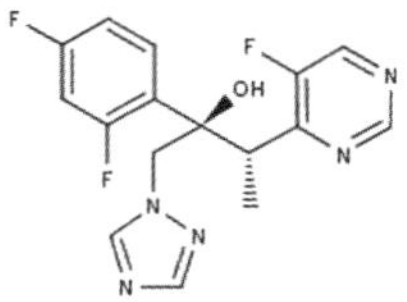	Pfizer New York, NY	Echinocandin: inhibits 1,3-β-D-glucan synthesis – a key cell wall component. Azoles are 14-alpha-demethylase inhibitors	Phase III
Isavuconazole 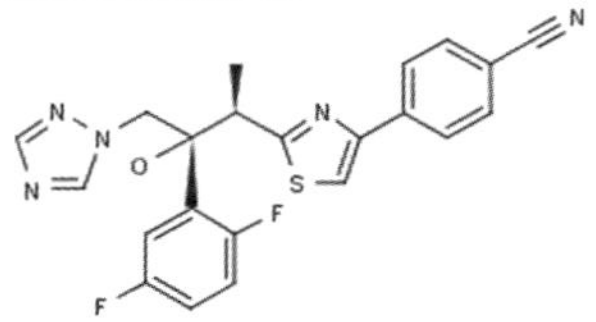	Basilea Pharmaceutica Basel, Switzerland Atellas Pharma US Deerfield, IL	14-alpha-demethylase inhibitor. Membrane intergrity and function is compromised	Phase III. Also highly soluble and orally bioavailable pro drug form

Table 13.9 (*Continued*)

Product Name	*Sponsor*	*Class/Mechanism*	*Development Status*
PAC-113 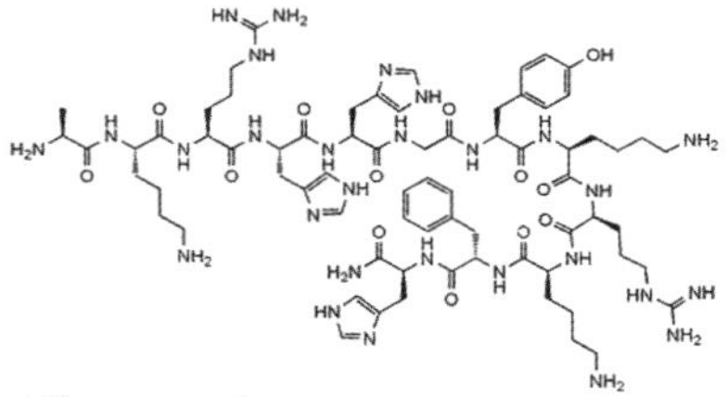	Pacgen Bio-pharmaceu- ticals Vancouver, Canada	12 amino-acid antimicrobial peptide, interacts with fungal cell membranes and, mitochondria causing production of reactive oxygen species	Phase II, candidal oral infection
Albaconazole 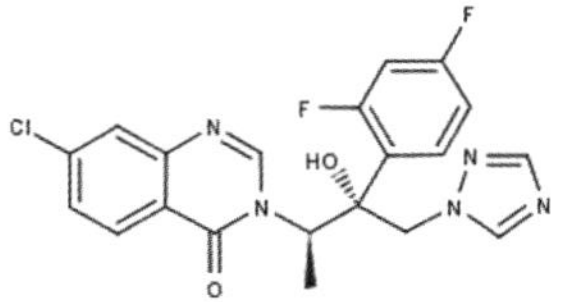	GlaxoSmithKline	14-alpha-demethylase inhibitor. Membrane intergrity compromised	Phase II
CB-182804 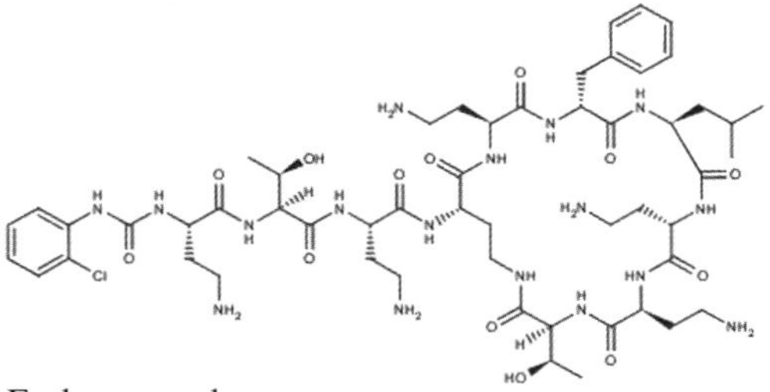	Cubist Pharmaceuticals Lexington MA	Polymyxcin-B analog Modifies membranes	Phase I
Embeconazole 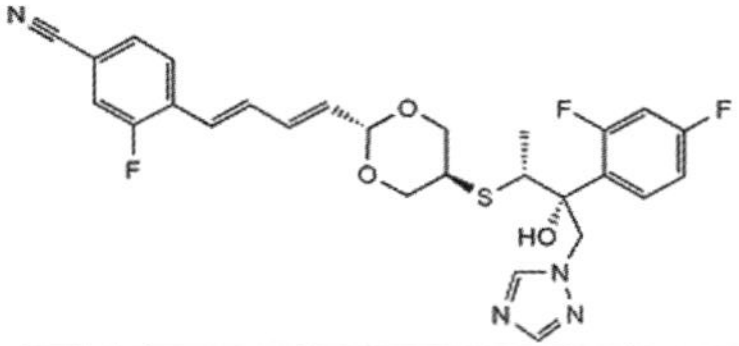	Daiichi Sankyo Tokyo Japan	14-alpha-demethylase inhibitor. Membrane intergrity compromised	Phase I discontinued

Neuraminidase inhibitors are a class of antiviral drugs targeting the influenza virus, and work by blocking the function of the viral neuraminidase protein. This prevents the virus from reproducing by budding from the host cell. Neuraminidase inhibitors are active against both influenza A and influenza B. Oseltamivir (Tamiflu) a prodrug, Zanamivir (Relenza), Laninamivir (Inavir), and Peramivir are members of this class.

Adamantanes – The currently available inhibitors working only against influenza A, are (amantadine and rimantadine). Adamantane-resistant isolates of influenza A are common.

Respiratory syncytial virus (RSV) is the most common cause of LRTIs worldwide, causing pneumonia and bronchiolitis, particularly in children. RSV604, a novel benzodiazepine with submicromolar anti-RSV activity is equipotent against both the A and B subtypes of the virus. RSV604 targets the nucleocapsid protein, is active post-RSV infection and is currently in Phase II.

MicroDose Therapeutx: MDT-637 is an inhalable small molecule anti-viral fusion inhibitor acquired from ViroPharma. The target is blocking the viral fusion protein and is more potent than ribavirin. MDT-637 was effective in reducing RSV viral count both pre- and post-infection giving it the potential to both prevent and treat RSV infection.

NexBio: Fludase blocks viral entry into respiratory cells by removing cell surface sialic acids. Fludase is a recombinant sialidase fusion protein composed of a sialidase catalytic domain derived from *Actinomyces viscosus*. Respiratory tract cell surface sialic acids act as host cell receptors for influenza A and B viruses. Fludase is highly active against viruses resistant to Tamiflu and effective against H5N1 strain. It is administered by oral inhalation.

Nitazoxanide is active against influenza A virus. The mechanism appears to be by selectively blocking the maturation of the viral hemagglutinin at a stage preceding resistance to endoglycosidase H digestion. This impairs hemagglutinin intracellular trafficking and insertion of the protein into the host plasma membrane.[239]

Development of antivirals for lower respiratory tract infection suffers from the same difficulties as anti-infectives in general. Large pharmaceutical companies have been more willing to invest in antiviral agents which can be used for chronic treatment in diseases such as hepatitis C. Table 13.10 summaries the antiviral medicines which are presently in development. Research into new antivirals would be of great value and since viral illness frequently precedes bacterial infections many would benefit from treatment of the LRTI.

13.3.5 Emerging Classes of Potential Antimicrobials

Besides the present classes of antimicrobials for which research is continuing, there are new agents which are being studied. Table 13.11 shows new agents which are under investigation. These include FabI inhibitors, peptide deformylase inhibitors, novel oxaborole compounds, diamines, and ATP synthase. The oxaboroles are an interesting new class, some of which act through the Leu-tRNA synthase. They are relatively small, and have good drug-like

Table 13.10 Antivirals with respiratory applications.

Product Name	Sponsor	Class/ Mechanism	Development Status
c 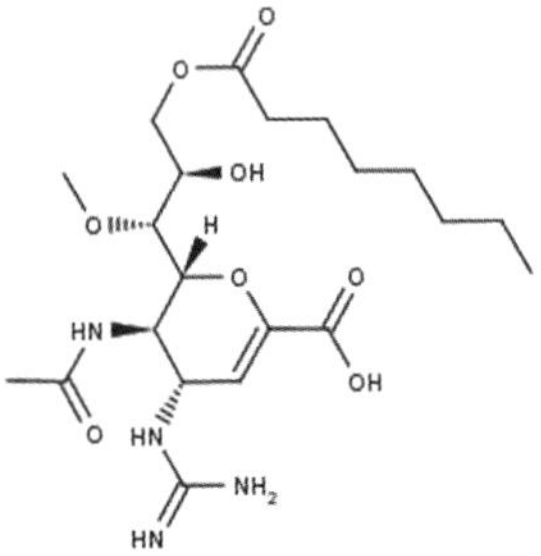	Biota Melbourne, Australia	Neuraminidase inhibitor targeting influenza Prodrug	Launched
BTA-798 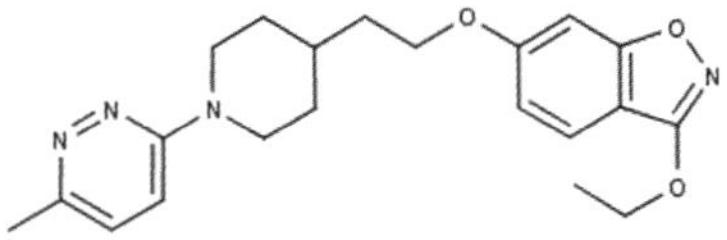	Biota Holdings Melbourne, Australia	Capsid binding inhibitor Targeting rhinovirus	Phase II
RSV604 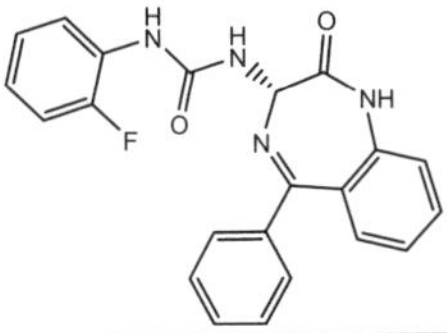	Novartis Pharmaceuticals East Hanover, NJ	Novel benzodiaze-pine nucleo-capside pro-tein inhibitor targeting RSV	Phase II

properties. Different members of the series are also being studied for their activity in human African trypanosomiasis and for malaria.

The sequencing of pathogen genomes brought the promise of many new interesting targets for killing the bacteria responsible for lower respiratory tract infection. Unfortunately, so far the results have proved to be somewhat disappointing.[240] The conclusion after many relatively fruitless target-based high-throughput screens is that target validation is at best an imprecise art, and that targets are perhaps best validated by pharmacological activity *in vivo*. The cycle of finding interesting molecules and then defining their molecular target has served us well in the first 50 years of antibiotic history, and continues to bear fruit, when the correct diversity is used, as the oxaborole case underlines. No doubt with time and a more integrated understanding of the organism we need to treat, new classes of drugs will be identified, but it should not be underestimated that the microorganisms have had millions of years to develop means of survival.

13.4 Affordable Medicines for Lower Respiratory Tract Infections in the Least Developed Countries

The key questions for the treatment of lower respiratory tract infections in the least developed countries are the availability and affordability of the medicines

Table 13.11 Antibacterials with diverse mechanisms of action in development.

Product Name	*Sponsor*	*Class/Mechanism*	*Development Status*
AFN-1252 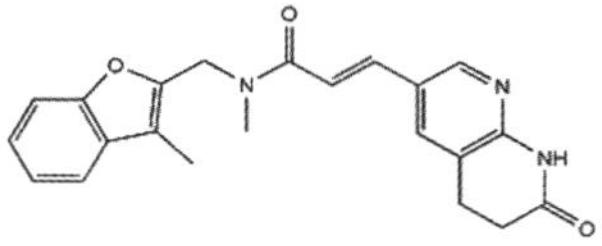	GlaxoSmithKline London, UK. Affinium Pharmaceuticals Austin, TX	Fabl (enol ACP reductase inhibitor)	Phase I
GSK1322322 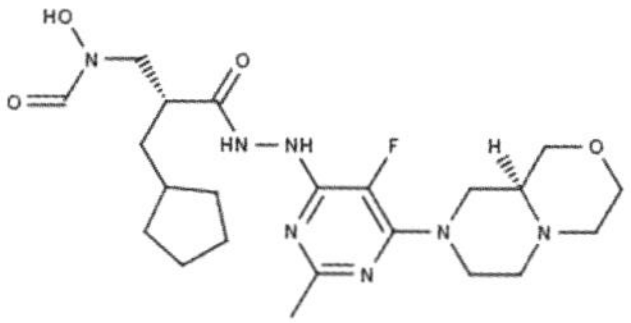	GlaxoSmithKline Research Triangle Park, NC	Peptide deformylase inhibitor	Phase II
GSK2251052 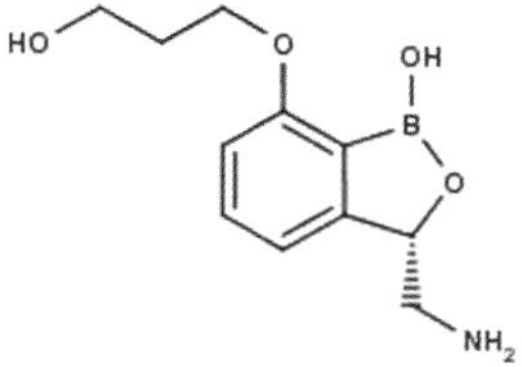	Anacor Pharmaceuticals Palo Alto, CA. GlaxoSmithKline London	Novel oxaborole Antibiotic Aminoacyl-tRNA Synthetase Inhibition	Phase I
MK-1682 Structure undisclosed	Merck Whitehouse Station, NJ	Hypothetical protein MK1682 From *M. kandleri*	Phase I Discontinued
PA-824 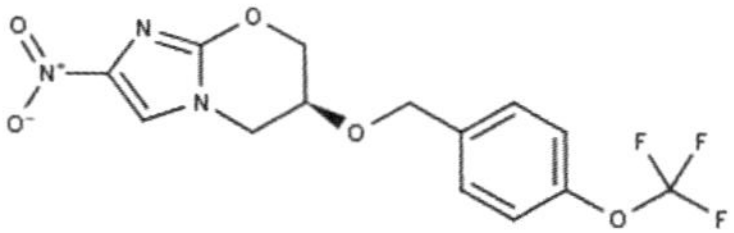	Global Alliance for TB Drug Development New York, NY. Novartis Basel	Nitroimidazole Intracellular NO release	Phase II For tuberculosis

Table 13.11 (*Continued*)

Product Name	*Sponsor*	*Class/Mechanism*	*Development Status*
SQ-109 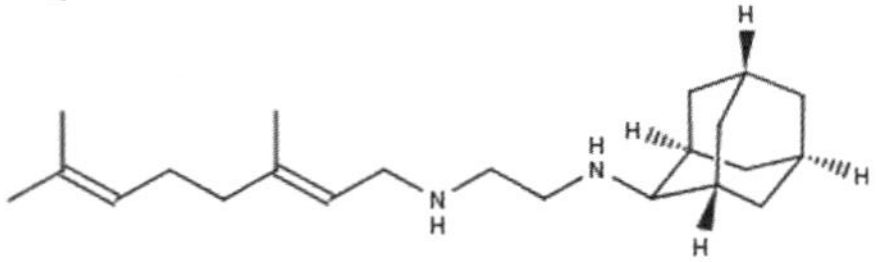	Sequella Rockville, MD	Diamine – ethambutol analog. Inhibits cell wall synthesis	Phase II for tuberculosis, Phase I for H. pylori
Bedaquiline TMC207 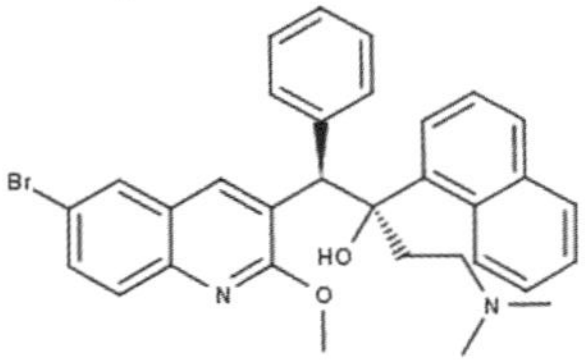	Global Alliance for TB Drug Development New York, NY Tibotech Yardley, PA	Diarylquinoline. Inhibiting mycobacterial ATP synthase	Phase II for tuberculosis
AZD9742 Structure undisclosed	AstraZeneca	BTGT4 undisclosed molecular target	Phase I

to the poorest populations. Most newly developed antibiotics are initially sold at a relatively high price into Western markets, and so can only be afforded by a relatively small sub-population in Africa or other developing countries. One consequence of this is that resistance to these antibiotics arrives much later. The question then becomes, how affordable can these medicines be made once the patent protection is removed? Insight into this question can be seen from the prices of medicines in Africa. For example, the data for azithromycin prices in the Management Sciences for Health/World Health Organization International Drug Price Indicator Guide gives prices of \$0.33–0.48 per 500 mg tablet. These prices for the public sector are much lower than those in the generic markets in Western countries – where prices are around \$2.50–\$10 per tablet. There is however scope for improvement: the price of bulk azithromycin in some places has been around \$200/kg, suggesting that a per tablet price of \$0.10 is eminently possible. The fact that for most patients price is an issue, means that more attention needs to be paid to making these medicines available at an affordable cost once the patent life in the Western markets is expired.

13.5 Conclusions

The need for new drugs to treat lower respiratory tract infections is clear. Resistance will continue to make those agents which still work go the way of the others. We know what organism we need to treat and what has worked in the past. Some research is being done, but mostly on drugs targeting Gram-positive organisms, because of concerns about methicilin-resistant *Stapholoccus aureus*. Concerningly, some Gram-negative organisms have become resistant to all presently available agents, and there is an urgent need to identify new classes of antibiotics. There is little research on drugs to treat multiple resistant organisms, and the ease by which bacteria can share resistance genes means that this is a serious concern for the future. There is really no difference between the developed world and the developing world: what affects the developed world today will impact on the developing world as soon as the antibiotics are rolled out in each of the countries. Lowering the drug prices in the developing world is only possible if the costs of research and development are already paid for, or can be written off. Then the problem becomes one of lowering the price of active ingredients, and ultimately one of supply chain. However, if new classes of medicines are to be developed, there will need to be a much closer alignment of the pharmaceutical discovery and development experts with government and philanthropic research funders, and with the purchasers. Only with such major commitments will we be able to prevent a return to the pre-antibiotic era.

References

1. R. Beaglehole, *The World Health Report 2004-Changing History*, World Health Organization, 2004, p. 120.
2. S. K. L. R. Kabra and R. M. Pandey, *Cochrane Database of Systematic Reviews*, 2006, issue 3.

3. Brad Spellberg, *Rising Plague*, Prometheus Books, Amherst New York, 2009.
4. Lewis Thomas, *Notes of a Medicine Watcher*, Viking Press, Published in Penquin Books, 1983.
5. H. W. Boucher, G. H. Talbot, J. S. Bradley, J. E. Edwards, D. Gilbert and L. B. Rice, *Clin. Infect. Dis.*, 2008, **47(S3)**, S249.
6. H. W. Boucher, G. H. Talbot, J. S. Bradley, J. E. Edwards, D. Gilbert, L. B. Rice, M . Scheld, B. Spellberg and J. Bartlett, *Clin. Infect. Dis.,* 2009, **48**, 1.
7. G. L. Armstrong, L. A. Conn and R. W. Pinner, *JAMA*, 1999, **28**, 61.
8. R. M. Klevens, A. M. Morrison, J. Nadle, S. Petit, K. Gershman, S. Ray, L. H. Harrison, R. Lynfield, G. Dumyati, J. M. Townes, A. S. Craig, E. R. Zell, G. E. Fosheim, L. K. McDougal, R. B. Carey and S. K. Fridkin, *JAMA*, 2007, **298**, 1763.
9. R. M. Klevens, A. M. Morrison, J. Nadle, S. Petit, K. Gershman, S. Ray, L. H. Harrison, R. Lynfield, G. Dumyati, J. M. Townes, A.S. Craig, E. R. Zell, G. E. Fosheim, L. K. McDougal, R. B. Carey and S. K. Fridkin, *JAMA*, 2007, **298**, 1803.
10. S. Blot, D. Vandijck, C. Lizy, L. Annemans, D. Vogelaers, P. D. Mauldin, C. D. Salgado and J. A. Bosso, *Antimicrob. Agents Chemother.*, 2010, **54**, 109.
11. R. R. Roberts, B. Hota, I. Ahmad, R. D. Scott 2nd, S. D. Foster and F. Abbasi, *Clin. Infect. Dis.*, 2009**, 49**, 1175.
12. M. S. Niederman, J. I. McCombs, A. N. Unger, A. Kumar and R. Popovian, *Clin. Ther.,* 1998, **20**, 820.
13. J. A. Dimasi, R. W. Hansen and H. G. Grabowski, *J. Health Polit. Policy Law,* 2008, **33**, 319.
14. J. A. DiMasi, R. W. Hansen and H. G. Grabowski, *J. Health Econ.*, 2003, **22**, 151.
15. L. M. Jarvis, Ketek Affect, *Chemical and Engineering News*, 2008, **86**, Number 15.
16. J. H. Bates, G. D. Campbell, A. L. Barron, G. A. McCracken, P. N. Morgan and E.B. Moses*, Chest*, 1992, **101**, 1005.
17. G. D. Fang, M. Fine, J. Orloff, D. Arisumi, V. L. Yu, W. Kapoor, J. T. Grayston, S. P. Wang, R. Kohler and R. R. Muder, *Medicine,* 1990, **69**, 307.
18. T. J. Marrie, H. Durant and L. Yates, *Rev. Infect. Dis.*, 1989, **11**, 586.
19. L. M. Mundy, P. G. Auwaerter, D. Oldach, M. L. Warner, A. Burton, E. Vance, C. A. Gaydos, J. M. Joseph, R. Gopalan, R. D. Moore, T. C. Quinn, P. Charache and J. G. Bartlett, *Am. J. Respir. Crit. Care Med.,* 1995, **152**, 1309.
20. B. J. Marston, J. F. Plouffe, T. M. File, B. A. Hackman, S. J. Salstrom, H. B. Lipman, M. S. Kolczak and R. F. Breiman, *Arch. Intern. Med.,* 1997, **157**, 1709.
21. M. Ruiz, S. Ewig, M. A. Marcos, J. A. Martinez, F. Arancibia, J. Mensa and A. Torres, *Am. J. Respir. Crit Care Med.,* 1999, **160**, 397.

22. J. Young, A. De Sutter, D. Merenstein, G. A. van Essen, L. Kaiser, H. Varonen, I. Williamson and H. C. Bucher, *Lancet*, 2008, **371**, 908.
23. FDA, Guidance for Industry Antibacterial Drug Products: Use of Non-inferiority Trials to Support Approval *http://www.fda.gov/Drugs/GuidanceComplianceRegulatoryInformation/Guidances/default.htm* (accessed 2 March 2010).
24. P. G. Ambrose, Clin. Infect. Dis., 2008, **47**: S225–231.
25. WHO: *The global burden of disease*, 2004. http://www.who.int/healthinfo/global_burden_disease/2004_report_update/en/index.html.
26. I. Korvula, *Chapter 2 Community-Acquired Pneumonia*, ed. Thomas J. Marrie, Plenum Publishers, New York, NY, 2001, p. 13.
27. J. L. Vincent, J. Rello, J. Marshall, E. Silva, E. Silva, C. D. Martin, R. Moreno, J. Lipman, C. Gomersall, Y. Sakr and K. Reinhart, *JAMA*, 2009, **302**, 2323.
28. Antibiotic Expert Group, Therapeutic guidelines: Antibiotics. 13th ed. North Melbourne Therapeutic Guidelines; 2006.
29. R. Guthri, *Chest*, 2001, **120**, 2021.
30. S. Weber, A. R. Wilkinson, D. Lindsell, P. L. Hope, S. R. Dobson and D. Isaac, *Archives of Disease in Childhood*, 1990, **65**, 207.
31. M. J. Abzug, A. C. Beam, E. A. Gyorkos and M. J. Levin, *The Pediatric Infectious Disease Journal*, 1990, **9**, 881.
32. L. Wubbel, L. Muniz, Ahmed A, M. Trujillo, C. Carubelli, C. McCoig, T. Abramo, M. Leinonen and G. H. McCracken, *The Pediatric Infectious Disease Journal*, **18**, 98.
33. M. S. Niederman, L. A. Mandell and A. Anzueto, *Am. J. Resp. Crit. Care Med.*, 2001,**163**, 1730.
34. K. McIntosh, *New Eng. J. Med.*, 2002, **346**, 429.
35. G. L. Mandell, J. E. Bennett and R. Dolin, *Mandell's Principles and Practices of Infectious Diseases 6th Edition*, Churchill Livingstone, 2004.
36. *The Oxford Textbook of Medicine 4th edition*, ed. D. A. Warrell, T. M. Cox, J. D. Firth and E. J. Benz, Oxford University Press, 2003.
37. *Harrison's Principles of Internal Medicine 16th Edition*, McGraw-Hill.
38. J. G. Bartlett, P. O'Keefe, F. P. Tally, T. J. Louie and S. L. Gorbach, *Arch. Intern. Med.*, 1986, **146**, 868.
39. J. Y. Fagon, J.Chastre, A. Vuagnat, J. L. Trouillet, A. Novara and C. Gibert, *JAMA*, 1996, **275**, 866.
40. R. M. Allen, W. F. Dunn and A. H. Limper, *Mayo Clinic Proc.*, 1994, **69**, 962.
41. I. Uçkayl, Q. A. Ahmed, H. Saxl and D. Pittet, *Clin. Infect. Dis.*, 2008, **46**, 557.
42. J. Pugin, R. Auckenthaler, N. Mili, J. P. Janssens, P. D. Lew and P. M. Suter, *Am. Rev. Respir. Dis.*, 1991, **143**, 1121.
43. F. Rodriguez de Castro, J. Sole Violan, B. Lafarga Capuz, J. Caminero Luna, B. Gonzalez Rodriguez and J.L. Manzano Alonso, *Crit. Care Med.*, 1991, **19**, 171.
44. M. Davidson, B. Tempest and D. L. Palmer. *JAMA*, 1976, **235**, 158.

45. J. Y. Fagon, J. Chastre, A.J. Hance, P. Montravers, A. Novara and C. Gibert, *Am. J. Med.*, 1993, **94**, 281.
46. J. H. Higuchi, J. J. Coalson, W. G. Johanson Jr., *Am. Rev. Respir. Dis.*, 1982, **125**, 53.
47. C. S. Bryan and K. L. Reynolds, *Am. Rev Respir. Dis.*, 1984, **129**, 668.
48. T. C. Horan, J. W. White, W. R. Jarvis, T.G. Emori, D. H. Culver, V. P. Munn, C. Thornsberry, D. R. Olson and J. M. Hughes, *Morbidity and Mortality Weekly Report*, 1986, **35(1SS)**, 17SS.
49. D. R. Schaberg, D. H. Culver and R. P. Gaynes, *Am. J. Med.*, 1991, **91**(suppl 3B), 72S.
50. J. Y. Fagon, J. Chastre, A. Hance, P. Montravers, A. Novara and C. Gibert, *Am. Rev. Respir. Dis.*, 1989, **139**, 877.
51. A. Torres, R. Aznar, J. M. Gatell, P Jiménez, J.González, A. Ferrer, R. Celis and R. Rodriguez-Roisin. *Am. Rev. Respir. Dis.*, 1990, **142**, 523.
52. A. Torres, J. Puig de la Bellacasa, R. Rodriguez-Roisin, M. T. Jimenez de Anta and A. Agusti-Vidal., *Am. Rev. Respir. Dis.*, 1988, **138**, 117.
53. J. Rello, E. Quintana, V. Ausina, J. Castella, M. Luquin, A. Net and A.I.Guillein, *Chest*, 1991, **100**, 439.
54. F. Espersen and J. Gabrielsen, *J. Infect. Dis.*, 1981, **144**, 19.
55. T. J. Inglis, L. J. Sproat, P. M. Hawkey and J.S. Gibson, *J. Hosp. Infect.*, 1993, **25**, 207.
56. G. U. Meduri, D. H. Beals, A. G. Maijub and V. Baselski, *Am. Rev. Respir. Dis.*, 1991, **143**, 855.
57. P. Reusser, W. Zimmerli, D. Scheidegger, G. A. Marbet, M. Buser and K. Gyr, *J. Infect. Dis.*, 1989, **160**, 414.
58. A. H. Limper, *Overview of Pneumonia in Cecil Textbook of Medicine. 23rd ed.*, ed. L. Goldman and D Ausiello, Saunders Elsevier; Philadelphia, Pa, 2007, chap 97.
59. P. E. Marik, *N. Engl. J. Med.*, 2001, **344**, 665.
60. I. Ben-Dov and Y. Aelony, *Postgrad. Med. J.*, 1989, **65**, 299.
61. H. Shigemitsu and K. Afshar, *Curr. Opin. Pulm. Med.*, 2007, **13**, 192.
62. M. H. DeLegge, MD, *Journal of Parenteral and Enteral. Nutrition*, 2002, **26**, S19.
63. N. A Metheny, R. E. Clouse, Y-H. Chang, B. J. Stewart, D. A. Oliver and M. H. Kollef, *Crit. Care Med.*, 2006, **34**, 1007.
64. S. Mukhopadhyay and A. L. Katzenstein, *American Journal of Surgical Pathology*, 2007, **31**, 752.
65. N. A. Metheny, *J. Parenter. Enteral. Nutr.*, 2002, **26**, S26.
66. J. G. Bartlett and S. M. Finegold, *Medicine (Baltimore)*, 1972, **51**, 413.
67. J. G. Bartlett, *Clin. Infect. Dis*,. 2005, **40**, 923.
68. J. G. Bartlett and S.M. Finegold, *Am Rev Respir Dis.*, 1974, **110**, 56.
69. J. L. Wang, K. Y. Chen, C. T. Fang, P. R. Hsueh, P. C. Yang and S. C. Chang, *Clin. Infect. Dis.*, 2005, **40**, 915.
70. H. C. Mwandumba and N. J. Beeching, *Curr. Opin. Pulm. Med.*, 2000, **6**, 234.

71. E. C. Pohlson, J. J. McNamara, C. Char and L. Kurata, *Am. J. Surg.*, 1985, **150**, 97.
72. J. G. Bartlett, *Chest*, 1987, **91**, 901.
73. R. P. Wenzel and A. A. Fowler, *N. Engl. J. Med.*, 2006, **355**, 2125.
74. W. J. Hueston, *The Journal of Family Practice*, 1997, **44**, 261.
75. T. D. Girard and G. R. Bernard, *Chest*, 2007, **131**,921.
76. S. B. Shaker, A. Dirksen, K. S. Bach and J. Mortensen, *COPD*, 2007, **4**, 143.
77. W. Schlick, *Wien. Med. Wochenschr.*, 1986, 31, 610.
78. P. B. Bach, C. Brown, S. E. Gelfand and D. C. McCrory, *Ann. Intern. Med.*, 2001, **134**, 600.
79. B. Burrows, *N. Engl J. Med.*, 1969, **280**, 397.
80. I. Tager and F. E. Speizer, *N. Engl. J. Med.*, 1975**, 292**,563.
81. T. F. Murphy and S. Sethi, *Am. Rev. Respir. Dis.* 1992**, 146**,1067.
82. C. M. Isada, *Semin. Respir. Infect.*, 1993, **8**, 243.
83. M. B. Nicotra and R. S. Kronenberg, *Semin Respir Infect.*, 1993**, 8**, 254.
84. D. W. Gump, C. A. Phillips and B. R. Forsyth*, Am. Rev. Respir. Dis.,* 1976, **113**, 465.
85. R. O. Buscho, D. Saxtan and P. S. Shultz, *J. Infect. Dis.*, 1978, **137**, 377.
86. C. B. Smith, C. Golden, R. Kanner, *Am. Rev. Respir. Dis.*, 1980, **121**, 225.
87. N. Soler, A. Torres and S. Ewig, *Am. J. Respir. Crit. Care Med.*, 1998, **157**, 1498.
88. F. Blasi, D. Legnani, V. M. Lombardo, G. G. Negretto, E. Magliano, R. Pozzoli, F. Chiodo, A. Fasoli and L. Allegra, *Eur. Respir. J.*, 1993, **6**, 19.
89. N. Miyashita, Y. Niki, M. Nakajima, H. Kawane and T. Matsushima, *Chest*, 1998, **114**, 969.
90. J. Fujita, N. L. Nelson, D. M. Daughton, C. A. Dobry, J. R. Spurzem, S. Irino and S. I. Rennard, *Am. Rev. Respir. Dis.*, 1990, **142**, 57.
91. S. Sethi, *Chest*, 2000, **117 (5 Suppl 2)**, 380S.
92. S. Sethi, K. Muscarella, N. Evans, K. L. Klingman, B. J. B. Grant and T.F. Murphy, *Chest*, 2000, **118**, 1557.
93. J. Diamond in *Guns, Germs, and Steel: The Fates of Human Societies*, ed. W. W. Norton, New York, 1997.
94. G. Majno, *The Healing Hand: Man and Wound in the Ancient World*, Harvard University Press, Cambridge, MA, 1975.
95. G. W. Hudler, *Magical Mushrooms, Mystical Molds*, Princeton University Press, NJ, 1998; R. C. Moelering Jr. and G. M. Eliopoulos, *Principles of Anti-Infective Therapy,* Elsevier, Philadelphia, 2005, p. 242.
96. S. H. Zinner and K. H. Mayer, *Choosing the Right Antibiotic in Ambulatory Care,* Elsevier, Philadelphia, Pa, 2005, p. 440.
97. A. Fleming, *British Journal of Experimental Pathology*, 1929, **10**, 226.
98. M. Wianwrightr and H. T. Swan, *Medical History*, 1986, **30**, 42.
99. R. Austrian and J. Gold, *Annuals of Internal Medicine*, 1964, **60**, 759.
100. R. A. Samson, R. A. Hadlok and A. C. Stolk, *Antonie van Leeuwenhoek,* 1977, **43**, 169.

101. J. Lederberg, R. E. Shope and S. C. Oaks Jr,, *Emerging Infections: Microbial Threats to Health in the United States*, National Academy Press, Washington DC, 1992.
102. P. J. Chesney, *Am. J. Dis. Child.*, 1992, **146**, 912.
103. CDC, Drug-resistant Streptococcus Pneumoniae-Kentucky and Tennessee 1993, *Morbidity and Mortality Weekly Report*, 1994, **43**, 23.
104. J. S. Spika, R. R. Facklam, B. D. Plikaytis and M. J. Oxtoby, *J. Infect. Dis.*, 1991, **163**, 1273.
105. M. C. Enright, D. A. Robinson, G. Randle, E. J. Feil, H. Grundmann and B. G. Spratt, *Proc. Natl. Acad. Sci.*, 2002, 99, **11**, 7687.
106. C. H. Nash and F. M. Huber, *Appl. Microbiol.*, 1971, **22**, 6.
107. L. C. Blaszczak, C. N. Eid, J. Flokowitsch, G. S. Gregory, S. A. Hitchcock, G. W. Huffman, D. R. Mayhugh, M. J. Nesler, D. A. Preston and M. Zia-Ebrahimi, *Bioorg. Med. Chem. Lett.*, 1997, **7**, 2261.
108. J. Birnbaum, F. M. Kahan, H. Kropp and J. MacDonald, *The American Journal of Medicine*, 1985, **78 (6A)**, 3.
109. N. Durán and C. F. Menck, *Crit. Rev. Microbiol.*, 2001, **27**, 201.
110. J. M. Ghuysen, *Annu. Rev. Microbiol.*, 1991, **45**, 37.
111. N. H. Georgopapadakou, *Antimicrob. Agents Chemother.*, 1993, **37**, 2045.
112. E. W. Brunskill and K. W. J. Bayles, *Bacteriol.*, 1996, **178**, 611.
113. R. P. Ambler, A. F. Coulson, J. M. Frere, J. M. Ghuysen, B. Joris, M. Forsman, R. C. Levesque, G. Tiraby and S. G. Waley, *Biochem. J.*, 1991, **276**, 269.
114. K. Bush, G. A. Jacoby and A. A. Medeiros, *Antimicrob. Agents Chemother.*, 1995, **39**, 1211.
115. B. A. Rasmussen and K. Bush, *Antimicrob. Agents Chemother.*, 1997, **41**, 223.
116. J. L. Mainardi, R. Villet, T. D. Bugg, C. Mayer and M. Arthur, *FEMS Microbiol. Rev.*, 2008, **32**, 386.
117. L. Martinez-Martinez, S. Hernandez-Alles, S. Alberti, J. Tomas, V. Benedi and G. Jacoby, *Antimicrob. Agents Chemother.*, 1996, **40**, 342.
118. K. K. Kumarasamy, M. A. Toleman, T. R. Walsh, J. Bagaria, F. Butt, R. Balakrishnan, U. Chaudhary, M. Doumith, C. G. Giske, S. Irfan, P. Krishnan, A.V. Kumar, S. Maharjan, S. Mushtaq, T. Noorie, D. L. Paterson, A. Pearson, C. Perry, R. Pike, B. Rao, U. Ray, J. B. Sarma, M. Sharma, E. Sheridan, M. A. Thirunarayan, J. Turton, S. Upadhyay, M Warner, W. Welfare, D. M. Livermore and N. Woodford, *Lancet Infect. Dis.*, 2010, **10**, 597.
119. Center for Disease Control and Prevention, *Morbidity and Mortality Weekly Report*, 2010, **59**, 750.
120. M. W. Olson, A. Ruzin, E. Feyfant, T. S. Rush, J. O'Connell and P. A. Bradford, *Antimicrob. Agents Chemother.*, 2006, **50**, 2156.
121. M. C. Roberts, *FEMS Microbiol. Rev.*, 1996, **19**, 1.
122. I. T. Paulsen, M. H. Brown and R. A. Skurray, *Microbiol. Rev.*, 1996, **60**, 575.
123. M. C. Roberts, *Trends in Microbiol.*, 1994, **2**, 353.
124. M. C. Roberts, *Ciba Found. Symp,.* 1997, **207**, 206.
125. D. E. Taylor and A. Chau, *Antimicrob. Agents Chemother.*, 1996, **40**, 1.

126. S. B. Levy, L. M. McMurry, T. M. Barbosa, V. Burdett, P. Courvalin, W. Hillen, M. C. Roberts, J. I. Rood and D. E. Taylor, *Antimicrob. Agents Chemother.*, 1999, **43**, 1523.
127. B. R. Kaplan and K. Weinstein, *Appl. Microbiol.*, 1969, **17**, 653.
128. C. Cocito, M. Di Giambattista, E. Nyssen and P. Vannuffel, *J. Antimicrob. Chemother.*, 1997, **39 (Suppl A)**, 7.
129. W. Namwat, Y. Kamioka, H. Kinoshita, Y. Yamada and T. Nihira, *Gene*. 2002, **286**, 283.
130. T. A. Mukhtar and G. D. Wright, *Chem. Rev.*, 2005, **105**, 529.
131. F. Van Bambeke in *J. Infectious Diseases*, ed. D. Armstrong and J. Cohen, Mosby, London, 1999, 7/1.1.
132. T. Tenson, M. Lovmar and M. Ehrenberg, *Journal of Molecular Biology*, 2003, 1005.
133. M. C. Roberts, J. Sutcliffe, P. Courvalin, L. B. Jensen, J. Rood and H. Seppala, *Antimicrob. Agents Chemother.*, 1999, **43**, 2823.
134. T. Tenson and A. S. Mankin, *Peptides*, 2002, **22**, 1661.
135. R. C. Goldman and S. K. Kadam, *Antimicrob. Agents Chemother.*, 1989, **33**, 1058.
136. V. D. Shortridge, G. V. Doern, A. B. Brueggemann, J. M. Beter and R. K. Flamm, *Clin. Infect. Dis.*, 1999, **29**, 1186.
137. A. Tait-Kamradt, T. Davies, M. Cronan, M. R. Jacobs, P. C. Appelbaum and J. Sutcliffe, *Antimicrob. Agents Chemother.*, 2000, **44**, 2118.
138. B. Weisblum, *Antimicrob. Agents Chemother.*, 1995, **39**, 577.
139. B. Vester and S. Douthhwaite, *Antimicrob. Agents Chemother.*, 2001, **45**, 1.
140. M. Gaynor and A. S. Mankin, *Frontiers in Medicinal Chemistry*, 2005, **2**, 21.
141. K. Nagai, P. C. Appelbaum, T. D. Davies, L. M. Kelly, D. B. Hoellman, A. T. Andrasevic, L. Drukalska, W. Hryniewicz, M. R. Jacobs, J. Kolman, J. Miciuleviciene, M. Pana, L. Setchanova, M. K. Thege, H. Hupkova, J. Trupl and P. Urbaskova, *Antimicrob. Agents Chemother.*, 2002, **46**, 371.
142. E. Rubinstein, *J. Antimicrobial Agents*, 2001, **18 (Suppl 1)**, 71.
143. D. Bertrand, S. Bertrand, E. Neveu and P. Fernandes, *Antimicrob. Agents Chemother.*, 2010, **54**, 5399.
144. C. Thomas, M. Stevenson and T. V. Riley, *J. Antimicrob. Chemother.* 2003, **51**, 1339.
145. M. P. Mingeot-Leclercq, Y. Glupczynski and P. M. Tulkens, *Antimicrob. Agents Chemother.*, 1999, **43**, 727.
146. D. Gilbert in Mandell, Douglas, and Bennett's Principles and Practice of Infectious Diseases. 5th ed., ed. G. L. Mandell, J. E. Bennett and R. Dolin R, Churchill Livingstone, Philadelphia, Pa, 2000, p. 307.
147. L. P. Kotra, J. Haddad and S. Mobashery, *Antimicrob. Agents Chemother.*, 2000, **44**, 3249.
148. J. Davies and G. Wright, *Trends in Microbiology*, 1997, **5**, 234.
149. J. Davies, *Science*, 1994, **264**, 375.
150. D. M. Livermore, T. G. Winstanley and K. P. Shannon, *Journal of Antimicrobial Chemotherapy*, 2001, **48 (Suppl S1)**, 87.

151. E. Durante-Mangoni, A. Grammatikos, R. Utili and M.E. Falagas, *International Journal of Antimicrobial Agents*, 2009, **33**, 201.
152. D. Levine, *Clin. Infect. Dis.*, 2006, **42**, S5.
153. R. C. Moellering Jr., *Clin. Infec.t Dis.*, 2006, **42 (Suppl 1)**, S3.
154. R. S. Griffith, *Rev. Infect. Dis.*, 1981, **3**, S2004.
155. S. A. Samel, M. A. Marahiel and L. O. Essen, *Mol. Biosyst.*, 2008, **4**, 387.
156. R. Quintiliani Jr, and P. Courvalin in *Manual of Clinical Microbiology 6th ed.*, ed. P. R. Murray, E. J. Baron, M. A. Pfaller, F. C. Tenover and R. H. Yolken, ASM Press, 1995, Washington DC, 1995, p. 1319.
157. J. E. Geraci and W. R. Wilson, *Rev. Infect. Dis.*, 1981, **3** (Suppl), S250.
158. C. A. Loffler and C. Macdougall, *Expert. Rev. Anti. Infect. Ther.*, 2007, **6**, 961.
159. T. L. Smith, M. L. Pearson, K. R. Wilcox, C. Cruz, M. V. Lancaster, B. Robinson-Dunn, F. C. Tenover, M. J. Zervos, J. D. Band, E. White and W. R. Jarvis, *New Engl. J. Med.*, 1999, **340**, 493.
160. T. Peláez, L. Alcalá, R. Alonso, M. Rodríguez-Créixems, J. M. García-Lechuz and E. Bouza, *Antimicrob. Agents Chemother.*, 2002, **46**, 1647.
161. C. Gonzalez, M. Rubio, J. Romero-Vivas, M. Gonzalez and J.J. Picazo, *Clin. Infect. Dis.* 1999, **29**, 1171.
162. B. F. Farber and R. C. Moellering Jr., *Antimicrob. Agents Chemother.*, 1983, **23**, 138.
163. F. Van Bambeke, *Curr. Opin. Investig. Drugs,* 2006, **7**, 740.
164. P. Kirkpatrick, A. Raja, J. LaBonte and J. Lebbos, *Nat. Rev. Drug Discov.*, 2003, **2**, 943.
165. M. Debono, M. Barnhart, C. B. Carrell, J. A. Hoffmann, J. L. Occolowitz, B. J. Abbott, D. S. Fukuda, R. L. Hamill, K. Biemann and W. C. Herlihy, *J. Antibiot. (Tokyo)*, 1987, **40**, 761.
166. P. E. Pertel, P. Bernardo, C. Fogarty, P. Matthews, R. Northland, M. Benvenuto, G. M. Thorne, S. A. Luperchio, R. D. Arbeit and J. Alder, *Clin. Infect. Dis.*, 2008, **46**, 1142.
167. J. A. Silverman, L. I. Mortin, A. D. Vanpraagh, T. Li and J. Alder, *J. Infect. Dis.*, 2005, **191**, 2149.
168. P. Margalith and G. Beretta, *Mycopatho. Mycol. Appl.*, 1960, **8**, 321.
169. V. Prelog and W. Oppolzer, *Helv. Chim. Acta.*, 1973, **56**, 2279.
170. A. F. Widmer, R. Frei, Z. Rajacic and W. Zimmerli, *J. Infect. Dis.*, 1990, **162**, 96.
171. C. Calvori, L. Frontali, L. Leoni and G. Tecce, *Nature*, 1965, **207**, 417.
172. H. G. Floss and T. Yu, *Chem. Rev.*, 2005, **105**, 621.
173. W. Wehrli and M. Staehelin, *Bacteriol. Rev.*, 1971, **3**, 290.
174. W. Achour, O. Guenni, M. Fines, R. Leclercq and A. Ben Hassen, *Antimicrob. Agents Chemother.*, 2004, **48**, 2757.
175. H. Aubry-Damon, C. J. Soussy and, P. Courvalin, *Antimicrob. Agents Chemother.*, 1998, **42**, 2590.
176. H. Aubry-Damon, M. Galimand, G. Gerbaud and P. Courvalin, *Antimicrob. Agents Chemother.*, 2002, **46**, 1571.

177. L. Herrera, C. Salcedo, B. Orden, B. Herranz, R. Martinez, A. Efstratiou and J. A. Saez Nieto, *Eur. J. Clin. Microbiol. Infect. Dis.*, 2002, **21**, 411.
178. B. I. Eisenstein and M. Schaechter in *Schaechter's Mechanisms of Microbial Disease,* Lippincott Williams & Wilkins, Hagerstwon, MD, 2007, p. 28.
179. V. O. Rotimi, M. Khoursheed, J. S. Brazier, W. Y. Jamal and F. B. Khodakast, *Clinical Microbiology and Infection*, 1999, **5**, 166.
180. P. Summanen, E. J. Baron, D. M. Citron, C. Strong, H. M. Wexler and S. M. Finegold in *Wadsworth Anaerobic Bacteriology Manual, 5th edn.*, Star Publishing, Belmont, CA, 1993, p. 111.
181. S. Trinh and G. Reysset, *Journal of Clinical Microbiology*, 1996, **34**, 2078.
182. J. S. Brazier, S. L. J. Stubbs and B. I. Duerden. *J. Antimicrob. Chemother.*, 1999, **44**, 580.
183. E. Hunt, *Drugs Future*, 2000, **25**, 1163.
184. C. Walsh, *Antibiotics: Actions, Origins, Resistance*, ASM Press, Washington, 2003, p. 237.
185. G. Brooks, W. Burgess, D. Colthurst, J. D. Hinks, E. Hunt, M. J. Pearson, B. Shea, A. K. Takle, J. M. Wilson, and G. Woodnutt, *Bioorg. Med. Chem.* 2001, **9**, 1221.
186. M. Pringle, J. Poehlsguard, B. Vester, and K. S. Long, *Mol. Microbiol.*, 2004, **54**, 1295.
187. R. J. Jones, T.R. Fritsche, H. S. Sader and J. E. Ross, *Antimicrob. Agents Chemother.*, 2006, **50**, 2583.
188. M. E. Falagas, A. P. Grammatikos and A. Michalopoulos, *Expert. Rev. Anti. Infect. Ther.*, 2008, **6**, 593.
189. J. McIntyre and I. Choonara, *Biol. Neonate,* 2004, **86**, 218.
190. A. Mulhall, J. de Louvois and R. Hurley, *Br. Med. J. (Clin. Res. Ed.)*, 1983, **287**, 1424.
191. V. Piñeiro-Carrero and E. Piñeiro, *Pediatrics*, 2004, **113 (4 Suppl)**, 1097.
192. H. C. Neu and T. D. Gootz, *"Antimicrobial Chemotherapy:Antimicrobial Inhibitors of Ribosome Function"*. in *Baron's Medical Microbiology (4th ed.)*, ed. S. Baron, Univ of Texas Medical Branch, 1996.
193. O. Jardetzky, *J. Biol. Chem.,* 1963, **238**, 2498.
194. R. Holt, *Lancet*, 1967, **1**, 1259.
195. *Harrison's Principles of Internal Medicine, 13th Ed.*. McGraw-Hill Inc. 1994. p. 604.
196. R. L. Then, *J. Chemother.*, 1993, **5**, 361.
197. S. A. Tilles, *Southern Medical Journal,* 2001, **94**, 817.
198. C. G. Slatore and S. Tilles, *Immunology and Allergy Clinics of North America*, 2004, **24**, 477.
199. C. C. Brackett, H. Singh and J.H. Block, *Pharmacotherapy,* 2004, **24**, 856.
200. C. L. Smith and K.R. Powell, *Pediatrics in Review,* 2000, **21**, 368.
201. J. M. Nelson, T.M. Chiller, J.H. Powers and F.J. Angulo, *Clin. Infect. Dis.,* 2007, **44**, 977.
202. D. V. Ivanov and S.V. Budanov, *Antibiot. Khimioter.*, 2006, **51**, 29.

203. M. P. Wentland in memoriam: G. Y. Lesher in *Quinolone Antimicrobial Agents ed. 2*, ed. D.C. Hooper and J.S. Wolfson, American Society for Microbiology XIII-XIV, Washington DC, 1993.
204. P. Ball, *J. Antimicrob. Chemother.*, 2000, **46 (Suppl T1, S3)**, 17.
205. S. Norris and G. L. Mandell, *"The Quinolones: History and Overview"*, San Diego: Academic Press Inc., 1988, p. 1.
206. S. J. Childs, *Infect. Urol.*, 2000, **13**, 3.
207. D. C. Hooper, *Emerg. Infect. Dis.*, 2001, **7**, 337.
208. F. J. Castora, F. F. Vissering and M. V. Simpson, *Biochim. Biophys. Acta.*, 1983, **740**, 417.
209. N. Kaplowitz, *Hepatology*, 2005, **41**, 227.
210. R. Schaumann and A. C. Rodloff, *Anti-Infective Agents in Medicinal Chemistry*, 2007, **6**, 49.
211. Y. H. Chang. H. K. Se and K. K. Young, *Bioorg. Med. Chem. Lett.*, 1997, **7**, 1875.
212. N. Mittmann, F. Jivarj, A. Wong and A. Yoon, *The Canadian Journal of Infectious Diseases*, 2002, **13**, 293.
213. D. E. Karageorgopoulos, K. P. Giannopoulou, A. P. Grammatikos, G. Dimopoulos and M. E. Falagas, *Canadian Medical Association Journal*, 2008, **178**, 845.
214. K. Z. Vardakas, I. I. Siempos, A. Grammatikos, Z. Athanassa, I. P. Korbila and M. E. Falagas, *Canadian Medical Association Journal*, 2008, **179**, 1269.
215. N. Le Saux, *Canadian Medical Association Journal*, 2008, **178**, 865.
216. A. De Sarro and G. De Sarro, *Curr. Med. Chem.*, 2001, **8**, 371.
217. R. C. Owens and P. G. Ambrose, *Clin. Infect. Dis.*, 2005, **41 (Suppl 2)**: S144.
218. L. Galatti, S. E. Giustini, A. Sessa, G. Polimeni, F. Salvo, E. Spina and A. P. Caputi, *Pharmacol. Res.*, 2005, **51**, 211.
219. R. C. Owens and P. G. Ambrose, *Clin. Infect. Dis.*, 2005, **41 (Suppl 2)**, S144.
220. P. B. Iannini, *Curr. Med. Res. Opin.*, 2007, **23**, 1403.
221. B. Rouveix, *Med. Mal. Infect.*, 2006, **36**, 697.
222. A. J. Mehlhorn and D. A. Brown, *Ann. Pharmacother.*, 2007, **41**, 1859.
223. R. J. Lewis and J. F. Mohr, *Drug Saf.*, 2008, **31**, 283.
224. E. Rubinstein, *Chemotherapy*, 2001, **47 (Suppl 3)**, 3.
225. S. F. Jones and R. H. Smith, *Brit. Med. J.*, 1997, **314**, 869.
226. F. Saint, G. Gueguen, J. Biserte, C. Fontaine and E. Mazeman, *Rev. Chir. Orthop. Reparatrice Appar. Mot.*, 2000, **86**, 495.
227. C. A. Muto, J. A. Jernigan, B. E. Ostrowsky, H. M. Richet, W. R. Jarvis, J. M. Boyce and B. M. Farr, *Infect. Control Hosp. Epidemiol.*, 2003, **24**, 362.
228. E. J. Kuijper, J. T. van Dissel and E. J. Wilcox, *Curr. Opin. Infect. Dis.*, 2007, **20**, 376.
229. D. B. Blossom and L. C. McDonald, *Clin. Infect. Dis.*, 2007, **45**, 222.
230. A. Robicsek, G. A. Jacoby and D. C. Hooper, *Lancet Infect. Dis.*, 2006, **6**, 629.

231. Y. Morita, K. Kodama, S. Shiota, T. Mine, A. Kataoka, T. Mizushima and T. Tsuchiya, *Antimicrob. Agents Chemother.*, 1998, **42**, 1778.
232. D. I. Diekema and R. N. Jones, *Drugs*, 2000, **59**, 7.
233. D. I. Zonios and J. E. Bennet, *Semin. Respir. Crit. Care Med.*, 2008, **29**, 198.
234. M. I. Morris and M. Villmann, *Am. J. Health Syst. Pharm.*, 2006, **63**, 1693.
235. M. I. Morris and M. Villmann, *Am. J. Health Syst. Pharm.*, 2006, **63**, 1813.
236. C. Wagner, W. Graninger, E. Presterl and C. Joukhadar, *Pharmacology*, 2006, **78**, 161.
237. J. Chapman, E. Abbott, D. G. Alber, R. C. Baxter, S. K. Bithell, E. A. Henderson, M. C. Carter, P. Chambers, A. Chubb, S. Cockerill, P. L. Collins, V. C. L. Dowdell, S. J. Keegan, R. D. Kelsey, M. J. Lockyer, C. Luongo, P. Najarro, R. J. Pickles, M. Simmonds, D. Taylor, S. Tyms, L. J. Wilson and K. L. Powell, *Antimicrob. Agents Chemother.*, 2007, **51**, 3346.
238. A. Moscona, *N. Engl. J. Med.*, 2005, **353**, 1363.
239. J. F. Rossignol, S. La Frazia, L. Chiappa, A. Ciucci and M.G. Santoro, *J. Biol. Chem.*, 2009 , **284**, 29798.
240. David Payne NRDD.

Subject Index